NORTH DAKOTA
STATE UNIVERSITY

AUG 2 4 2007

SERIALS DEPT.
LIBRARY

WITHDRAWN

Geophysical Monograph Series

Including
IUGG Volumes
Maurice Ewing Volumes
Mineral Physics Volumes

Geophysical Monograph Series

134 **The North Atlantic Oscillation: Climatic Significance and Environmental Impact** *James W. Hurrell, Yochanan Kushnir, Geir Ottersen, and Martin Visbeck (Eds.)*

135 **Prediction in Geomorphology** *Peter R. Wilcock and Richard M. Iverson (Eds.)*

136 **The Central Atlantic Magmatic Province: Insights from Fragments of Pangea** *W. Hames, J. G. McHone, P. Renne, and C. Ruppel (Eds.)*

137 **Earth's Climate and Orbital Eccentricity: The Marine Isotope Stage 11 Question** *André W. Droxler, Richard Z. Poore, and Lloyd H. Burckle (Eds.)*

138 **Inside the Subduction Factory** *John Eiler (Ed.)*

139 **Volcanism and the Earth's Atmosphere** *Alan Robock and Clive Oppenheimer (Eds.)*

140 **Explosive Subaqueous Volcanism** *James D. L. White, John L. Smellie, and David A. Clague (Eds.)*

141 **Solar Variability and Its Effects on Climate** *Judit M. Pap and Peter Fox (Eds.)*

142 **Disturbances in Geospace: The Storm-Substorm Relationship** *A. Surjalal Sharma, Yohsuke Kamide, and Gurbax S. Lakhima (Eds.)*

143 **Mt. Etna: Volcano Laboratory** *Alessandro Bonaccorso, Sonia Calvari, Mauro Coltelli, Ciro Del Negro, and Susanna Falsaperla (Eds.)*

144 **The Subseafloor Biosphere at Mid-Ocean Ridges** *William S. D. Wilcock, Edward F. DeLong, Deborah S. Kelley, John A. Baross, and S. Craig Cary (Eds.)*

145 **Timescales of the Paleomagnetic Field** *James E. T. Channell, Dennis V. Kent, William Lowrie, and Joseph G. Meert (Eds.)*

146 **The Extreme Proterozoic: Geology, Geochemistry, and Climate** *Gregory S. Jenkins, Mark A. S. McMenamin, Christopher P. McKay, and Linda Sohl (Eds.)*

147 **Earth's Climate: The Ocean–Atmosphere Interaction** *Chunzai Wang, Shang-Ping Xie, and James A. Carton (Eds.)*

148 **Mid-Ocean Ridges: Hydrothermal Interactions Between the Lithosphere and Oceans** *Christopher R. German, Jian Lin, and Lindsay M. Parson (Eds.)*

149 **Continent-Ocean Interactions Within East Asian Marginal Seas** *Peter Clift, Wolfgang Kuhnt, Pinxian Wang, and Dennis Hayes (Eds.)*

150 **The State of the Planet: Frontiers and Challenges in Geophysics** *Robert Stephen John Sparks and Christopher John Hawkesworth (Eds.)*

151 **The Cenozoic Southern Ocean: Tectonics, Sedimentation, and Climate Change Between Australia and Antarctica** *Neville Exon, James P. Kennett, and Mitchell Malone (Eds.)*

152 **Sea Salt Aerosol Production: Mechanisms, Methods, Measurements, and Models** *Ernie R. Lewis and Stephen E. Schwartz*

153 **Ecosystems and Land Use Change** *Ruth S. DeFries, Gregory P. Anser, and Richard A. Houghton (Eds.)*

154 **The Rocky Mountain Region—An Evolving Lithosphere: Tectonics, Geochemistry, and Geophysics** *Karl E. Karlstrom and G. Randy Keller (Eds.)*

155 **The Inner Magnetosphere: Physics and Modeling** *Tuija I. Pulkkinen, Nikolai A. Tsyganenko, and Reiner H. W. Friedel (Eds.)*

156 **Particle Acceleration in Astrophysical Plasmas: Geospace and Beyond** *Dennis Gallagher, James Horwitz, Joseph Perez, Robert Preece, and John Quenby (Eds.)*

157 **Seismic Earth: Array Analysis of Broadband Seismograms** *Alan Levander and Guust Nolet (Eds.)*

158 **The Nordic Seas: An Integrated Perspective** *Helge Drange, Trond Dokken, Tore Furevik, Rüdiger Gerdes, and Wolfgang Berger (Eds.)*

159 **Inner Magnetosphere Interactions: New Perspectives From Imaging** *James Burch, Michael Schulz, and Harlan Spence (Eds.)*

160 **Earth's Deep Mantle: Structure, Composition, and Evolution** *Robert D. van der Hilst, Jay D. Bass, Jan Matas, and Jeannot Trampert (Eds.)*

161 **Circulation in the Gulf of Mexico: Observations and Models** *Wilton Sturges and Alexis Lugo-Fernandez (Eds.)*

162 **Dynamics of Fluids and Transport Through Fractured Rock** *Boris Faybishenko, Paul A. Witherspoon, and John Gale (Eds.)*

163 **Remote Sensing of Northern Hydrology: Measuring Environmental Change** *Claude R. Duguay and Alain Pietroniro (Eds.)*

164 **Archean Geodynamics and Environments** *Keith Benn, Jean-Claude Mareschal, and Kent C. Condie (Eds.)*

165 **Solar Eruptions and Energetic Particles** *Natchimuthukonar Gopalswamy, Richard Mewaldt, and Jarmo Torsti (Eds.)*

166 **Back-Arc Spreading Systems: Geological, Biological, Chemical, and Physical Interactions** *David M. Christie, Charles Fisher, Sang-Mook Lee, and Sharon Givens (Eds.)*

167 **Recurrent Magnetic Storms: Corotating Solar Wind Streams** *Bruce Tsurutani, Robert McPherron, Walter Gonzalez, Gang Lu, José H. A. Sobral, and Natchimuthukonar Gopalswamy (Eds.)*

168 **Earth's Deep Water Cycle** *Steven D. Jacobsen and Suzan van der Lee (Eds.)*

169 **Magnetic ULF Waves: Synthesis and New Directions** *Kazue Takahashi, Peter J. Chi, Richard E. Denton, and Robert L. Lysak (Eds.)*

170 **Earthquakes: Radiated Energy and the Physics of Faulting** *Rachel Abercrombie, Art McGarr, Hiroo Kanamori, and Giulio Di Toro (Eds.)*

Geophysical Monograph 171

Subsurface Hydrology: Data Integration for Properties and Processes

David W. Hyndman
Frederick D. Day-Lewis
Kamini Singha
Editors

American Geophysical Union
Washington, DC

Published under the aegis of the AGU Books Board

Jean-Louis Bougeret, Chair; Gray E. Bebout, Cassandra G. Fesen, Carl T. Friedrichs, Ralf R. Haese, W. Berry Lyons, Kenneth R. Minschwaner, Andrew Nyblade, Darrell Strobel, and Chunzai Wang, members.

Library of Congress Cataloging-in-Publication Data

Subsurface hydrology : data integration for properties and processes / David W. Hyndman, Frederick D. Day-Lewis, Kamini Singha, editors.
    p. cm. -- (Geophysical monograph ; 171)
  ISBN 978-0-87590-437-5
  1. Groundwater flow--Mathematical models. I. Hyndman, David W. II. Day-Lewis, Frederick D. III. Singha, Kamini. IV. American Geophysical Union.
  GB1197.7.S84 2007
  551.49--dc22
                        2007017693

ISBN 978-0-87590-437-5

ISSN 0065-8448

Front cover image: Spectral analysis of the stream discharge hydrograph (top) for the Muskegon River in central-northern Michigan, USA, reveals a rich time-varying power spectrum (bottom). Direct comparison of the discharge power spectrum to that of precipitation or water table fluctuations can provide significant insight into watershed processes. *Courtesy of David W. Hyndman.*

Copyright 2007 by the American Geophysical Union
2000 Florida Avenue, N.W.
Washington, DC 20009

Figures, tables and short excerpts may be reprinted in scientific books and journals if the source is properly cited.

Authorization to photocopy items for internal or personal use, or the internal or personal use of specific clients, is granted by the American Geophyscial Union for libraries and other users registered with the Copyright Clearance Center (CCC) Transactional Reporting Service, provided that the base fee of $1.50 per copy plus $0.35 per page is paid directly to CCC, 222 Rosewood Dr., Danvers, MA 01923. 0065-8448/07/$01.50+0.35.

This consent does not extend to other kinds of copying, such as copying for creating new collective works or for resale. The reproduction of multiple copies and the use of full articles or the use of extracts, including figures and tables, for commercial purposes requires permission from the American Geophysical Union.

Printed in the United States of America.

CONTENTS

Preface
David W. Hyndman, Frederick D. Day-Lewis, and Kamini Singha vii

Introduction
Kamini Singha, David W. Hyndman, and Frederick D. Day-Lewis 1

I. Approaches to Data Integration

A Review of Geostatistical Approaches to Data Fusion
Clayton V. Deutsch 7

On Stochastic Inverse Modeling
Peter K. Kitanidis 19

II. Data Integration for Property Characterization

A Comparison of the Use of Radar Images and Neutron Probe Data to Determine the Horizontal Correlation Length of Water Content
Rosemary J. Knight, James D. Irving, Paulette Tercier, Gene J. Freeman, Chris J. Murray, and Mark L. Rockhold 31

Integrating Statistical Rock Physics and Sedimentology for Quantitative Seismic Interpretation
Per Avseth, Tapan Mukerji and Gary Mavko, and Ezequiel Gonzalez 45

A Geostatistical Approach to Integrating Data From Multiple and Diverse Sources: An Application to the Integration of Well Data, Geological Information, 3d/4d Geophysical and Reservoir-Dynamics Data in a North-Sea Reservoir
Jef Caers and Scarlet Castro 61

A Geostatistical Data Assimilation Approach for Estimating Groundwater Plume Distributions From Multiple Monitoring Events
Anna M. Michalak and Shahar Shlomi 73

A Bayesian Approach for Combining Thermal and Hydraulic Data
Allan D. Woodbury 89

Fusion of Active and Passive Hydrologic and Geophysical Tomographic Surveys: The Future of Subsurface Characterization
Tian-Chyi Jim Yeh, Cheng Haw Lee, Kuo-Chin Hsu, and Yih-Chi Tan 109

III. Data Integration to Understand Hydrologic Processes

Evaluating Temporal and Spatial Variations in Recharge and Streamflow Using the Integrated Landscape Hydrology Model (ILHM)
David W. Hyndman, Anthony D. Kendall, and Nicklaus R.H. Welty 121

Integrating Geophysical, Hydrochemical, and Hydrologic Data to Understand the Freshwater Resources on Nantucket Island, Massachusetts
Andee J. Marksamer, Mark A. Person, Frederick D. Day-Lewis, John W. Lane, Jr., Denis Cohen, Brandon Dugan, Henk Kooi, and Mark Willett 143

Integrating Hydrologic and Geophysical Data to Constrain Coastal Surficial Aquifer Processes at Multiple Spatial and Temporal Scales
Gregory M. Schultz, Carolyn Ruppel, and Patrick Fulton 161

Examining Watershed Processes Using Spectral Analysis Methods Including the Scaled-Windowed Fourier Transform
Anthony D. Kendall and David W. Hyndman 183

Integrated Multi-Scale Characterization of Ground-Water Flow and Chemical Transport in Fractured Crystalline Rock at the Mirror Lake Site, New Hampshire
Allen M. Shapiro, Paul A. Hsieh, William C. Burton, and Gregory J. Walsh 201

IV. Meta Analysis

Accounting for Tomographic Resolution in Estimating Hydrologic Properties from Geophysical Data
Kamini Singha, Frederick D. Day-Lewis, and Stephen Moysey 227

A Probabilistic Perspective on Nonlinear Model Inversion and Data Assimilation
Dennis McLaughlin 243

PREFACE

Groundwater is the principal source of drinking water for over 1.5 billion people. With increasing demands for potable water, continued threats to water quality, and growing concerns about climate change, the processes controlling groundwater availability are of paramount concern. There are also considerable concerns about the sustainability of groundwater supplies, given that much of the water withdrawn from aquifers today was recharged thousands of years ago. Data about hydrologic properties controlling flow and transport are needed to predict and simulate water-resources management practices, aquifer remediation, well-head protection, ecosystem management, and geologic isolation of radioactive waste. As the study of fundamental processes moves forward, we find that the physical processes of flow are complex at all scales, and furthermore are coupled with chemical and biological processes. In the 21st century, hydrologic scientists increasingly find themselves considering a diverse range of processes, data types, and analytical tools to help unravel processes controlling subsurface dynamics.

Quantifying the nature of hydrogeologic processes such as fluid flow, contaminant transport, or groundwater-surface-water interactions is difficult due to poor spatial sampling, heterogeneity at multiple scales, and time-varying properties. This book provides a series of examples where multiple data types have been integrated to better understand subsurface hydrology. We hope it serves to stimulate discussion and research on ways to improve our understanding on hydrologic processes, which are increasingly relevant as societal needs for clean water become more pressing. We thank the authors and reviewers of the chapters contained within this monograph and Allan Graubard, our AGU acquisitions editor.

David W. Hyndman
Frederick D. Day-Lewis
Kamini Singha
Editors

INTRODUCTION

Kamini Singha

Department of Geosciences, The Pennsylvania State University, University Park, Pennsylvania, USA

David W. Hyndman

Department of Geological Sciences, Michigan State University, Michigan, USA

Frederick D. Day-Lewis

U.S. Geological Survey, Office of Ground Water, Bureau of Geophysics, Storrs, Connecticut, USA

Understanding the processes that control water movement in the subsurface has been recognized as a "grand challenge" in environmental science [*National Research Council*, 2001b]. Research into methods to estimate hydrologic parameters that control water movement extends at least back to *Theis* [1935], who worked simultaneously on methods to predict (forward model) aquifer response to pumping, and also to estimate (using an inverse model) the controlling hydrologic parameters—transmissivity and storativity. Seventy years after Theis' pioneering work, hydrologists continue to use pumping tests and slug tests to characterize heterogeneous aquifers. Despite advances in modeling tools and inverse methods, aquifer characterization remains an extremely difficult problem due to spatial heterogeneity, temporal variability, and coupling between chemical, physical, and biological processes.

The concept of data integration (also called data fusion or data assimilation) involves merging multiple data types to develop more reliable predictive models, and to answer basic and applied science questions. In many applications, combinations of complementary data types has been shown to yield more information than analysis of more abundant data of a single type [*National Research Council*, 2000; *National Research Council*, 2001a]. Ideally, this would involve a seamless connection of field data across broad ranges of data types, temporal scales, and spatial scales from pores to watersheds and beyond. In practice, hydrologic measurements tend to be either sparse, local, and representative of only small volumes of the subsurface, or integrated over large volumes making it difficult to characterize heterogeneous hydrologic parameters. As a result, there remains a need for cost-effective data sources, and novel approaches to integrate multiple data types that consider coupled processes across multiple scales. Data integration is thus critical to improve our understanding of complex, multi-scale hydrologic processes, which often have feedbacks with other physical, chemical, and biological processes at multiple scales.

Reliable predictions of future system behavior depend on our ability to develop models that accurately represent field conditions based on collected data, while simulating key processes with a sparse set of parameters. With limited data, the problem of model identification is generally poorly constrained; as additional data types are considered, however, the intersection between viable sets of models becomes smaller (Figure 1) and estimates of parameters and rates of processes in the field improve. Recognition of this synergy is evidenced by the increasing number of integrated analyses of multiple data types, and a growing realization that simultaneous consideration of multiple data types, provides improved ways to characterize and monitor subsurface hydrologic properties and processes [e.g., *Hubbard and Hornberger*, 2006].

There are a wide range of data types that can be used to improve our understanding of hydrologic processes, ranging from direct estimates of hydrologic parameters (e.g., perme-

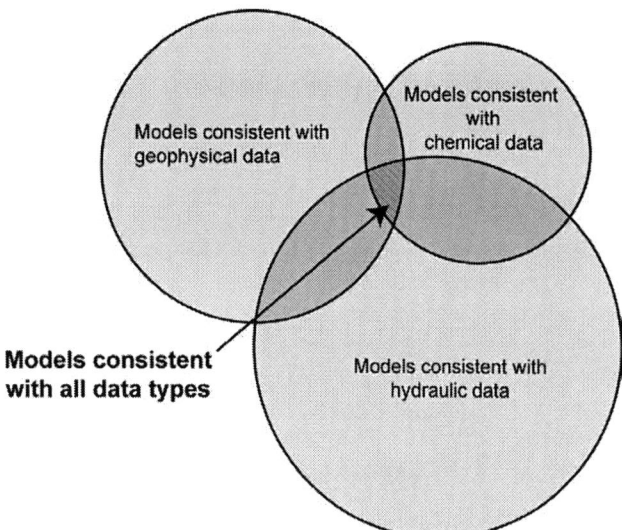

Figure 1. Sets of models can explain different data types. The intersection identifies the "best" model or models that represent the system across the integrated range of available data types.

ameter measurements on cores or flowmeter measurements of hydraulic conductivity) to indirect information from geologic maps, geophysical tomography, or quantities related to parameters of interest through physical models such as heat or solute transport. Table 1 provides a list of representative references where the listed data type is used to estimate parameters in subsurface models. This list is by no means exhaustive, but indicates the diversity of information sources used in hydrology. While data integration is increasingly implemented in hydrologic studies, it is also an active area of research due to the complexities of scale and measurement support volume, data weighting, model parameterization, realistic representation of geology in numerical models, and implementation of coupled-process numerical models.

This volume provides a broad sampling of papers that represent the current state of the science of data integration for subsurface hydrology. The premise underlying the collected work in this volume is that simultaneous consideration of multiple data types allows for an improved understanding of subsurface hydrology. The monograph is divided into four sections: (1) approaches to quantitative data integration; (2) data integration for characterization of hydrologic properties; (3) data integration for understanding hydrologic processes; and (4) meta analysis.

The first section includes papers on approaches to hydrologic data integration, which range from qualitative interpretation of multiple data types to rigorous non-linear inversion of coupled-process numerical models. In the last few decades, non-linear regression models that estimate subsurface properties based on groundwater data [e.g., *Neuman and Yakowitz*, 1979; *Gorelick*, 1990; *Gailey et al.*, 1991; *Wagner*, 1992; *Poeter and Hill*, 1997] have been developed and are built into commercially available modeling software. Software packages such as PEST [*Doherty*, 2002] and UCODE [*Poeter et al.*, 2005] allow for automated model calibration that includes multiple datasets (e.g., hydraulic heads and tracer concentrations). Often, regression modeling requires that the inverse problem be overdetermined; hence only a handful of parameters can be estimated, or zonal patterns of heterogeneity need to be defined a priori. Stochastic inversion methods provide alternatives to conventional non-linear regression by seeking to identify multiple models that match a given dataset, thus yielding additional information on parameter uncertainty and how this translates into uncertainty in model predictions. Although papers on stochastic inversion abound in the hydrologic literature [e.g., *Ginn and Cushman*, 1990; *Harvey and Gorelick*, 1995; *McLaughlin and Townley*, 1996; *Gomez-Hernandez et al.*, 1997; *Capilla and Gomez-Hernandez*, 2003], widespread use of such methods has been hampered by the perceived complexity of these tools. In this volume, *Deutsch* provides an overview of common geostatistical approaches that were originally developed for petroleum and mineral problems but are applied with increasing frequency in subsurface hydrology. The author discusses practical aspects of geostatistical methods that range from estimation with sparse data and declustering, to integration of secondary data and complex geological structures. *Kitanidis* provides a review of a Bayesian framework for inversion of groundwater data,

Table 1. A partial list of information sources used for estimating hydrologic parameters or processes.

Data Type	Representative papers or books
Stratigraphic/sedimentologic information	*Weissmann and Fogg* [1999], *Koltermann and Gorelick* [1992]
Temperature	*Anderson* [2005], *Stonestrom and Constantz* [2003]
Geophysics	*Vereecken et al.* [2006], *Rubin and Hubbard* [2006]
Isotopes	*Clark and Fritz* [1997], *Kaufmann et al.* [1984]
Geochemistry and microbiology	*Chappelle* [2000], *Kendall and McDonnell* [1998]
Hydraulic head	*Hill and Tiedeman* [2007], *Kitanidis* [1997]
Solute concentrations	*Rubin* [2003], *Harvey and Gorelick* [1995]
Remote sensing	*Hoffmann et al.* [2003], *Houser et al.* [1998]

with emphasis on estimating hydraulic parameters using head data; this paper describes a linear Gaussian stochastic inverse approach (often referred to as geostatistical inversion) including the underlying concepts, mathematics, and applications.

The second section of this monograph includes papers that use data integration methods to characterize hydrologic properties such as hydraulic conductivity, porosity, or fracture connectivity as well as parameters representing boundary conditions and contaminant release histories. The interest in estimating hydrologic properties is many-fold, including development of models that can be used to assess the risks that contamination poses to potential receptors or to evaluate rates of natural processes including recharge. The papers collected here represent work with different data across a range of settings.

For vadose-zone applications, spatially variable water content controls flow in the subsurface. Extrapolation of these data to large spatial scales is complicated, however, given only direct measurements of water content. *Knight et al.* integrate neutron-probe and ground-penetrating radar data to assess specific geostatistical characteristics of water content data from Hanford, Washington, USA. This work moves toward more quantitative integration of surface GPR data in hydrologic studies, and offers insights into issues with the measurement support volume.

The hydrologic community has long benefited from shared interests and cross-pollination with petroleum engineering and exploration geophysics. This monograph includes two crossover papers from the petroleum community. *Avseth et al.* present a data integration method developed to characterize lithologic facies in reservoirs. Their approach combines geologic and seismic information using petrophysical relations within a Bayesian framework, while *Caers and Castro* present an application of a probabilistic approach to integrate geologic, facies, seismic, and well production data to characterization of a North Sea reservoir. To estimate geologic facies and match water and oil production data, they analyze static and dynamic data with multipoint geostatistical and perturbation methods. The work they present is applied to basin-scale fluid flow and reservoir dynamics; the methodologies, however, have direct application for hydrologic data integration. Multipoint geostatistics for data integration is still not commonly used in hydrology, despite work such as this that indicates its promise [e.g., *Feyen and Caers*, 2006].

Michalak and Shlomi contribute a theoretical framework for estimating the spatial and temporal evolution of solute plume distributions. This framework is based on geostatistical inverse modeling and multiple monitoring events, given knowledge about geological variability and other factors affecting solute transport, but without knowing the source location or release history. In their approach, concentration data can be used to reconstruct past plume distributions that are consistent with all available information. *Woodbury* presents the generalized inverse problem for heat and groundwater, as an example of how the Bayesian framework can be used for data integration. The paper includes two examples, one focusing on inversion of heat conduction for paleoclimate reconstructions, and the second focusing on groundwater flow within the Edwards aquifer.

In a vision paper, *Yeh et al.* discuss state-of-the-art tomographic approaches including both hydraulic tomography and electrical resistivity tomography. Several examples illustrate the benefits of combining multiple data types, such as hydraulic and tracer data. The authors then propose tomographic approaches to basin-scale hydrologic characterization; they suggest that natural hydrologic, geologic, and climatic stimuli (e.g., river-stage fluctuations, earthquakes, and lightning) can serve as hydrologic or geophysical perturbations needed for regional-scale tomographic surveys (i.e., hydraulic, seismic, or electrical).

In addition to characterizing physical or chemical properties that affect hydrologic processes, data integration methods are used to shed light on the processes themselves. The third section of the monograph is a collection of papers that demonstrate the use of diverse types of data to elucidate processes spanning subsurface-hydrologic research, from paleohydrology to watershed response to modern coastal aquifer dynamics. A range of data types (e.g., geochemical, isotopic, hydraulic, geophysical) and integration methods (i.e., spectral analysis, physically based numerical modeling, etc.) are considered. This range of topics is timely as we attempt to identify the influence of human activities associated with land use and climate change on hydrologic and ecological systems.

Hyndman et al. illustrate the use of the new Integrated Landscape Hydrology Model (ILHM), which was developed to predict spatial and temporal variations in groundwater recharge at the watershed scale. This code simulates the redistribution of precipitation through the vegetation canopy, sediment surface, soil and sediment layers, and snow pack to various surface and subsurface pathways using a process-based description of the water balance, based on GIS data and minimal use of site-specific parameters. A process-based simulation for a watershed in western Michigan, USA, illustrates the region's strong seasonality in recharge rates; most of the precipitation and snowmelt becomes groundwater recharge from September through March, while virtually none of the precipitation during the growing season is recharged.

The dynamics of coastal and island aquifers remain important basic- and applied-science topics. Understanding interactions between aquifers, estuaries, and the coastal ocean requires consideration of many different data types collected over a range of temporal and spatial scales. Saltwater intrusion is a potential threat to many coastal and island aquifers, many of which are sole-source supplies of potable water. *Marksamer et al.* investigate the Nantucket Island aquifer in Massachusetts, USA, which extends deeper than expected given the current climate and water-table configuration. The authors use numerical modeling and multiple lines of evidence to test alternative paleohydrologic hypotheses to explain anomalous offshore freshwater and Nantucket's deep freshwater lens. Working in the coastal region of the southeastern USA, *Schultz et al.* combine groundwater monitoring, geochemical, electrical, electromagnetic, and vegetation mapping data to examine multi-scale, spatial and temporal coastal-aquifer dynamics. Target processes include saltwater intrusion, submarsh groundwater discharge, salinity gradients at the ocean boundary, and possible pore-water free convection.

The spectral content of hydrologic time series can provide insight into the time-scales of, and linkages between, important natural processes. *Kendall and Hyndman* demonstrate how spectral analysis of hydrologic datasets can be used to better understand linkages between precipitation, streamflows, and groundwater levels for watersheds in northern lower Michigan, USA. This analysis shows non-stationary behavior in these hydrologic systems, including the large reductions in summer streamflows due to canopy interception and evapotranspiration.

Fractured rock is, perhaps, the most complicated hydrologic setting [*National Research Council*, 1996]. Fluid concentration data from many fractured rock sites do not follow standard advective-dispersive behavior, and new data integration approaches are needed to identify dominant processes and understand the role of permeability heterogeneity [*National Research Council*, 2000, 2001b]. *Shapiro et al.* present an example of data integration from the U.S. Geological Survey's Fractured-Rock Hydrology research site, near Mirror Lake, New Hampshire, USA. The authors investigate anomalous solute-transport behavior at a variety of spatial scales using tracer and hydraulic testing as well as chemical sampling. Detailed borehole information and fracture mapping was integrated with the hydrologic data to clarify the geologic controls on flow and transport at each scale.

The collection of studies in this volume clearly demonstrates the value of data integration for hydrology; important limitations, however, remain. Recent work has underscored pitfalls and limitations of certain approaches or strategies used to combine data of different types. For example, additional work is needed to address the problems arising from model identification, non-linear feedbacks, uncertainty assessment, realistic characterization of geological variability, and discrepancies between the support volumes of different measurement types. The monograph's fourth section focuses on meta analysis and includes papers that reflect on opportunities for further research. *Singha et al.* discuss problems in the conversion of geophysical tomograms to hydrologic properties of interest. Although tomograms may provide qualitative information about hydrologic properties, the images have limited resolution and tend to be blurry versions of reality. The authors compare an analytical approach with a numerical approach to evaluate and address this problem. *McLaughlin* also discusses limitations associated with environmental data assimilation, in particular, problems that arise from the assumptions of linearity and normality on which most current approaches are based. He proposes that robust, rather than optimal, estimates should be sought, and that nonlinearity should be accepted and addressed.

Given current attention to coupled physical and chemical processes, and the increasing importance of groundwater as a resource, there is a strong need for novel data integration methods in hydrology. With continued advances in computational resources and rapidly evolving software for numerical modeling and inversion, future data integration methods will be better able to resolve both the nature of subsurface heterogeneities and the rates of critical processes across the range of hydrologic scales. Such developments will provide tools to help scientists address questions that arise through interdisciplinary research, where the measurements and models incorporate a host of processes that were typically studied individually within single disciplines. We believe that the integration of data and methods from hydrology with those from other sciences will be an active area of future research as hydrologic problems are increasingly recognized as being complex and dynamic. Data integration methods can provide important advances in the study of water quality and quantity, which will both be imperative for future decision-making in water resources.

REFERENCES

Anderson, M. P., Heat as a groundwater tracer, *Ground Water*, 6(43): 951–968, 2005.

Capilla, J. E. and J. J. Gomez-Hernandez, Stochastic inversion in hydrogeology, *Journal of Hydrology*, 281(4): 326 pp., 2003.

Chapelle, F. H., *Ground-water Microbiology and Geochemistry*, John Wiley and Sons, 468 pp., 2000.

Clark, I. D. and P. Fritz, *Environmental Isotopes in Hydrogeology*, CRC, 352 pp., 1997.

Doherty, J., *PEST—Model independent parameter estimation, version 6*, Queensland, Australia, Watermark Numerical Computing., 2002.

Feyen, L. and J. Caers, Quantifying geological uncertainty for flow and transport modeling in multi-modal heterogeneous formations, *Advances in Water Resources,* 29(6): 912–929, 2006.

Gailey, R. M., S. M. Gorelick and A. S. Crowe, Coupled process parameter estimation and prediction uncertainty using hydraulic head and concentration data, *Advances in Water Resources,* 14(5): 301–314, 1991.

Ginn, T. R. and J. H. Cushman, Inverse methods for subsurface flow; a critical review of stochastic techniques, *Stochastic Hydrology and Hydraulics,* 4(1): 1–26, 1990.

Gomez-Hernandez, J. J., A. Sahuquillo and J. E. Capilla, Stochastic simulation of transmissivity fields conditional to both transmissivity and piezometric data; I, Theory, *Journal of Hydrology,* 203(1–4): 162–174, 1997.

Gorelick, S. M., Large scale nonlinear deterministic and stochastic optimization: Formulations involving simulation of subsurface contamination, *Mathematical Programming,* 48(1–3): 19–39, 1990.

Harvey, C. F. and S. M. Gorelick, Mapping hydraulic conductivity; sequential conditioning with measurements of solute arrival time, hydraulic head, and local conductivity, *Water Resources Research,* 31(7): 1615–1626, 1995.

Hill, M. C. and C. R. Tiedeman, *Effective Groundwater Model Calibration: With Analysis of Data, Sensitivities, Predictions, and Uncertainty,* Wiley, 455 p., 2007.

Hoffmann, J., D. L. Galloway and H. A. Zebker, Inverse modeling of interbed storage parameters using land subsidence observations, Antelope Valley, California, *Water Resources Research,* 39(2): 1031, doi:10.1029/2001WRR001252, 2003.

Houser, P. R., W. J. Shuttleworth, J. S. Famiglietti, H. V. Gupta, K. H. Syed and D. C. Goodrich, Integration of soil moisture remote sensing and hydrologic modeling using data assimilation, *Water Resources Research,* 34(12): 3405–3420, 1998.

Hubbard, S. and G. Hornberger, Introduction to special section on Hydrologic Synthesis, *Water Resources Research,* 42: W03S01, doi:10.1029/2005WR004815, 2006.

Kaufmann, R., A. Long, H. Bentley and S. Davis, Natural chlorine isotope variations, *Nature,* 309: 338–340, 1984.

Kendall, C. and J. J. McDonnell, *Isotope tracers in catchment hydrology,* Elsevier, 1998.

Kitanidis, P. K., *Introduction to geostatistics; applications to hydrogeology,* Cambridge, Cambridge University Press, 249, 1997.

Koltermann, C. and S. M. Gorelick, Paleoclimatic signature in terrestrial flood deposits, *Science,* 256: 1775–1782, 1992.

McLaughlin, D. and L. R. Townley, A reassessment of the groundwater inverse problem, *Water Resources Research,* 32(5): 1131–1161, 1996.

National Research Council, *Rock Fractures and Fluid Flow: Contemporary Understanding and Applications,* Washington D.C., National Academy Press, 551 pp., 1996.

National Research Council, *Seeing into the Earth: Noninvasive Characterization of the Shallow Subsurface for Environmental and Engineering Application,* Washington D.C., National Academy Press, 129 pp., 2000.

National Research Council, *Basic Research Opportunities in Earth Science,* Washington D.C., National Academy Press, 168 pp., 2001a.

National Research Council, *Grand Challenges in Environmental Sciences,* Washington D.C., National Academy Press, 2001b.

Neuman, S. P. and S. Yakowitz, A Statistical Approach to the Inverse Problem of Aquifer Hydrology, 1. Theory *Water Resources Research,* 15(4): 845–860, 1979.

Poeter, E. P. and M. C. Hill, Inverse models; a necessary next step in ground-water modeling, *Ground Water,* 35(2): 250–269, 1997.

Poeter, E. P., M. C. Hill, E. R. Banta, S. Mehl and S. Christensen, *UCODE_2005 and Six Other Computer Codes for Universal Sensitivity Analysis, Calibration, and Uncertainty Evaluation,* U.S. Geological Survey Techniques and Methods. 6-A11: 283 pp., 2005.

Rubin, Y., *Applied Stochastic Hydrogeology,* Oxford University Press, 416 pp., 2003.

Rubin, Y. and S. S. Hubbard, Eds., *Hydrogeophysics (Water and Science Technology Library).* Netherlands, Springer, 2006.

Stonestrom, D. A. and J. Constantz, *Heat as a tool for studying the movement of ground water near streams,* USGS Circular 1260: 105 pp., 2003.

Theis, C. V., The relation between the lowering of the piezometric surface and the rate and duration of discharge of a well using ground-water storage, *American Geophysical Union Transcript,* 16: 519–524, 1935.

Vereecken, H., A. Binley, G. Cassiani, A. Revil and K. Titov, Eds., *Applied Hydrogeophysics,* NATO Science Series: IV: Earth and Environmental Sciences, Springer-Verlag, 2006.

Wagner, B. J., Simultaneous parameter estimation and contaminant source characterization for coupled groundwater flow and contaminant transport modelling, *Journal of Hydrology,* 135: 275–303, 1992.

Weissmann, G. S. and G. E. Fogg, Multi-scale alluvial fan heterogeneity modeled with transition probability geostatistics in a sequence stratigraphic framework, *Journal of Hydrology,* 226(1–2): 48–65, 1999.

A Review of Geostatistical Approaches to Data Fusion

Clayton V. Deutsch

University of Alberta, Edmonton, Alberta, Canada

Geostatistics has evolved to a mature discipline with a well understood theoretical framework and a standard set of tools. The tools have been applied with many geospatial variables in many different contexts. This paper provides a brief review of geostatistical approaches to problems involving multiple data types in subsurface hydrology. The random function paradigm of geostatistics is presented. Bayes Law is the engine that permits multivariate spatial and remotely sensed data to be integrated. The required multivariate probabilities are often fit with the Gaussian distribution. There are many implementation decisions and practicalities of geostatistics. These include declustering, inference in presence of sparse data, dealing with many secondary data, and modeling complex geological features. Subjects of practical importance are reviewed.

1. INTRODUCTION

The word *geostatistics* commonly refers to the theory of regionalized variables and the related techniques that are used to predict rock properties at unsampled locations. Georges Matheron formalized this theory in the early 1960's (Matheron, 1971). The development of geostatistics was led by engineers and geologists faced with real problems. They were searching for a consistent set of numerical tools that would help them with ore reserve estimation, reservoir performance forecasting, and site characterization.

At any instance in geological time, there is a single true distribution of rock properties over each study area. This true distribution is inaccessible with limited data and the chaotic nature of certain aspects of geological processes. Geostatistics strives to create numerical models that mimic the physically significant features of property variations.

Conventional mapping algorithms were devised to create smooth maps to reveal large-scale geologic trends; they are low pass filters that remove high frequency property variations. For practical problems of flow prediction, however, this variability has a large affect on the predicted response. Geostatistical simulation techniques, conversely, were devised with the goal to reproduce a realistic amount of variability, that is, create maps or realizations that are neither unique nor smooth. Although the small-scale variability of these realizations may mask large-scale trends, geostatistical simulation is more appropriate for predictions of subsurface flow.

Geostatistics is primarily concerned with constructing high-resolution 3-D models of categorical variables such as facies and continuous variables such as porosity and permeability. It is necessary to have *hard* truth measurements at some volumetric scale. All other data types including geophysical data are called *soft* data and must be calibrated to the hard data. It is neither possible nor optimal to construct models at the resolution of the hard data. Models are generated at some intermediate geological modeling scale, and then scaled to an even coarser resolution for flow modeling. An important goal of geostatistics is the creation of detailed numerical 3-D geologic models that simultaneously account for a wide range of relevant data of varying degrees of resolution, quality, and certainty. Much of geostatistics relates to data calibration and reconciling data types at different scales. This data integration or *fusion* is the focus of this review paper.

Geostatistical techniques allow alternative realizations to be generated. These realizations are often combined in a model of uncertainty, that is, they are processed through a numerical model of the response and the different outcomes are assembled in a distribution of response uncertainty. Uncertainty is becoming an important goal of geostatistical studies.

Numerical models are rarely built in one step. A hierarchical framework is followed with different techniques and tools at each level. A typical scenario consists of (1) mapping large scale bounding surfaces with conventional or geostatistical techniques, (2) mapping trends of facies proportions within each major stratigraphic layer, (3) creating high resolution facies models within each layer reproducing the mapped trends, (4) assigning continuous rock properties such as porosity and permeability within each facies, and (5) post processing and upscaling the resulting high resolution models for flow simulation. The classical random function model formalism of geostatistics is presented first, then some of the practical implementation aspects are described.

2. RANDOM FUNCTION FORMALISM

We start by considering a regionalized variable such as a subsurface elevation, formation thickness, facies proportion, facies indicator, porosity or permeability. We denote a specific value as z. The uncertainty about an unsampled value z is modeled through the probability distribution of a random variable (RV) Z. The probability distribution of Z after data conditioning is usually location-dependent; hence the notation $Z(\mathbf{u})$, with $\mathbf{u}$ being the coordinate location vector. A random function (RF) is a set of RVs defined over some field of interest, e.g., $Z(\mathbf{u})$, $\mathbf{u} \in$ study area A. Geostatistics is concerned with inference of statistics related to a random function (RF).

Inference of any statistic requires some repetitive sampling. For example, repetitive sampling of the variable $z(\mathbf{u})$ is needed to evaluate the cumulative distribution function: $F(\mathbf{u};z) = \text{Prob}\{Z(\mathbf{u}) \leq z\}$ from experimental proportions. However, in most cases, at most one sample is available at any single location $\mathbf{u}$; therefore, the paradigm underlying statistical inference processes is to trade the unavailable replication at location $\mathbf{u}$ for replication over the sampling distribution of z-samples collected at other locations within the same general area.

This trade of replication corresponds to the decision of stationarity. Stationarity is a property of the RF model, not of the underlying regionalized variable. Thus, it cannot be checked from data. The decision to pool data into statistics across facies is not refutable a priori from data; however, it can be shown inappropriate a posteriori if differentiation per facies is critical to the study.

The first and most important aspect of stationarity is the decision to pool data together for common processing. Another aspect of stationarity is a decision regarding the location-dependency of statistical parameters. A common practical approach is to assume that key statistical parameters do not depend on location within reasonably defined geological populations.

The statistical paradigm faced by geostatisticians is one of multivariate statistics: the same variable at multiple locations and multiple secondary data. We could denote the secondary data as $Y(\mathbf{u})$ and index Y if required to be clear regarding the number of secondary data. This is illustrated schematically in Plate 1.

The two wells and gridded seismic response on Plate 1 illustrate the multivariate aspect of the problem faced by geostatisticians. We are interested in the uncertainty at a location that has not been drilled. The nearby data (n) consist of well and seismic data:

$$(n) = \{z(\mathbf{u}_\alpha), \alpha=1,\ldots,n_w\}, \{y(\mathbf{u}_\beta'), \beta=1,\ldots,n_s\} \quad (1)$$

The uncertainty at a particular unsampled location must be inferred in light of the (n) conditioning data. A best estimate can be retrieved from the conditional distribution or it could be sampled by Monte Carlo simulation for alternative realizations. The standard approach to estimate conditional probabilities is Bayes Law, which has been used for more than 200 years. Bayes Law provides the arithmetic to infer the conditional distribution of the unsampled value $z(\mathbf{u})$:

$$F_{Z(\mathbf{u})|(n)}(z) = \frac{F_{Z(\mathbf{u}),(n)}(z_0, z_1, \ldots, z_n)}{F_{(n)}(z_1, \ldots, z_n)} \quad (2)$$

The numerator on the right side is an $n+1$ variate distribution of the unknown and the n data. The denominator on the right side is the n variate distribution of the conditioning data. The univariate distribution on the left side is what we are after–the conditional distribution of the unsampled value given the set of conditioning data (n).

Inference of the required multivariate distributions is virtually impossible. There are no replications of the unsampled value with the data values and there are unlikely to be replications of the precise data configuration (n). Nevertheless, those multivariate probabilities are required for inference of the conditional distribution.

The required multivariate probabilities are calculated from either an analytical distribution model or from a large set of analogue data deemed representative (sometimes referred to as a training image). The conventional paradigm of geostatistics is to use analytical distributions with parameters inferred from the available data. The multivariate Gaussian distribution will

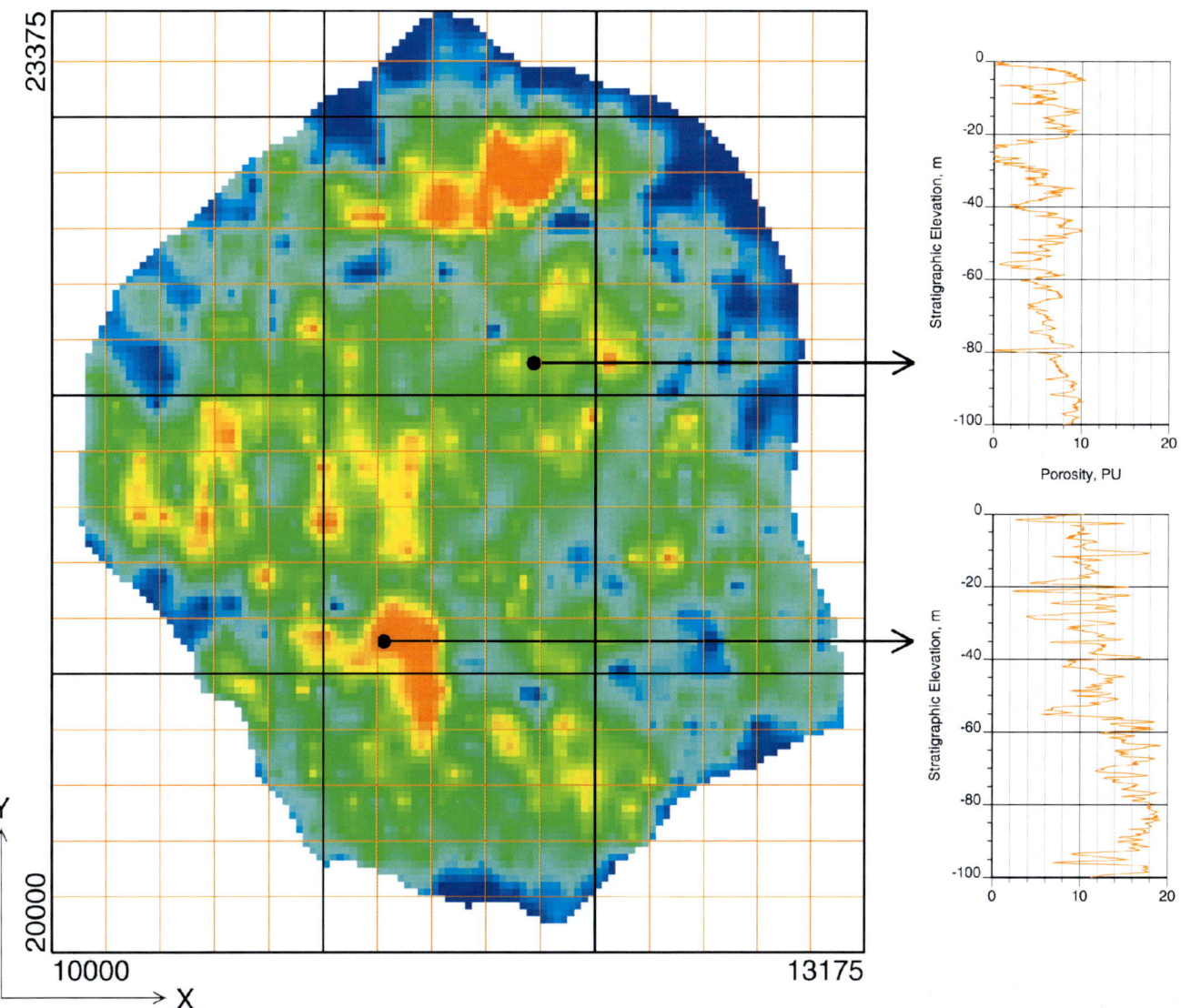

Plate 1. Illustration of the typical case faced by geostatisticians: there are a limited number of locations with precise measurements (the two wells in this case) and secondary variables that are often on grids (one variable shown).

be explained in the next section. The classical approach of variograms and kriging will be explained now.

Subsurface variables are heterogeneous. Their spatial variability is quantified by the variogram function:

$$2\gamma(\mathbf{h}) = E\left\{\left[Z(\mathbf{u}) - Z(\mathbf{u}+\mathbf{h})\right]^2\right\} \quad (3)$$

2γ is variability and is in the units of variance, $\mathbf{h}$ is a vector distance, $Z(\mathbf{u})$ is the random variable. The expected value is approximated by a discrete sum over the available pairs. The available data do not permit estimation of 2γ for many distance and direction lags. The function is fit to interpolate 2γ for $\mathbf{h}$-values that cannot be calculated.

Estimation can be formulated as an optimization problem. The linear estimate at unsampled location $\mathbf{u}_0$ is written:

$$\left[z^*(\mathbf{u}_0) - m(\mathbf{u}_0)\right] = \sum_{i=1}^{n} \lambda_i \cdot \left[z(\mathbf{u}_i) - m(\mathbf{u}_i)\right] \quad (4)$$

$m(\mathbf{u})$ is the location-dependent mean and the λs are weights that are calculated to minimize the expected error variance. The equations that lead to the optimal weights are referred to as the normal equations or the simple kriging equations:

$$\sum_{j=1}^{n} \lambda_j \cdot C(\mathbf{u}_i - \mathbf{u}_j) = C(\mathbf{u}_i - \mathbf{u}_0), \quad i = 1, \dots, n \quad (5)$$

The $C(\mathbf{h})$ covariance values are derived from the variogram through the relation $C(\mathbf{h}) = \sigma^2 - \gamma(\mathbf{h})$, which is valid with the assumption of stationarity. Relatively straightforward modifications are necessary if the decision of stationarity is relaxed. Constraints may be added to ensure unbiasedness without specifying the location-dependent mean; the modifications are straightforward. We can calculate the minimized error variance, but it has no practical meaning outside of the Gaussian context (see Section 3 below).

Estimates from Equation 4 are useful. They provide a useful means to construct a grid of estimates. These estimates are used for resource assessment and visualization of geologic trends. The kriging formalism of Equations 4 and 5 may be extended to multiple correlated variables. The result is cokriging. The estimate follows the same form; however, the covariance values must come from a mathematically valid model of coregionalization.

Kriging was state of the art in the late 1970s and early 1980s. Geostatisticians have come to expect more from their numerical models: local and joint uncertainty. Equation 2 is valid. A multivariate distribution is required. The multivariate Gaussian distribution is a remarkably tractable model that has come to be relied upon in geostatistical calculations.

3. MULTIVARIATE GAUSSIAN DISTRIBUTION

The multivariate probabilities required for inference of continuous variable uncertainty cannot be directly inferred from data. A multivariate Gaussian model is systematically adopted. The continuous variable is transformed to a Gaussian distribution, and then all multivariate distributions are assumed to be Gaussian. We would wish for alternative probabilistic models to choose from; however, the multivariate Gaussian probability distribution is remarkably tractable and used almost exclusively. Figure 1 illustrates transformation of a continuous variable from an arbitrary distribution to a Gaussian distribution. The distributions are shown as cumulative distribu-

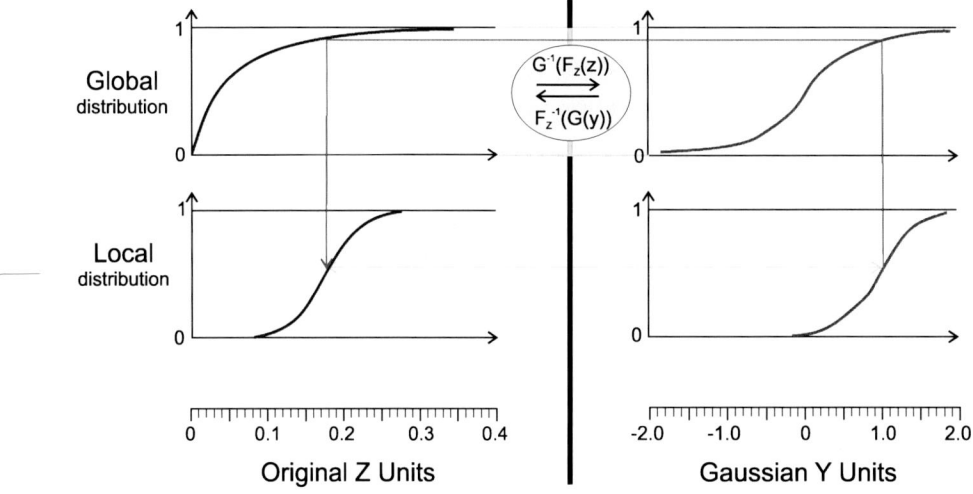

Figure 1. Schematic illustration of normal score transform. The original Z- data are on the left and the Gaussian Y-values are on the right. The top figures are the global CDFs and the bottom figures represent local CDFs. Quantiles are transformed using the global distribution (the three part blue line).

tions. In Gaussian units (the right side), all distributions are Gaussian in shape. The uncertainty in original units must be established by back transformation. The transformation and back transformation are written as:

$$y = G^{-1}(F(z)) \text{ and } z = F^{-1}(G(y)) \qquad (6)$$

Figure 1 reveals an important point. All conditional distributions in Gaussian units are non-standard Gaussian, see the lower right. The quantiles of such distributions can be back transformed via the global transformation. Conditional distributions in original units are not Gaussian, but we can establish their shape numerically, that is, back transformation of many quantiles. The 99 percentiles would be a good start; more are required for a stable estimate of the variance.

The mean and variance of each conditional non-standard Gaussian distribution are calculated with the normal equations that are identical to the kriging equations given in equations 4 and 5. The stationary mean is set to 0.0 in Gaussian units and the variogram/covariance are calculated from the normal score transforms of the data. The variance of estimation has particular meaning in the Gaussian case; it is the variance of the conditional distribution:

$$\sigma^2 = 1 - \sum_{i=1}^{n} \lambda_i \cdot C(u_i - u_0) \qquad (7)$$

A small example will be developed at the expense of some space. This example is a classic illustration of modern geostatistical tools used to assess uncertainty.

3.1 Small Example

Consider a square grid of 101–16m grid cells that cover just over one regular Section of land. Let's directly model porosity. The global representative distribution will be taken as lognormal with a mean $m=0.15$ and a standard deviation $\sigma=0.075$. The global representative distribution would be obtained by declustering and/or debiasing using the available well and seismic data. Consider an average data of 0.15 in the northwest corner of the area and a high data of 0.25 in the southeast corner of the area.

Uncertainty is characterized in Gaussian units. The transformation to a standard Gaussian distribution is defined analytically in this case:

$$y = \frac{\log(z) - \alpha}{\beta}$$
$$\beta = \sqrt{\log\left(1 + \frac{\sigma^2}{m^2}\right)} \text{ and } \alpha = \log(m) - \beta^2/2 \qquad (8)$$

In our case $\alpha = -2.01$ and $\beta = 0.472$. The back transform is also defined analytically: $z = exp(y\beta + \alpha)$. The porosity data values of 0.15 and 0.25 are transformed to 0.236 and 1.317, respectively.

A fitted variogram model of the Gaussian transformed values is required. This would be obtained from the available data and analogue information. The variogram will be taken as an exponential function with an effective range of 2000m: $\gamma(\mathbf{h}) = 1 - exp(3h/2000)$. In fact, $\gamma(\mathbf{h})$ is the semivariogram or one half of the variogram. Under a decision of stationarity, the covariance function is $C(\mathbf{h}) = 1 - \gamma(\mathbf{h}) = exp(3h/2000)$.

Local conditional distributions are defined everywhere by a local conditional mean and variance that are computed by simple kriging. Plate 2 shows these results. The locations of the wells are evident on the conditional variance map–the conditional variance is zero at the two well locations. These results are in Gaussian units. We back transform these conditional distributions to original units by back transforming a large number of quantiles, say 200. Plate 3 shows maps of the conditional mean, conditional variance, P_{90} low value and P_{10} high value in original units. Note how the conditional variance in original units is higher in the south and east because the mean is higher; the conditional variance in original units depends on the data as well as the data configuration.

Simulated realizations are required for two reasons. Firstly, they provide numerical models of heterogeneity for process evaluation. Secondly, they permit input uncertainty to be transferred to output uncertainty, for example, calculating uncertainty in resources or transport. There are a number of implementations that generate multiple realizations. Sequential methods such as sequential Gaussian simulation are popular.

Multiple realizations of porosity are generated by Gaussian simulation. Five realizations are shown on the left of Plate 4. The two well data are reproduced by all realizations. The pore volume was calculated on each realization assuming a thickness of 10m. The distribution of pore volume is shown at the right of Plate 4. These realizations allow us to visualize heterogeneity as well as assess uncertainty. The realizations could be ranked by their pore volume and select realizations (say the ones with the P_{90}, P_{50} and P_{10} outcomes) could be input to flow simulation.

This little example shows a hint of what geostatistics is aimed at. In practice, we must consider multiple stratigraphic layers, multiple facies, multiple data types, and multiple variables such as residual saturation and permeability. Some practicalities are addressed below.

3.2 Block Cokriging

An important practical reality of geostatistics is the presence of data with different type, noise content, and volume

12 A REVIEW OF GEOSTATISTICAL APPROACHES TO DATA FUSION

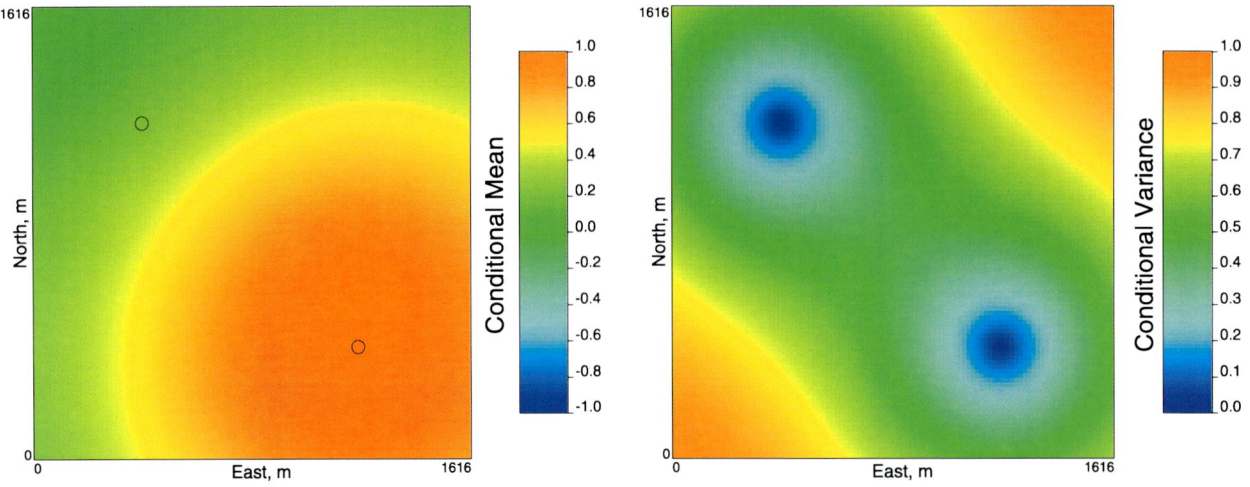

Plate 2. Map of the conditional mean (left side) and conditional variance (right side) for the Small Example.

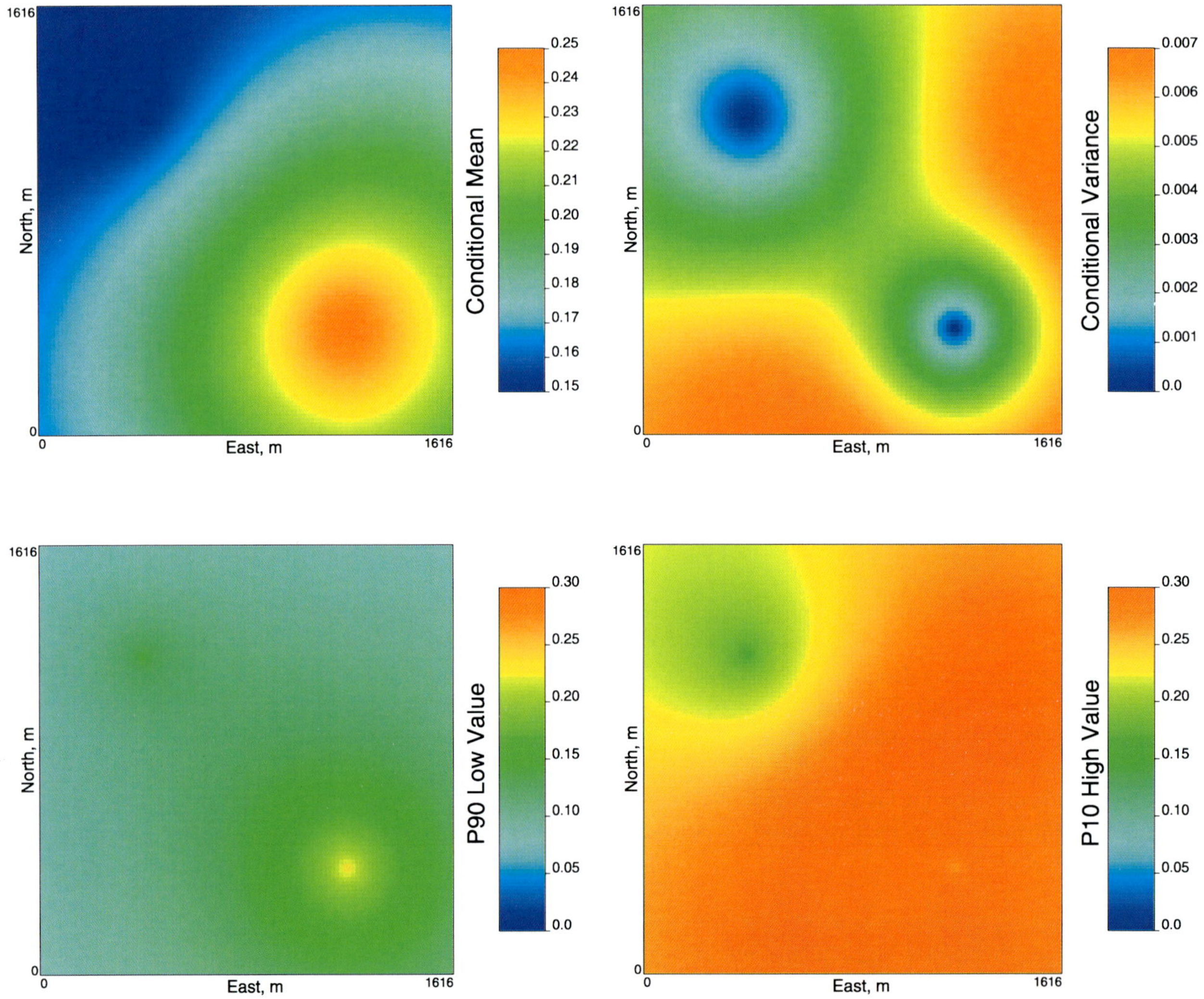

Plate 3. Map of the conditional mean (upper left) and conditional variance (upper right) in original units. Maps of the P90 low value and P10 high values are shown in the lower left and right.

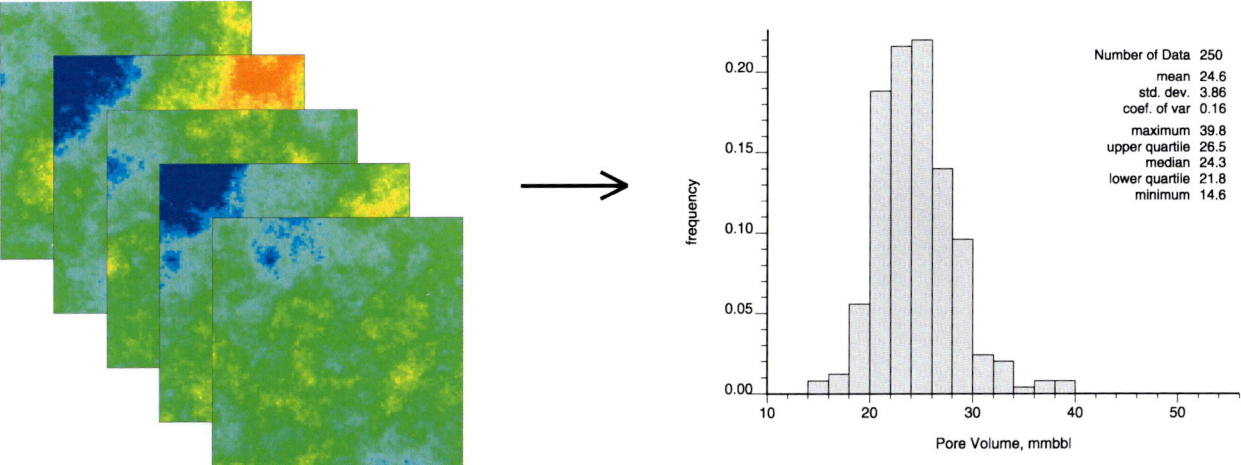

Plate 4. Multiple realizations (5 out of 250) are illustrated on the left and a histogram of the OOIP for the 250 realizations is shown to the right.

scale. We could invoke block cokriging to address these three critical issues. Different data types are handled with a cokriging and a model of coregionalization. Different volume scales of measurement are handled by block cokriging, that is, the use of volume averaged covariances. There are a number of inference problems and challenges with this approach: (1) linear averaging is assumed in Gaussian units, which is only correct if the original variable histograms are Gaussian in shape, (2) the point-scale statistics including histograms and variograms must be known, and (3) the noise content of each data source must also be known. This approach is valid and manageable in many cases. Nevertheless, these assumptions are serious and often lead practitioners to consider some simplifications. A number of practical implementation issues will now be discussed. These are unquestionably important for reasonable results in the combination of data with geostatistics.

4. PRACTICAL IMPLEMENTATION

4.1 Representative Statistics

Wells are not drilled to be statistically representative of the site; they are often intended as locations for production. Even in preliminary appraisal, there is a desire to delineate interesting areas of the site. It is critical to establish a representative distribution for each variable being modeled. This includes facies proportions and the histograms of porosity and permeability within each facies type. *Declustering* techniques weight the data such that wells drilled close together are given less weight. Wells drilled farther apart are given more weight. Declustering is suitable when there are sufficient data to sample areas of high and low quality. Sometimes there are too few wells. There may be areas of relatively poor reservoir quality that have not been drilled. *Debiasing* techniques are used to establish representative distributions based on a secondary variable such as seismic or a geologic trend. The results of declustering and debiasing include representative facies proportions and representative histograms of each continuous variable under consideration. A large-scale trend model may have been built for debiasing–this trend model will also come into subsequent geostatistical calculations.

An essential feature of geostatistics is inference in presence of sparse data. We are faced with a paradox. A lack of data is precisely when a geostatistical model of uncertainty is warranted; however, it is also the case when inferring required parameters is difficult. Limiting ourselves to statistics we can infer from the available data would be a mistake. We must often use analogue information related to spatial continuity, particularly in the vast interwell region. The spatial continuity in the vertical direction is relatively easy to infer even with limited well data. Horizontal to vertical anisotropy ratios based on the geologic setting can be useful to infer the horizontal continuity. The vertical variogram shape is used, but scaled according to a ratio. Figure 2 shows some typical ratios (Deutsch, 2003).

4.2 Hierarchical Modeling

A sequential approach is often followed for reservoir modeling. The large-scale features are modeled first followed by smaller, more uncertain, features:

(1) Establish the stratigraphic layers to model, that is, define the geometry of the *container* being modeled. A conceptual model for the large scale continuity of facies and petrophysical properties within each major layer is chosen.
(2) The bounding surfaces are mapped. They may be simulated with geostatistical techniques if they are associated with considerable uncertainty.
(3) The facies rock types are modeled by cell-based or object-based techniques within each stratigraphic layer (see below). Multiple realizations represent uncertainty in facies.
(4) The porosity and other petrophysical variables are modeled on a by-facies basis. These may be modeled one after another or all together. Multiple realizations are used to represent uncertainty.
(5) The models are revised to match dynamic data such as pumping tests and flow history. Knowledge gained from trying to match this data may be coded as spatial constraints and the modeling repeated.
(6) These set of multiple realizations are input to flow and transport modeling or simply visualized to aid in decision making and resource assessment.

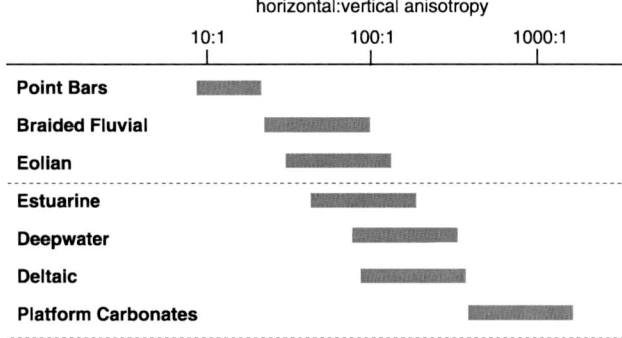

Figure 2. Some typical horizontal-to-vertical anisotropy ratio conceptualized from available literature and experience. Such generalizations can be used to verify actual calculations and supplement very sparse data.

A detailed description of these steps is beyond the scope of this review paper, but some of the references are suitable. The book by Chiles and Delfiner (1999) is a comprehensive overview of geostatistical techniques. The book by Cressie (1991) presents a statistical perspective on this approach. The book by David (1977) is a practical mining approach. The two books by Deutsch (1998 and 2003) provide a software and petroleum perspective, respectively. Goovaerts (1997) provides another comprehensive overview of geostatistical techniques. Isaaks and Srivastava (1989) provide a nice introduction to basic concepts. Journel and Huijbregts (1978) provide a comprehensive theoretical presentation from a mining perspective. Kitanidis (2000) provides an introduction from a hydrogeologic perspective.

4.3 Facies Modeling

Facies are often important in reservoir modeling because the petrophysical properties of interest are highly correlated with facies type. Facies are distinguished by different grain size or different diagenetic alteration. The facies must have a significant control on the porosity and other properties of interest; otherwise, modeling the 3-D distribution of facies will be of little benefit since uncertainty will not be reduced and the resulting models will have no more predictive power. An additional constraint on the choice of facies is that they must have straightforward spatial variation patterns. The distribution of facies should be at least as easy to model as the direct prediction of petrophysical properties. Once the facies are defined, relevant data must be assembled and a 3-D modeling technique selected.

The alternatives are (1) cell-based geostatistical modeling, (2) object-based stochastic modeling, or (3) deterministic mapping. Deterministic mapping is always preferred when there is sufficient evidence of the facies distribution to remove any doubt of the 3-D distribution. In many cases, there is evidence of geologic trends, which should be included in stochastic facies modeling.

Cell-based techniques are commonly applied to create facies models. The popularity of cell-based techniques is understandable: (1) local data are reproduced by construction, (2) the required statistical controls (variograms) may be inferred from limited well data, (3) soft seismic data and large-scale geological trends are handled straightforwardly, and (4) the results appear realistic for geological settings where there are no clear geologic facies geometries, that is, when the facies are diagenetically controlled or where the original depositional facies have complex variation patterns. Of course, when the facies appear to follow clear geometric patterns, such as sand-filled abandoned channels or lithified dunes, object-based facies algorithms should be considered.

From a geological perspective, it is convenient to view reservoirs and aquifers from a chrono-stratigraphic perspective. The sedimentary architecture is considered in light of a hierarchical classification scheme. We consider modeling this genetic hierarchy of heterogeneities by surfaces and objects representing facies associations.

Despite the realism of object-based modeling, many reservoirs show very complicated architectural element configurations developed during meander migration punctuated by avulsion events. It is becoming increasingly common to attempt facies modeling in a manner that mimics original deposition and alteration. Like object-based modeling, there is a perception that these process-based models are difficult to condition to well data.

Image analysis based techniques using multiple point statistics have evolved to use the models generated by object- and process-based models as training images. The features of such models are imposed on 3-D geocellular models with multiple point statistics (Guardiano and Srivastava, 1992).

4.4 Secondary Data

The block cokriging approach mentioned above has limited applicability in presence of many secondary data at different scales. Inference of the required statistics is virtually impossible. Collocated cokriging simplifies the process to consider collocated secondary variables; however, there is no simple way to consider a large number of secondary variables simultaneously.

Many different variables must be considered: small scale well data, large-scale remotely sensed variables, interpreted trend-like variables, and other response variables. These data often cover different areas, provide data at different scales, and are variably correlated together. Conventional geostatistical techniques, such as the block cokriging mentioned above, incorporate the spatial structure but these techniques are cumbersome in the presence of many secondary variables. An increasingly common approach is to merge all secondary data into a single variable that contains all of the secondary variable information; this provides a conditional distribution. The spatial distribution of each variable is mapped with data of the same type of information; this provides a second conditional distribution. The two conditional distributions are merged to provide updated posterior distributions. This merging is done in Gaussian units and the variables must be back transformed for final analysis.

Two Gaussian conditional distributions may be merged to an updated Gaussian distribution assuming conditional independence of the two distributions. This type of Markov model is very common. The parameters of the updated Gaussian distribution are given by:

$$m_U = \frac{m_1 \cdot \sigma_2^2 + m_2 \cdot \sigma_1^2}{\sigma_1^2 - \sigma_1^2 \sigma_2^2 + \sigma_2^2} \text{ and } \sigma_U^2 = \frac{\sigma_1^2 \cdot \sigma_2^2}{\sigma_1^2 - \sigma_1^2 \sigma_2^2 + \sigma_2^2} \quad (9)$$

This simple result is at the heart of much data integration. An important notion of data integration is that corroborating data cause the updated distributions to be non-convex. Figure 3 shows three examples. In the first case, both distributions are high (m_1=0.8, σ^2_1=0.6, m_2=1.0, σ^2_2=0.4), that is, the mean values are greater than the global mean; therefore, the updated distribution is quite high (m_u=1.21, σ^s_u=0.316). In the second case, one distribution is high and the other low (m_1=-0.5, σ^2_1=0.6, m_2=0.5, σ^2_2=0.6); therefore, the result is in the middle (m_u=0.0, σ^s_u=0.429). In the third case, both distributions are low (m_1=-0.8, σ^2_1=0.6, m_2=-1.0, σ^2_2=0.4); therefore, the updated distribution is even lower (m_u=-1.21, σ^s_u=0.316).

Recall that all of our data are transformed to Gaussian units, uncertainty is assessed, and the resulting uncertainty is back transformed to original units. Quantiles of local distributions can be back transformed or entire simulated realizations are back transformed. The multivariate Gaussian distribution is used routinely in geostatistics because it is straightforward to infer the required parameters with few data. Data integration and uncertainty prediction is relatively easy. Moreover, it is common that data within reasonably defined facies are often Gaussian. Nevertheless, we often seek an alternative to the Gaussian model to handle more complex features. The most common alternative is the indicator formalism (Journel, 1983).

4.5 Indicator Formalism

Indicators are applied to both continuous and categorical variables. A series of threshold values z_c are used to discretize the range of variability of the continuous Z-variable. The indicator coding of continuous variables:

$$i(\mathbf{u}; z_c) = \begin{cases} 1, \text{if } z(\mathbf{u}) \leq z_c \\ 0, \text{otherwise} \end{cases} \quad (10)$$

This amounts to coding the continuous data as a series of cumulative probability values. It is common to consider between 9 and 20 threshold values; less than 9 leads to poor resolution and greater than 20 leads to difficult inference of the required parameters and no significant increase in precision of calculated conditional distributions.

Variogram analysis is conducted for each threshold. This permits the continuity of the low and high values to be modeled differently. The variograms should be consistent since they are based on the same underlying continuous variable; however, they are more flexible than the simplistic Gaussian model.

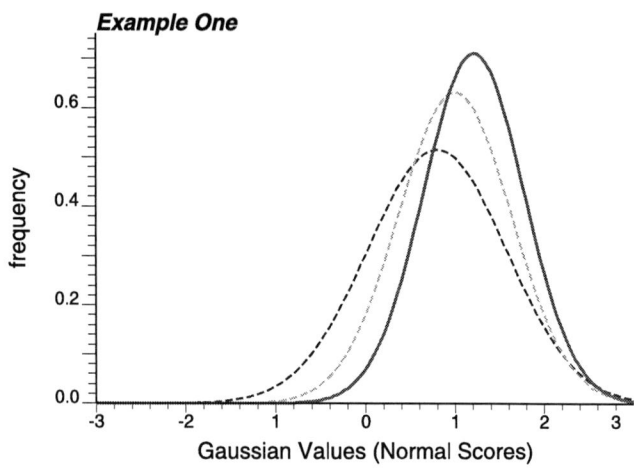

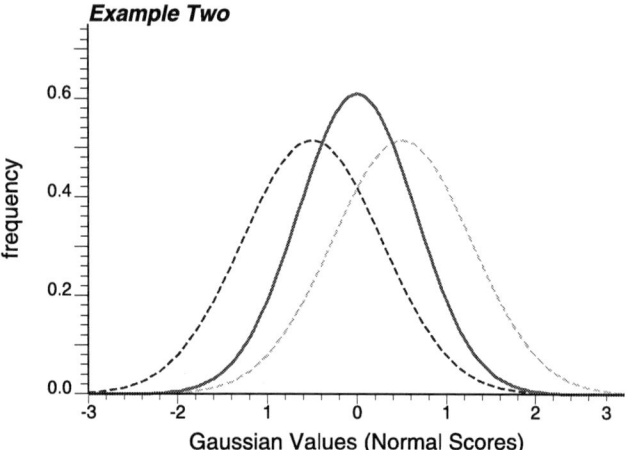

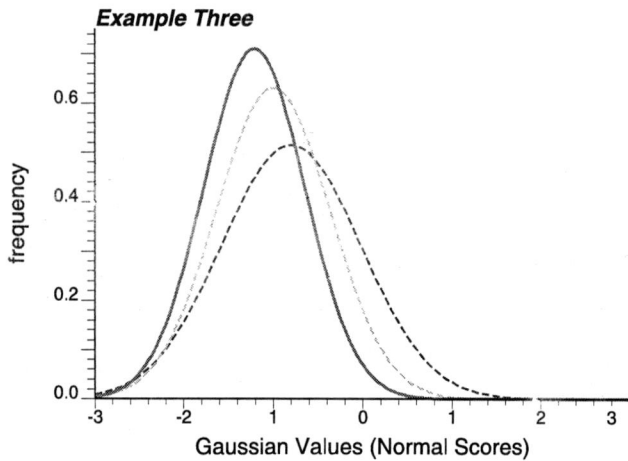

Figure 3. Three examples of updating two conditional Gaussian distributions into updated distributions.

Kriging is applied at each threshold with the corresponding indicator variogram to directly calculate an estimate of the CDF value at the threshold values. This leads to a direct estimate of the conditional distribution. Figure 4 shows a schematic example. It is necessary to ensure that the estimated CDF values form a valid distribution (non decreasing between 0 and 1) and to interpolate and extrapolate the CDF beyond the values predicted at the thresholds. These distributions can be used directly for uncertainty assessment or used in simulation to assess joint uncertainty.

The indicator coding for categorical variables is similar. Consider K facies. The data are coded as the probability of occurrence:

$$i(\mathbf{u};k) = \begin{cases} 1, \text{if facies } k \text{ at location} \\ 0, \text{otherwise} \end{cases} \quad (11)$$

As with continuous variables, variograms are constructed for each of the K indicators. Kriging can be applied to predict the probability of each facies at an unsampled location. The probability estimates are corrected if necessary to ensure that they are non negative and sum to one. They are then used for uncertainty assessment or the simulation.

The hierarchical scheme described in Section 4.2 leads to multiple realizations of the study area under investigation. A variety of techniques, including indicator techniques, are used at different steps to arrive at a set of realizations that quantify the uncertainty. Each realization is a full specification of the study area: location, geometry, internal facies and petrophysical properties. These realizations must be post processed.

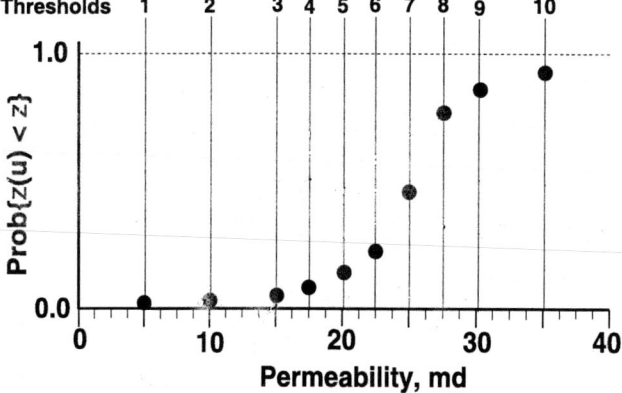

Figure 4. Example of a probability distribution derived from the indicator formalism. Each probability estimate is derived from kriging the data coded at that threshold.

4.6 Post Processing

Geostatistical models are useful for many purposes. The estimates at unsampled locations can be used directly for some decisions. The local uncertainty, that is, uncertainty at one location at a time is easily assembled from the multiple realizations. Maps can be made of P10 low values (the 0.1 quantiles of the local distributions) and the P90 high values (the 0.9 quantiles). These maps reveal two important features: (1) when the P10 value is high, then the actual value is surely high–there is a 90% probability to be even higher, and (2) when the P90 value is low, then the actual value is surely low–there is a 90% probability to be even lower. The local conditional variance could also be mapped to summarize local uncertainty. Another summary statistic that can be useful is the probability for the true value to be within a percentage (say 15%) of the estimate. There is little uncertainty when this probability is high.

Local estimates and local uncertainty are useful; however, they do not tell us the uncertainty at multiple locations simultaneously. Whenever uncertainty at many locations is required, then simulated realizations must be used.

Estimates are smooth and often inappropriate for direct input into flow simulation; flow predictions are biased because the connectivity of high permeability flow conduits and low permeability flow baffles is not accounted for. Simulated realizations are more appropriate. They also carry a measure of uncertainty. The paradigm of probabilistic analysis is that the set realizations are processed through a transfer function to assess uncertainty in response variables, see Figure 5.

Some response variables are straightforward such as volumetric calculation of resources. In fact, the response of smooth estimated models should match the average of the simulated realizations. In most cases, the transfer function is non-linear. Resources above a critical threshold and the response of flow and transport modeling are non-linear transfer functions. The response of an average is not the same as the average response.

All realizations are processed through the transfer function. This provides a distribution of uncertainty in the response variables. There are times when the transfer function (flow simulation) is very CPU-demanding. Moreover, many different scenarios of the transfer function must be considered. It is intractable to process all of the realizations through all scenarios.

The realizations are ranked according to some easy to calculate statistic such as the connected resource. Then, selected *low*, *median*, and *high* realizations are processed through the full transfer function. The ranking measure may be as simple as pore volume or as complex as the results of a fast flow

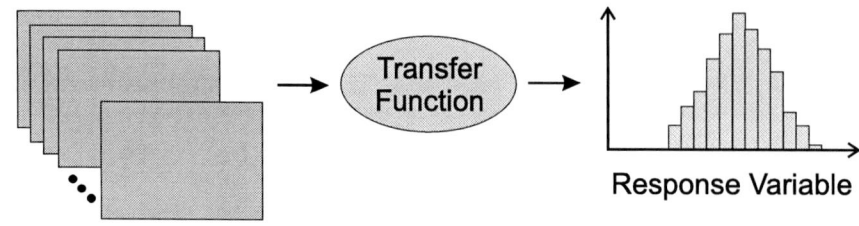

Figure 5. Schematic of how multiple realizations are processed through a transfer function to calculate uncertainty in a response variable (or multiple response variables).

simulation such as streamlines. A ranking measure of intermediate complexity often suffices: the connected volume to well locations is a good intermediate measure.

The preceding discussion has focused on uncertainty. Another aspect of post processing is sensitivity analysis, that is, determining how sensitive the response variables are to each of the input parameters/variables. This is done by holding some parameters constant or with experimental design techniques.

5. FUTURE DIRECTIONS

Geostatistical techniques for data fusion are applied in subsurface hydrology and other areas of geological modeling. A number of alternatives exist; however, the classical geostatistical paradigm presented here has had a history of successful prediction, is applied regularly and will provide unquestioned value in future applications.

The main problems with the geostatistical approach are that (1) it is poorly constrained by geological knowledge and processes, and (2) many statistical parameters must be inferred. No approach is perfect and people often want a change from the tried, true and boring applications with conventional techniques Alternatives are under consideration. Process-mimicking geological modeling, multiple point statistics, and data integration techniques provide interesting future directions.

Acknowledgments. Canada's NSERC organization and the industry sponsors of the Centre for Computational Geostatistics at the University of Alberta are gratefully acknowledged for financial assistance.

REFERENCES

Chiles, J. P., and Delfiner, P. (1999). "Geostatistics : Modeling Spatial Uncertainty (Wiley Series in Probability and Statistics. Applied Probability and Statistics.)," Wiley, New York.

Cressie, N. (1991). "Statistics for Spatial Data." Wiley, New York.

David, M. (1977). "Geostatistical Ore Reserve Estimation." Elsevier, Amsterdam.

Deutsch, C. V., and Journel, A.G. (1997). "GSLIB: Geostatistical Software Library." Second Edition, Oxford University Press, New York.

Deutsch, C. V., (2003). "Geostatistical Reservoir Modeling," Oxford University Press, New York.

Goovaerts, P. (1997). "Geostatistics for Natural Resources Evaluation." Oxford University Press, New York.

Guardiano, F. and Srivastava, R. M. (1992), Multivariate Geostatistics: Beyond Bivariate Moments, in *Proceedings of Fourth International Geostatistics Congress*, Kluwer, New York.

Isaaks, E. H., and Srivastava, R. M. (1989). "An Introduction to Applied Geostatistics." Oxford University Press, New York.

Journel, A. G., and Huijbregts, C. (1978). "Mining Geostatistics." Academic Press, London.

Journel, A. G. (1983). Non-parametric estimation of spatial distributions, *Mathematical Geology*, 15(3), PAGES 445–469.

Kitanidis, P. (1997), *Introduction to Geostatistics: Applications in Hydrogeology*, Cambridge University Press, New York.

Matheron, G. (1971). The theory of regionalized variables and its applications. Les cahiers du CMM. Fasc. No. 5, Ed. Ecole Nationale Superieure des Mines de Paris, Paris.

―――――――

C. V. Deutsch, Department of Civil and Environmental Engineering, 3-40 NREF Building, University of Alberta, Edmonton, Alberta, CANADA T6G 2W2

On Stochastic Inverse Modeling

Peter K. Kitanidis

Civil and Environmental Engineering, Stanford University, Stanford, California, USA

I review key concepts and equations related to the stochastic approach to inverse modeling. After going over the equations for linear inversing using geostatistical models, I discuss the challenge of striking a proper balance between resolution and noise suppression. Specifically the following questions are addressed: How closely to reproduce data; how much resolution to strive for; and how to evaluate realistic confidence intervals. It is argued that the geostatistical approach provides rigorous and systematic ways to deal with these important issues.

INTRODUCTION

Hydrogeologists, geophysicists, and other practitioners in the earth sciences rely heavily on data to infer properties of geologic formations with methods of analysis that fall under the general rubric of *inverse methods*. Each field has developed its own methods that reflect its special needs as well as the training and biases of its researchers. For example, someone trained in the solution of partial differential equations tends to see inverse problems differently from someone trained in statistics or linear algebra.

Hydrogeologists have traditionally dealt with sparse data and, consequently, many of them have embraced and employed stochastic methods, such as regression or least squares, geostatistical, and Bayesian. Geophysicists, on the other hand, have focused on the analysis of large data sets usually through suitable deterministic approximations, such as inferring media properties from seismic or ground penetrating radar data. Both fields recognize that there is uncertainty in what hydrologists call the estimate and geophysicists call the model. However, with notable exceptions, the majority of geophysicists have been reluctant to use stochastic methods to quantify uncertainty or to select among the many possible solutions. Many of them believe that proponents of the stochastic approach are overeager to apply stochastic methods when what is really needed is a good deterministic approximation. Some of those who identify stochastic methods with classical statistics wonder how one can find a probability distribution, when information is insufficient to identify even a single solution.

Stochastic inverse methods have probably not been part of the curriculum of most hydrogeophysicists. The basic motivation for this chapter is to advocate the use of stochastic methods and encourage readers to incorporate them in their work. I believe that those most familiar with the data must be given the tools to analyze them. In the words of Birnbaum [1962], "... each scientist and interpreter of experimental results bears responsibility for his own concepts of evidence and his interpretation of results..." Within the unavoidable space constraints, we will discuss the issue of uncertainty and the role of stochastic methods in a practical context. In particular, we will review the geostatistical inverse method [Kitanidis and Vomvoris, 1983, Dagan, 1985, Hoeksema and Kitanidis, 1984, 1985, 1989, Rubin and Dagan, 1987a, 1987b, Wagner and Gorelick, 1989, Hoeksema and Clapp, 1990, and others] in the framework of empirical Bayes methods, the meaning of the equations, and the practical significance of the results.

It must be emphasized that this is not, by any means, a review of methods that have appeared in the literature or an *apologia* for the stochastic approach. It is simply a tutorial on select important concepts that could be useful to practitioners.

UNCERTAINTY AND STOCHASTIC METHODS

Mathematical models are valuable tools in describing the hydrogeology of a site, understanding groundwater flow and transport processes, evaluating the effectiveness of management schemes, and making predictions. However, the parameters of these models must be inferred from data. For example, before using a numerical flow model, the conductivity and specific storage must be estimated at every block in the finite-difference mesh. A rich source of hydrogeologic data is piezometric head measurements. The literature on the subject of model calibration, parameter identification, history matching, or inverse modeling is voluminous (for example, see the reviews by Yeh, 1986, Carrera et al., 2005, or the book by Sun, 1994). The crucial issue is that the problem of specifying transmissivity in every block from sparse head observations is underdetermined, i.e., there are many solutions that are consistent with the data. The ambiguity is largely due to the scarcity of the data but is also inherent in the mathematics of typical inverse problems: a small range of values in the observed head is consistent with a larger range of transmissivity values. This characteristic is known as ill-posedness and results in non-uniqueness in the solution of the inverse problem. How does one pick the right answer? Isn't one solution as good as another?

Certain formalisms [e.g., Tikhonov and Arsenin, 1977, Tikhonov et al., 1997, Backus and Gilbert, 1967, 1968, Guidici et al, 1995] focus on finding *a* solution, a task that may be achieved by:

1) imposing restrictions on the solution, such as assuming that the domain is homogeneous, or consists of a few homogeneous zones, or that the solution is flat or smooth; or
2) making assumptions about available information, such as assuming that head is measured without error at every node of the model.

Such methods have their appeal in practice. However, the single answer they yield should not be confused with the "real thing", which can be determined only when sufficient information is truly available.

An alternative approach is to take up willingly and deliberately the multiplicity of solutions. Let us first see what we would like to achieve:

1) We would like to obtain a representative solution that contains features that are, in a sense, common to all possible solutions. For example, the many transmissivity functions that reproduce head observations vary in the small-scale characteristics but share the same large-scale features. The representative solution should have only the common features and be free of details that, though consistent with the data, are not necessarily valid. The representative solution should be relatively simple, flat, or smooth, not because the actual transmissivity function is necessarily so, but in view of the lack of information to resolve the small-scale characteristics. The representative solution takes the place of the single answer of the deterministic methodologies but does not claim to be the unique and true solution. It should be viewed as a good estimate. (Note that the term model, which is often used in geophysics to indicate an estimate, here is used to indicate a conceptual or mathematical representation of a system.)

2) We would like to evaluate the range of possible values. For example, we would like to bracket between two values the possible value of the transmissivity at a location. The idea is that although the exact value is not known, one should be able to identify an interval that contains the true value with a high degree of assurance. Being able to identify such a confidence or credible interval is useful in faithfully representing what is really known about the transmissivity, in contrast to maintaining the illusion of a unique answer. A large interval is indicative of more uncertainty and may be a justification to seek more data. Very small intervals indicate the ideal case of sufficient information to obtain an almost unique solution.

3) A more ambitious objective is to identify many solutions that are equally plausible candidates to be the actual transmissivity and collectively represent fairly the range of possible solutions. Each of these solutions is consistent with observations and other information, and is also sufficiently different from the others. By computing more and more of these solutions, one of them will approximate, in some sense, the actual solution. A set of such solutions can be used to perform risk analysis, which is roughly defined as the assessment, characterization, communication, and management of hazards. For example, one could evaluate the likelihood that the capture zone of a municipal well contains a neighboring contaminant plume and evaluate the chance that the contaminant levels in the extracted water may violate a water quality standard.

Stochastic methods are well suited and developed to meet these three objectives. Obviously, we do not have in mind classical statistics where probabilities are approximations to frequencies and are computed from repeated measurements of the same or similar quantities. Instead, we refer to Bayesian statistics, where the probabilities represent state of knowledge or available information. The idea is that the unknown function, such as transmissivity over a region, is modeled as a random function exactly because there is insufficient information to model it as deterministic, rather than because repeated measurements indicate a statistical regularity. For example, if we are asked to guess the porosity of an exotic material of which we know nothing, we may treat it as a random variable distributed uniformly in the interval 0 to 1. This probability was not obtained from the analysis of a large data set but reflects that we know nothing other than, by definition, the porosity must be between 0 and 1. If additional information is given, the probability distribution must change but, at every stage, the probability distribution must represent the available information.

The distinction between probability and frequency has been made by, among others, Jaynes [2003] who emphasized that probability theory can be regarded as the "calculus of inductive reasoning", or making inferences on the basis of incomplete information. This is not necessarily the same as the analysis of the frequencies of actual or (in the case of classical statistics) hypothesized repeated experiments. Furthermore, Jaynes [2003, p. 314] pointed out that "... consideration of random experiments is only one specialized application of probability theory, and not even the most important one; for probability theory as logic solves far more general problems of reasoning which have nothing to do with chance or randomness, but a great deal to do with the real world."

There are, however, skeptics about the validity of probability values thus obtained. In fact, one sometimes hears that this approach is inappropriate because certain events are not probabilistic or that there is not enough information to assign probabilities. I believe that these criticisms are largely due to inadequate understanding of the objectives of Bayesian methods. These methods make no claim of prescience but evaluate probabilities that reflect available information. A Bayesian method to solve an inverse problem should be judged on the basis that it is a logical, systematic, and largely objective way to explore the range of possible solutions based on whatever data one considers.

BAYES THEOREM

Let us now look at some mathematics, primarily for the purpose of making obvious the logical underpinnings of this approach. Bayes theorem is the centerpiece in the implementation of this approach. Let $\mathbf{s}$ be the unknown, an m-dimensional vector that is typically obtained from the discretization of an unknown function. In the source-identification example to be discussed later, $\mathbf{s}$ is the vector obtained from the discretization of the function that represents the pumping rate in a certain well over a period of interest. We distinguish between the prior probability density function (pdf) $p'(\mathbf{s})$ and the posterior pdf $p''(\mathbf{s})$, meaning prior and posterior in reference to some new data $\mathbf{y}$, which is, for mathematical convenience, represented as an n-dimensional vector. In our example, $\mathbf{y}$ represents measurements in a well. The posterior is computed from the prior by weighting by the likelihood, which is the pdf of $\mathbf{y}$ given $\mathbf{s}$, $p(\mathbf{y} \mid \mathbf{s})$:

$$p''(\mathbf{s}) = C p(\mathbf{y} \mid \mathbf{s}) p'(\mathbf{s}) \qquad (1)$$

The normalization factor C is required to make the left-hand side an appropriate probability density function (the integral should equal 1).

The simplicity of this formula belies some computational and conceptual challenges. Regarding computations, implementation of Bayes theorem in real-world inverse problems is more difficult than it appears because m, the length of $\mathbf{s}$, can be very large. Thus, brute-force integrations are out of the question. However, in some special cases, the most prominent being the linear-Gaussian one that we will discuss later, efficient methods of solution are available.

The most serious conceptual difficulty is the choice of the prior distribution because one may criticize it as subjective. However, it is possible to implement the approach in ways that leave little room for such criticism:

1. We may select distributions that are appropriately diffuse (or have large entropy) so that the results are primarily affected by the data, through the likelihood function. Those who criticize the use of Gaussian prior on the basis that the Gaussian distribution has high entropy fail to see that this feature is precisely what makes the Gaussian distribution appropriate. There may indeed be many cases that alternative distributions should be used; however, these distributions should also have the feature of large entropy while satisfying certain constraints. The topic is discussed in, for example, Christakos (1990).
2. We can select the prior on the basis of the data in what is known in a broader statistical context as empirical Bayes methodology (e.g., see Carlin and Louis, 2000). This approach is consistent with geostatistics (Matheron, 1971), as previously discussed (Kitanidis, 1986), as well as other applied statistical modeling methods. In geostatistics, the semivariogram, which is involved in the parameterization of the prior pdf, is determined from information that includes the data. (Thus, the often heard statement that the estimation variance depends on the data configuration and not the data themselves is patently false. The variance depends on the semivariogram which is selected from data.)
3. Last but not least, we should remind ourselves that the skillful analysis of data is an art guided by sound scientific principles and directed to the solution of practical problems. The selection of the prior, perhaps more than anything else, must be addressed in the context of an actual problem and of practical issues.

We will apply some of these ideas in a specific example.

LINEAR INVERSION

For illustration purposes, we will focus on linear inversion. The prevalence of linear inversion methods is justified by, among others, their computational efficiency and versatility, although they should not monopolize the analysis because they are certainly not applicable to all cases. It is worth noting that most "nonlinear" inversion methods used in practice involve application of the linear inversion equations, combined with successive linearization [e.g., Kitanidis, 1995]. We will summarize the key assumptions and results.

Consider that $s(\mathbf{x})$ is a function, such as the log-conductivity over a three-dimensional continuum or the pumping rate at a well over a period of time, to be estimated. Its structure is represented through the prior pdf $p'(\mathbf{s})$. Our basic model is that $s(\mathbf{x})$ can be represented as

$$s(\mathbf{x}) = \sum_{k=1}^{K} f_k(\mathbf{x})\beta_k + \varepsilon(\mathbf{x}) \quad (2)$$

The first term is the prior mean, where $f_k(\mathbf{x})$ are known functions and β_k are unknown coefficients, $k = 1,...K$; the second term is a function characterized through a mean that is zero and a covariance function. In a sense, the first part is deterministic (precisely specified) and the second part is stochastic (specified through probabilities or averages). This model is popular in geostatistics and in other data analysis approaches.

This representation, eq. (2), is quite versatile. For example, the zonation/regression approach is included as a special case: K is the number of zones,

$$f_k = \begin{cases} 1, & \text{if in zone } k \\ 0, & \text{otherwise} \end{cases},$$

and we set the covariance of ε equal to zero. In the regression approach, the variability is described through deterministic functions. In geostatistical approaches, one resolves variability mainly through the stochastic part and the simplest deterministic model is used, meaning $K = 1$ and $f(x) = 1$. After discretization (e.g., in the application of finite-difference and finite-element models), $s(\mathbf{x})$ is represented through an m by 1 vector $\mathbf{s}$. The mean of $\mathbf{s}$ is

$$\mu = E[\mathbf{s}] = \mathbf{X}\beta \quad (3)$$

where $\mathbf{X}$ is a known $m \times K$ matrix, $X_{ij} = f_j(x_i)$, and β are K unknown drift coefficients. The covariance of $\mathbf{s}$ is

$$E\left[(\mathbf{s} - \mathbf{X}\beta)(\mathbf{s} - \mathbf{X}\beta)^T\right] = \mathbf{Q} \quad (4)$$

In the geostatistical context, $\mathbf{Q}$ is parameterized through a semivariogram or generalized covariance function. In summary, Eq. (2) can be written, where ε is has zero mean and covariance matrix $\mathbf{Q}$,

$$\mathbf{s} = \mathbf{X}\beta + \varepsilon \quad (5)$$

The prior pdf is thus modeled as Gaussian with mean $\mathbf{X}\beta$ and covariance matrix $\mathbf{Q}$.

The observation vector $\mathbf{y}$ is related to the unknown vector $\mathbf{s}$ by the linear relation

$$\mathbf{y} = \mathbf{H}\mathbf{s} + \mathbf{v} \quad (6)$$

where $\mathbf{H}$ is an n by m given matrix; $\mathbf{v}$ is a random vector of deviations between observations and model, probabilistically independent from $\mathbf{s}$ (or ε), with mean zero and covariance matrix $\mathbf{R}$. The $\mathbf{H}\mathbf{s}$ part is the values that would be predicted from the mathematical model for the data. In the absence of the $\mathbf{v}$ term, the model should reproduce the observations exactly. However, by introducing the $\mathbf{v}$ terms, we recognize that the model predictions may deviate from the observations because neither observations nor mathematical models are flawless. Examples, including application to affine and nonlinear problems, can be found in the literature, e.g., Snodgrass and Kitanidis (1997).

For now, let us consider $\mathbf{Q}$ and $\mathbf{R}$ as given. The posterior distribution of $\mathbf{s}$ is Gaussian with mean and covariance matrix that can be computed through the solution of linear systems. The negative logarithm of the posterior pdf is given by:

$$\frac{1}{2}(\mathbf{y} - \mathbf{H}\mathbf{s})^T \mathbf{R}^{-1}(\mathbf{y} - \mathbf{H}\mathbf{s}) + \frac{1}{2}(\mathbf{s} - \mathbf{X}\beta)^T \mathbf{Q}^{-1}(\mathbf{s} - \mathbf{X}\beta) \quad (7)$$

The posterior mean in this case is the same with the value that maximizes the posterior pdf and can then be computed by minimizing (7) with respect to $\mathbf{s}$ and β. Note that the first term represents a penalty for not reproducing the data and the second term a penalty for deviating from the mean.

We will summarize two solutions, which should yield identical results.

The ξ Form

The posterior mean is

$$\hat{\mathbf{s}} = \mathbf{X}\hat{\beta} + \mathbf{Q}\mathbf{H}^T \xi \quad (8)$$

where the n by 1 vector ξ and the K by 1 vector $\hat{\beta}$ are found from the solution of:

$$\begin{bmatrix} \Psi & \Phi \\ \Phi^T & 0 \end{bmatrix} \begin{bmatrix} \xi \\ \hat{\beta} \end{bmatrix} = \begin{bmatrix} \mathbf{y} \\ 0 \end{bmatrix} \quad (9)$$

Thus, the solution is obtained by solving a system of $n + K$ linear equations. The coefficients are:

$$\Psi = \mathbf{H}\mathbf{Q}\mathbf{H}^T + \mathbf{R}, \quad \Phi = \mathbf{H}\mathbf{X} \quad (10)$$

The procedure to generate a conditional realization is as follows:

Generate an unconditional realization $\tilde{\mathbf{s}}_u$ with zero mean and covariance matrix $\mathbf{Q}$ and a realization of the measurement error $\mathbf{v}$ with zero mean and covariance matrix $\mathbf{R}$.

Then compute the conditional realization $\tilde{\mathbf{s}}_c$:

$$\tilde{\mathbf{s}}_c = \tilde{\mathbf{s}}_u + \mathbf{X}\hat{\boldsymbol{\beta}} + \mathbf{Q}\mathbf{H}^T\boldsymbol{\xi} \quad (11)$$

Where: $\boldsymbol{\xi}$ and $\hat{\boldsymbol{\beta}}$ are obtained from the solution of a system of $m + K$ linear equations with the same number of unknowns:

$$\begin{bmatrix} \boldsymbol{\Psi} & \boldsymbol{\Phi} \\ \boldsymbol{\Phi}^T & 0 \end{bmatrix} \begin{bmatrix} \boldsymbol{\xi} \\ \hat{\boldsymbol{\beta}} \end{bmatrix} = \begin{bmatrix} \mathbf{y} + \mathbf{v} - \mathbf{H}\tilde{\mathbf{s}}_u \\ 0 \end{bmatrix} \quad (12)$$

The Λ Form

The posterior mean is

$$\hat{\mathbf{s}} = \boldsymbol{\Lambda}\mathbf{y} \quad (13)$$

where Λ is m by n, which functions as a pseudo-inverse of $\mathbf{H}$, and can be found, in conjunction with the K by m matrix $\mathbf{M}$ from the following system:

$$\begin{bmatrix} \boldsymbol{\Psi} & \boldsymbol{\Phi} \\ \boldsymbol{\Phi}^T & 0 \end{bmatrix} \begin{bmatrix} \boldsymbol{\Lambda}^T \\ \mathbf{M} \end{bmatrix} = \begin{bmatrix} \mathbf{HQ} \\ \mathbf{X}^T \end{bmatrix} \quad (14)$$

These equations are basically the cokriging equations of geostatistics.

To generate a conditional realization, one uses

$$\tilde{\mathbf{s}}_c = \tilde{\mathbf{s}}_u + \boldsymbol{\Lambda}\left(\mathbf{y} + \mathbf{v} - \mathbf{H}\tilde{\mathbf{s}}_u\right) \quad (15)$$

where Λ is the same as the one used to find the posterior mean.

Additionally, the posterior covariance matrix $\mathbf{V}$ may be computed as follows:

$$\mathbf{V} = -\mathbf{XM} + \mathbf{Q} - \mathbf{QH}^T\boldsymbol{\Lambda}^T \quad (16)$$

ILLUSTRATIVE EXAMPLE

For illustration, we will consider a relatively simple source identification problem. Consider two-dimensional flow in a homogeneous and isotropic aquifer. We consider an extraction well at a location with coordinates $(0,0)$, i.e., the origin of the coordinate system. The hydraulic head satisfies the following partial differential equation.

$$S\frac{\partial \phi}{\partial t} - T\left(\frac{\partial^2 \phi}{\partial x_1^2} + \frac{\partial^2 \phi}{\partial x_2^2}\right) = -s(t)\delta(x_1)\delta(x_2) \quad (17)$$

where t is time $[T]$; (x_1, x_2) are spatial coordinates $[L]$; ϕ is the (depth-averaged) head $[L]$; S is the storativity (or storage coefficient) $[\,]$; T the transmissivity $[L^2/T]$; $s(t)$ is the extraction rate $[L^3/T]$; and $\delta(x)$ is the Dirac delta function $[L^{-1}]$. Using the hydraulic diffusivity, $D = \frac{T}{S}$, the equation can also be written

$$\frac{\partial \phi}{\partial t} - D\left(\frac{\partial^2 \phi}{\partial x_1^2} + \frac{\partial^2 \phi}{\partial x_2^2}\right) = -\frac{1}{S}s(t)\delta(x_1)\delta(x_2) \quad (18)$$

The solution, for initial head ϕ_0 and for an unbounded domain with level ϕ_0 away from the well is:

$$\phi(t, x_1, x_2) = \phi_0 - \frac{1}{4\pi T}\int_0^t \frac{1}{(t-\tau)}\exp\left(-\frac{x_1^2 + x_2^2}{4D(t-\tau)}\right)s(\tau)d\tau \quad (19)$$

This can also be written in terms of the drawdown,

$$\begin{aligned} Y(t, x_1, x_2) &= \phi_0 - \phi(t, x_1, x_2) \\ &= \frac{1}{4\pi T}\int_0^t \frac{1}{(t-\tau)}\exp\left(-\frac{x_1^2 + x_2^2}{4D(t-\tau)}\right)s(\tau)d\tau \end{aligned} \quad (20)$$

Consider the use of a monitoring well to determine whether a groundwater user has exceeded the water allocation. Our task is to estimate the pumping history at the extraction well, $s(t)$, from head observation at a nearby monitoring well. Aquifer data and observation-well coordinates are:

$$T = 0.01\frac{m^2}{\min}, \; S = 0.01, \; x_1 = 2m, \; x_2 = 0m \quad (21)$$

The drawdown observations are at 5-minute intervals starting at $t = 5$ min and ending at 1000 min, for a total of 200 observations.

To solve, we will discretize the unknown pumping rate into 1-minute intervals and approximate the integral through a Riemann sum, with uniform partition $\tau_1 < \tau_2 < ... < \tau_m$ with increment $\Delta\tau$

$$Y(t_i, x_1, x_2) = \frac{1}{4\pi T}\sum_{j=1}^{J}\frac{1}{(t_i - \tau_j)}\exp\left(-\frac{x_1^2 + x_2^2}{4D(t_i - \tau_j)}\right)s(\tau_j)\Delta\tau \quad (22)$$

where J is the j for the maximum τ_j that does not exceed τ_i. Thus,

$$y_i = Y(t_i, x_1, x_2)$$
$$s_j = s(\tau_j),$$
$$H_{ij} = \begin{cases} \dfrac{1}{4\pi T} \dfrac{1}{(t_i - \tau_j)} \exp\left(-\dfrac{x_1^2 + x_2^2}{4D(t_i - \tau_j)}\right) \Delta\tau, & \tau_j < t_i \\ 0, & \text{otherwise} \end{cases} \quad (23)$$

for $i = 1,..,n$, $j = 1,...,m$.

It is reasonable to presume that the pumping should have a degree of continuity, since it is impractical to start and stop the pumps all the time. Even if the actual pumping schedule were intermittent and erratic, we are primarily interested in trends and not in catching every zigzag. This information is captured by the model:

$$X_j = 1, \quad Q_{jl} = -\theta_1 |\tau_j - \tau_l|, \quad j, l = 1,...,m \quad (24)$$

This is the linear semivariogram (or generalized covariance function, Matheron, 1973) model of geostatistics which conveys the information that the s(t) function varies gradually over time. (More information and background on these models can be found in Kitanidis, 1997.) This model is, in many applications, a good starting point.

The measurement errors are supposed to have the same variance and be independent of each other. Thus, the measurement error covariance matrix is

$$\mathbf{R} = \theta_2 \mathbf{I}, \text{ or } R_{kl} = \begin{cases} \theta_2, & k = l \\ 0, & \text{otherwise} \end{cases} \quad (25)$$

The two parameters θ_1 and θ_2 are called structural parameters. We have now parameterized the stochastic inverse problem and it is straightforward to apply the formulas of the previous section.

As parameter values for the basic case we will use $\theta_1 = 3.22E-7$ and $\theta_2 = 1.05E-4$ (we will discuss later the procedure that led to these particular values). Figure 1 shows the best estimate, represented through the posterior mean, and the 95% confidence interval computed from the mean plus or minus two times the standard deviation. Note that, since this is a Bayesian procedure, the term credible interval might be preferable to the term confidence interval that originates in classical statistics. In any case, the intuitive meaning is that we anticipate that the actual values of the unknowns should be included in the interval with probability 95%. The estimate reproduces the data with mean square deviation between observed and predicted value equal to $4.95E-5$.

Since this is a synthetic case, we know the actual pumping rate that we used to generate the data (which were contaminated with random error with mean 0 and variance $1.0E-4$) and we plot it on figure 1 so that we can evaluate the effectiveness of the inverse method. The best estimate follows quite nicely the actual pumping rate. The confidence interval contains the actual pumping without being overly broad and thus provides a fair graphical representation of the uncertainty in the estimate. Thus, the overall performance of the inverse methodology is satisfactory, despite the fact that the model of structure that we adopted is not a particularly good representation of the temporal structure of the actual function. The model with the linear semivariogram model that we adopted corresponds to a random walk while the actual function was piecewise constant. The reason the method worked despite the less than optimal choice of a model is that the adopted prior distribution is broadly inclusive (i.e., it has high entropy) and thus the data, through the likelihood function, were not hindered from identifying a satisfactory solution.

Let us see how the results are affected if θ_1 were to increase by three orders of magnitude over the base case. The new estimate reproduces the data more closely, the mean square deviation between observed and predicted value dropping to $1.0E-9$. However, the estimates are worse because they fluctuate more, as shown in figure 2. That better reproduction of data goes with worse estimates may appear counterintuitive. However, consider that: (a) although data should be reproduced within their perceived accuracy, perfect reproduction

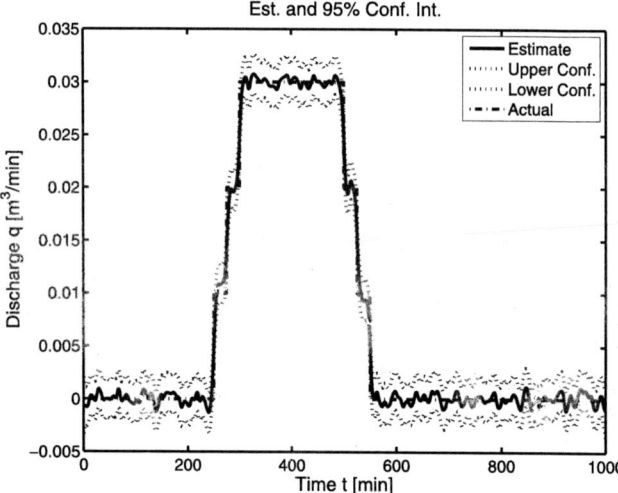

Figure 1. Best estimate (solid line), confidence interval (dotted lines), and actual pumping rate (dash-dotted line) for base case of structural parameters.

of imperfect data serves no useful purpose; (b) due to the ill-posedness in the mathematics of this inverse problem, one must change the estimate drastically to reproduce details in the data that are insubstantial in comparison to the observation error. Thus, overfitting of the data results in excessive fluctuations in the estimates and deterioration in the performance of the inverse method.

Next, consider the effects of decreasing θ_1 by three orders of magnitude over the base case, see figure 3. The estimate now reproduces the data less faithfully, the mean square deviation between observed and predicted value increasing to $2.9E-9$. The lower value in θ_1 has the effect of suppressing fluctuations in the estimate but also sacrificing some of the resolution.

The ratio θ_1/θ_2 affects the shape of the best estimate and degree of data fitting: the smaller this ratio, the flatter or less variable the solution but the less satisfactory the reproduction of the data. Using a large ratio allows good reproduction of the data but this does not necessarily mean a satisfactory solution. In fact, in this as well as most other applications, one can find many solutions that reproduce the data but most of them are not satisfactory because they are mostly noise.

Finally, consider the effect of parameter scaling. If both parameters are multiplied by the same factor α, the posterior variances are multiplied by α and the confidence interval is broadened by a factor equal to $\sqrt{\alpha}$ (this follows directly from the equation that gives the posterior covariance, **V**), but the best estimate remains unaffected. Thus, the scaling of the two parameters fully controls the breadth of the confidence intervals, but has no consequence whatsoever on the reproduction of the observations.

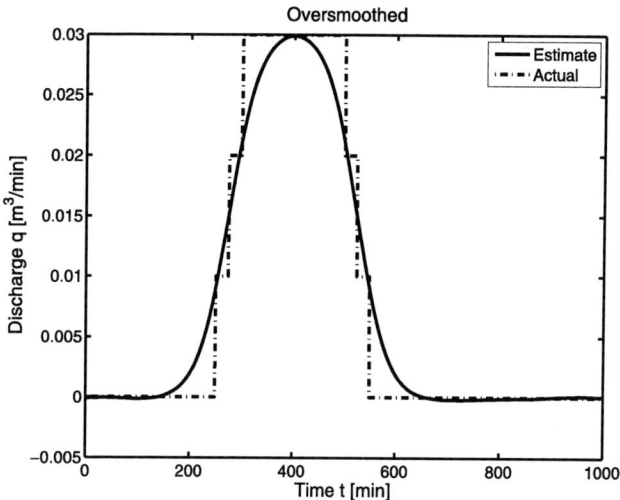

Figure 3. Estimate (solid line) and actual (dash-dotted line) for over-regularized case.

SELECTION OF PARAMETERS

As the example has illustrated, the results of linear inversion are affected by the structural parameters that are used to weigh data versus prior information or regularization. Selecting the right parameters is one of the most interesting and challenging aspects of inverse methods.

In practice, the most common method for parameter selection is probably *ad hoc*. For example, one may select θ_2 based on one's judgment of measurement error. Then, the other parameter θ_1 may be selected to regularize the estimate, i.e., to suppress erratic fluctuations or *noise* in the estimate while hopefully maintaining sufficient resolution in the *signal*, i.e., the variability in the unknown function. This procedure is obviously subjective but it may work at the practical level if one knows how a solution should look.

In selecting the ratio θ_1/θ_2, one seeks a good trade-off between data fitting and resolution. The degree of meeting these conflicting objectives can be represented, for example, through the following metrics (other metrics could also be defined as appropriate in every case):

$$J_f = \sum_{i=1}^{n}\left(y_i - \sum_{j=1}^{m} H_{ij}\hat{s}_j\right)^2 \qquad (26)$$

$$J_r = \sum_{i=1}^{m-1}\left(\hat{s}_{i+1} - \hat{s}_i\right)^2 \qquad (27)$$

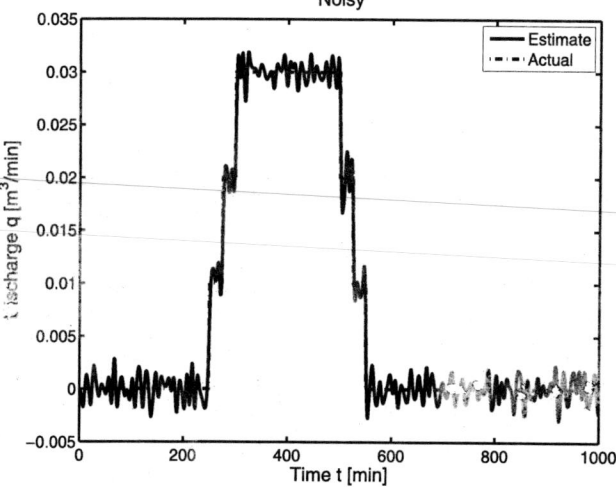

Figure 2. Estimate (solid line) and actual (dash-dotted line) for under-regularized case.

The first is small when the data is reproduced faithfully, which occurs when θ_1/θ_2 is large. The second represents the flatness of the estimate, when s is defined in one dimension and a uniform grid; it is small when the estimate is relatively uniform and without many lumps, which is achieved when θ_1/θ_2 is small. Thus, selection of θ_1/θ_2 can be viewed as seeking a good trade-off between two objectives: reproducing data and suppressing excessive fluctuations in the estimate.

Fortunately, statistical theory provides a simple, straightforward, objective, and general approach to selecting the parameters. The idea is to find the parameters that maximize the probability of the data. Using the same probabilistic model that we used to find the posterior of s, we can derive the pdf of y given the vector of structural parameters θ. Then, using a well-established approach, known as restricted maximum a posteriori probability, we can find the parameters that maximize this expression for the actual data.

This method of finding the structural parameters is known and appreciated in statistics (see Edwards, 1992). Here, we must stress its relation to cross-validation and residual-examination approaches (see Kitanidis, 1991, 1997, for interpolation methods) because it makes it easier to understand the meaning and value of the approach. This issue cannot be covered comprehensively here but we will take a peek at what residuals can tell us.

Consider an $(n-K) \times n$ matrix $\mathbf{P}$ so that $\mathbf{P}\Phi = 0$ and $\mathbf{P}\Psi\mathbf{P}^T$ is a diagonal matrix. Then, the transformed data

$$\delta = \mathbf{P}\mathbf{y} \qquad (28)$$

have zero mean and covariance matrix $\mathbf{P}\Psi\mathbf{P}^T$ which is diagonal with entries σ_i^2 equal to the variances of δ. Then ε values are found through normalization, $\varepsilon_i = \frac{\delta_i}{\sigma_i}$. For example, the procedure to find $\mathbf{P}$ can be summarized in the following MATLAB commands, which is given here in lieu of pseudocode:

```
T = null(PHI')'; Pyy = T'*inv(T*PSI*T')*T;
P = (orth(Pyy))';
```

Note that $\mathbf{P}$ should be $(n-K)$ by n and verify that the imposed conditions are met. The significance of the generated transformed data (which play a role similar to that of residuals in regression) is that they allow us to test the model and fit its parameters. The ε values are supposed to have zero mean and variance 1. The restricted maximum likelihood approach can be seen as a method to select parameters so that the variance,

$$Q_2 = \frac{1}{n-K}\sum_{k=1}^{n-K}\varepsilon_i^2 \qquad (29)$$

is indeed near 1. By doing so, we select the right scaling for the parameters and thus find proper confidence intervals.

Also the method of restricted maximum likelihood finds the ratio of the parameters that optimizes the predictive ability of the model, in some average sense. Such a measure is cR (Kitanidis, 1991, 1997):

$$cR = Q_2 \exp\left(\frac{1}{n-K}\sum_{k=1}^{n-K}\ln\sigma_i^2\right) \qquad (30)$$

This is the geometric mean of the normalized variances of the delta residuals. The idea is that a good set of θ parameters should give small delta residuals, which can be measured by the value of cR. We will see how this method works by continuing with the example.

EXAMPLE CONTINUED

Application of the maximum likelihood methodology of the previous section yielded the parameter that we used as our base case, figure 1. This procedure is equivalent to finding the ratio θ_1/θ_2 that minimizes cR and scaling the two parameters so that $Q_2 = 1$.

In the upper half of figure 4 we show the value of cR as a function of θ_1/θ_2 and indicate by a small circle the point from the maximum likelihood procedure, where cR is minimum. In the lower half, we show how the actual mean square error (average of square differences between the actual and estimated pumping rate)

$$MSE = \frac{1}{m}\sum_{k=1}^{m}(s_i - \hat{s}_i)^2 \qquad (31)$$

varies with θ_1/θ_2, where s_i is actual value, $\hat{s}_i$ is the estimate (the posterior mean). The smaller the actual MSE, the better the estimate. This figure illustrates that cR is an excellent indicator of the accuracy of the estimate, because there is a strong correlation between cR and actual MSE. Of course, in practice one cannot evaluate the actual MSE, since the actual function is unknown, but one can always evaluate cR because it depends only on the data. This example verifies the usefulness of cR as a guide for selecting θ_1/θ_2. The flatness of the actual MSE vs. θ_1/θ_2 function near its minimum indicates that there is a range of θ_1/θ_2 values that yield satisfactory results and the minimum-cR method has identified one such value.

Next, we maintain θ_1/θ_2 at its maximum-likelihood value, which is $3.06E-3$, but vary each of the parameters individually. In the upper half of figure 5 we show Q_2 as a function of θ_2. The maximum-likelihood values give $Q_2 = 1$, indicated on the graph by a small circle. In the lower half, we plot the mean square actual normalized error. In other words, for each value of θ_2, we compute

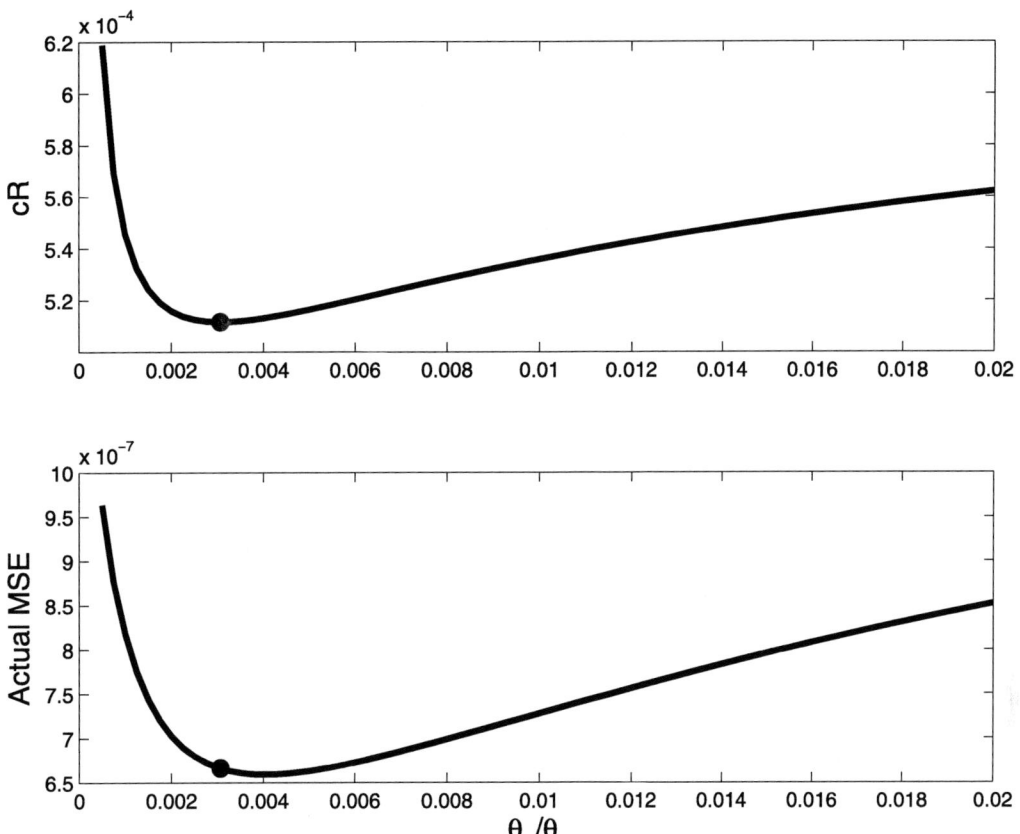

Figure 4. Validation criterion cR (upper) and actual mean square error (lower) expressed as functions of the ratio of structural parameters, θ_1/θ_2.

$$MSNE = \frac{1}{m}\sum_{k=1}^{m}\frac{(s_i - \hat{s}_i)^2}{\sigma_i^2} \qquad (32)$$

where s_i is actual value, $\hat{s}_i$ is posterior mean, and σ_i^2 is posterior variance. Ideally, MSNE should be near 1. If this ratio is much larger than 1, the confidence intervals are optimistically too narrow; if it is much smaller than 1, the confidence intervals are pessimistically too broad. Comparison of the two figures show that the MSNE is near 1 where Q_2 is 1, indicating that Q_2 is a good guide to whether the confidence intervals are properly scaled. In real-world applications, one cannot evaluate MSNE because the actual s is not known but one can be guided by the value of Q_2 which depends only on the data.

In the author's opinion, this example illustrates that the mean square error computed from the inverse method (estimation variance) can be a satisfactory measure of the reliability of the estimate, provided that a good method has been used to select the parameters of the underlying model. The results of this example challenge the misguided notion that, unfortunately, has gained currency among geostatisticians, that the variance is necessarily highly subjective or even useless as a measure of reliability. Such a notion is perhaps a consequence of neglecting to use sound methods of parameter estimation and excessive preoccupation with the evils of "multigaussianity" and other red herrings. Of course, the estimation error is affected by modeling choices, but then so is the best estimate and everything else that we do. At a deeper level, many misunderstandings are caused by the fact that the mainstream of the geostatistics community either discards or only grudgingly accepts the Bayesian viewpoint that estimation, including inverse modeling, is all about management of available information.

It is worth mentioning that a method used in practice to determine the proper weighting is to plot a measure of roughness, Eq. (27) versus a measure of misfit, Eq. (26) and to select a point that gives a reasonable trade-off. This plot, known as the L-curve, is shown in figure 6, where the result from the cR-method is also shown as a small circle. It is hard

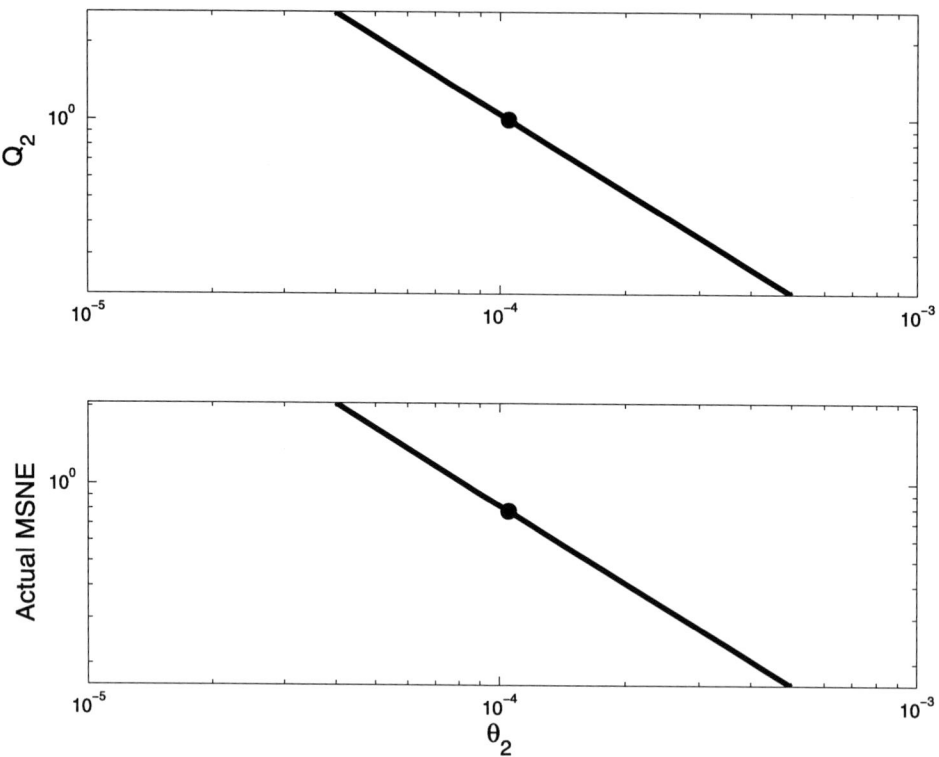

Figure 5. Validation criterion Q_2 (upper) and actual mean square normalized error (lower) expressed as functions of the measurement-variance parameter θ_2.

to select parameters on the basis of the L-curve because, as figure 6 illustrates, the curve may look more like a "C" than an "L".

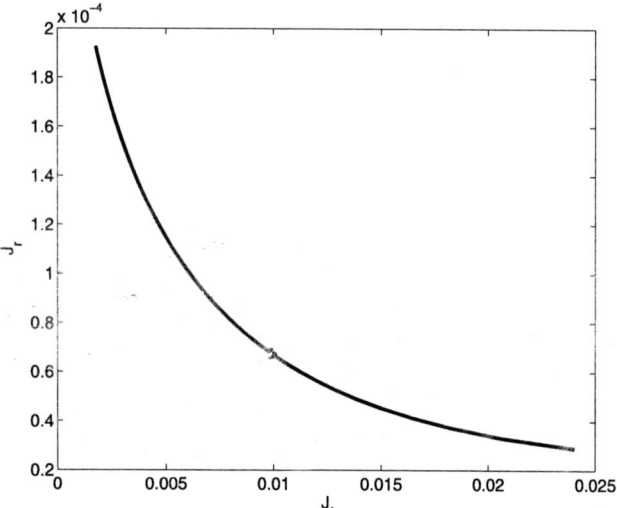

Figure 6. The tradeoff curve between a measure of roughness and a measure of misfit for a range of θ ratios. The solution from the cR method is shown as a circle.

CONCLUDING REMARKS

Stochastic methods are well suited for the solution of algebraically underdetermined and mathematically ill-posed inverse problems. Probabilities should be interpreted within a Bayesian context as encoding available information, rather than frequencies. We have focused on linear inversion, where the relation between measurements and unknowns is linear and all probability distributions are Gaussian, but the basic ideas apply to other cases.

In virtually all methods used in practice, including the geostatistical approach, inverse modeling is set up as the minimization of an objective function that consists of two parts: the first part is a penalty for not reproducing the data and the second part is a penalty for not being consistent with prior information (or a *regularization* criterion). Prior information is often expressed as a notion of uniformity, flatness, or smoothness of the estimate. The focus of inverse modeling is in the solution of this optimization problem, which can be quite challenging.

However, choices on the regularization scheme and the relative weighting of the two terms in the objective function have profound effects on the results of the inverse problem, when the data are information-poor as is usually the case in hydrogeologic applications. The regularization and the relative weighting control the level of reproduction of detail in the obtained image and also the closeness of the estimated to the true values. By overweighting the data-reproduction penalty, the data are reproduced more closely and more details appear in the image that is estimated from the optimization but the image is also more affected by spurious features. The challenge is to find weights that will allow the method to extract useful information from the data without being overly affected by errors that may produce artifacts. Schemes that are often employed by practitioners, like selecting a point on the L-curve that shows the trade-off between meeting the two objectives, can be inconclusive. Another challenge is to evaluate confidence intervals that in an objective sense reflect the range of possible error in the estimates. In inverse problems, which are algebraically underdetermined and mathematically ill-posed in the sense of Hadamard, the degree of data reproduction is a poor indicator of the accuracy of estimates—which is a big difference from classical least squares regression methods.

Within the context of stochastic methods, the geostatistical inverse method provides a rigorous and objective approach to meet these challenges. The prior information (or regularization) and the measurement error are parameterized using 2–3 *structural* parameters (*or hyper-parameters*) which are determined guided by the data. The method can be described in terms of two criteria. The first is an index of the actual error and thus needs to be minimized; this minimization determines the relative weighting of the two penalty terms (for not reproducing data and for not being consistent with prior information). The second criterion is an index of actual squared errors versus mean squared errors computed by the inverse problem so this ratio must be unity; this criterion scales the confidence intervals so that they are more realistic.

Acknowledgments. This work was partially funded by the National Science Foundation under Grant No. NSF/EAR 0001441-002.

REFERENCES

Backus, G. E., and J. F. Gilbert, Numerical application of a formalism for geophysical inverse problems, *Geophys. J. Royal Astron. Soc.*, 13, 247–276, 1967.

Backus, G. E., and J. F. Gilbert, The resolving power of gross earth data, *Geophys. J. Royal Astron. Soc.*, 16, 169–205, 1968.

Birnbaum, A., On the foundations of statistical inference, *J. Am. Stat. Assoc.*, 57, 269–326, 1962.

Carlin, B. P., and T. A. Louis, *Bayes and Empirical Bayes Methods for Data Analysis*, Chapman & Hall/CRC, Boca Raton, 2000.

Carrera, J., A. Alcolea, A. Medina, J. Hidalgo, and L. Slooten, Inverse problem in hydrogeology, *Hydrogeology Journal*, 13, 206–222, 2005.

Christakos, G., A Bayesian/maximum-entropy view to the spatial estimation problem, *Math. Geology*, 22, 763–777, 1990.

Dagan, G., Stochastic modeling of groundwater flow by unconditional and conditional probabilities: The inverse problem, *Water Resour. Res.*, 21(1), 65–72, 1985.

Edwards, A. W. F., *Likelihood*, pp. 275, The Johns Hopkins Univ. Press, 1992.

Guidici, M., G. Morossi, G. Parravicini, and G. Ponzini, A new method for the identification of distributed transmissivities, Water Resour. Res., 31(8), 1969–1988, 1995.

Hoeksema, R. J., and P. K. Kitanidis, An application of the geostatistical approach to the inverse problem in two-dimensional groundwater modeling, *Water Resour. Res.*, 20(7), 1003–1020, 1984.

Hoeksema, R. J., and P. K. Kitanidis, Comparison of Gaussian conditional mean and kriging estimation in the geostatistical solution of the inverse problem, *Water Resour. Res.*, 21(6), 825–836, 1985.

Hoeksema, R. J., and P. K. Kitanidis, Prediction of transmissivities, heads, and seepage velocities using mathematical models and geostatistics, *Adv. in Water Resour.*, 12(2), 90–102, 1989.

Hoeksema, R. J., and Clapp, R. B., Calibration of groundwater flow models using Monte Carlo simulations and geostatistics, in ModelCARE 90: Calibration and Reliability in Groundwater Modelling (pp. 33–42). IAHS Publ. No 195, 1990.

Jaynes, E. T., *Probability Theory*, Cambridge Univ. Press, 2003.

Kitanidis, P. K., Parameter uncertainty in estimation of spatial functions: Bayesian analysis, *Water Resources Research*, 22, 499–507, 1986.

Kitanidis, P. K., Orthonormal residuals in geostatistics: model criticism and parameter estimation, *Math. Geol.*, 23(5), 741–758, 1991.

Kitanidis, P. K., Quasilinear geostatistical theory for inversing, *Water Resour. Res.*, 31, 2411–2419, 1995.

Kitanidis, P. K., *Introduction to Geostatistics*, pp. 249, Cambridge University Press, 1997.

Kitanidis, P. K., and E. G. Vomvoris, A geostatistical approach to the inverse problem in groundwater modeling (steady state) and one-dimensional simulations, *Water Resour. Res.*, 19(3), 677–690, 1983.

Matheron, G., *The Theory of Regionalized Variables and its Applications*, 212 pp., Ecole de Mines, Fontainbleau, France, 1971.

Matheron, G., The intrinsic random functions and their applications, *Adv. Appl. Prob.*, 5, 439–468, 1973.

Rubin, Y., and G. Dagan, Stochastic identification of transmissivity and effective recharge in steady groundwater flow, 1. Theory, *Water Resour. Res.*, 23(7), 1185–1192, 1987a.

Rubin, Y., and G. Dagan, Stochastic identification of transmissivity and effective recharge in steady groundwater flow, 2. Case study, *Water Resour. Res.*, 23(7), 1193–1200, 1987b.

Snodgrass, M. F., and P. K. Kitanidis, A geostatistical approach to contaminant source identification, *Water Resour. Res.*, 33, 537–546, 1997.

Sun, N.-Z., *Inverse Problems in Groundwater Modeling*, Kluwer Publ., Norwell, MA, 1994.

Tikhonov, A. N., and V. Y. Arsenin, *Solutions of Ill Posed Problems*, Halsted Press/Wiley, New York, 1977.

Tikhonov, A., I. Leonov, A. N. Tikhonov, A. G. Yagola, and A. S. Leonov, *Non-linear ill-posed problems*, CRC Press, Boca Raton, FL, 1997.

Wagner, B. J., and Gorelick, S. M., Reliable aquifer remediation in the presence of spatially variable hydraulic conductivity: From data to design, *Water Resour. Res.*, 25(10), 2211–2225, 1989.

Yeh, W. W.-G., Review of parameter identification procedures in groundwater hydrology: The inverse problem, *Water Resour. Res.*, 22(1), 95–108, 1986.

A Comparison of the Use of Radar Images and Neutron Probe Data to Determine the Horizontal Correlation Length of Water Content

Rosemary J. Knight[1], James D. Irving[1], Paulette Tercier[2], Gene J. Freeman[3], Chris J. Murray[3], and Mark L. Rockhold[3]

Surface-based ground-penetrating radar data were collected at the Hanford Site in Washington, U.S.A. to assess the use of radar reflection images as a means of quantifying the spatial variability of subsurface water content. Available at the selected test site were two sets of water content data derived from neutron probe measurements that had been made to a depth of ~18 m in 32 wells in 1980 and 1995. The comparison of probe-derived water content data, synthetic radar data, and the acquired radar data indicated a good correspondence between the changes in probe-derived water content and the location of reflections in the radar data. Geostatistical analysis was conducted on the two sets of probe-derived water content values and the amplitudes of the reflections in the radar reflection image to determine the horizontal correlation length of water content. The experimental semivariograms for the water content data were fit with a single exponential model with a correlation length of 10 m. The semivariogram for the radar data was fit with a nested structure containing a dominant long-range structure with a correlation length of 14 m, and a smaller-scale structure with a correlation length of 0.3 m. Quantifying the scale triplet—the spacing, extent, and support—for the two forms of measurement provided a framework for comparing and assessing the derived correlation structures. This approach also highlighted the importance of identifying methods for properly determining the scale of radar measurements required for the imaging of subsurface water content.

INTRODUCTION

One of the challenges in modeling the transport of contaminants in the vadose zone is acquiring the data required to adequately characterize the spatial variability of subsurface hydraulic properties. One approach is to determine the spatial and temporal variability in water content. The water content distribution can be an excellent indicator of soil texture, which strongly influences the unsaturated hydraulic properties controlling vadose zone contaminant transport behavior. Field-measured water content data can be used as soft data from which estimates of soil hydraulic properties can be inferred (Rockhold, 1999), and can provide very useful information about possible migration pathways.

When the subsurface region of interest extends tens of meters below the surface, water content data can be obtained by drilling wells and using neutron probe data to estimate water content with probe-specific relationships between neutron counts and water content. While such estimates are

[1] Geophysics Department, Stanford University, Stanford, California, USA
[2] Burnaby, British Columbia, Canada
[3] Pacific Northwest National Laboratory, Richland, Washington, USA

Subsurface Hydrology: Data Integration for Properties and Processes
Geophysical Monograph Series 171
Copyright 2007 by the American Geophysical Union.
10.1029/171GM05

considered to be reasonably accurate, the lateral spacing and number of wells is rarely sufficient to adequately constrain the lateral variation in water content. Additional problems with a reliance on information acquired in wells are the costs of drilling and casing, and the health, safety and environmental risks associated with all forms of invasive sampling if information is required in regions with subsurface contamination. The purpose of this study was to assess the use of surface-based ground-penetrating radar (GPR) to obtain information about the spatial variation in water content in the vadose zone.

Surface-based GPR data are acquired by sending a pulse of high-frequency (1 MHz to 1 GHz) electromagnetic (EM) energy into the earth and recording energy reflected back to the surface from interfaces across which there are changes in electrical properties, most notably the dielectric constant κ. After processing, the location of these interfaces in the subsurface and the magnitude of reflected energy are displayed as reflections in the radar image. Our hypothesis is that the correlation structure seen in the radar image can be used to determine the correlation structure of the subsurface property determining κ.

The use of GPR data to obtain the lateral correlation length of a subsurface region was investigated in a field experiment conducted by Rea and Knight (1998). In this study the geostatistical analysis of a binary digital photograph of a cliff-face was compared with the geostatistical analysis of the corresponding radar reflection image, acquired along the top of the cliff. A very similar lateral correlation structure was obtained for the two data sets (a nested structure with ranges of 0.5 m and 2.1 m for the photographic image, 0.8 m and 2.0 m for the radar image), suggesting that the radar image was capturing information about the spatial variability seen in the photograph. In the study, the black and white pixels in the binary image were taken as corresponding to grain size, with the black indicating finer grained material. Given the fact that the cliff-face was unsaturated, the separation between black and white most likely corresponded to differences in water content as the finer grained materials would preferentially retain water. This suggests that the lateral correlation structure seen in the radar reflections was capturing information about the spatial variability in water content.

A more recent study by Dafflon et al. (2004), using a very similar approach, again found that the lateral correlation lengths derived from GPR data were consistent with those derived from the corresponding digital outcrop images. These authors note that the correlation lengths of the radar data tended to be 10–30% longer than those obtained for the digital images, but conclude, based on their field experiment and theoretical analysis, that "the geostatistical analysis of surface georadar data offers an easy, robust and reliable way to estimate the average lateral correlation structure of the shallow subsurface".

In this study, we further explored the concept that geostatistical analysis of a GPR image can yield estimates of the lateral correlation length of water content, by working at a site where detailed information was available about subsurface water content. In May 2000, we collected a GPR profile, 30 m long and ~12 m deep, at a test site located within the U.S. Department of Energy Hanford Site in south-central Washington, U.S.A. Available at the test site were water content data derived from neutron probe measurements made in an array of wells in the years 1980 and 1995. The 32 wells penetrating to a depth of ~18 m over a 16 m by 16 m grid, with thousands of probe-derived water content measurements, provided an ideal data set for comparison with the radar data. Also available at the site were results from studies employing crosswell radar (Majer et al., 2001) which provided information about the magnitude of, and the controls on, subsurface electromagnetic properties. The first objective of our study was to determine whether the locations and amplitudes of the radar reflections corresponded to changes in water content. Our second objective was to compare the estimates of lateral correlation length obtained from analysis of the radar reflection image and neutron probe data. This comparison focused on the effect of the scale of the measurement on the determined correlation length.

A rigorous assessment of the effect of measurement scale on the correlation structure obtained from our two forms of measurement would have required a full treatment of the spatial characteristics of each measurement and the interaction of the measurement with the spatial variability of the sampled system; an analysis that required far more information than was available in this study. We instead adopted the concept of the "scale triplet", defined by Bloschl and Sivapalan (1995) to include the spacing, extent, and support of a measurement, and considered our results in light of observations made in earlier publications by Journel and Huijbregts (1978), Gelhar (1993) and Western and Bloschl (1999). The extent is defined as the overall coverage of the measurements; the spacing is defined as the distance between samples; the support is defined as the volume sampled by the measurement. In general it has been found that the correlation length will be underestimated if the extent is too small, and will be overestimated if the spacing or support is too large. While the specific form of dependence of the determined correlation length on measurement scale will depend on the spatial distribution of the parameter, the analysis of soil moisture data presented in Western and Bloschl (1999) suggests that significant bias will be introduced if 1) the extent is smaller than about 5 times the correlation length; 2) the spacing is greater than about twice the correlation length; and 3) the support is greater than about 20% of the correlation length. In our study, the subsurface is stratified to some extent, so the correlation structure is anisotropic. For the analysis of data from such a system, the support dimension of the measurement in both the horizontal and vertical direction will affect the derived horizontal correlation length (Journel and Huijbregts, 1978; Gelhar, 1993). The previous studies using geostatistical analysis of radar data to determine lateral correlation lengths did not address the impact of the scale of the radar measurement, but there is evidence that the radar-derived correlation length is very sensitive to the vertical resolution of the data (Knight et al., 2004). We found in this study that the scale triplet provided a very useful framework for quantitatively describing the scale of measurement and for comparing the correlation structures derived from the two, very different, forms of measurement.

DESCRIPTION OF THE FIELD SITE

The field site selected for this study is referred to as the Sisson and Lu Injection Test Site and was designed to monitor the movement of water from a liquid injection point source in the unsaturated conditions at Hanford (Sisson and Lu, 1984; Fayer et al., 1995). The site is located in the south-eastern corner of the region referred to as the 200 East Area. Annual precipitation at Hanford is 16 cm/yr, which is representative of a semi-arid climate. The depth to the water table is approximately 80 m. Plant cover at the site is predominantly a shallow-rooted cheat grass with scattered, deep-rooted sagebrush and Russian thistle.

The test site consists of 32 metal-cased wells, 18 to 19 m deep, surrounding a central injection well that is 5 m deep. A plan view of the well locations is shown in Figure 1. Each well is referred to with a letter and number, with the letter designating the radial arm (A through H) and the number designating the approximate distance in meters from the central well. Also shown on this figure is the location of the line along which the GPR data were collected. The GPR profile, referred to as SISREF1, starts at well D8 and extends 30 m in the northwest direction. We were unable to acquire radar data over the wells due to the metal casing.

A detailed discussion of the lithology of the test site and surrounding area is presented in the thesis by Smoot (1995). The sediments in the area consist of up to a few meters of Quaternary age aeolian silty sand overlying Pleistocene medium- to coarse-grained fluvial sands with interbedded gravelly sand and silt-sand layers. The test site is heterogeneous, consisting of layers and lenses of alluvial sediments ranging in grain size from silt to gravel that fall within the sand-dominated facies of the Hanford formation (DOE, 2002). The spatial variation in the texture of the sediments

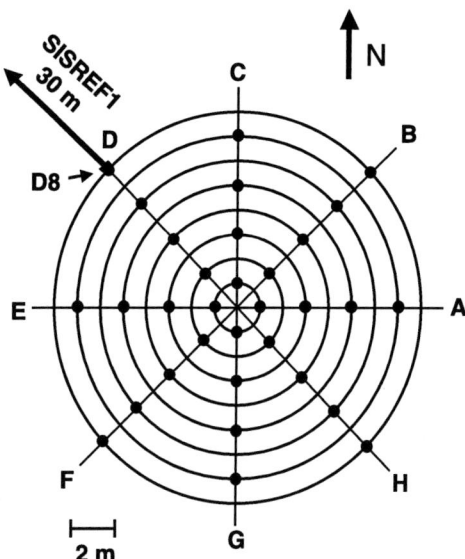

Figure 1. A plan view of the well locations (solid circles) at the Sisson and Lu Injection Test Site. Thirty-two wells surround a central injection well. GPR data were acquired along a 30 m line labeled SISREF1 that starts at the well labeled D8.

can be expected to be a major control on the distribution of water in the vadose zone.

WATER CONTENT DATA FROM NEUTRON PROBE MEASUREMENTS

Volumetric water content data are available from neutron probe measurements made in two field tests conducted at the Sisson and Lu site in June 1980 and February 1995. During each of the tests, neutron probe measurements were made both before and after an injection test. In the report by Fayer et al. (1995), the comment is made that "at a majority of depths the differences [in water content between the 1980 and 1995 pre-injection tests] are less than 1 vol%, which is smaller than the calibration error of the probe". We therefore presume, given the dry conditions at Hanford, that the data acquired before the injection tests in 1980 and 1995 are representative of the ambient water content distribution in the vadose zone during the radar survey, made in May 2000, prior to an injection test. Detailed descriptions of the 1980 measurements are in Sisson and Lu (1984) and Fayer et al. (1995); descriptions of the 1995 measurements are in Fayer et al. (1995). More recent neutron probe measurements were made at the site in the summer of 2000, but data were not acquired in the top 4 m; we chose not to use these data due to the limited region of overlap with the GPR data.

The 1980 pre-injection neutron probe measurements were made in the 32 observation wells using a Campbell-Pacific neutron probe. Measurements were made from a depth of 0.3

m to 18 m, with vertical sampling every 0.3 m. Neutron counts were recorded for 15 seconds at each sampling location. In order to obtain estimates of volumetric water content θ_w from the neutron probe data we used the calibration curve for the 1980 probe that was determined by Fayer et al. (1995):

$$\theta_w = (0.0182\, C_{15} - 3.82) / 100 \qquad (1)$$

where C_{15} represents the probe count over 15 seconds; Fayer et al. (1995) reported a correlation coefficient of 0.829 and a standard error in volumetric water content of ~2 to 3%. This calibration curve was developed using three calibration standards with volumetric water contents of 5%, 12% and 20%. The material used for the standards was a mixture of sand and alumina trihydrate, placed in a container 1.5 m in diameter and 1.8 m high. The probe was run in a 0.15 m diameter steel casing to simulate the casing at the Sisson and Lu site. Full details of the procedure are given in Engelman et al. (1995) and in Fayer et al. (1995).

In 1995 a Schlumberger compensated neutron tool was used to log the 32 wells prior to an injection experiment. In most wells the maximum depth of the measurements was 18.29 m, but in a few of the wells the measurements were made to slightly greater depths up to 18.75 m. Measurements were made in all of the wells with vertical sampling every 6 in (~0.15 m) and the data reported as volumetric water content. The calibration curve was not provided by the operators but is reported by Fayer et al. (1995) to have produced excellent results in the dry range and a 1.4% error for a volumetric water content of 30%. In the 1995 data there is an unusually large spike in water content values in the top 1 m in a number of the wells. We attribute this to probe malfunction, so in our analysis used only the data below that interval.

In Figure 2 are shown the water content data (as % volumetric water content) from the 1980 and 1995 neutron probe measurements in well D8, the well that lies at the southeast end of the GPR profile. The water content values determined in 1980 and 1995 are very close; except for the spike in water content at the surface in the 1995 data. The main features in these two data sets are the pronounced increases in water content at depths of approximately 7.5 m and 11.5 m.

GROUND PENETRATING RADAR DATA

A radar reflection image is a compilation of the radar traces recorded as transmitter and receiver antennas are moved across the surface of the earth. The arrival times and relative amplitudes of reflections in a single radar trace can be represented by the convolution of the source EM pulse (or wavelet) with a series of reflection coefficients, defined at each subsurface interface as:

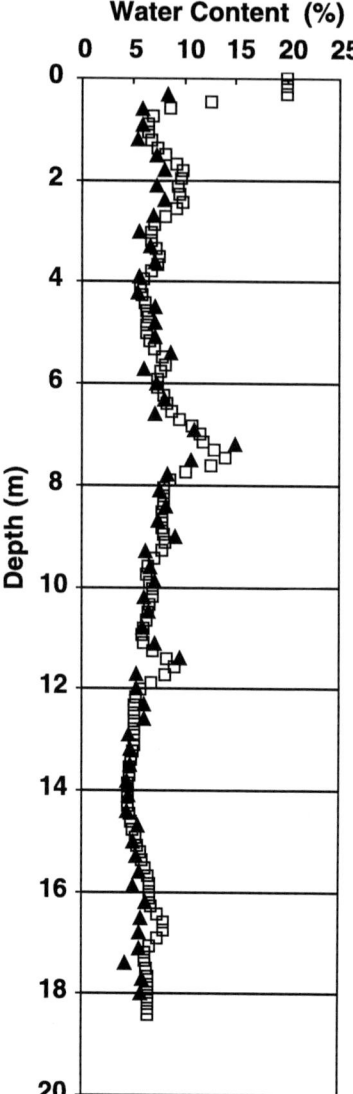

Figure 2. Water content values in well D8 obtained from neutron probe measurements made in 1980 (triangles) and 1995 (squares).

$$R = \frac{\sqrt{\kappa_1} - \sqrt{\kappa_2}}{\sqrt{\kappa_1} + \sqrt{\kappa_2}} \qquad (2)$$

where subscripts 1 and 2 refer to the materials above and below the interface, respectively. Here, R represents the ratio of the amplitudes of reflected energy to incident energy for a normally incident EM wave at a planar interface, where the incident wave's electric field is polarized perpendicular to the plane of incidence (referred to as TE mode). In order to convert the reflection amplitude data, recorded as a function of time, to a radar reflection image displayed in terms of depth, the velocity at which an EM wave travels through the subsurface must be known. The EM wave velocity v is equal to $c/\sqrt{\kappa}$ where c is the speed of light (3.0×10^8 m/s). Note that both this expression for v and Equation 2 assume EM wave propagation through a material with relatively low electrical conductivity and with magnetic permeability equal to its value in free space. We believe this to be a valid assumption for our study given the dry conditions and reported composition of the sediments at the Hanford test site.

The resolution of a radar image is determined by the frequency f of the antennas and the subsurface EM velocity v. The vertical resolution of a radar measurement is commonly taken to be one quarter of the dominant wavelength λ of the transmitted EM pulse, where $\lambda = v / f$. The horizontal resolution is the Fresnel zone of the measurement, the width (W) of which can be approximated by the following expression (Yilmaz, 1987):

$$W = v\sqrt{\frac{2d}{vf}} \quad (3)$$

where d is the depth from which the energy is returned. Detailed discussion of the fundamental principles of GPR can be found in the publications by Daniels et al. (1988) and Davis and Annan (1989).

The GPR data (SISREF1) were collected along a 30 m survey line running northwest from well D8, as shown in Figure 1. We used a Sensors and Software Pulse EKKO IV GPR system with 100 MHz antennas. For the survey, the transmitter and receiver antennas were kept 0.5 m apart, and were moved down the survey line at a station spacing of 0.1 m. Each recorded trace was stacked 64 times to ensure a high signal-to-noise ratio, and was sampled in time every 0.8 ns.

A number of signal processing steps were taken in order to prepare the GPR data for interpretation and analysis. The first step involved shifting the traces in the data set to align them on the first arrival. This corrects for any time drift caused by temperature changes in the system electronics during the survey. Next, a 25-point residual median filter was applied to each trace in the data set to remove the low frequency transient (the "wow") that underlies the GPR reflection signal. Trace amplitudes were then corrected using two methods. The first method involved applying a spherical and exponential compensation (SEC) gain to the data set to correct for the geometrical spreading of energy and attenuation losses during propagation. The amount of exponential boosting was chosen such that the amplitudes of the reflections were roughly balanced in time. We also used automatic gain control (AGC) to balance the trace amplitudes so that the average signal strength in a time window was constant.

We chose a window of 25-points and a maximum gain of 300 for our AGC function. Following the procedure described in Rea and Knight (1998), we used the AGC-gained image in the geostatistical analysis of the reflection image. The SEC-gained section, however, is more useful for comparison with the probe-derived water content data because the relative amplitudes of reflections are preserved.

The SEC-gained radar section is shown in Figure 3. The position marked along the horizontal axis is the distance from well D8. The radar data contain the amplitude and arrival time of received energy, and are therefore displayed as a time section with time along the left vertical axis. The energy arriving in the first 20 ns contains the high amplitude direct air and ground wave, and was therefore not considered in our analysis. Below this are a number of other continuous reflections, close to horizontal in orientation, which we interpret as imaging variation in the water content of the subsurface; a detailed analysis of these data is given in the following sections. There are two types of features seen in Figure 3 that have no relationship to subsurface properties: 1) The dipping reflection that cuts across the lower right region of the radar section was caused by energy reflecting off a fence post approximately 12 m northeast of well D8 (to the right of the line as shown in Figure 3). 2) The multiple reflections in the lower half of the plot are a result of internal ringing within the GPR system or antennas. In addition, we also observe a difference in the character of the radar section at late times between 25 m and 27 m; the data are noisier and the dominant frequency appears to be lower. We have no explanation for this, as there is no known surface or subsurface structure or disturbance that could cause this change.

LINK BETWEEN WATER CONTENT AND THE RADAR REFLECTION IMAGE

The radar reflection image contains information about the way in which κ varies spatially throughout the sampled subsurface region. The key question for our study: Is this variation in κ, imaged in the radar data, directly related to variation in water content? A review of numerous laboratory studies, as provided in Knight (2001), shows that the dominant factors controlling κ of a material are the water content, the volume fraction of high surface area materials (such as clays), and the geometry of the solid phase. However, given the large contrast between κ of water (80) and that of air (1) and commonly occurring solid components (typically 5–12) it is often assumed that variation in κ is primarily due to variation in water content.

Of specific relevance to our study are the results from the work of Majer et al. (2001) that show six images of EM slowness ($1/v$) obtained from crosswell radar measurements made

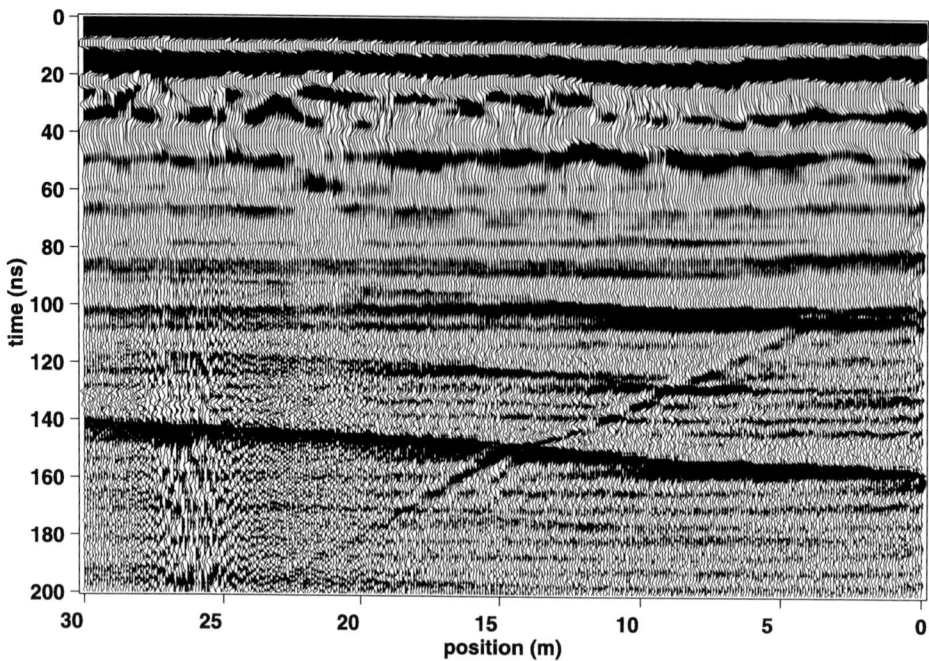

Figure 3. SEC-gained radar section. Well D8 is located at the right end of this section, with position from D8 given along the bottom. The vertical axis is in time, as in the form of the collected data.

between four wells, approximately 4 m apart, at our study site at Hanford. The EM slowness in each pixel was used to determine κ and a good correlation was found between κ-values near the wells and θ_w derived from the neutron probe data. This indicates that θ_w is the dominant control on κ at the Sisson and Lu site, an important result for our study. With an established link between κ and θ_w at a site, the acquisition of surface radar reflection data provides a way to obtain information about the variation in subsurface water content.

We further examined the link between water content and the radar reflection data at the field site by comparing the variation in water content in well D8, derived from the 1995 neutron probe data, to the adjacent radar image. This requires converting the time section, shown in Figure 3 to a depth section. The standard approach to making this conversion is to use estimates of subsurface velocity obtained from common midpoint (CMP) data. The CMP data that we acquired at the Sisson and Lu site were of such poor quality that we took an alternate approach. We used the water content data in D8 to determine the variation in subsurface EM wave velocity as a function of depth.

The first step was to convert θ_w to κ. To accurately model the relationship between θ_w and κ in any multi-component system requires using a theoretical approach that can incorporate information about volume fractions, geometries, and dielectric constants of all the solid and fluid components, and can account for physical and chemical interactions at the interfaces between the components. To do this we would need far more information than we had available. As an approximation, we used the empirically-derived Topp relationship (Topp et al., 1980):

$$\kappa = 3.03 + 9.30\,(\theta_w) + 146.00\,(\theta_w)^2 - 76.70\,(\theta_w)^3, \quad (4)$$

obtaining values of κ that ranged from 3.8 to 7.0. We then used these κ-values to determine a model of the EM velocity at well D8 ($v = c/\sqrt{\kappa}$). The velocity values were found to range from 0.11 to 0.16 m/ns, the same range reported by Majer et al. (2001), with an average velocity of 0.14 m/ns. The 1-D (i.e. varying with depth) velocity model at D8 was used to convert the radar time section to the depth section shown in Figure 4, with the 1995 water content data from well D8 repeated for comparison. As shown in Equation (2), large changes in water content should cause high-amplitude reflections because of the corresponding large changes in κ. This can be seen in Figure 4 at depths of 7.5 and 11.5 m. A qualitative comparison of the radar image and water content data suggests that there is also a correlation at other depths between changes in water content and the presence of radar reflections.

As a more rigorous way of comparing radar reflections and changes in water content, we generated the synthetic radar trace that would be produced by the κ-values (calculated using the Topp equation) at the location of well D8. We assumed a 100 MHz zero phase Ricker wavelet and

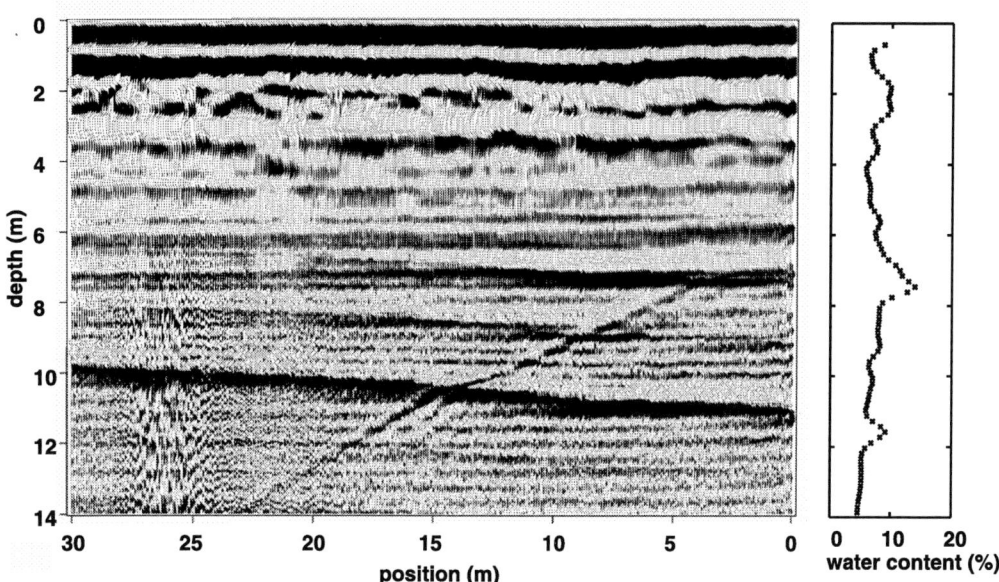

Figure 4. SEC-gained radar data converted to a depth section using the velocity model derived from water content measurements in well D8. The water content data acquired in 1995 are shown to the right of the radar section.

used a program written by Steve Cardimona (University of Missouri-Rolla) to simulate the 1-D propagation of an EM plane wave. The synthetic trace predicts the way in which changes in θ_w alone would be imaged in radar data. In Figure 5 we compare the synthetic trace to the radar trace acquired adjacent to well D8; both traces have been repeated 10 times. As can be seen in the results shown in Figure 5, many of the features in the synthetic trace are present in the real data indicating that the amplitudes and locations of radar reflections correspond primarily to changes in subsurface θ_w; a result we expected given the findings of Majer et al. (2001). The fact that we do not see perfect agreement between the synthetic data and the acquired data is largely due to the inability of our convolution model to accurately represent all aspects of radar wave propagation.

The observed correspondence between the radar reflection image and the changes in water content led to the conclusion that it would be appropriate to continue with the second part of this study, which involved the geostatistical analysis of the radar reflection image to obtain an estimate of the correlation length in water content.

GEOSTATISTICAL ANALYSIS OF PROBE-DERIVED WATER CONTENT AND RADAR DATA: RESULTS AND DISCUSSION

We adopted a geostatistical framework as a means of quantifying the correlation structure in the lateral direction in the radar image and in the probe-derived water content values. Specifically we obtained experimental semivariograms and addressed the question: How do the experimental semivariograms and lateral correlation lengths obtained for the two forms of measurement compare?

The experimental semivariogram is described by the following equation (Journel and Huijbregts, 1978):

$$\hat{\gamma}(\mathbf{h}) = \frac{1}{2N} \sum_{i=1}^{N} [z(\mathbf{x_i} + \mathbf{h}) - z(\mathbf{x_i})]^2, \qquad (5)$$

where $\mathbf{h}$ is the lag, or separation vector, between two data points $z(x_i+h)$ and $z(x_i)$, and N is the number of data pairs in each summation. As the separation distance increases the data points often tend to become less correlated and the semivariogram may be seen to flatten as γ reaches the sill or variance of the dataset.

Semivariogram models are used to provide an analytic description of the experimental semivariogram. While the positive-definite functions commonly used in semivariogram modeling are usually selected based on an empirical fit of the data, a discussion of the theoretical basis for some of the functions used to fit semivariogram models is given by McBratney and Webster (1986). For modeling the data we used either a single or nested exponential model. The exponential model can be shown to describe the semivariogram which will result from a variety of processes such as first-order Markov processes (Agterberg, 1970; McBratney and Webster, 1986) and is the one that is often assumed by researchers in stochastic hydrology (Woodbury and Sudicky,

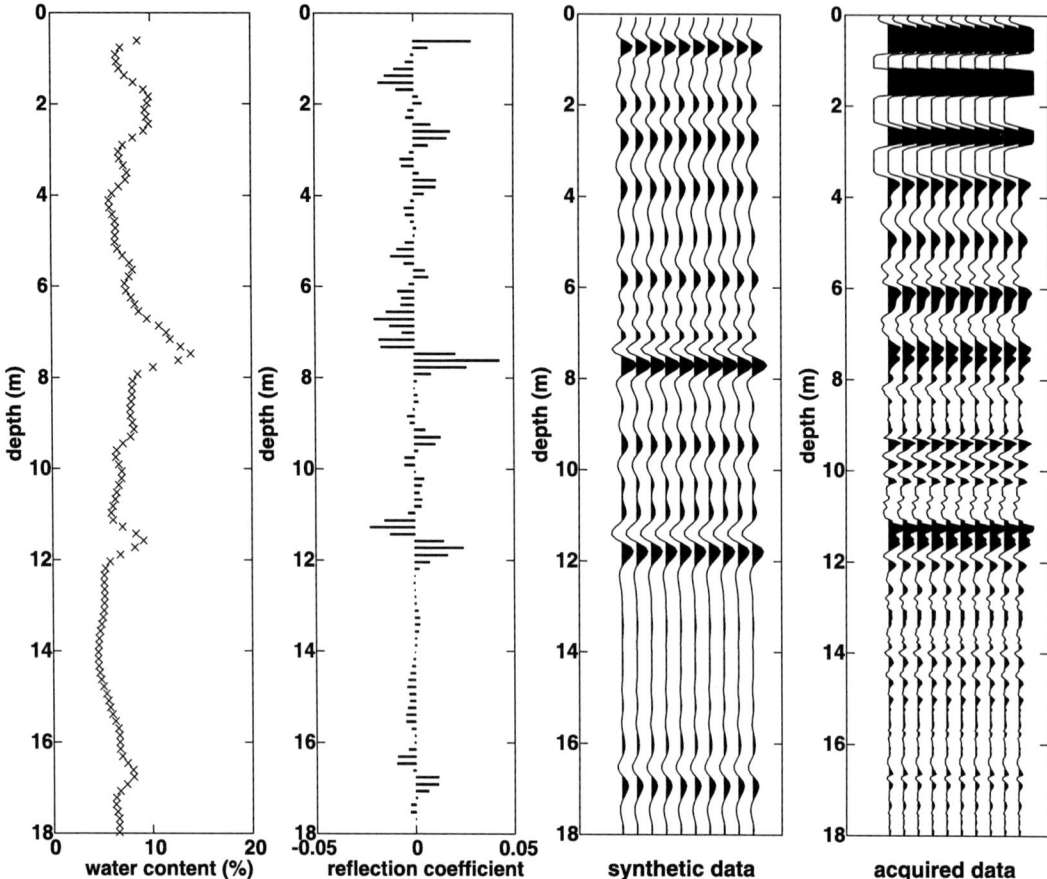

Figure 5. The water content data from well D8 were used to produce a synthetic radar trace using calculated dielectric constant values and a model of 1-D EM wave propagation. Shown are the water content data, reflection coefficients determined using Equation (2), a synthetic trace, and an acquired trace. The synthetic and acquired traces have been repeated 10 times. The comparison of the synthetic data to the acquired radar data illustrates the way in which the radar data capture the subsurface variation in water content.

1991). The nested exponential model is given by the following equation:

$$\gamma(h) = C_1(1-e^{-\frac{h}{\lambda_1}}) + C_2(1-e^{-\frac{h}{\lambda_2}}), \quad \text{if } h > 0 \quad \gamma(0) = 0 \quad (6)$$

where C_1 and C_2 are weighting factors for λ_1 and λ_2, the two correlation lengths (or integral scales) of the data set. The single exponential model is given by the above expression with $C_2 = 0$. The correlation length is the parameter that we used to describe the spatial structure of the data.

Semivariogram analysis was completed on the two water content data sets (which include all measured values of water content in the wells) using the program gamv found in GSLIB (Deutsch and Journel, 1998). The lag vector was in the horizontal plane for all sampled depths and set parallel to the northwest direction of the radar line. As an example of the data, we show in Figure 6 the transect from well D8 to well H8, displaying the 1980 and 1995 water content data as a function of depth in each well. We used a minimum length of 2 m for the lag vector. Due to the well arrangement and spacing, lags shorter than 2 m would have sampled a limited region with very few data points near the injection well. The maximum lag used was approximately one half the extent of the measurements.

Table 1 contains the scale triplet for the neutron probe measurements used in the geostatistical analysis. The neutron probe measurements were made in wells separated by 2 m in a radial pattern. Given an analysis limited to the northwest direction, we define the horizontal spacing as the minimum value of 2 m. The vertical spacing is 0.3 m in the 1980 data and ~0.15 m in the 1995 data. The horizontal extent of both sets of neutron probe measurements is the 16 m of the line of wells; the vertical extent is the ~18 m depth of the wells. We did not have an accurate measure of the support for either the

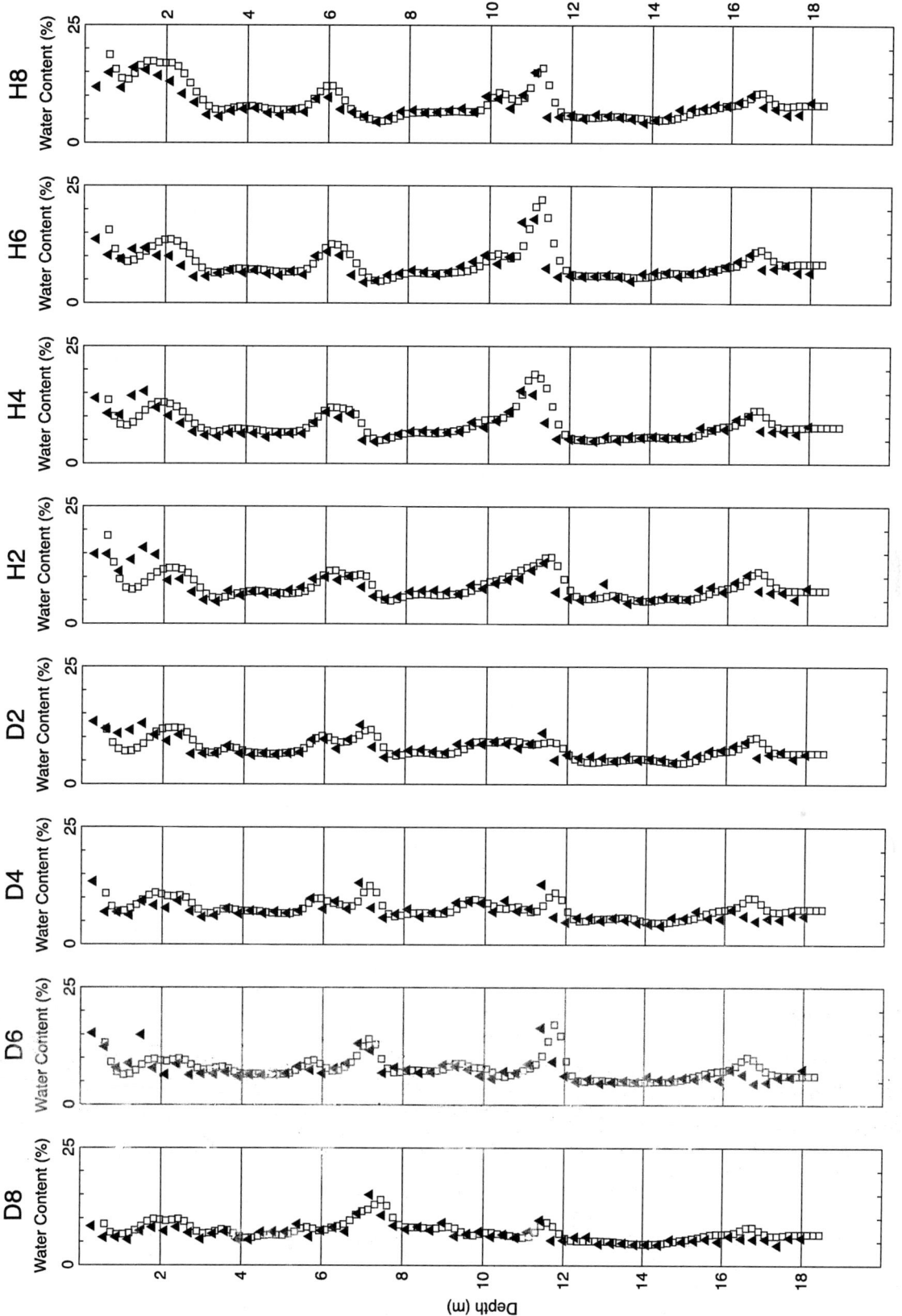

Figure 6. The transect from well D8 to well H8, displaying the 1980 and 1995 water content data as a function of depth in each well.

Table 1. The horizontal and vertical spacing, extent and support volume for the neutron probe and radar measurements.

	1980 Neutron Probe	1995 Neutron Probe	Radar	Radar-migrated
Spacing				
horizontal	2 m	2 m	0.1 m	0.1 m
vertical	0.3 m	~0.15 m	0.1 m	0.1 m
Extent				
horizontal	16 m	16 m	30 m	30 m
vertical	~18 m	~18 m	9.87 m	10 m
Support				
horizontal	~0.1–0.3 m	~0.1–0.3 m	~2-6 m	~1.3 m
vertical	~0.1–0.3 m	~0.1–0.3 m	0.35 m	0.35 m

Campbell-Pacific probe used in 1980 or the Schlumberger tool used in 1995. It is commonly assumed that a neutron probe measurement is sensitive to an approximately spherical volume that may range from 0.1 m to 0.3 m in diameter, depending on soil water content (Haverkamp et al., 1984). We have therefore used an estimate of 0.1 to 0.3 m for the horizontal and vertical dimensions of the support volume for both of the probe-derived water content data sets. It is very likely that the two systems had different support volumes.

The experimental and model semivariograms are shown in Figure 7a for the 1980 data and in Figure 7b for the 1995 data. Each experimental semivariogram was normalized by the variance of the data set on which it was calculated. The water content semivariograms were both fit with single exponentials with a correlation length of 10 m. The results of the modeling for all of the data are given in Table 2.

Let us consider the result of a horizontal correlation length of 10 m in terms of the horizontal dimensions of the scale triplet, given in Table 1, for the neutron data. We conclude that both the horizontal spacing and the horizontal support are more than adequate; these measurement scales are much less than the determined correlation lengths. The horizontal extent of the neutron data, however, is not much greater than the determined correlation length so it might not be large enough. We do not see either semivariogram flattening as it would when reaching the sill. It is also important to note that the neutron probe data could not be used to accurately characterize shorter-range structure (on the order of a few meters or less) due to the horizontal spacing of the measurements (determined by the spacing of the wells).

Semivariogram analysis in the horizontal direction was completed on the AGC-gained radar data using the program gam found in GSLIB (Deutsch and Journel, 1998). The

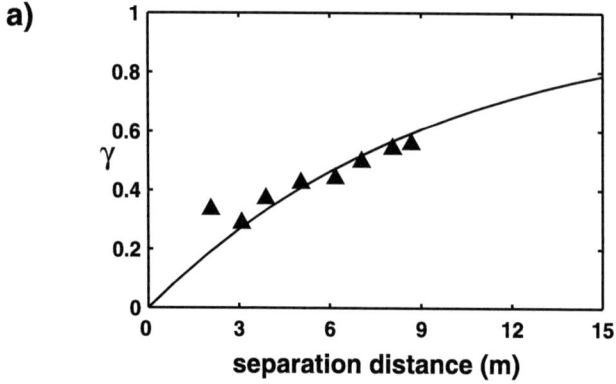

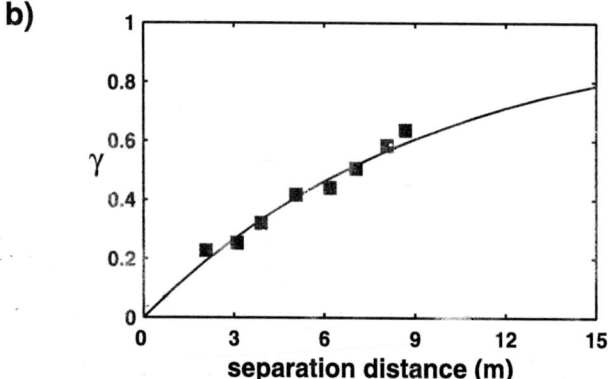

Figure 7. The experimental (symbols) and modeled (line) semivariograms calculated along the northwest direction for the probe-derived water content data from (a) 1980 and (b) 1995. Both experimental semivariograms were modeled with a single exponential model with a correlation length of 10 m.

Table 2. The parameters obtained in modeling of the water content and radar experimental semivariograms with single (for the water content data) and nested (for the radar data) exponential models.

Dataset	C_1	λ_1 (m)	C_2	λ_2 (m)
1980 θ_w	1	10	----	----
1995 θ_w	1	10	----	----
radar	0.2	0.3	0.80	14
radar-migrated	0.13	0.6	0.87	14

data values that were used were the recorded amplitudes of received energy, as done in the study by Rea and Knight (1998); more details on the geostatistical analysis of radar data can be found in that reference. The semivariogram analysis was conducted over the 30 m length of the radar section and across the depth interval of 1.76 to 11.63 m. The lag vector was horizontal, with zero tolerance on the vector direction. As suggested by Journel and Huijbregts (1978), we limited the lag vector to one half the extent of the measurements, so the maximum lag used was 15 m. Semivariogram analysis of radar images, conducted with a lag vector in the vertical direction, contains no useful information about subsurface properties because the length scale associated with the radar wavelet itself dominates the results. This is discussed in the paper by Rea and Knight (1998).

Table 1 contains the scale triplet for the analyzed radar data. The horizontal spacing of the radar measurements is the trace spacing of 0.1 m. The vertical sampling interval is set in terms of time and is 0.8 ns; this converts to a vertical spacing of 0.1 m using an EM velocity at the site of 0.14 m/ns. The horizontal and vertical extents of the radar measurements used in the geostatistical analysis are 30 m and 9.87 m respectively. The values given for the support of the radar measurements are the horizontal and vertical resolution of the data. These values were calculated using the expressions, given earlier, that are commonly used but are simplified representations of the support of the radar measurement. Accurately determining the support of any geophysical measurement is a highly non-linear problem, as the support depends on the measured subsurface properties. Using the average value for v of 0.14 m/ns found at the Sisson and Lu site, the vertical resolution with the 100 MHz system is 0.35 m (similar to the support of the neutron measurement); the horizontal resolution (Fresnel zone) ranges from ~2 m at a depth of 1 m to ~6 m at a depth of 12 m.

The radar experimental semivariogram is shown in Figure 8, with the semivariograms for the water content data included for comparison. It was fit with a nested exponential model (shown as the solid line in Figure 9) containing a dominant long-range structure with a correlation length of 14 m, and a smaller-scale structure with a correlation length of 0.3 m. Although the GPR profile suggests that some reflections have even longer spatial correlation lengths, the 14 m correlation length represents the entire profile rather than any particular depth-discrete layer or zone.

Let us now consider the results from geostatistical analysis of the radar data, in light of the horizontal spacing, extent and support of the measurement, given in Table 1. The relatively small horizontal spacing and large horizontal extent of the radar data should be sufficient to determine, without bias, a horizontal correlation length of 0.3 m; but the

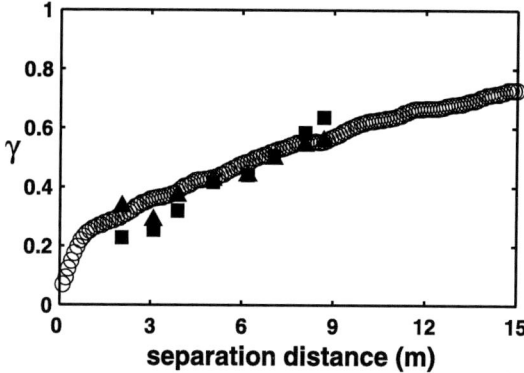

Figure 8. The experimental semivariogram for the radar data (circles). The water content semivariograms for 1980 (triangles) and 1995 (squares) are repeated for comparison.

horizontal support dimension is much too large. While it is possible that there is short-range subsurface structure that was imaged with the radar data, we conclude that we could not have accurately quantified it. If we compare the other determined horizontal correlation length of 14 m to the radar measurement scales, it is clear that the horizontal spacing of the radar data is adequate. As with the neutron probe data, however, we suspect that the horizontal extent is too limited to accurately determine the correlation length that we have estimated. Given the expected form of dependence of apparent correlation length on extent, the radar data, with an extent larger than that of the neutron data, should provide a correlation length that is greater than that determined from the neutron data, as observed.

The horizontal support dimension of the radar data varies from approximately 2 to 6 m; whether this is adequate is questionable, but can be further assessed by processing of the radar data so as to reduce the support. We migrated the data using phase shift migration (Gazdag, 1978) and the velocity model from well D-8. The migration process serves to reposition reflection events to their true subsurface locations, thereby improving the resolution of the image. With seismic data, migration can reduce the horizontal resolution to a theoretical limit of $\lambda/2$. The radiation pattern of the radar antennas is such that the resolution cannot reach this limit but rather approaches the theoretical limit given by the expression in Berkhout (1984; page 175) for limited aperture seismic data. Using this relationship along with the expression for the antenna radiation pattern given in Annan and Cosway (1994) we calculated an improved horizontal resolution in our migrated image that ranges from approximately 1.0 m to 2.1 m with an average value of 1.3 m.

With this reduced horizontal support dimension, we repeated the geostatistical analysis of the radar image and obtained the model results given in Table 2. We found no

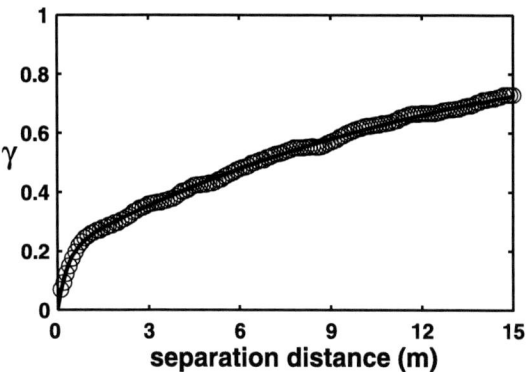

Figure 9. The experimental (circles) and modeled (line) semivariogram for the radar data. The data were modeled with a nested exponential model with two correlation lengths of 0.3 m and 14 m.

change in the longer correlation length of 14 m, suggesting that the horizontal dimension of the support of the radar data both before and after migration was sufficient to accurately determine the 14 m correlation length. (If the 14 m estimate had been biased due to support we should have seen a variation in the apparent value with a change in the support.) In contrast, we did find a change in the shorter correlation length. We conclude that neither of the radar data sets has the horizontal support dimension required to accurately estimate the parameters of the sub-meter structure. However, the radar data suggest that a sub-meter structure is present, which cannot be detected at all in the semivariogram of the probe-derived water content data due to the coarser horizontal spacing of the probe data.

The final measurement scale that we consider is the vertical support. As discussed in Knight et al. (2004), what has been observed in practice in anisotropic systems is a radar-derived lateral correlation length that is very sensitive to changes in the vertical resolution of the radar data; as the dominant frequency of the transmitted EM pulse decreased (resulting in a loss of vertical resolution or increase in vertical support), the estimated correlation length increased. In our study, the vertical supports for the neutron probe and radar measurements are very similar; we therefore do not anticipate significant differences in the determined correlation structure due to differences in vertical support.

The methodology developed by Rea and Knight (1998) for the geostatistical analysis of radar data has a critical underlying assumption: the correlation structure seen in the radar reflection image represents the correlation structure of the subsurface property governing the variation in κ; in this study, water content. It is important to emphasize that the radar reflection image does not correspond directly to an image of water content, but to an image of *changes* in water content. The reflection image can be represented (as an approximation) by the convolution of the radar wavelet with a subsurface model of reflection coefficients. The transform to reflection coefficients from subsurface water content acts like a vertical differencing filter, disrupting the spatial continuity of θ_w in the horizontal direction. The extent to which the correlation structure of the radar reflections represents the correlation structure of θ_w will therefore depend on the extent to which the convolution with the radar wavelet recovers, through spatial averaging, the continuity in θ_w. This is the link to the vertical resolution of the measurement.

An alternate approach would be to determine the horizontal correlation structure of the radar stack velocity (rather than the radar reflections), as done by Oldenborger et al. (2003). While there is a more direct link from subsurface properties to radar velocity than to radar reflections, it is extremely time-consuming to acquire and analyze radar data to obtain velocities, there can be a high level of uncertainty in the estimates of velocity in many geologic environments, and this approach results in a subsurface image that has much poorer resolution than is present in a radar reflection image. We therefore chose to focus in this study, and in our ongoing research, on the use of radar reflections alone to represent the variation in subsurface water content.

CONCLUSIONS

The Sisson and Lu test site, with the large volume of probe-derived water content data, is an ideal site for developing ways of using geophysical data to address characterization needs at Hanford and contaminated sites elsewhere. Our study using surface-based GPR, and the study of Majer and others (2001) using crosswell radar, indicate that there is a close link at Hanford between water content and κ of subsurface sediments. This results in a correspondence between the location of changes in θ_w and the location of reflections in a radar image. We therefore conclude that surface-based GPR can be very useful at Hanford as a way to map changes in θ_w, observed in well measurements, away from the well locations. This alone could provide valuable information about the spatial distribution in θ_w, a key factor in predicting the fate and transport of contaminants in the unsaturated zone at the Hanford site. In order to use GPR in this way at other sites, an essential step is to first determine that changes in θ_w are the dominant cause of the reflections in the radar data. While this is commonly assumed, given the strong dependence of κ on θ_w seen in laboratory data, the use of θ_w data from a site to produce a synthetic radar section for comparison with acquired data is an excellent way to check this critical assumption.

In the second part of our study, we determined the horizontal correlation length of the radar reflection image and compared this to the horizontal correlation length of the probe-derived water content data. We found that geostatistical analysis of the radar reflection image yielded a correlation structure similar to that obtained from neutron probe-derived θ_w values, and provided additional evidence of structure at the sub-meter scale. As a framework for comparing the results from the two forms of measurement, we focused on the effects of the scale of measurement. We conclude that the derived horizontal correlation lengths are likely to be too low, given the limited horizontal extent of both the neutron probe data and the radar data. In general, however, it is likely that the measurement scales that can be achieved with GPR data can lead to more accurate estimates of correlation length than can be obtained with other forms of well-based measurements.

We recognize that further work is required in order to fully understand the general applicability of the geostatistical analysis of radar images for characterizing the spatial distribution in subsurface water content. One of the key challenges is developing an improved understanding of the link between the correlation structure of the reflections, which correspond to changes in water content, and the correlation structure in water content. While it has been shown that the spatial averaging of the GPR wavelet can lead to a close correspondence between the reflection image and the distribution of water content (Knight et al., 2004), we lack a basis for predicting the accuracy of the radar-derived correlation length.

What do the results of this study suggest about the use of radar reflection images for characterizing the horizontal correlation length of subsurface water content? We conclude that although radar data clearly contain information about the spatial distribution of water content, they, like all forms of measurement, are highly sensitive to the scale (spacing, extent, support) of the measurement. Thus critical questions that remain to be answered are: How can we define the "correct" radar measurement scales to use in order to recover the desired correlation structure? How can these measurement scales be determined for a specific field site? We are confident that continued research at well-characterized field sites will ultimately allow us to use radar reflection images to describe the correlation structure of the subsurface.

Acknowledgments. The study was supported primarily by funding to R. Knight under Grant No. DE-FG07-96ER14711, Environmental Management Science Program, Office of Science and Technology, Office of Environment Management, United States Department of Energy (DOE). However, any opinions, findings, conclusions, or recommendations expressed herein are those of the authors and do not necessarily reflect views of DOE. The field work was funded by the Immobilized Low-Activity Waste project through U.S. DOE contract number DEAC06-99RL14047. We would like to thank Tapan Mukerji, Biondo Biondi and Jerry Harris for helpful discussions regarding the resolution and migration of radar data. We also wish to thank Andrew Binley and an anonymous reviewer for their thoughtful reviews.

REFERENCES

Agterberg, F. P., Autocorrelation functions in geology, in *Geostatistics*, edited by D. F. Merriam, pp. 113–141, Plenum, New York, 1970.

Annan, A.P. and S.W. Cosway, GPR frequency selection, Int. Conf. Ground Penetrating Radar, 5th, Waterloo, Ontario, Canada, Waterloo Centre for Groundwater Research, pp. 747–760, 1994.

Berkhout, A. J., Seismic Resolution, Geophysical Press, London, 228 p., 1984.

Bloschl, G and M. Sivapalan, Scale issues in hydrological modelling—a review, *Hydrological Processes, 9*, 251–290

Dafflon, B., Tronicke, J., and K. Holliger, Inferring the lateral subsurface correlation structure from georadar data: Methodological background and experimental evidence, Proc. GEOENV2004, 2004.

Daniels, D. J., D. J. Gunton, and H. F. Scott, Introduction to subsurface radar. *IEEE, 135*, (F4), 278–320, 1988.

Davis, J. L. and A. P. Annan, Ground penetrating radar for high resolution mapping of soil and rock stratigraphy, *Geophys. Prospecting, 37*, 531–551, 1989.

Deutsch, C. V. and A. G. Journel, *GSLIB Geostatistical Software Library and User's Guide*, 2nd edition, Oxford Univ. Press, New York, 369 p., 1998.

DOE (U.S. Department of Energy), Standardized stratigraphic nomenclature for Post-Ringold-Formation sediments within the Central Pasco Basin, DOE/RL-2002-39, (http://www.erc.rl.gov/pgs/readroom/doerl/rl02–39.pdf),U.S. Department of Energy, Richland, Washington, 2002.

Engelman, R. E., R. E. Lewis, D. C. Stromswold, and J. R., Hearst, Calibration Models for Measuring Moisture in Unsaturated Formations by Neutron Probe, PNL-10801, Pacific Northwest Laboratory, Richland, Washington, 1995.

Fayer, M. J., R. E. Lewis, R. E. Engelman, A. L. Pearson, C. J. Murray, J. L. Smoot, R. R. Randall, W. H. Wegner, and A. H. Lu, Re-Evaluation of a Subsurface Injection Experiment for Testing Flow and Transport Models, PNL-10860, Pacific Northwest Laboratory, Richland, Washington, 1995.

Gazdag, J., Wave equation migration with the phase shift method, *Geophysics, 43*, 1342–1351, 1978.

Gelhar, L.W., *Stochastic subsurface hydrology*, Prentice-Hall, Englewood Cliffs, NJ, 390 p., 1993.

Haverkamp, R., M. Vauclin, and G. Vachaud, Error analysis in estimating soil water content from neutron probe measurements: 1. Local standpoint, *Soil Sci., 137*, 78–90, 1984.

Journel, A. G. and Ch. J. Huijbregts, Mining Geostatistics, Academic Press, San Diego, 600 p., 1978.

Knight, R., Ground penetrating radar for environmental applications, *Annu. Rev. Earth Planet. Sci., 29*, 229–255, 2001.

Knight, R., P. Tercier, and J. Irving, The effect of vertical measurement resolution on the correlation structure of a ground penetrating radar reflection image, *Geophysical Research Letters, 31*, L21607, doi:10.1029/2004GL021112, 2004.

Majer, E. L., K. H. Williams and J. E. Peterson, High Resolution Imaging of Vadose Zone Transport Using Crosswell Methods, Lawrence Berkeley National Laboratory Report, 2001.

McBratney, A. B. and R. Webster, Choosing functions for semi-variograms of soil properties and fitting them to sampling estimates, *Journal of Soil Science, 37*, 617–639, 1986.

Oldenborger, G. A., R. A. Schincariol, and L. Mansinha, Radar determination of the spatial structure of hydraulic conductivity, *Groundwater, 41*(1), 24–32, 2003.

Rea, J. and R. J. Knight, Geostatistical analysis of ground-penetrating radar data: A means of describing spatial variation in the subsurface, *Water Resour. Res., 34*, 329–339, 1998.

Rockhold, M.L., Parameterizing flow and transport models for field-scale applications in heterogeneous, unsaturated soils. In Corwin, D. L., K. Loague, and T. R. Ellsworth (eds.) Assessment of Non-Point Source Pollution in the Vadose Zone, Geophysical Monograph 106, American Geophysical Union, Washington, D.C., 1999.

Sisson, J. B., and A. Lu, Field calibration of Computer Models for Application to Buried Liquid Discharges: A Status Report, RHO-ST-46 P, Rockwell Hanford Operations, Richland, Washington, 1984.

Smoot, J. L., *Development of a Geostatistical Accuracy Assessment Approach for Modeling Water Content in Unsaturated Lithologic Units*, Ph.D. thesis, University of Idaho, Moscow, 1995.

Topp, G. C., J. L. Davis and A. P. Annan, Electromagnetic determination of soil water content: Measurements in coaxial transmission lines, *Water Resour. Res., 16*, 574-582, 1980.

Western, A. W. and G. Bloschl, On the spatial scaling of moisture content, Journal of *Hydrology, 217*, 203–224, 1999.

Woodbury, A. D., and E. A. Sudicky, The geostatistical characteristics of the Borden aquifer, *Water Resour. Res., 27*, 533–546, 1991.

Yilmaz, O., Seismic Data Processing, Society of Exploration Geophysicists, Tulsa, 526 p., 1987.

Rosemary Knight, Mitchell Building, Geophysics Department, Stanford University, Stanford, CA, USA; email: rknight@pangea.stanford.edu

Integrating Statistical Rock Physics and Sedimentology for Quantitative Seismic Interpretation

Per Avseth

Rock Physics Technology, Bergen, Norway

Tapan Mukerji and Gary Mavko

Stanford Rock Physics Laboratory, Stanford University, Palo Alto, California, USA

Ezequiel Gonzalez

Shell Exploration and Production, Houston, Texas, USA

This paper presents an integrated approach for seismic reservoir characterization that can be applied both in petroleum exploration and in hydrological subsurface analysis. We integrate fundamental concepts and models of rock physics, sedimentology, statistical pattern recognition, and information theory, with seismic inversions and geostatistics. Rock physics models enable us to link seismic amplitudes to geological facies and reservoir properties. Seismic imaging brings indirect, noninvasive, but nevertheless spatially exhaustive information about the reservoir properties that are not available from well data alone. Classification and estimation methods based on computational statistical techniques such as nonparametric Bayesian classification, Monte Carlo simulations and bootstrap, help to quantitatively measure the interpretation uncertainty and the mis-classification risk at each spatial location. Geostatistical stochastic simulations incorporate the spatial correlation and the small scale variability which is hard to capture with only seismic information because of the limits of resolution. Combining deterministic physical models with statistical techniques has provided us with a successful way of performing quantitative interpretation and estimation of reservoir properties from seismic data. These formulations identify not only the most likely interpretation but also the uncertainty of the interpretation, and serve as a guide for quantitative decision analysis. The methodology shown in this article is applied successfully to map petroleum reservoirs, and the examples are from relatively deeply buried oil fields. However, we suggest that this approach can also be carried out for improved characterization of shallow hydrologic aquifers using shallow seismic or GPR data.

1. INTRODUCTION

In petroleum geophysics our main goal is to discover more oil or gas, to optimize new well locations or to improve petroleum recovery. Within hydrology and hydrogeophysics, major goals include mapping the best aquifers, optimizing water production, or attempting to detect non-aqueous phase liquid contamination. The bottom line in both disciplines is to quantify and reduce uncertainties in subsurface exploration, characterization, and management.

Both petroleum geologists and geohydrologists recognize that subsurface heterogeneity delineation is a key factor in reliable reservoir or aquifer characterization and subsurface remediation. Heterogeneities occur at various scales, and can include variations in lithology, pore fluids, clay content, porosity, pressure and temperature. Increased global demand for petroleum and water has resulted in a deliberate search for these vital resources in more complex and subtle reservoirs. Heightened environmental awareness has increased the needs for effective site remediation. Accordingly, there has been a quest for more quantitative seismic methods and improved understanding of seismic data, beyond the conventional geometric structural and stratigraphic interpretations. The quantitative information in seismic amplitudes opens up new gates for reservoir characterization and monitoring including the predictability of pore fluid types, fluid saturation, lithologies, and pore pressure [e.g., *Lumley*, 2001; *Castagna et al.*, 1998; *Ursin et al.*, 1996]. Some of the methods used in seismic reservoir characterization are purely statistical, based on multivariate techniques [e.g., *Fournier*, 1989]. Others are deterministic, based on physical models (theoretical, laboratory). Each group of techniques can have some degree of success depending on the particular study. The optimum strategy is to combine the best of each method to generate results much more powerful than would be possible from purely statistical or purely deterministic techniques alone. Some of the pioneering work that combined rock physics with statistical techniques includes *Doyen* [1988], *Lucet and Mavko* [1991], *Doyen and Guidish* [1992], and *Mukerji et al.* [1998]. Combined applications of geophysics, rock physics, and statistical techniques for hydrological investigations include the early works of *Rubin et al.* [1992], and *Copty et al.* [1993], followed amongst others by *Poeter et al.* [1997], *Ezzedine et al.* [1999], *Hyndman et al.* [2000], *Hubbard and Rubin* [2000], *Chen et al.* [2001], and *Tronicke and Holliger* [2005]. Some recent works where combined methodologies have been used for petroleum reservoir characterization, include among others, *Avseth* [2000], *Avseth et al.* [2001], *Mukerji et al.* [2001b], *Caers et al.* [2001], *Eidsvik et al.* [2004], and *Bachrach and Dutta* [2004]. *Bachrach and Mukerji* [2005] used shallow seismic reflections with crosswell radar tomography and geostatistics for aquifer characterization. *Gonzalez* [2005] combined rock physics with seismic inversion and multipoint geostatisics to predict lithofacies from seismic data. In this paper we will concentrate on statistical rock physics methodology. Other companion papers in the volume describe modern geostatitical methods including multipoint geostatistics for subsurface characterization.

Subsurface property estimation from remote geophysical measurements is always subject to uncertainty because of many inevitable difficulties and ambiguities in data acquisition, processing, and interpretation. It is therefore necessary to express quantitatively the information content, and uncertainty in rock property estimation from seismic data. Statistical probability density functions (pdfs) give us one way to describe quantitatively the state of our knowledge about the targeted rock properties, and the relations between rock properties and seismic signatures, including their inherent uncertainty. The pdfs may be estimated from available training data. The training set often has to be extended or enhanced using physical models to derive pdfs for situations not sampled in the original training data.

Additional data of a different kind can sometimes (but not always) bring in information that can help to reduce the uncertainty. For example, studies have shown that knowing shear wave velocities (Vs), in addition to pressure wave velocities (Vp), can help to resolve ambiguities in lithofacies versus fluids identification. From seismic data aquired with varying reflection angle, which implies varying offset between seismic source and receivers, one can extract reflection amplitudes as a function of offset (AVO). These AVO attributes implicitly include shear information even though the recorded wave is a pressure wave, i.e., when only one component is acquired. The AVO gradient (proposed by *Shuey*, 1985) and the far-offset elastic impedance [*Connolly*, 1998; *Mukerji et al.*, 1998] are examples of such "physical attributes" that indirectly contain shear wave information. Shear wave information may also be obtained more explicitly from multi-component surveys, which, however, are costlier than conventional single component surveys. The use of seismic AVO technology is widely used in the petroleum industry to detect hydrocarbons, but has been rarely applied to shallow environmental characterizations. *Waddell et al.* [2002] used high-resolution seismic reflection data and AVO analysis to locate high concentrations of dense nonaqueous phase (DNAPL) contaminants at a naval waste location near Charleston, SC. Though the examples described here all use attributes extracted from seismic data, the principles of the methodology can be applied to interpret other geophysical measurements such as ground-penetrating radar or electrical resistivity. *Baker* [1998] applied AVO analysis to

GPR data in order to differentiate NAPL from stratigraphic changes. *Bradford and Deeds* [2006] demonstrated how a better understanding of the offset dependent reflectivity from GPR could improve the detectability and accuracy in GPR subsurface characterization. Our integrated, probabilistic approach could be complementary to the workflows suggested in these and other GPR studies.

When using statistical data-mining techniques it is wise to keep in mind some of the myths and pitfalls of these methods. It is a myth is that the more attributes we throw in, the more effective will be the statistical effort. More attributes are useful only if they can contribute more information about the goal of the data-mining exercise. Otherwise they can do more harm than good. No statistical data-mining technique is so powerful that it can substitute for 'domain knowledge' and expertise in reservoir analysis and physical modeling, whether it is within petroleum geoscience or within hydrology.

Below, we go through the essential steps of the methodology, suggesting a workflow that we have found useful in our efforts to perform seismic reservoir characterization. Next, we go into more details of each step, and illustrate these with representative examples from North Sea oil fields. In spite of a lack of hydrologic case examples, for each of the steps we attempt to keep a link to hydrologic challenges and aquifer characterization, both with reference to other work as well as simple suggestions and recommendations.

2. WORK FLOW

The statistical rock physics methodology [*Mukerji et al.*, 2001a] described in this paper may be broadly divided in to four phases or steps (see flow chart, Plate 1):

Briefly, first, the well log information is analyzed to identify and define characteristic *facies*. These are classes of lithologies or sedimentary facies saturated with various pore fluid types. Prior to and during this step, it is important to do appropriate quality control and corrections of the petrophysical log data. Basic rock physics relations such as velocity-porosity, and Vp-Vs are defined for the facies. (If using other geophysical measurements, the appropriate rock-physics relations such as resistivity, and dielectric properties have to be defined for the facies.) This step is fundamental because it links the rock and fluid properties to the seismic parameters. Without this link we will not be able to obtain a deterministic and physical understanding of the seismic signatures. Since we want to exploit the shear wave velocity (Vs) information in the seismic data, we also need to establish the link between facies and Vs. If this information is not available from well log data, there exist empirical formulas to estimate Vs from Vp. If an expected pore fluid scenario is not encountered in the wells, one can perform fluid substitution using Gassmann theory.

After the rock physics analysis of well log data, we perform Monte Carlo simulation of the rock physics properties (Vp, Vs, density) and computation of the facies dependent statistical pdfs for various possible seismic attributes of interest (e.g., reflectivity, AVO gradient, near and far offset impedance, anisotropy parameters etc.) Rock physics modeling is used to extend the pdfs to situations not encountered in the wells (e.g., different fluid saturations, lithologies, presence of fractures).

Following this, the seismic data (attributes) from seismic inversion or analyses (e.g., AVO analyses, impedance inversion, etc.) are used in a Bayesian classification technique to extend the facies defined in the wells to all voxels within the seismic attribute cube. Calibrating the attributes with the probability distributions defined at well locations allows us to obtain a measure of the probability of occurrence of each facies.

Finally, geostatistics is used to include the spatial correlation, represented by the variograms or multipoint statistics, and the small scale variability, which is not captured in seismic data because of its limited resolution. Geostatistics can also be used at the initial stage to create spatial realizations of lithofacies consistent with the spatial correlation.

3. THE ROCK PHYSICS LINK BETWEEN SEISMIC AND GEOLOGY

3.1. Facies Definition

Usually the information from wells is the most directly available observation of a reservoir or an aquifer. The well log data can help us to calibrate the band-limited seismic data to a local background trend, and to better understand the expected contrasts between different lithologies and pore fluid zones.

For that reason, in many reservoir characterization projects, the first step is to define and to identify the facies that *a priori* we would like to delineate in the reservoir. In this paper we will use the term *facies* for categorical groups, not necessarily only by the lithology type, but also by some property or a collection of properties, as for example the combination of lithology and pores fluids. In an environmental site characterization project, one facies could for instance be sands contaminated with trichloroethane, which potentially could be separated from water-filled sand facies [e.g., *Waddell et al.*, 2002].

Using the available information at the well—cores, thin sections, geology, logs, production data—a facies indicator is assigned to each depth. It is convenient to do this process

Logs + Rock Physics + Geology

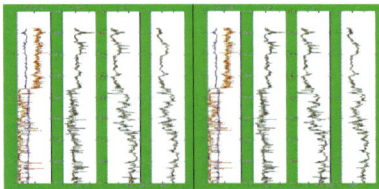

Monte Carlo - Probability distributions

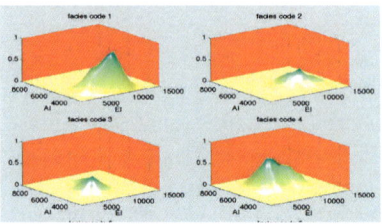

Seismic Inversions - near and far offset attributes

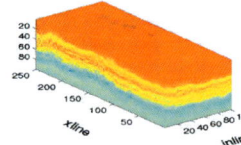
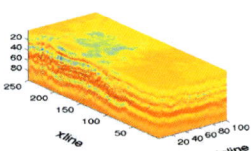

Integrated statistical classification - facies probability maps

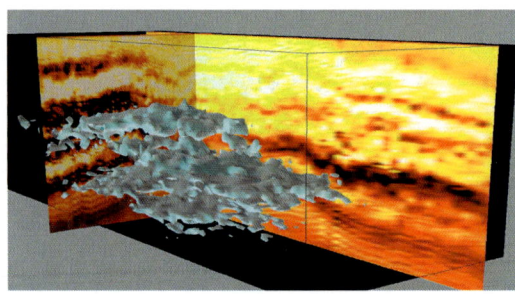

Plate 1. Schematic workflow for seismic reservoir characterization constrained by statistical rock physics and facies analysis of well log data.

with one or a few key wells where the data and interpretation are most complete and reliable. In petroleum fields, normally a complete suite of logs including both compressional and shear sonics as well as density logs are available at the key wells. This allows us to establish a link between the facies and the seismic properties. However, in lack of sonic velocity logs, seismic velocities can also be derived from high resolution vertical seismic profiling (VSP) where the seismic source is placed in the well and the receiver is located at the surface. Empirical relationships can then be used to derive densities from velocities [*Gardener et al.*, 1974]. This is exactly what *Waddell et al.* [2002] did in their environmemtal characterization of shallow sediments at an old naval site. They derived Vp and Vs for different facies from high resolution vertical seismic profiles (VSP) at well locations, and densities from empirical Vp-density relationsships. The criteria to define the facies depend on the targeted objective. In petroleum geophysics, for example, it could be to map different lithologies (sands-shales), to delineate fractures, to identify hydrocarbons, or changes in pressures and/or temperatures in a reservoir. Similar objectives are relevant within hydrogeophysics, where for instance contaminations are very important to map. It is a common practice to initiate the facies definition with exploratory crossplots between the logs looking for cluster separation. If there are poor separations in elastic properties for various facies at well log or VSP scale, there is likely no detectability at the seismic scale. In addition to elastic log data, other types of geologic information can also help to define useful facies categories, for instance the gamma ray log, core samples and thin section images. Plate 2 shows the result of the facies definition in a set of well logs from a North Sea oil field. As can be seen, each depth point has been assigned to a particular facies.

The different physical conditions or facies of interest that we would like to identify may not always be adequately sampled in the initial well training data. It may be necessary, using rock physics concepts, to extend the training data, modeling the reservoir properties after simulating changes in fluids, saturation, sorting, clay content, etc. For instance, *Waddell et al.* [2002] used Gassmann theory (see below) to estimate the expected seismic properties of sands filled with contaminants away from the VSP measurements where sands were water-filled.

3.2. Fluid Subsititution

Undoubtedly, the Gassmann theory is the most important and most frequently applied theory in rock physics [*Mavko et al.*, 1998]. Gassmann's equations allow us to predict the seismic properties of rocks saturated with hydrocarbons [e.g., *Smith et al.*, 2003], or contaminants [e.g., *Waddell et al.*, 2002], if we have only measured the properties of water saturated rocks. Seismic fluid sensitivity is determined by the combination of porosity and pore space stiffness. A softer rock will have a larger sensitivity to fluids than a stiffer rock at the same porosity. Gassmann's relations simply and reliably describe these effects:

$$\frac{K_{sat}}{K_{mineral} - K_{sat}} = \frac{K_{dry}}{K_{mineral} - K_{dry}} + \frac{K_{fluid}}{\phi(K_{mineral} - K_{fluid})}, \quad (1)$$

where $K_{mineral}$, K_{sat}, K_{dry} and K_{fluid} are the solid mineral, saturated rock, dry rock and pore fluid bulk moduli, respectively, and ϕ is the rock or sediment porosity. The shear modulus is insensitive to pore fluids, hence the companion result:

$$\mu_{sat} = \mu_{dry}, \quad (2)$$

where μ_{sat} and μ_{dry} are the saturated and dry rock shear moduli, respectively. Gassmann's equations (1) and (2) predict that for an isotropic rock, the rock bulk modulus will change if the fluid changes, but the rock shear modulus will not.

These dry and saturated moduli, in turn, are related to P-wave velocity $V_P = \sqrt{(K+(4/3)\mu)/\rho}$ and S-wave velocity $V_S = \sqrt{\mu/\rho}$, where ρ is the bulk density given by

$$\rho = \phi \rho_{fluid} + (1-\phi)\rho_{mineral} \quad (3)$$

The fluid sensitivity is not uniquely related to porosity, but to the rock stiffness [*Mavko et al.*, 1998]. Consequently, high-porosity sands can be much stiffer than low-porosity sands due to cementation. More often, however, high porosity sands are softer than sands with lower porosity. Hence, the potential to apply Gassmann theory for fluid substitution in shallow, unconsolidated sediments is normally greater than what is the case for deeply buried or uplifted consolidated rocks that are commonly encountered in petroleum reservoirs. In any case, lithology substitution is as important as fluid substitution.

3.3. Lithology Subsititution

In addition to fluid changes, we need to understand the expected changes in texture and lithology of the rocks or sediments on which we perform fluid substitution [*Smith et al.*, 2003]. For instance, diagenetic cement and clay lamination can have drastic effects on the dry rock frame as well as the fluid saturation pattern. In particular, clay effects may be very important in reservoir sandstones [*Avseth et al.*, 2005].

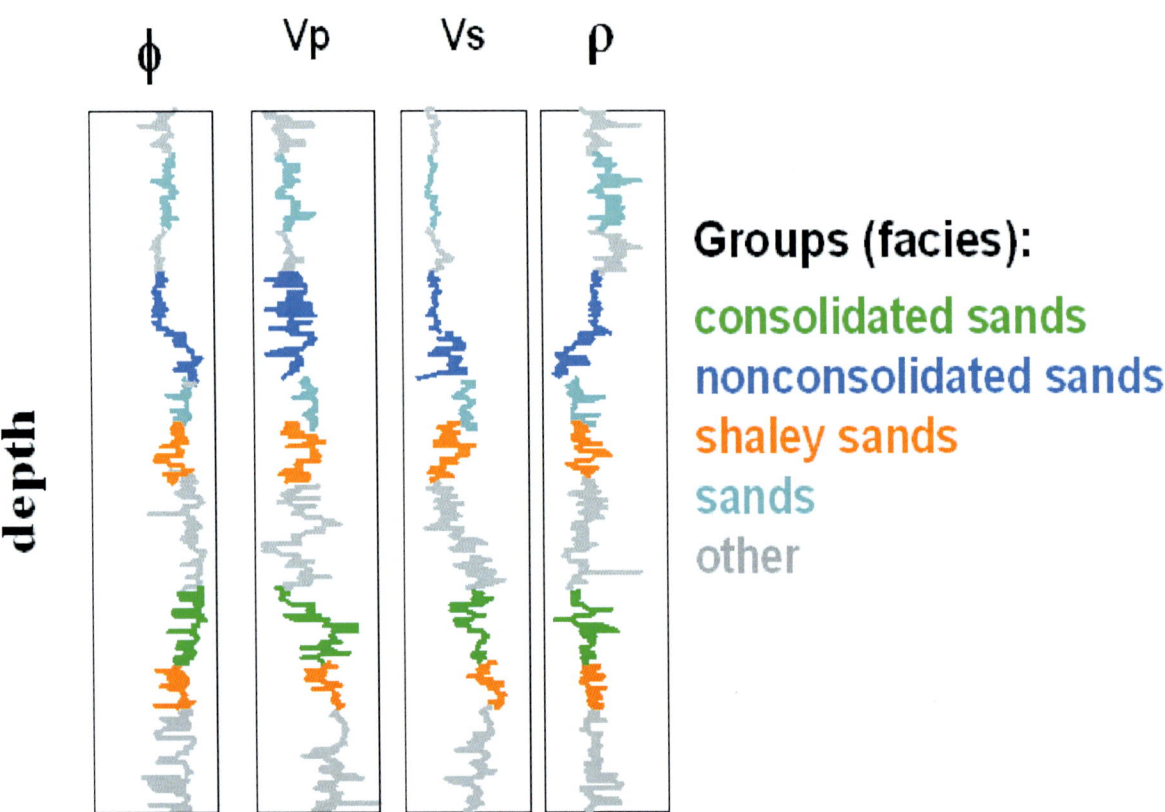

Plate 2. "Classified" well logs (each depth level has been identified as belonging to a particular facies). ϕ is total porosity and ρ is bulk density.

The most important reason for a rock physics—lithology link is to be able to calculate a correct dry frame in the Gassmann modelling (i.e., the correct relationship between stiffness and porosity). Furthermore, such models can be used for porosity prediction and lithology substitution. If we observe one type of sand at a well location, we may want to ask "what if" we have a different type of sand away from this well. A water aquifer with contaminants can have a different seismic response at two locations which is not related to the fluid properties, but to the change in for example sediment texture.

In our attempt to link seismic properties to reservoir properties, rock physics models can be useful complements to well log data. There is a large collection of different models that can be applied as tools for this purpose [Mavko et al., 1998]. For instance, the modified Hashin-Shtrikman upper and lower bounds have been found to be appropriate models to predict the seismic properties of sands and shales where the rock texture plays an important role in addition to the porosity [Avseth et al., 2005]:

$$K^{HS\pm} = K_1 + \frac{f_2}{(K_2 - K_1)^{-1} + f_1\left(K_1 + \frac{4}{3}\mu_1\right)^{-1}}$$

$$\mu^{HS\pm} = \mu_1 + \frac{f_2}{(\mu_2 - \mu_1)^{-1} + \frac{2f_1(K_1 + 2\mu_1)}{5\mu_1\left(K_1 + \frac{4}{3}\mu_1\right)}} \quad (4)$$

where

K_1, K_2 bulk moduli of individual phases
μ_1, μ_2 shear moduli of individual phases
f_1, f_2 volume fractions of individual phases

These formulas give us the effective bulk (K^{HS}) and shear moduli (μ^{HS}), respectively, as a function of volume fractions of porous rock versus mineral. Upper and lower bounds are computed by interchanging which material is subscripted 1 and which is subscripted 2. Generally, the expressions give the upper bound when the *stiffest* material (for instance the solid mineral) is subscripted 1 in the expressions above, and the lower bound when the *softest* material (for instance the porous sediment) is subscripted 1.

The modified Hashin-Shtrikman upper and lower bounds serve as very useful interpolators between the mineral point (i.e. zero porosity) and the high-porosity end member, normally given by the critical porosity (i.e., the physical upper porosity limit for a grain assemblage; for sands approximately 0.4). The lower bound of this model is found to give a very good representation of friable sand with varying sorting, where the stiffest material (i.e. the solid grains) is located passively inside the softest material (i.e. the pore space of the sediment). The upper bound is found to be more representative of diagenesis, where the stiffest material is added at grain contacts, causing a larger stiffening effect on the rock frame. However, for initial grain cement, the Dvorkin-Nur contact cement model [Dvorkin and Nur, 1996] has been found to work better than the Hashin-Shtrikman upper bound [see Avseth et al., 2005]. The elastic moduli estimated from equation 4, together with densities, allow us to estimate seismic velocities as a function of rock or sediment texture, using equation 3.

Figure 1 summarizes the diagnostic rock physics models which relate rock microstructure of sands to elastic properties. These models allow us to predict the geometrical arrangement of grains and pore space in sands from seismic velocities and densities. For more detailed descriptions of various rock models, see Avseth et al. [2005].

4. UNCERTAINTY IN ROCK PROPERTY ESTIMATION

4.1. Monte Carlo Simulation

A key point in the methodology is the concept of the *extended training data* and *derived distributions*. Using deterministic rock physics models in conjunction with statistical techniques allows us to extend the training data beyond

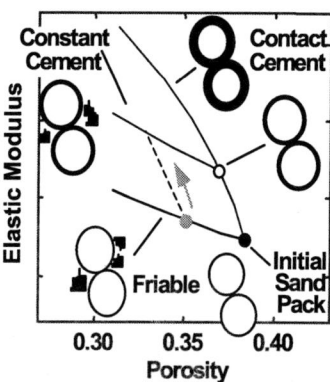

Figure 1. Rock physics models relating porosity and rock texture to seismic properties. The vertical axis could be any elastic moduli (bulk, K, or shear, μ) or seismic velocity (Vp or Vs), hence we have excluded any reference to units or absolute values. However, the elastic properties will increase from base to top. The figure includes only the high porosity range valid for unconsolidated or poorly consolidated sands/sandstones. The initial sand pack represents the high porosity end member, representative for a clean sand at deposition, also referred to as critical porosity. The reduction in porosity can either happen via sorting (pore filling material), via cementation around grains, or a combination of these two effects. In the former case, the elastic stiffness of the rock will increase more slowly with decreasing porosity than in the latter case. For more details on these diagnostic models, see Avseth et al. [2005].

what is just observed at the well, and derive the distribution of properties for scenarios not sampled in the original training data. For example, the well logs may have data for brine-saturated sandstones. Using Gassmann's equation (described above) we can compute the Vp, Vs and density for the same sandstones saturated with a different fluid (say air). Assuming that the well log data extended by rock physics modeling are sampling most possible values of Vp, Vs, and density for the study area, it is possible to fill the Vp-Vs-density space by generating additional points using *correlated Monte Carlo simulation*. Usually sequential simulation steps are used to generate correlated samples. For example, one strategy is to take Vp as the "base property", and use the available data to derive the Vp-Vs and Vp-density regressions. Then, Monte Carlo simulation is applied drawing values of Vp from the data derived non-parametric cumulative distribution function; then using the derived regressions, the corresponding Vs and density are simulated (allowing Gaussian variations around the regressions). This gives a realization of a correlated (Vp, Vs, density) sample. Instead of using regressions, a better approach is to draw Vs from the conditional distributions of Vs for each given Vp sample simulated in the first step. Given sufficient training data, the conditional distributions of Vs for different Vp bins can be pre-computed. Either way, a large number of points spanning the intrinsic variability (which gives rise to uncertainty) can be generated, respecting the Vp-Vs, and Vp-density data derived correspondence, as well as the distribution of the original data. This implicitly relies on Walther's law in geology that relates vertical variability to lateral variabilty within conformable stratigraphic sequences. At this step we have a non-parametric estimate of the multivariate distribution of Vp, Vs, and density for each group or facies of interest. Again, if the geophysical measurements are non-seismic, we need to estimate the distribution of the relevant property, e.g., dielectric or resistivity, using log data and appropriate rock physics models.

Next, to establish the link with the seismic information, seismic observables and attributes are theoretically calculated using the "extended" (through rock physics models and Monte Carlo simulation) log-based training data. An attribute is any characteristic that can be extracted from the seismic data. Although the methodology that we are presenting is completely general, in this paper only seismic attributes with some "physical meaning" are considered. This type of seismic attribute has a well defined physical relation with the reservoir properties, and can be either calculated using the well logs (Vp, Vs, density) or extracted from the seismic (e.g., with inversion, or AVO techniques).

Not all seismic attributes respond equally to different reservoir properties. Therefore the optimum seismic attribute or combination of seismic attributes to be used depends on the particular reservoir and the targeted facies or pore fluid classification problem. Maybe the easiest (but not the most rigorously objective) way to select the "best" attributes (when there are only a few of them) is by doing a visual inspection of color-coded comparative histogram plots of each attribute or cross plots of possible combinations between them, color-coding the points based on the facies to which they belong. A more quantitative approach is described in the section on information theory.

Plate 3a shows an example of a crossplot of two different seismic attributes, acoustic impedance (AI) vs. elastic impedance at 30 degrees (EI) calculated with well logs. Acoustic impedance (near–offset impedance) is the product of density and Vp, and is the impedance seen by a vertically propagating P-wave normal to the layers. Elastic impedance (far-offset impedance) is the approximate effective impedance seen by the wave traveling at non-normal incidence, and is a function of Vp, density, Vs, as well as the angle of incidence [*Connolly*, 1998; *Mukerji et al.*, 1998]. These impedance attributes can be computed from well logs (Vp, Vs, and density logs), and can also be extracted from seismic data that have been stacked at different angles. Inverting the near angle stack gives the acoustic impedance while inverting a far-angle stack gives the elastic impedance at the corresponding angle. As can be seen in Plate 3a, there are three color-coded groups: oil sandstones, brine sandstones, and shales, clearly well separated in this AI-EI plane. On the other hand, if a single attribute is used (equivalent to projecting the points over one of the axes) it is not possible to completely discriminate the three groups. The computation of seismic attributes and their pdfs from log data serves as a feasibility check to decide which attributes should be extracted from the field seismic data. In the initial exploration stages, this kind of feasibility study may also be used as a guide for designing the right survey that would be suitable for extracting the most promising attributes.

During this process of computing attributes it may be possible to find that not all the *a priori* defined facies, based on petrophysical and log data, can actually be separated in the seismic attributes space. In that case it is necessary to consider the union or division of some of the facies. Looking carefully at Plate 3a, we can identify different symbol shapes (triangles, circles, etc.) within each color-coded group. *A priori*, eight groups were defined, but it is clear that not all were separable with the proposed seismic attributes. In practice splitting or combining categories is done quantitatively using cluster analysis techniques. However, completely unsupervised cluster analysis usually gives poorer results than supervised learning, where clusters are defined based on expert knowledge (i.e., petrophysical and geologic expertise). When splitting or combining facies, it is not enough to analyze the attribute crossplots; it is also necessary to justify the decisions with geologic or production observations in order to attempt to avoid problems with the data (acquisition, processing, noise, etc.) that may drive the analysis to wrong conclusions.

4.2. Probability Density Functions (PDFs)

From the point distribution in the seismic attribute space, the *probability density functions (pdfs)*, [univariate (one attribute) or multivariate (combinations of attributes)], for each defined facies are estimated. In the simplest sense, an empirical *pdf* can be thought of as a normalized and smoothed histogram. In practice, to obtain the *pdfs* it is necessary to discretize the space where they will be calculated, and use a kernel (window) function for smoothing. Plate 3b shows the bivariate example of this process. In the *pdf* estimation, there has to be a compromise between the discretization and the smoothing. With too many cells, the pdfs would be too specific to the particularities of the input sample, and would not generalize to other data. With too much smoothing, the data variability would not be captured, and the discrimination between groups would be washed away. To choose these two parameters, a set of classification tests has to be done with a data validation subgroup. In spaces with few dimensions (attributes), the pdf calculation is not very difficult, although there are some non-trivial details in the process (smoothness, grid definition, limits extrapolation, etc.) that have to be carefully handled. On the other hand, in a space with high dimensionality, non-parametric pdf estimation is computationally highly demanding, and may not be very reliable due to sparse data. Other classification methods, such as K-nearest neighbors, neural networks, or classification trees have to be used in such situations.

4.3. Information Theory and Attribute Selections

Statistical information theory gives us simple yet powerful tools to quantify the information that each attribute can bring to discriminate the different facies [*Mavko and Mukerji*, 1998]. Using Shannon's information entropy concepts [e.g., *Cover and Thomas*, 1991] it is possible to select the "best" attributes as the one (or more) that most reduce the uncertainty in the reservoir properties identification. The quantity of information of a reservoir property "**X**", that an attribute "**A**" has, can be defined as:

$$I(X|A) = H(X) - H(X|A) \qquad (5)$$

where $H(\mathbf{X})$ is the *information entropy*, a statistical parameter that quantifies the intrinsic variability of **X**, without knowing the attribute **A**. $H(\mathbf{X})$ can be computed from the pdf, $p(\mathbf{X})$, of **X** [*Cover and Thomas*, 1991]:

$$H(X) = -\Sigma p(x_i) \log[p(x_i)] . \qquad (6)$$

$H(\mathbf{X}|\mathbf{A})$ is the conditional mean entropy of **X** given **A**, that is, the average uncertainty on **X** after observing **A**. The concept of information entropy which originated in statistics and communication theory has found applications in diverse fields such as computational chemistry, linguistics, bioinformatics and genetics.

The information $I(\mathbf{X}|\mathbf{A})$ can be interpreted as the reduction in the uncertainty of the reservoir property **X**, due to observing the attribute **A**. Therefore, a quantitative criterion to select the best attribute (or combination of attributes) is to choose the one (or ones) that maximize $I(\mathbf{X}|\mathbf{A})$ [*Takahashi*, 2000]. The reduction in information entropy and uncertainty by additional data can be shown by the following example. The relationships between porosity, Vp, and Vs of a particular reservoir are described by the trivariate pdf shown in Plate 4a. Conditioning of porosity information by velocities is summarized in Plate 4b. The unconditional prior pdf of porosity (blue curve) changes to narrower and taller conditional pdfs, p(porosity|Vp), and p(porosity|Vp, Vs) by velocity information. The velocity observation decreases the spread and variability (and hence uncertainty) about the porosity. This decrease in uncertainty is quantified by the information entropy. The prior information entropy about the porosity, computed from its unconditional pdf is 3.44. This decreases to 3.06 with Vp alone, and to 2.89 with both Vp, and Vs.

5. SEISMIC INFORMATION

Seismic attributes, which include reflectivities, velocities, impedances, and others, are derived from seismic data using different processing, analyses, or inversion techniques. Ways to obtain attributes from seismic data are topics of ongoing research and discussion. There are different algorithms for seismic inversion, each with its *pros* and *cons*. The common, fundamental goal of any inversion algorithm is is to estimate elastic parameters from the seismic data that will minimize the difference between the observed seismic data and a forward seismic model, and these results are used in the following reservoir characterization. However, there are many pitfalls: In order to build a good forward model, one needs to know a lot about the subsurface prior to the seismic inversion and parameter estimation. Here, well log data and information about local and regional geology are essential. The seismic inversion procedure suffers from a poor *a priori* earth model. There are many different scenarios that

54 QUANTITATIVE SEISMIC INTERPRETATION

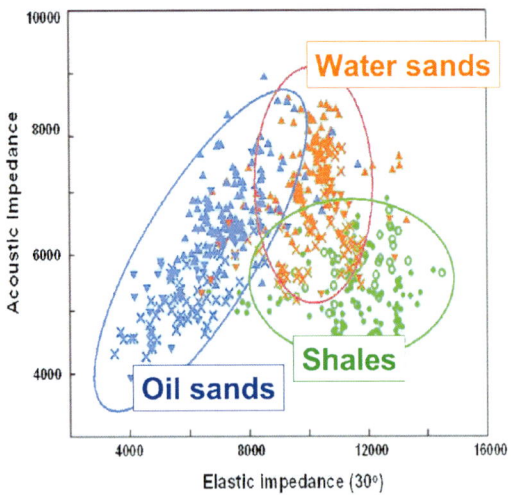

a)

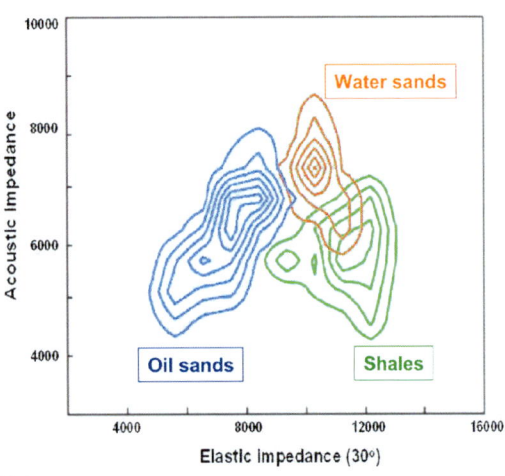

b)

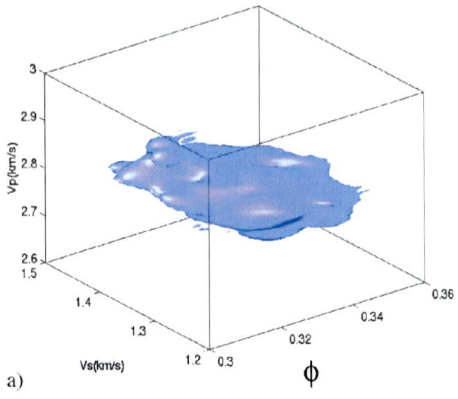

a)

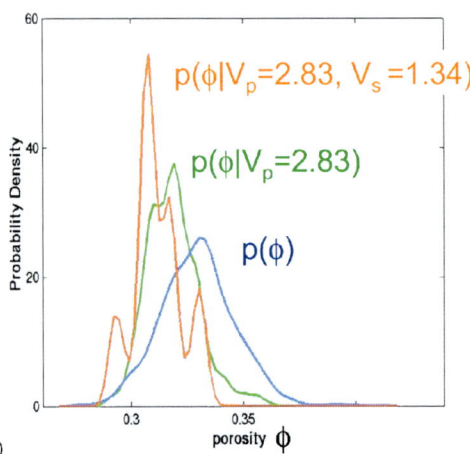

b)

Plate 4. Iso-surface of trivariate nonparametric pdf estimate for porosity, Vp, and Vs (a). Conditioning of porosity pdf (b) by Vp, and Vs information, corresponding to the trivariate pdf.

Plate 3. (a) Acoustic and elastic (30°) impedance (m/s.g/cm3) calculated theoretically from well logs. The color of each point corresponds to the facies to which it belongs. The ellipses are drawn by eye, to approximately represent the clusters. (b) Probability density function (pdf) contours generated with the data of (b) extended by Monte Carlo simulation. Notice that on the crossplot (a) the density of points can be obscured since points can overlap. The density in (b) is computed by smoothing a binned, normalized, 2-d histogram of the points generated by Monte-Carlo simulation.

can give the same seismic signature (i.e., non-uniqueness). Also, we need to have a good estimation of the seismic wavelet in order to obtain a reliable layer inversion from the seismic data. Moreover, the quality of the seismic data is critical for a robust inversion. Seismic processing seeks to increase the quality and the signal-to-noise ratio of the data. If this process is done poorly, the seismic inversion can fail to provide reliable results. In some cases, seismic data can show acquisition or processing footprints that may hide the reservoir reality. In other cases, these effects can influence the absolute values of seismic amplitudes but maintain their relative variations, which in the end could still be of real interest for discrimination and classification of reservoir properties. In general terms, having "good data" increases the probability of deriving reliable interpretations.

The seismic attributes derived from elastic seismic inversion respond to the reservoir interval properties (e.g., acoustic impedance, elastic impedance). However, we can also extract seismic attributes that respond to interface properties—contrasts between layers (e.g., AVO attributes). This technique is less time consuming, since it does not involve wavelet estimation and only minimal forward seismic modeling. The amplitudes are picked directly from seismic horizons in common depth point (CDP) gathers or from near and far stack seismic sections. Then, these amplitudes are used to calculate appropriate AVO attributes which are then calibrated to the corresponding Monte-Carlo simulated AVO pdfs from the well log data [*Avseth et al.*, 2001; *Houck*, 2002]. Plate 5a presents an example of physical seismic attributes of contrast at an interface: the AVO attributes defined by *Shuey* [1985], normal incidence P-to-P reflectivity R_0 (intercept) and G (gradient). The topography follows the traveltime interpretation of the seismic horizon along which the attributes were estimated from seismic AVO analyses. Plate 5b, a different example, shows acoustic and elastic impedance volumes resulting from inversion of near-offset and far-offset seismic partial stacks [*Mukerji et al.*, 2001].

As mentioned above, a detailed analysis of well log data and other geologic information is essential prior to the seismic data analysis, both in order to build a realistic prior model for the seismic inversion as well as to create realistic training data for the following classification. However, the attribute values derived from the seismic data are not always equal to the attribute values derived from the well logs. The reasons for those differences include the simplifications of the models used to derive the analytical expressions; imperfections in the data processing; and arbitrary scaling of the field amplitudes. Additionally, an important issue is that the measurement scales of the seismic and well logs are very different. The seismic responds to reservoir property averages that are not always well approximated by upscaling from the well logs. Due to these discrepancies, it is not possible (in general) to use directly the pdfs calculated with the well logs for classifying the attributes extracted from the seismic. In order to avoid the differences between the attributes computed with the well logs and the attributes extracted from the seismic, the classification system has to be generated with the traces nearest to the wells (taking into account deviations). Another option, when there are few available well data, is to recalibrate the pdfs derived from the seismic with the corresponding pdfs calculated from the well logs.

6. STATISTICAL CLASSIFICATION

When we have calibrated pdfs and seismically derived attributes, we can classify the volume or horizon of seismic attributes. These classes which we refer to as facies (defined in the first step) depend on the target—i.e., the classes could represent different lithologies, fluid types, uncontaminated versus contaminated sediments, fractures versus unfractured rock, etc. In other words, we want to "convert" the elastic parameters estimated from the seismic data into reservoir or aquifer properties, and create facies probability maps. There are many statistical methods to do pattern recognition or attributes classification [e.g., *Fukunaga*, 1990; *Duda et al.*, 2000; *Hastie et al.*, 2001; *Bishop*, 2006]. For example: linear and quadratic discriminant analysis (this only considers the mean and covariances of the reference pdfs), application of neural network or decision trees, or the use of the Bayes criterion with the obtained pdfs. With the Bayes classification method, the conditional probability of each group given one or a combination of attributes is calculated, and the sample is classified as belonging to the group that has the highest probability. Bayesian classification provides a maximum a posteriori (MAP) estimate of class as well as the uncertainty of the classification represented by the probabilities for each facies. When dealing with a few attributes (less than 5 or 6) Bayesian classification amounts to a table look up into the multivariate joint probability table computed in the previous step, followed by a normalization of the values so that they sum to one. This non-parametric Bayes classification works well with a few attributes, as is the case here where we have two seismic attributes. When dealing with a large number of attributes (more than 5 say), other parametric or semi-parametric methods of Bayesian classification have to be used. Plate 6 shows examples of results of applying the non-parametric Bayesian classification procedures to seismic attributes for two different cases. Plate 6a is the result of classifying the R_0 and G AVO attributes shown in Plate 5a, while Plate 6b shows iso-probability surfaces obtained after Bayesian classification of the near and far offset impedances shown in Plate 5b. It is important to keep in mind that such

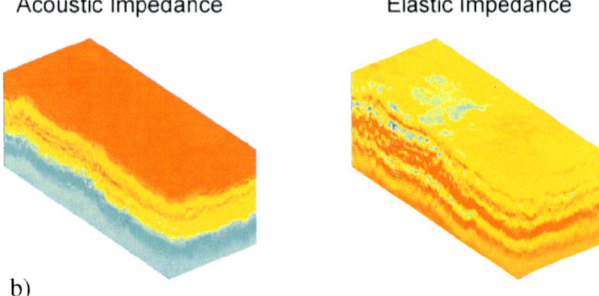

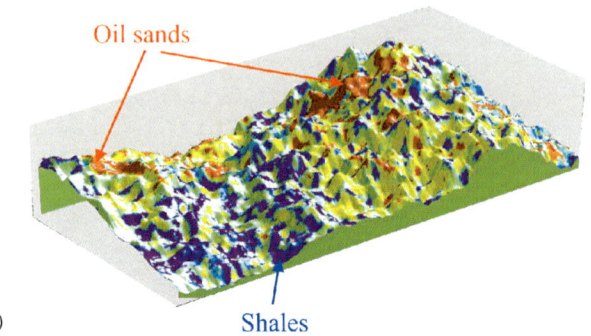

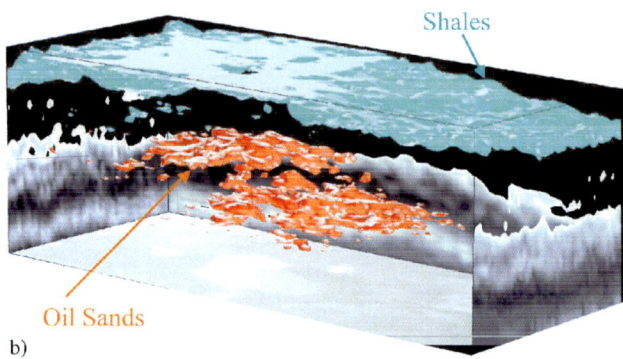

Plate 5. (a) P wave Shuey's AVO attributes (colors): To the left is R_0 (intercept), that is the zero-offset reflectivity of P-waves propagating and reflecting at vertical incidence, when the seismic source and the receiver are at the same location. Blue colors are representative of relatively high impedance values, while yellow is relatively low values. To the right is the AVO gradient, G, that is a scaled difference between the far-offset reflectivity (when the wave propagates and reflects at around 30 degrees incidence) and the zero-offset reflectivity. The yellow colours are relatively high negative gradients, whereas blue are relatively weak gradients. The topography follows the traveltime interpretation of the seismic horizon along which the reflectivity and gradient were estimated from AVO analyses of prestack data. (b) Acoustic and elastic impedance (at 30°) volumes. These two attributes respond to the reservoir interval properties. The far-offset elastic impedance implicitly contains shear wave information. These were estimated by impedance inversion of partial stacks. Red colors represent relatively low impedance, yellows represent intermediate values, while cyan and blue represent relatively high impedances.

Plate 6. (a) Areas with more probability of find oil sands (red) and shales (blue), resulting from the Bayesian classification using the P wave AVO attributes R_0 and G in Plate 5a. The topography follows the interpretation (travel time) of the seismic horizon (amplitudes) used to calculate the attributes. (b) Isoprobability surfaces resulting from applying a statistical classification process (nonparametric Bayesian) using the acoustic and elastic impedance in Plate 5b.

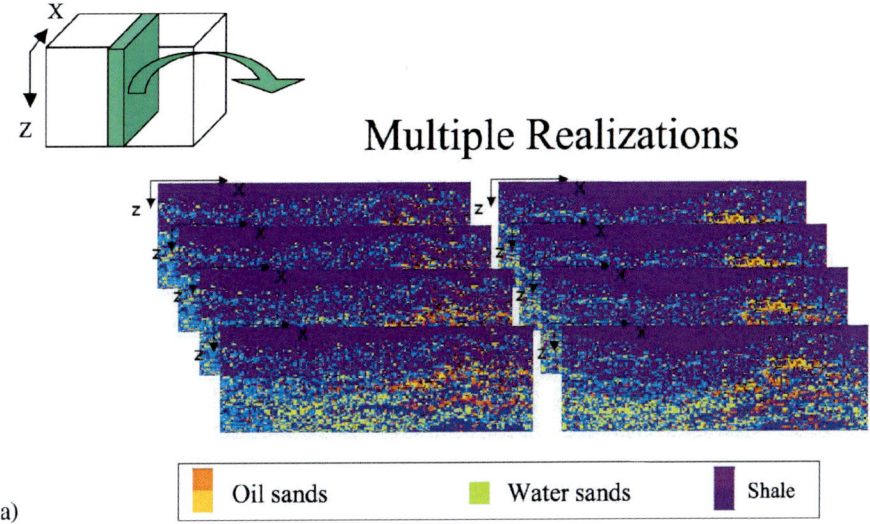

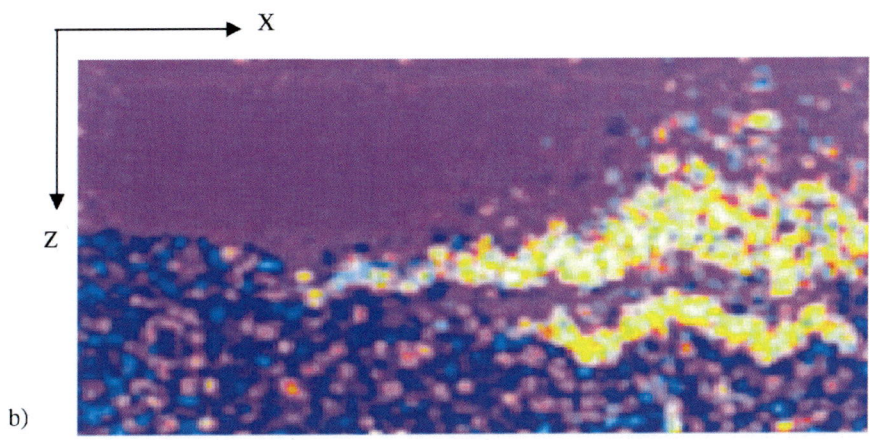

Plate 7. (a) The same vertical section (position) taken from different indicator stochastic simulation realizations. The red colors correspond to the oil sand facies. (b) Vertical section of the probability volume of finding oil sands obtained after geostatistical simulations. The yellow color indicates areas with higher probabilities. The geostatistical simulation updates the seismically derived probability (e.g., Plate 6) by accounting for the spatial correlation and small-scale variability in well logs.

probability surface visualizations do not show the actual sand (or shale) bodies but show the probability of the bodies having that spatial location and distribution.

By including geostatistics techniques of stochastic simulation in the analysis, we can take into account the spatial correlation (through the variogram) of reservoir properties. It can also attempt to reproduce the expected small scale variability, that cannot be detected with only the seismic data, but are seen in the well log data. Geostatistical analyses require estimation of spatial variograms which measure how different reservoir properties are correlated in space. Modern geostatistical techniques use not only the traditional two-point spatial correlation, but can also incorporate multi-point spatial statistics. As an example of the traditional two-point geostatistics, we show results from the geostatistical technique of indicator simulation. This technique generates multiple equiprobable realizations of facies in the reservoir and includes the seismic attribute classification results as soft indicators. Plate 7a presents a particular vertical section of the multiple equiprobable volumes (realizations) generated with indicator simulation. The figure clearly shows the characteristic variability of the stochastic process. For this example, the seismic attributes acoustic and elastic impedance volumes of Plate 5b were used as soft indicators, and two wells within the cube were used as hard indicators. The Markov-Bayes indicator formalism [*Deutsch and Journel*, 1998] was used to obtain the *posterior conditional pdfs*, including the facies spatial correlation through the indicator variograms. Plate 7b shows the result of this updating of the *prior pdfs*, P(facies | attributes), i.e. the probability of a facies given the attributes, to the *posterior pdfs*, P(facies | attributes, indicators information), i.e. the probability of a facies given attributes and the indicators data obtained from the well logs. In other words, the sections shown in Plate 7b correspond to the probability of each point to belong to a particular facies, oil sands in this case. It is calculated from the statistics of a large number of geostatistical realizations. This is an empirical Bayes approach without any priors on the parameters of the indicator variograms. As was mentioned, this type of result is an extension of the facies classification process described before, where the spatial correlation and small variability were included, in two and three dimensions. Modern methods go beyond the two-point methods and combine rock physics with multiple point geostatistics simulations [*Gonzalez*, 2005].

Some applications will require the pdfs (e.g., risks assessment for well placement) while others will need the stochastic realizations of actual reservoir properties drawn from the pdfs (e.g., reservoir flow simulations).

7. CONCLUSIONS

We presented in this paper concepts and methodologies that combine techniques of rock physics modeling, statistical pattern recognition and Bayesian classification, seismic AVO and impedance inversion, and geostatistics to quantify and reduce uncertainties in the reservoir characterization. The steps are summarized in the flow scheme in Plate 1 and include:

- Linking rock physics properties to observed and expected geologic facies and pore fluid scenarios (Plate 2 and Figure 1).
- Statistical rock physics and estimation of facies conditioned pdfs of seismic attributes from well log data (Plate 3)
- Selection of seismic attribute or attribute combinations based on information content for the target (Plate 4).
- Estimation of attributes from seismic data using various inversion methods (Plate 5).
- Bayesian classification of the volumes of seismic attributes into facies categories based on facies-conditioned, calibrated pdfs (Plate 6).
- Integrating spatial variability estimated from well-logs and training images using geostatistics (Plate 7).

The final products of this integrated technique are the spatial distribution of probabilities of reservoir fluids and facies, and stochastic realizations of the reservoir properties (Plate 7). In this way, not only do we obtain the most probable facies, but we can also quantify the uncertainty of the interpretation.

We have applied this integrated approach to do seismic reservoir characterization and shown how we can successfully predict hydrocarbons from seismic data. The same methodology could be applied to seismic, GPR, or electrical resistivity data for hydrologic aquifer characterization. The economic infrastructure is very different in the two disciplines, and large 3D seismic surveys are not common in hydrogeophysics. Nevertheless, the goals of hydrologists are very similar to those of petroleum geophysicists, which is to predict and map occurances of resources essential (or dangerous) to human life. Hence, we hope this effort to transfer technology across the discipline boundaries may motivate increased use of seismic data in hydrology.

Acknowledgments. We are grateful to Isao Takahashi, and Jack Dvorkin for their contributions to this work. In addition, we would like to thank Norsk Hydro and Statoil for permission to publish the data examples used in this study. Finally, we thank Stanford Rock Physics Laboratory for financial support, and Prof. Amos Nur for his early visions and devoted enthusiasm in creating the field of Rock Physics.

REFERENCES

Avseth, P., T. Mukerji, and G. Mavko (2005), *Quantitative Seismic Interpretation; Applying Rock Physics Tools to Reduce Interpretation Risk*, p. 378, Cambridge University Press, Cambridge.

Avseth, P. (2000), *Combining Rock Physics and Sedimentology for Seismic Reservoir Characterization in North Sea Turbidite Systems*, Ph.D. Dissertation, Stanford University.

Avseth, P., T. Mukerji, A. Jørstad, G. Mavko, and T. Veggeland (2001), Seismic reservoir mapping from 3-D AVO in a North Sea turbidite system, Geophysics, 66, 1157–1176.

Bachrach, R., and N. Dutta (2004), Joint estimation of porosity and saturation and of effective stress and saturation for 3D and 4D seismic reservoir characterization using stochastic rock physics modeling and Bayesian inversion: 74th Annual International Meeting, SEG, Expanded Abstracts, 1515–1518.

Bachrach, R. and T. Mukerji (2005), Analysis of 3D High-Resolution Seismic Reflection and Crosswell Radar Tomography for Aquifer Characterization: A Case Study, in Near-Surface Geophysics: Soc. of Expl. Geophys., 607–620.

Baker, G. S. (1998), Applying AVO analysis to GPR data: Geophysical Research Letters, 25, 397–400.

Bishop, C. (2006), *Pattern Recognition and Machine Learning*, Springer.

Bradford, J. H. and J. C. Deeds (2006), Ground-penetrating radar theory and application of thin-bed offset-dependent reflectivity; Geophysics, 71, K47–K57.

Caers, J., P. Avseth, and T. Mukerji, (2001), Geostatistical integration of rock physics, seismic amplitudes, and geologic models in North Sea turbidite systems, The Leading Edge, 20, 308–312.

Castagna, J., H. W. Swan, and D.J. Foster (1998), Framework for AVO gradient and intercept interpretation; Geophysics, 63, 948–956.

Chen, J., S. Hubbard, and Y. Rubin (2001), Estimating hydraulic conductivity at the South Oyster site from geophysical tomographic data using Bayesian techniques based on the normal regression model: Water Resources Research, 37, 1603–1613.

Connolly, P. (1998), Calibration and inversion of nonzero offset seismic, SEG Expanded Abstract.

Copty, N., Y. Rubin, and G. Mavko (1993), Geophysical-hydrogeological identification of field permeabilities through Bayesian updating: Water Resources Research, 29, 2813–2825.

Cover and Thomas (1991), *Elements of Information Theory*, Wiley and Sons.

Deutsch, C., and A. Journel (1998), GSLIB: Geostatistical Software Library and User's Guide, Oxford.

Doyen, P. (1998), Porosity from seismic data: A geostatistical approach, Geophysics, 53, 1263–1275.

Doyen, P., and T. M. Guidish (1992), Seismic discrimination of lithology: A Monte Carlo approach, in Reservoir Geophysics, Investigations in Geophysics, no. 7, ed. Sheriff, R. E., SEG, Tulsa.

Duda, R. O., P. E. Hart, and D. G. Stork (2000), *Pattern Classification*, John Wiley and Sons, New York.

Dvorkin, J., and A. Nur (1996), Elasticity of high-porosity sandstones: Theory for two North Sea datasets; Geophysics, 61, 559–564.

Eidsvik, J., P. Avseth, H. Omre, T. Mukerji, and G. Mavko (2004), Stochastic reservoir characterization using prestack seismic data, Geophysics, 69, pp. 978–993.

Ezzedine, S., Y. Rubin, and J. Chen (1999), Hydrological-geophysical Bayesian method for subsurface site characterization: Theory and application to LLNL Superfund Site: Water Resources Research, 35, 2671–2684.

Fournier, F. (1989), Extraction of quantitative geologic information from seismic data with multidimensional statistical analysis, SEG 59th Ann. Mtg. Exp. Abstr., 726–733.

Fukunaga (1990), *Introduction to Statistical Pattern Recognition*, Academic Press.

Gardener, G. H. F., L. W. Gardener, and A. R. Gregory. (1974), Formation velocity and density—the diagnostic basics for stratigraphic traps; Geophysics, 39, 770–780.

Gonzalez, E. (2005), *Physical and quantitative interpretation of seismic attributes for rock and fluids identification*, Ph.D. dissertation, Stanford University.

Hastie, T., R. Tibshirani, and J. Freidman (2001), *The Elements of Statistical Learning: Data Mining, Inference, and Prediction*, Springer-Verlag, New York.

Houck, R. (2002), Quantifying the uncertainty in an AVO interpretation; Geophysics, 67, 117–125.

Hubbard, S. S., and Y. Rubin (2000), Hydrogeological parameter estimation using geophysical data: A review of selected techniques: Journal of Contaminant Hydrology, 45, 3–34.

Hyndman, D. W., J. M. Harris, and S. M. Gorelick (2000), Inferring the relation between seismic slowness and hydraulic conductivity in heterogeneous aquifers: Water Resources Research, 36, 2121–2132.

Lumley, D. (2001), Time–lapse seismic reservoir monitoring; Geophysics, 66, 50–53.

Lucet, N., and G. Mavko (1991), Images of rock properties estimated from a crosswell tomogram, SEG, 61st Ann. Mtg. Exp. Abstr., 363–366.

Mavko, G., T. Mukerji, and J. Dvorkin (1998), *The Rock Physics Handbook: Tools for Seismic Analysis in Porous Media*, Cambridge Univ. Press.

Mavko, G., and T. Mukerji (1998), A rock physics strategy for quantifying uncertainty in common hydrocarbon indicators, Geophysics.

Mukerji, T., A. Jorstad, G. Mavko, and R. Granli (1998), Near and far offset impedances: Seismic attributes for identifying lithofacies and pore fluids, Geophysical Research Letters.

Mukerji, T., P. Avseth, G. Mavko, I. Takahashi, and F. Gonzalez (2001a), Statistical rock physics: Combining rock physics, information theory, and geostatistics to reduce uncertainty in seismic reservoir characterization: The Leading Edge, 20, 313–319.

Mukerji, T., A. Jørstad, P. Avseth, G. Mavko, and J.R. Granli (2001b), Mapping lithofacies and pore-fluid probabilities in a North Sea reservoir: Seismic inversions and statistical rock physics; Geophysics, 66, 988–1001.

Mukerji, T., A. Jørstad, G. Mavko, and J.R. Granli (1998), Applying statistical rock physics and seismic inversions to map lithofacies and pore fluid probabilities in a North Sea reservoir, SEG Technical Program Expanded Abstracts, pp. 894–897.

Poeter, E., W. L. Wingle, and S. A. McKenna, (1997), Improving groundwater project analysis with geophysical data: The Leading Edge, 16, 1075–1681.

Rubin, Y., G. Mavko, and J. Harris, (1992), Mapping permeability in heterogeneous aquifers using hydrological and seismic data: Water Resources Research, 28, 1192–1800.

Shuey (1985), A simplification of the Zoeppritz equations, Geophysics.

Smith, T. M., C. H. Sondergeld, and C. S. Raiz (2003), Gassmann fluid substitutions: A tutorial; Geophysics, 68, 430–440.

Takahashi, I. (2000), Quantifying information and uncertainty of rock property estimation from seismic data, Ph.D. Thesis, Stanford University.

Tronicke, J., and K. Holliger (2005), Quantitative integration of hydrogeophysical data: Conditional geostatistical simulation for characterizing heterogeneous alluvial aquifers, Geophysics, H1–H10.

Ursin, B., B. O. Ekren, and E. Tjåland (1996), Linearized elastic parameter sections; Geophysical Prospecting, 44, 427–455.

Waddell, M., W. Domoracki, and T. Temples (2002), Detection of DNAPLs using ultra high-resolution seismic data and AVO analysis at Charleston Naval Weapons Station, South Carolina; SEG Expanded Abstract, 72nd Annual Convention, Salt Lake City.

Per Avseth, Repslagergaten 20, N-5033 Bergen, Norway. E-mail: per@rpt.info

Ezequiel Gonzalez, Shell, Houston, USA. E-mail: Ezequiel.Gonzalez@shell.com

Gary Mavko, Mitcehll Bldg. Panama Mall, Geophysics Department, Stanford University, Stanford, CA94305, USA. E-mail: mavko@stanford.edu

Tapan Mukerji, Mitchell Bldg. Panama Mall, Geophysics Department, Stanford University, Stanford, CA 94305, USA. E-mail: mukerji@pangea.stanford.edu

A Geostatistical Approach to Integrating Data From Multiple and Diverse Sources: An Application to the Integration of Well Data, Geological Information, 3d/4d Geophysical and Reservoir-Dynamics Data in a North-Sea Reservoir

Jef Caers and Scarlet Castro

Stanford University, Department of Energy Resources Engineering, Stanford, California, USA

INTRODUCTION

Modeling the subsurface is an inherently difficult task due to limited access and lack of direct observation of the complex medium under investigation. Nevertheless, practical engineering questions often call for a full 3D modeling of subsurface heterogeneity, whether the task is to maximize production of an oil reservoir or to optimize storage of water during dry seasons in an aquifer storage and recovery process. While the goal of modeling and the nature of fluid flow may be different between the field of petroleum and hydrogeology, each deals with a similar heterogeneous medium and faces similar questions in model building.

Modeling aquifers or reservoirs requires integrating diverse sources of information into a single model (e.g., Deutsch, 2003, Caers, 2005). One faces many challenges in doing so, most related to the issue of scale, since the unit grid cell size of the model is different from the scale of information provided by each source of information. Each such source informs the aquifer or reservoir at a different scale of observation. Secondly, models contain several geological building blocks, such as a structural model (fault/horizons), 3D distribution of facies types, petrophysical properties (porosity and permeability) per facies, fluid distributions and fluid properties, etc.; each building block needs to be constrained to the available data.

Under such complex conditions, a probabilistic approach is desired for several reasons; first and foremost because the data gathered do not uniquely and deterministically determine the geological heterogeneity of the subsurface. Probabilistic models allow a more flexible integration of data than deterministic ones. In fact, deterministic modeling can be regarded as a special case of probabilistic modeling where all probabilities have been set to 1 or 0. Instead, we will present a method where the information content of each data source is coded into a probability value, then, a probabilistic framework for combining these elementary probabilities into a joint probabilistic statement based on all data sources is applied. From this joint probability, several realizations (reservoir models) are drawn. We will demonstrate the flexibility of our approach that may include a wide variety of prior geological models (no assumptions of multi-Gaussianity are needed) and can include a variety of diverse data sources, including time-varying data.

The ensemble of such reservoir models reflects the lack of data and knowledge to exactly and uniquely quantify the subsurface, hence quantify uncertainty. Recognizing that there are many aspects to subsurface modeling, many of them covered in this volume, this paper focuses on an application of this probabilistic data integration approach demonstrated on modeling complex fluvial channels in a prominent North Sea reservoir case. The paper can be seen as one of the state-of-the-art approaches to modeling oil & gas reservoirs (see also Caers et al., 2006).

SUBSURFACE DATA AND SPATIAL MODELING

Several sources of information may be available to quantify the subsurface heterogeneity and model the spatial distribution of rock properties. We will divide them into two groups: static data and dynamic data. Two major sources of static data are (1) data derived from wells, either from drilling, such as cores or from logs and (2) data obtained via geophysical surveys. Dynamic data are data that vary over time. In the context of producing reservoirs these consist of well and interference (pumping) tests, flow meter data, data from permanent downhole gauges, historical pressure and production data and time-lapse seismic (4D seismic) obtained from the reservoir.

The Prior Geological Scenario

The ensemble of such data allows geologists to interpret the nature of the depositional system present. Such interpretation is crucial since it will determine the level of heterogeneity of the resulting models. For example, one may calculate and model a variogram of permeability from the available well data (if possible) and use it as a measure of spatial continuity in building permeability realizations by means of geostatistical methods. In the petroleum industry, kriging for stand-alone interpolation is no longer used to map petrophysical or facies properties, due to the well-documented effect that produces overly smooth permeability and porosity models, and consequently biased flow predictions (Journel and Alabert, 1990; Srivastava, 1992, Deutsch, 2003, Caers, 2005). Instead, we rely on stochastic simulation to generate multiple reservoir models, each reproducing the variogram modeled/interpreted from the field. However, it should be understood that while some geostatistical simulation algorithms only require a variogram (and histogram,) as input, these algorithms/models make additional model assumptions (often hidden to the less informed user), such as that of multi-Gaussianity of the spatial variable under study. The multi-Gaussian assumption constrains considerable the type of spatial continuity of the simulated spatial variable. Reservoir models drawn under multi-Gaussian assumptions exhibit "maximum entropy", i.e. the extremes are maximally uncorrelated for the given variogram or spatial covariance. This property is unrealistic for real geological depositional systems.

Hence, modern spatial modeling techniques rely on more geologically realistic and explicit representations of heterogeneity by means of a 3D training image. In the practice of building models, it has been determined that traditional variogram-based methods are largely inadequate to model realistic geological heterogeneity, particularly when such heterogeneity is dominated by the spatial distribution of facies (see Feyen and Caers, 2006). In such cases, the distinct difference in petrophysical properties of facies bodies with specific geometries (e.g. curvi-linear), connectivities and mutual associations (overlap, erosion) can not be captured by a simple two-point statistical measure such as the variogram. In fact, it has been shown by means of examples (see Strebelle, 2000, Caers, 2005; Feyen and Caers, 2006) that strongly different geological depositional systems yield the exact same variogram.

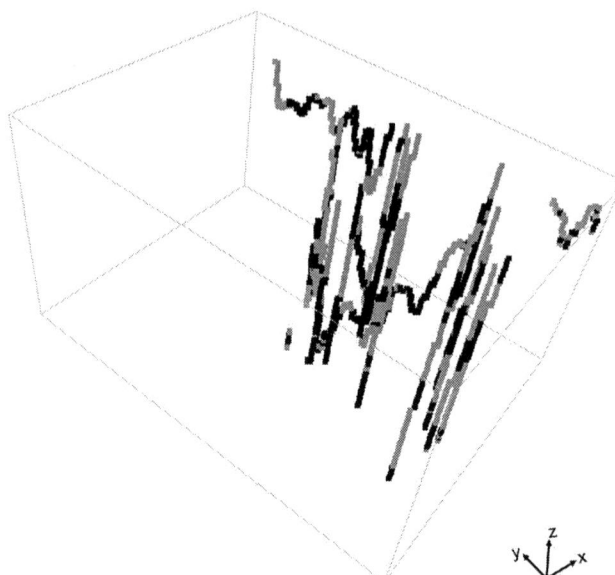

Figure 1. Facies succession along the various vertical and horizontal wells (black = sand, grey = shale). (x = North, y = East, z = Up). All Figures (1–7) are shown in depositional coordinates, i.e. after flattening the originally faulted and folded reservoir structure.

In the field of multiple-point geostatistics, one no longer represents spatial continuity by an explicit parametric model (e.g. variogram parameters and multi-Gaussian law) but by means of a 3D geological analog, or training image. A training image is an explicit 3D conceptual representation of the subsurface geological scenario reflecting, at least conceptually, all available geological information on facies distribution deemed relevant (see Caers and Zhang, 2004). The training image is a conceptual model, which means that it is not a reservoir model as such. It need not be constrained locally to well or seismic data. Instead it needs to reflect the type of geological patterns that are thought to be present in the subsurface (channels, elliptical lenses, barchains, etc.). Training images can be created from outcrop data, by means of deterministic process-based techniques, stochastic object-based techniques, or a geologists' conceptual renditions of subsurface properties. The idea of multiple-point geostatistics is to generate facies realizations by anchoring the geological patterns of the training image to subsurface data, be it static and/or dynamic data as will be explained in more detail further on.

It should be noted that any spatial continuity model, whether variogram or training image is the result of an interpretation or choice. While such interpretation/choice is always subjective, it represents information/expertise that needs to be integrated together with other reservoir data. In general, all methods of data integration and model building rely on such subjective interpretation whether this is done explicit (e.g. training image) or implicit (e.g. variogram with the implicit multi-Gaussian assumptions). It is often incorrectly stated that variogram-based modeling provides a more objective way to model than training images, since the former are directly inferred from data. Even if one could get a permeability variogram from very sparse well-data, any 3D model building requires more than a variogram, since the construction of any 2D or 3D "image" calls for the specification of all higher-order statistics present in the image. For example, when building a 3D model using a variogram, the higher order moments are frozen in the assumption of multi-Gaussianity when a Gaussian-type method is applied. In mapping kriging estimates, the higher moments are enforced through the smoothing implicit to kriging (or any other interpolation scheme). There is no escape to assuming the full joint distribution at all spatial locations of the variable being modeled, even in a deterministic approach. With training image-based modeling one simply states these higher-order statistics more explicitly as training image patterns, instead of hiding them in mathematical model that may be geologically unrealistic.

In the case of the North Sea reservoir (NSR), several well-logs sets are available providing an interpretation of the facies succession (sand/no-sand) along the well as well as porosity and permeability measurement, and several key rock physics properties such as P-wave and S-wave velocities (Fig. 1). A 3D training image was built by means of an unconstrained object simulation technique which was used to generate fluvial-type channels with dimensions derived from the well-log data as well as from analog outcrop information deemed representative for the reservoir under investigation (Fig. 2). Note that all models and data are shown in a depositional coordinate system transformed from the original faulted and folded reservoir structure.

Seismic Data

Geophysical data provides an exhaustive but indirect quantification of the 3D geological variability be it of the reservoir structure or of the reservoir rock property variations. While current geophysical surveys (seismic mostly in oil and gas reservoirs) are now fully 3D, they still only provide indirect information in terms of amplitude variations.

Moreover, the vertical and horizontal scale of information provided by seismic data is usually larger than the unit grid cell size on which one builds a detailed model of facies and petrophysical properties. For example in the NSR (reservoir unit between 25–40m thickness) the seismic data provides

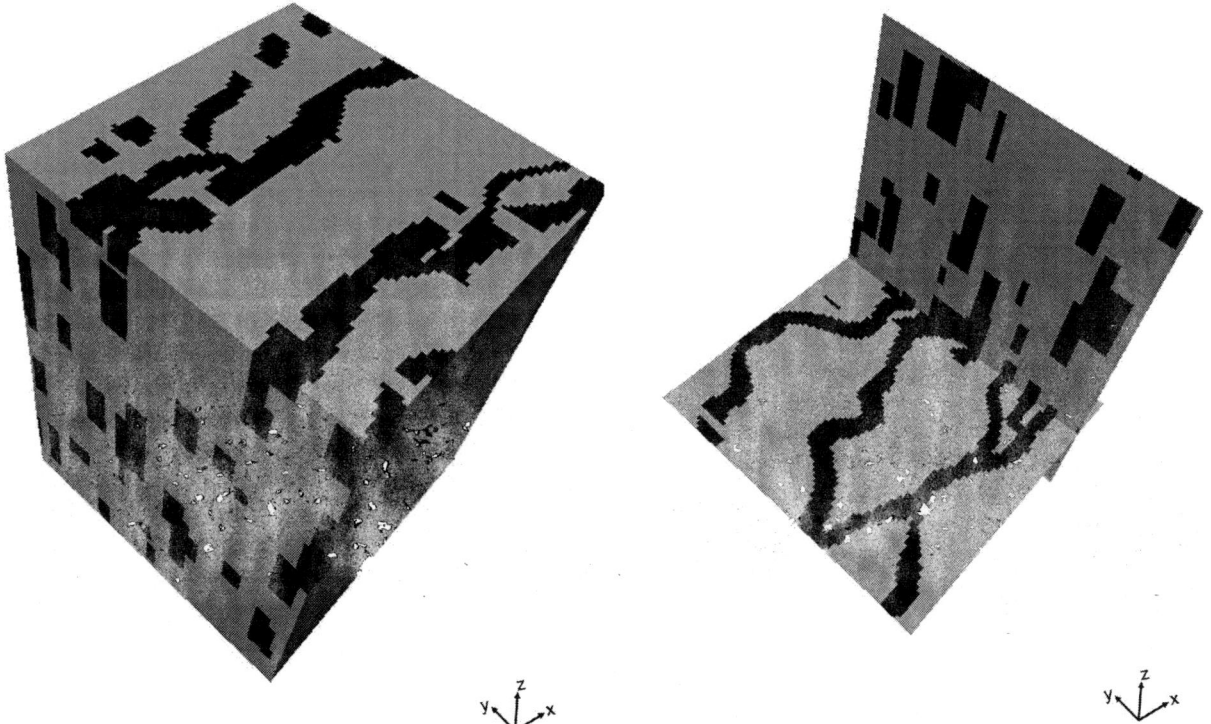

Figure 2. 3D training image of a fluvial channel system reflecting channels with varying length, thickness and width—right image shows selected cross sections. (black = channel sand, grey = background shale)

information at a scale much larger than the unit grid cell of the high-resolution model (see next section) which is 25x25x0.8 meter. The seismic surveys for this reservoir provides a vertical scale of information of roughly 25m and horizontal scale of observation of 500m (based on the Fresnel zone). The small-scale channel geometries that are known to affect reservoir flow, as well as the detailed facies variation interpreted from wells (Fig. 1) need to be included by building high-resolution models (higher resolution than the scale of information provided by seismic). Due to the limited resolution of geophysical data, one often calibrates the seismic data into a facies-probability cube using a rock-physics and/or statistical calibration method of choice. This cube contains in each grid cell of the high-resolution model the probability of occurrence of each facies type. In general terms, such calibration needs to account for the indirectness of information provided by the seismic as well as the scale difference between seismic and higher resolution well data. Several statistical regression techniques can be used to obtain such a probability cube. More specifically, for the NSR, a sand probability cube was created by calibrating both 3D and 4D (3D + time) information with sand data from wells (see Andersen et al., 2006 for how this was done at NSR). Fig. 3 shows this cube.

Dynamic Data

Next to static data, information on reservoir dynamics through pressure and flow data of various kinds are available and grouped under the term "dynamic data". Such data provide direct information of the actual process of interest: subsurface flow. The NSR has been in production since 1993; since then several injector wells (gas and water) have been drilled to sustain reservoir pressure. However, such injection also leads to breakthrough of water in oil producing wells. The breakthrough times and subsequent ratio of water produced compared to the total volume of fluid produced, or water-cut, as well as the cumulative oil produced are excellent indicators of reservoir connectivity (or lack thereof). However, the integration in the reservoir model jointly with the well-log, seismic and geological interpretation calls for the iterative solution of an inverse problem as elaborated below.

INTEGRATION OF STATIC AND DYNAMIC DATA

Building a High-Resolution Geo-Cellular Model

The task is now to integrate the well-log data, facies probability cube derived from 3D/4D seismic, the geological interpretation depicted in the training image and the production data obtained from the various producing wells to create several alternative reservoir models. These models will be termed "equiprobable" referring to the fact that they are all consistent with the prior geological information (training image) and match the available data. To achieve this we propose the workflow depicted in Fig. 4 where each step is elaborated in more detail.

In this NSR, the reservoir structure (horizon and fault network) is fairly well-known from seismic. The major uncertain driver for subsurface flow is the stratigraphic position of the sand channels and their mutual connectivity. A high-resolution model of sand channels is built that integrates all available static data. We will show step-wise how this is done, each time integrating more data into the model.

To build a high-resolution facies model, the multi-point geostatistical algorithm "snesim" (Strebelle, 2002) is used. This algorithm is a particular implementation of a larger family of stochastic simulation algorithms known as "sequential simulation algorithms". In sequential simulation one simulates the facies type or petrophysical property one grid cell at a time, whereby the next grid cell simulation is constrained/conditioned by all previously simulated values as well as any "hard" data (hard data in this case are the facies observations from wells). To simulate each cell, the sequential simulation algorithm calls for the inference of the conditional probability $P(A|B)$ where A is the unknown variable to be simulated, e.g. "sand facies occurrence" or "permeability less than 50 millidarcies" and B are the hard data (from wells) and previously simulated nodes. In snesim, this

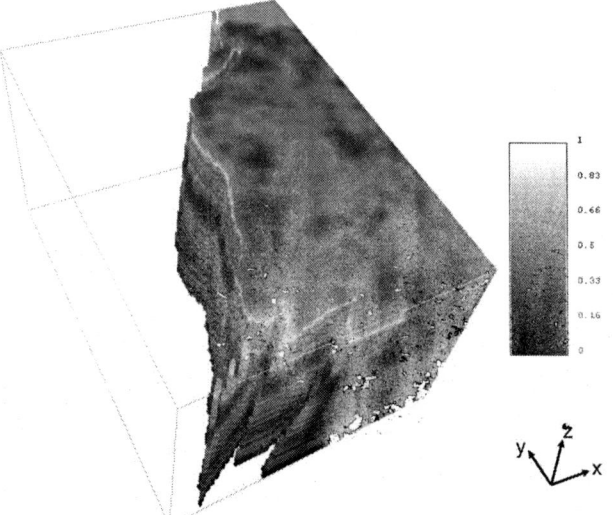

Figure 3. sand channel probability cube derived from 3D and 4D seismic data (from Andersen et al., 2006). The cube is shown in depositional coordinates (obtained after flattening the reservoir structure), not in actual reservoir coordinates.

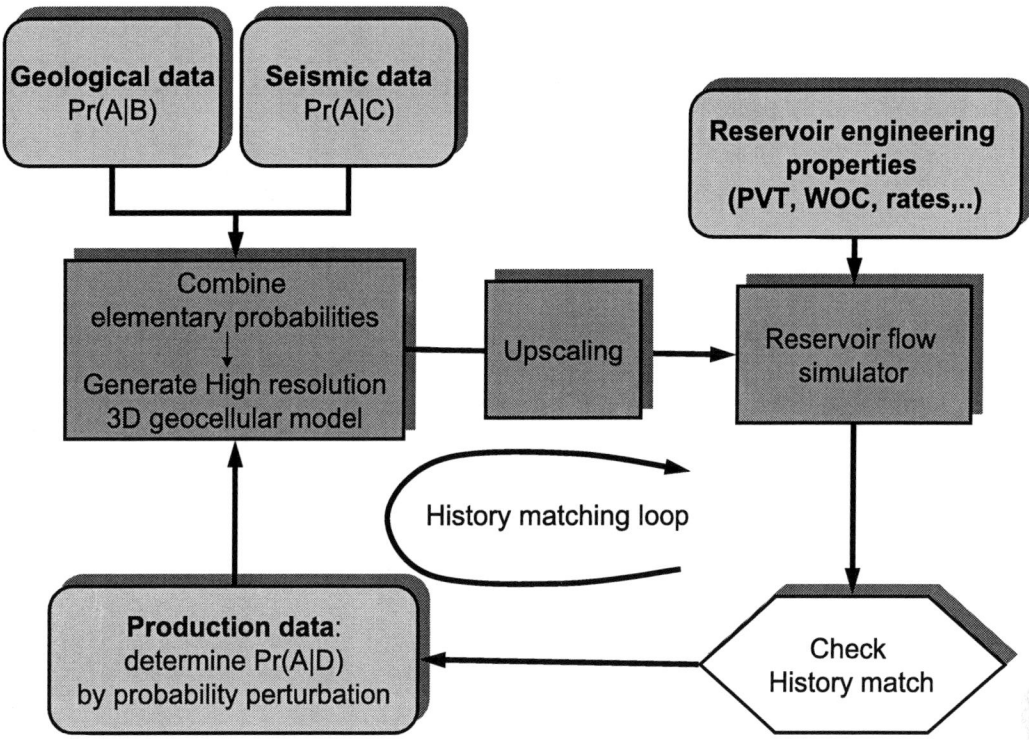

Figure 4. workflow to integrate static and dynamic data (after Caers et al., 2006).

conditional probability is inferred directly from the training image (see Strebelle, 2002 for a detailed description). In more traditional sequential simulation methods, such as sequential Gaussian simulation, this conditional probability would be derived by means of kriging, which requires the specification of a variogram model and the assumption of a multi-variate Gaussian spatial law.

Fig. 5 shows several high-resolution realizations of sand distribution generated with snesim. These realizations are constrained to the facies observations from wells and reflect fairly accurately the type of channels depicted in the 3D training image. In addition, the models are constrained to a vertical proportion variation with increasing occurrence of sand channels towards the bottom of the reservoir, as evident from Fig. 5.

Next, we show how this facies model can be further constrained by the facies probability cube derived from 3D/4D seismic. In a similar notation as above, we denote this facies probability as P(A|C), i.e. the cube specifies the probability of sand occurrence at each location given the seismic data C. C could denote either the co-located seismic datum or any set of values near the location where variable A needs to be simulated. The sequential simulation is extended as follows to include the facies probability cube. At each location to be simulated one has now two probabilities, P(A|B) as inferred from the training image, and P(A|C) obtained by calibration with the seismic data. These two probabilities are then combined into a single probability P(A|B,C) from which then a simulated facies type is drawn in sequential simulation. To combine both probabilities we rely on Journel's tau model (Journel, 2003). The tau model states that P(A| B,C) can exactly be decomposed into P(A|B) and P(A|C) using the following equations

$$\frac{x}{a} = \left(\frac{b}{a}\right)^{\tau_1} \left(\frac{c}{a}\right)^{\tau_2} \quad (1)$$

where $x = \dfrac{1-P(A|B,C)}{P(A|B,C)}$, $a = \dfrac{1-P(A)}{P(A)}$,

$b = \dfrac{1-P(A|B)}{P(A|B)}$, $c = \dfrac{1-P(A|C)}{P(A|C)}$

The tau-parameters in (1) model the mutual redundancy of both data sources B and C in predicting A. For example if $\tau_1 = 0$, then Pr(A|B) is ignored in the calculation of P(A|B,C); when $\tau_1 = \tau_2 = 1$ the data are equally redundant. Expression (1) is exact in the sense that any joint conditional probability can be decomposed as shown in expression (1), without approximation. In practice the tau-values are not known a-priori and may require a tedious estimation procedure (Krishnan, 2005). However, for the application in question

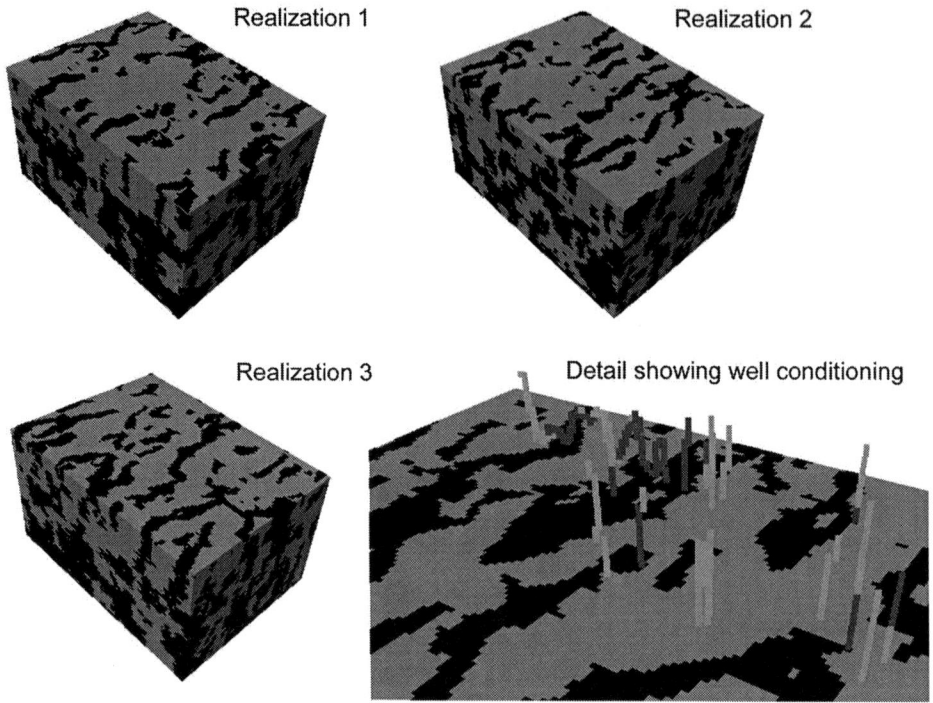

Figure 5. three realizations on a high resolution grid of 96x128x70 with unit cell size of 25x25x0.8m. The models are constrained to well facies data only. Realizations are in depositional coordinates, not in actual reservoir thicknesses. (black = channel sand, grey = background shale)

it has been shown that setting both values to unity is often a robust choice (Caers and Hoffman, 2006, Strebelle et al., 2003).

The strength of the tau model over more traditional Bayesian approaches that rely specification of prior and likelihood densities is that it divides the problem of data integration into easier sub-problems, i.e. the separate specification of P(A| a single data source). P(A| a single data source) can be any distribution function serves as an explicit quantification of the information content of each data source on the unknown variable A being modeled. It would be too difficult to directly state P(A| all data sources) through Bayes' rule unless one relies on unrealistic assumptions of multi-Gaussianity (often in both prior and likelihood) and conditional independence between data sources or data errors.

Fig. 6 shows several realizations of the facies distribution constrained now to the well-log data, seismic derived probability cube and reflecting the channel objects of the training image. Note how, compared to Fig. 5, these models are constrained by probability cube.

Once the facies geometry is defined, a constant permeability and porosity value is assigned to the two facies. While internal variations in petrophysical properties per facies may exist, they have little impact on the actual flow response of the reservoir model. These constant values were derived from core data.

Iterative Calibration With Dynamic Data

Unlike geophysical data, production data are less spatially exhaustive in the sense that their coverage does span as much of 3D space as geophysical data do. Moreover, they may vary considerably over time. Hence, it would be difficult to turn production data directly into a "facies probability cube" as was done for seismic data. A second problem relates to flow simulation. In order to evaluate how well the high-resolution models generated in Fig. 6 match the production data from the field, a flow simulation model is required. For flow simulation, one has to specify the various initial, boundary and well conditions, the fluid properties as well as the permeability and porosity in each grid cell. Current-day flow simulators can handle 105 cell models, typically, depending on the complexity of the flow problem. In most cases, a flow simulator can handle models with grid cells much less than the typical high resolution models which contain 10^6–10^7 cells. This means that the model needs to be coarsened (upscaled), (Fig. 4). Such coarsening is often difficult since one has to preserve as much as possible the connectivity of the high resolution model. For the NSR a coarsening ratio of 2:2:5 was deemed accurate enough for the purpose at hand.

The coarsened model is subject to flow simulation (Fig. 4), from which a mismatch between simulated production and field production data can be calculated. In this case we calculate the mismatch with the water and oil production in the producing wells which has been evaluated to be sensitive to the position of the channels in the model. In most cases there will be a considerable mismatch, as shown in Fig. 8. The latter is not necessarily due to the fact that the high-resolution model is wrong, but because of the large remaining uncertainty in building such high-resolution model with limited data. The next step is to adjust the model to match production data. Such adjustment in the petroleum literature is also termed "history matching". This process of history matching cannot be done arbitrarily. Indeed, one cannot randomly or arbitrarily change the flow simulation models as this could potentially destroy the data conditioning and geological interpretation of the original high-resolution model (see Caers, 2005 for an example). The ill-posedness of the inverse problem may make it possible to obtain an excellent match at the cost of geological realism of the model. Instead any adjustment should be consistent with the geological interpretation (prior geological scenario) in the sense that one should still end up with a history match model that contains patterns similar to the training image. Moreover, any adjustment should not violate the well data constraints or become incompatible with the seismic probability cube by generating consistently sand channels where a low seismic derived probability occurs. Models that are more realistic in terms of geology tend to predict better future performance which is after all the ultimate goal of model building.

We therefore consider the history matching problem as a search problem as follows. Amongst all possible realizations of the high-resolution facies model, termed "the prior model space" one needs to find those that match the production data, i.e. the "posterior model space". Formulating the problem as a search problem will by construction result in history matched models that also honor the static data. However, the search needs to be efficient (not a simple acceptance/rejection type search) since the prior model space can be large (contains millions of possible reservoir models) and since flow simulation is CPU expensive (approximately one hour for the NSR for a single flow simulation on a typical PC). To achieve such efficient search we employ a relatively new search method termed probability perturbation method" (PPM), see Fig. 4 (Caers, 2003 for the original idea and, Caers and Hoffman, 2006, for a discussion of this method

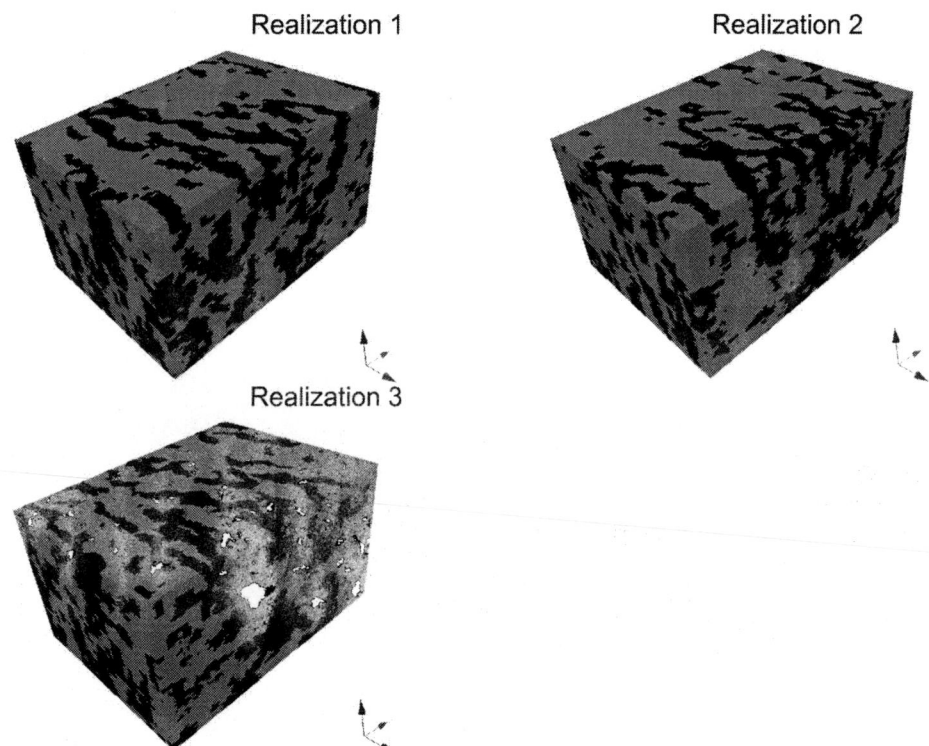

Figure 6. Three realizations constrained to the facies probability cube as well as well facies data. (black = channel sand, grey = background shale)

within a Bayesian context and example demonstrations on the efficiency).

To explain briefly the PPM method, consider a binary spatial variable (sand vs no-sand)

described by an indicator random function model is given as:

$$I(\mathbf{u}) = \begin{cases} 1 & \text{if sand occurs at location } \mathbf{u} \\ 0 & \text{else} \end{cases} \quad (2)$$

where $\mathbf{u} = (x, y, z) \in$ reservoir, is the spatial location of the node, and $i(\mathbf{u}) = 1$ means that

channel sand occurs at location $\mathbf{u}$, while $i(\mathbf{u}) = 0$ indicates non-channel occurrence. An initial non-history matched model constrained to well-log and seismic data of Fig. 3 is then denoted as $i^{(0)}(\mathbf{u})$. To perturb this initial model one introduces a new probability termed P(A|D), where D represents the set of production data.

$$P(A|D) = (1-r_D) i^{(0)}(\mathbf{u}) + r_D P(A) \in [0,1] \quad (3)$$

The expression is a function of the initial realization, $i^{(0)}(\mathbf{u})$, a free parameter r_D and the prior probability, P(A). To create a perturbation of the initial model, one simply runs the snesim algorithm anew, now with two probability cubes as input, namely, P(A|C) from 3D/4D seismic and P(A|D) from the above equation with a given dimensionless parameter, $r_D \in [0, 1]$, which control how much the initial realization will be perturbed. The algorithm combines at each node to be simulated three difference probabilities each related to a different source of information using Journel's tau model (tau's equal to one)

$$x = \frac{bcd}{a^2} \quad (4)$$

where $x = \dfrac{1-P(A|B,C)}{P(A|B,C)}$, $b = \dfrac{1-P(A|B)}{P(A|B)}$,

$c = \dfrac{1-P(A|C)}{P(A|C)}$, $d = \dfrac{1-P(A|D)}{P(A|D)}$

To better understand why a perturbation is created and assess the impact of the parameter r_D, consider the two limiting cases when $r_D = 1$ and $r_D = 0$. When $r_D = 0$, $P(A|D) = i^{(0)}(u)$, hence according to the tau-equation (4), $P(A,B,C,D) = i^{(0)}(u)$, no perturbation is made and when $r_D = 1$, $P(A|D) = P(A)$ which will result in, $i^{(1)}(u)$, which is another realization that is equally probably to be drawn as $i^{(0)}(u)$ (essentially another reservoir model out of the search space is selected). The parameter r_D, therefore, defines a perturbation of an initial realization into another equiprobable realization. It is shown (Caers, 2003) that regardless of the value of r_D, each realization $i^{(rD)}(u)$ is consistent with the static data and training image, in other words, each $i^{(rD)}(u)$ is a sample of the prior search space in Bayesian context (see Caers and Hoffman, 2006).

There may exist a value of r_D, such that $i^{(rD)}(u)$ will match the production data better than the initial realization. Finding the optimum realization, $i^{(rDopt)}(u)$, is a problem parameterized by only one free parameter, r_D; therefore, the optimum realization is selected using a simple one-dimensional optimization routine. The parameter r_D can be made spatially

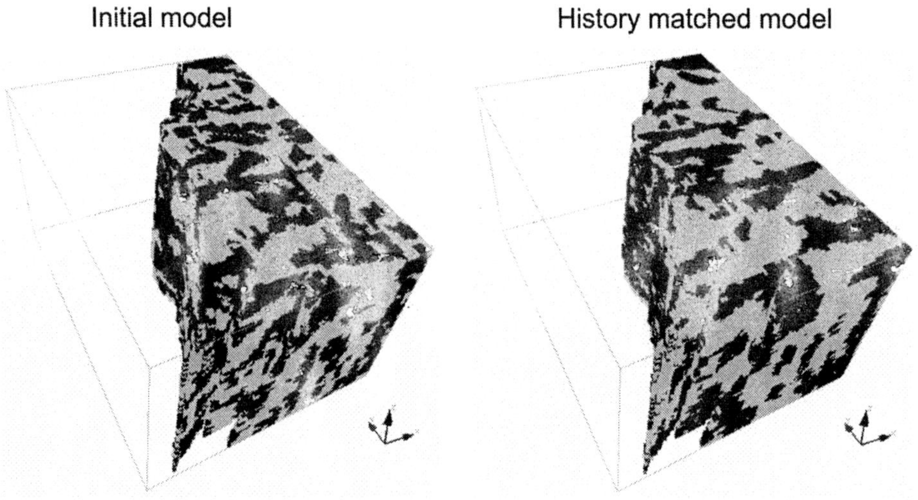

Initial model   History matched model

Figure 7. Initial model and history matched model both cut along a fault. (black = channel sand, grey = background shale)

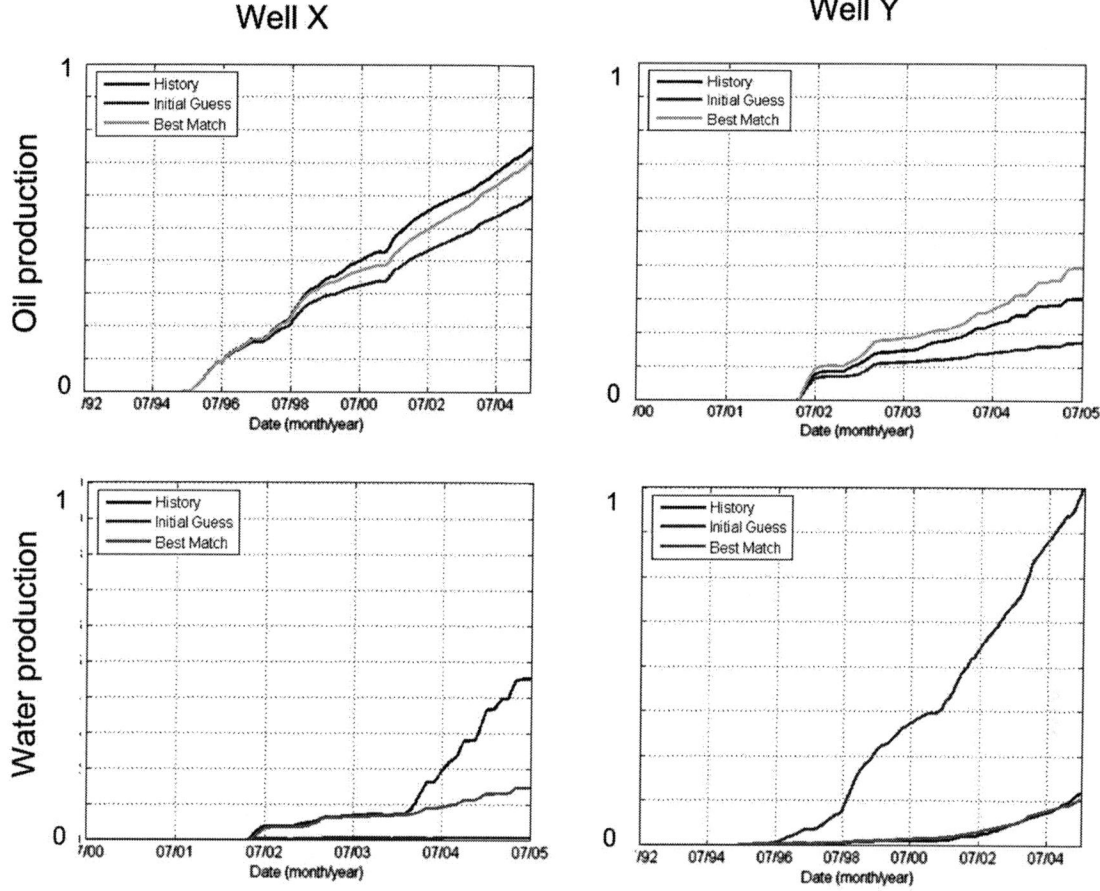

Figure 8. History match results showing a considerable improvement in matching both water and oil production for the dedicated wells for NSR. Actual rates (y-axis) cannot be shown and are scaled appropriately.

varying which results in the regional probability perturbation methods (Hoffman and Caers, 2005). An iterative procedure to find a history matched model is then created by in each step setting the optimal $i^{(rDopt)}(u)$ as the next initial model and iterate by changing the random seed.

The PPM was applied to NSR resulting in a perturbation of the simulated channel bodies until they are positioned such that the production history (water-cut and cumulative oil production) is matched. Prior to starting this procedure, it was established, by means of a simple sensitivity study that the position of channel bodies had a considerable impact on production response. Fig. 7 shows an initial realization and a history matched realization. As is obvious, both reflect the same geological concept. Several realizations that match static and dynamic data can be generated simply by starting the iterative process each time with a different initial model. It took on average 50 flow simulations to create one matched model. Fig. 8 shows that the history match was successful. Fig. 9 show the history match of 5 realizations.

CONCLUSIONS

This paper illustrates the versatility of probabilistic modeling to integrate various diverse sources of data in modeling subsurface formations. A probabilistic modeling approach is convenient in integrating data that inform the reservoir at different scale because probabilities are essentially unit-free values that inform how accurately each data source informs the variable being modeled. As evidenced in this paper and others case studies in the reference list, this approach is not merely an academic exercise on a small synthetic case; rather, it is practical and can be applied to the modeling of large subsurface reservoirs.

Key features of this method that made such practical application possible are:

- It does not rely on variogram-based models since they are inadequate in modeling distinct facies geometries. A variogram-based approach for modeling permeability fields would never be able to model, nor predict the complex

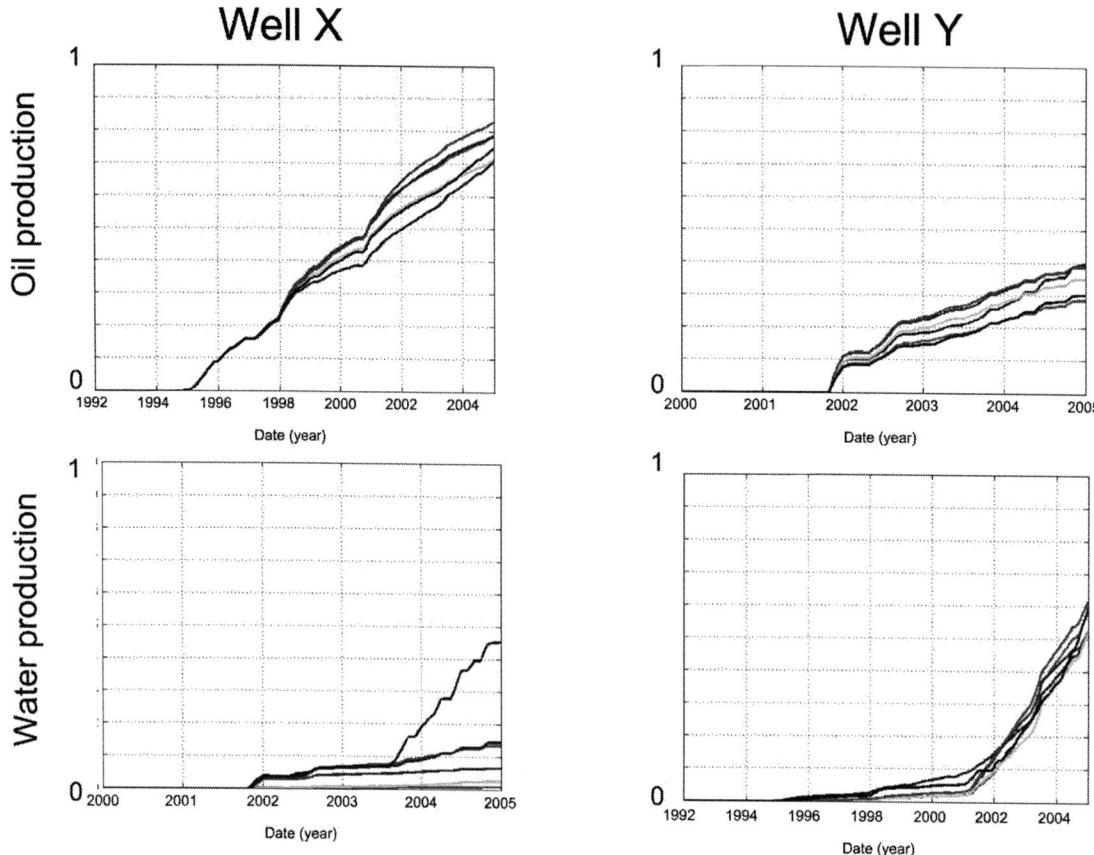

Figure 9. Flow response of five history matched models. Actual rates (y-axis) cannot be shown and are scaled appropriately.

flow through channels in the subsurface. Therefore, a 3D training image approach is used that allows the geologist to explicitly state the prior geological scenario (conceptual model), compatible with his/her interpretation of the actual subsurface heterogeneity, without being bounded by the limitations of parsimonious mathematical models. Many training images may be constructed reflecting uncertainty in the interpretation of such concept.
- It does not rely on any geological and physically unrealistic multi-Gaussian assumptions or independence assumptions, neither in prior models or likelihood. Instead, a divide-and-conquer approach is taken via the tau-model that requires the explicit specification of the information content of each data source through any type of probability distribution. The tau-model provides a solution for combining these diverse sources into a single joint probabilistic statement about the unknown. However, in the end one is interested in the samples (reservoir models) simulated from these probability distributions as they are the ones used in flow simulation for reservoir performance prediction, reservoir engineering and management planning.
- It relies on a flexible and efficient search method, termed probability perturbation method (PPM) that allows searching for model realizations in the prior model space that match the time-varying dynamic data. PPM does not require the specification of derivatives (sensitivity coefficients) which are specific to the forward model being used and may be difficult to obtain. Moreover the latter methods only apply to continuous variables (derivates must exist) and have no bearing on discrete systems such as the case for fluvial system in NSR and many other types of depositional systems.

While the technique is shown for modeling facies geometries, current research focuses on extending the same principles to other reservoir properties, most importantly the reservoir structural model which often has an order-one impact on reservoir flow behavior.

Acknowledgements. We thank Norsk Hydro for their time, data and expertise in modeling the NSR used as illustration for this paper.

REFERENCES

Andersen, T., Zachariassen, E., Hoye, T., Meisinget, H.C., Otterlei, C, Van Wijngaarden, A.J., Hatland, K., Mangeroy, F., 2006. Method for conditioning the reservoir model on 3D and 4D elastic inversion data applied to a fluvial reservoir in the North Sea, SPE 100190.

Caers, J., 2003. History matching under a training image-based geological model constraint. SPE Journal, SPE # 74716, p. 218–226

Caers, J. and Zhang, T., 2004, Multiple-point geostatistics: a quantitative vehicle for integrating geologic analogs into multiple reservoir models. In: ""Integration of outcrop and modern analog data in reservoir models"" AAPG memoir 80, p. 383–394

Caers, J., 2005. Petroleum Geostatistics, Society of Petroleum Engineers, 104p.

Caers, J., Hoffman, B.T., Strebelle, S. and Wen, X-H., 2006, Probabilistic integration of geological information, seismic and production data, The Leading Edge, 25: 240–24.

Caers, J. and Hoffman, T.B., 2006. The probability perturbation method: a new look at Bayesian inverse modeling. Math. Geol., v 38, no 1, 81–100.

Deutsch, C.V., 2003. Geostatistical reservoir modeling. Oxford University Press, 376p.

Feyen, L. and Caers, J., Quantifying geological uncertainty for flow and transport modeling in multi-modal heterogeneous formations, Advances in Water Resources; June 2006; v.29, no.6, p.912–929.

Hoffman, B.T, and Caers, J., 2005, Regional probability perturbations for history matching, Journal of Petroleum Science and Engineering, 46, 53–71.

Journel, A.G., and Alabert, F.G., 1990. New methods for reservoir mapping. Journal of Petroleum Technology, 212–224. SPE# 18324.

Journel, A.G. 2002. Combining knowledge from diverse data sources: an alternative to traditional data independence hypothesis. Math. Geol., v34, no 1, 573–596.

Krishnan, S., 2005. The tau model to integrate prior probabilities. Doctoral Dissertation, Stanford University, Stanford, California, USA.

Srivastava, R.M., 1992. Reservoir characterization with probability field simulation. SPE# 24753. in Proceedings to the SPE Annual Technical conference and Exhibition, Oct 4–7, Washington D.C.

Strebelle, S. 2000, Sequential Simulation Drawing Structures from Training Images, PhD dissertation, Stanford University, California, USA.

Strebelle, S., (2002). Conditional simulation of complex geological structures using multiple-point geostatistics. Math. Geol., v34, p1–22.

Strebelle, S., Payrazyan, K. and Caers, J. Modeling of a deepwater turbidite reservoir conditional to seismic data using multiple-point geostatistics, SPE Journal, 227–235

A Geostatistical Data Assimilation Approach for Estimating Groundwater Plume Distributions From Multiple Monitoring Events

Anna M. Michalak and Shahar Shlomi

Department of Civil and Environmental Engineering, University of Michigan, Ann Arbor, Michigan, USA

Knowledge of the distribution of groundwater contaminant plumes is needed to avoid pumping contaminated water, assess past exposure to contamination, and design remediation schemes to contain or treat the contaminated area. In most field cases, however, contamination is discovered by a small number of fortuitously located wells, and the full distribution of the plume is never known. This paper presents a stochastic geostatistical data assimilation approach capable of estimating the plume distribution at any time before, during or after the monitoring history of a site. The approach uses concentration data from all available monitoring events and results in plume estimates that are also consistent with groundwater flow and transport at the affected site. One of the unique features of the approach is that measurements taken at times subsequent to the time for which the plume is to be estimated can be used as an additional constraint on the plume distribution. The method is demonstrated using two hypothetical examples. In the first example, the distribution of a plume is estimated based on multiple sampling events from a sparse monitoring network. In the second example, the plume distribution is recovered using temporal breakthrough curves from downgradient monitoring wells.

1. INTRODUCTION

The recognition of the risk to human health and the environment associated with groundwater contamination has increased dramatically over the past several decades. Groundwater management in proximity to contaminated areas aims to avoid or minimize pumping of contaminated water, contain the plume within a specified area, assess past exposure to compromised water supplies, and/or treat the contaminated groundwater to reduce contaminant levels to within an acceptable threshold. These tasks require knowledge of the spatial and temporal distribution of chemical concentrations in the plume. In most practical situations, however, groundwater contamination is detected by a small number of fortuitously-located groundwater wells or monitoring locations, and the full spatial distribution of the plume is almost never known. To overcome this data limitation, interpolation tools are often used to estimate the plume distribution using the available concentration measurements. In typical cases where concentration data are limited, however, plume distributions estimated through interpolation often do not represent the true plume distribution adequately (e.g. *Shlomi and Michalak* [2007]).

In field applications, other forms of data that can inform the plume distribution are often also collected. In many cases, data available at field sites include information on the flow and transport in the affected aquifer and concentration data taken either before or after the time at which the plume distribution is to be estimated. Historical concentration measurements and transport information theoretically provide a strong constraint on possible plume shapes and distributions. As will be described in more detail in Section 2, however, interpolation methods are not well equipped to assimilate this additional data into the estimation of the plume distribution. Existing data assimilation approaches can overcome this limitation by allowing transport information to be used, but often require prior information on contaminant source locations and/or release histories and cannot make use of measurements taken after the time at which the plume is to be estimated.

The availability of improved methods for plume estimation would have a substantial impact on the management of contaminated groundwater resources, by improving our ability to avoid pumping contaminated water, amending the design of groundwater remediation alternatives, and providing a framework for monitoring remediation progress. The high cost associated with groundwater monitoring and the high risks posed by these contaminants contribute to the importance of developing robust estimation techniques that take into account diverse types of data for plume tracking and delineation.

In this paper, we present a data assimilation approach for estimating the full spatiotemporal distribution of a contaminant plume. The method integrates concentration measurements taken at different times and locations as well as knowledge of the flow and transport in the affected aquifer. Key features of the proposed methods are that the plume estimate at any given time is conditioned on both earlier and subsequent measurements and the source location or timing need not be known, thereby extending the applicability of the proposed approaches. At present, the method assumes that transport can be represented by a linear transport model, and that flow and transport at the site are either known or that transport model errors can be described using an error covariance structure. Also, the method as formulated in this paper is applicable to conservative tracers.

2. CURRENT APPROACHES TO PLUME ESTIMATION

There is a vast body of literature on the development and application of estimation and data assimilation approaches for subsurface applications. The majority of work has focused on the estimation of the physical properties of the subsurface, such as, for example, hydraulic conductivity distributions. Although the immediate aim of these approaches is not to estimate plume distributions, the ultimate goal is to help predict tracer transport, and tracer concentration data are often used as a constraint in the estimation. A brief discussion of these approaches is presented in Section 2.1. A wide range of data assimilation tools have also been used for various hydrological applications including plume estimation and monitoring network design, and some of the methods have strong analogies to the tools proposed here. These are briefly described in Section 2.2. Section 2.3 presents a short description of geostatistical kriging tools and their variations proposed specifically for plume estimation. The methods proposed here are based on inverse modeling tools originally developed for contaminant source identification, and a short description of these tools is provided in Section 2.4.

2.1. Parameter Estimation in Subsurface Hydrology

The estimation of subsurface parameter distributions has been the focus of substantial research. These approaches aim to provide optimal representations of subsurface parameter distributions and their associated uncertainty, given limited observations of related quantities. For example, quantities such as hydraulic conductivity or transmissivity can be estimated from any combination of sparse measurements of these same quantities, hydraulic head distributions, tracer concentrations, breakthrough curves, etc. These methods have evolved over the last three decades from simple interpolation of measured property values, to data assimilation methods that account for other measured parameters such as hydraulic head, and later, tracer concentrations. Linear and nonlinear methods have been proposed for assimilating the various sources of information into the estimation process. Reviews of such applications in subsurface hydrology and petroleum engineering are available in *McLaughlin and Townley* [1996] and de *Marsily et al.* [1999], among others. An intercomparison of seven linear and nonlinear geostatistically based inverse approaches for estimating the transmissivity distribution in an aquifer is presented in *Zimmerman et al.* [1998].

Although the ultimate goal of these approaches is to improve predictions of tracer transport, the primary focus has most often been on the uncertainty associated with the physical subsurface parameter distribution, and not on estimating the distribution of existing plumes. The methods proposed in this paper can also contribute to the prediction of contaminant transport, but the primary objective is to estimate the plume at one or multiple specific points in time during the monitoring history of a site, given a known flow and transport field.

2.2. Kalman Filtering and Other Data Assimilation Approaches in Hydrology

McLaughlin [2002] provides an introduction to, and review of, the application of statistical interpolation, filtering and smoothing to hydrological applications. Kalman filtering applications in groundwater flow modeling are reviewed in *Eigbe et al.* [1998]. In general terms, for the problem examined here, interpolation approaches use measurements taken at a single time to estimate the spatial distribution of a parameter. Filtering approaches are applied when historical measurements are used the inform the current or future distribution of a parameter. Smoothing aims to assimilate data gathered throughout the monitoring history of a site to estimate the parameter distribution at one historical point in time. The methods proposed in this work are closely related to the least-squared filtering and smoothing methods described in *McLaughlin* [2002], but the priors are defined based on the spatial autocorrelation of the plume distribution, instead of representing a prior estimate of the concentration distribution within the domain.

A related topic was examined in *McLaughlin et al.* [1993] and *Graham and McLaughlin* [1991], who applied an extended Kalman filter developed by *Graham and McLaughlin* [1989a,b] to sequentially update estimates of plume distributions at field sites by using hydraulic conductivity, hydraulic head and concentration measurements. This approach involved the simultaneous estimation of the subsurface parameters and the most recent plume distribution. The plume distribution is informed by measurements taken at or before the time at which the plume is to be estimated.

Kalman filtering and related approaches have also been applied for groundwater monitoring network design, where the goal is often to minimize the uncertainty associated with the current or future distribution of a tracer plume. These approaches have the ability to use information on the flow and transport in contaminated aquifers in a data assimilation framework to estimate plume distributions. *Loaiciga* [1989] used transport information to quantify the covariance between plume concentrations at different times and locations, and used this information to select optimal monitoring locations. In this approach, historical concentration measurements are used to inform future plume distributions. Similarly, *Herrera and Pinder* [2005] recently proposed a Kalman filtering approach that maps information from past measurements to the current time in a method aimed at locating optimal sampling locations. This recent study also made assumptions about the location and timing of the initial contaminant release, and incorporated uncertainty in the transport model in the form of a spatial covariance of transport properties such as hydraulic conductivities. *Chang and Jin* [2005] also examined the impact of model uncertainty on sequential estimates of the plume distribution in a Kalman filtering framework, but assumed that the source of the contamination was exactly known. As such, all uncertainty was the result of errors in the transport model, which were parameterized using a system error covariance structure which was assumed known. *Zou and Parr* [1995] also applied a Kalman filter starting with a known initial plume distribution. The transport model error was represented as a white Gaussian noise process. All of these past works implemented various forms of a linear Kalman filter (e.g. *Gelb* [1974]), a Bayesian approach where a prior distribution is defined by forecasting the plume distribution from a previous time at which measurements were taken, and the posterior distribution represents a compromise between this prior and information supplied by new observations at the current time.

All of the above monitoring network design approaches except that of *Loaiciga* [1989] require knowledge of either the contaminant source location and release history, or an initial plume distribution. In the cases of *Zou and Parr* [1995] and *Chang and Jin* [2005], this distribution was exactly known, whereas *Herrera and Pinder* [2005] assumed that the source locations were known, but presented a statistical model for the uncertainty associated with the release history of the sources. In addition, all the above methods allow information to propagate forward in time through the Kalman filtering procedure, but no information propagates from later to earlier times. As such, if the contaminant distribution at a particular time is to be estimated, measurements taken after that time cannot be used to constrain the estimate.

2.3. Interpolation Approaches

Many current methods for estimating plume distributions rely on interpolation of concentration measurements taken at the time at which the plume needs to be estimated. Geostatistical kriging is a popular method for estimating plume distributions that can use measurements at sampled locations, incorporate trends, and take advantage of measurements of certain other related variables (e.g. in a cokriging framework). Spatial analysis can be performed to identify spatial trends and variogram structures to be used in the estimation process. Variants and extensions of kriging can be used to introduce some additional information into geostatistical analyses (e.g. *Diggle et al.* [1998]; *Figueira et al.* [2001]; *Kitanidis and Shen* [1996]; *Saito and Goovaerts* [2001]), and a review of these approaches is presented in *Shlomi and Michalak* [2007].

However, many types of supplementary physical data, such as knowledge about the groundwater flow and transport in the aquifer or concentration data taken at different times cannot be used directly in kriging. Data taken at different times must either be ignored, or the plume must be assumed to be at a relative steady state in order to incorporate different measurement periods within an interpolation framework. Alternatively, the differential equations describing contaminant flow and transport can potentially be used to define spatio-temporal covariance functions for interpolation (e.g. *Kolovos et al.* [2004]).

Overall, current interpolation methods for estimating the spatial and/or temporal distribution of contaminant plumes do not take into account all information available about the transport properties of the affected aquifer, and are most easily applicable with measurements taken at a single time.

2.4. Geostatistical Inverse Modeling and Extensions

The method proposed in this work leverages recent advances in geostatistical contaminant source identification to develop novel tools aimed at estimating plume distributions. Inverse methods applied to contaminant source identification use modeling and statistical tools to determine the historical distribution of observed contamination, the location of contaminant sources, or the release history from a known source. Reviews of existing methods are available in *Atmadja and Bagtzoglou* [2001b] and *Michalak and Kitanidis* [2004].

One subset of inverse methods focuses on determining the values of a small number of parameters describing the source of a contaminant such as, for example, the location and magnitude of a steady-state point source, and may include additional parameters such as the times at which the release began and ended.

More directly related to the proposed methods, a second subset of existing contaminant source identification tools uses a function estimate to characterize the historical contaminant distribution, source location, or release history. In this case, the contaminant distribution or source description is not limited to a small set number of fixed parameters, but can instead vary in space and/or in time. This category includes methods that use a deterministic approach and others that offer a stochastic approach to the problem. Deterministic approaches include Tikhonov regularization [*Skaggs and Kabala*, 1994, 1998; *Liu and Ball*, 1999; *Neupauer et al.*, 2000], quasi-reversibility [*Skaggs and Kabala*, 1995; *Bagtzoglou and Atmadja*, 2003], non-regularized non-linear least squares [*Alapati and Kabala*, 2000], the progressive genetic algorithm method [*Aral et al.* 2001], and the Marching-Jury Backward Beam Equation method [*Atmadja and Bagtzoglou*, 2001a; *Bagtzoglou and Atmadja*, 2003]. Although these methods provide an estimate of a source location or release given certain assumptions, they cannot be used to directly quantify the uncertainty associated with that estimate. In stochastic approaches, parameters are viewed as jointly distributed random fields, and estimation uncertainty is recognized and its importance can be determined. Two stochastic approaches that offer a function estimate have been proposed to address the problem of source identification: geostatistical inverse modeling [*Snodgrass and Kitanidis*, 1997; *Michalak and Kitanidis*, 2002, 2003, 2004a,b; *Butera and Tanda*, 2003] and the minimum relative entropy method [*Woodbury and Ulrych*, 1996; *Woodbury et al.*, 1998; *Neupauer et al.*, 2000].

These stochastic approaches are particularly useful for potential application to improving plume estimation, because they allow the information content of concentration measurements taken at different times to be directly evaluated. This information can, in turn, be used to determine the precision with which the plume distribution can be estimated. In addition, these methods are not limited to instantaneous or point releases. The applicability of geostatistical inverse methods to multidimensional solute transport has already been demonstrated for heterogeneous media [*Michalak and Kitanidis*, 2004a; *Shlomi and Michalak*, 2007] and is similar in form to geostatistical kriging, making it amenable to the development of a stochastic method for identifying the spatiotemporal distribution of groundwater contamination. The approach aims to assimilate knowledge of the transport properties of the affected aquifer and concentration measurements from any number of times and locations. Of particular relevance to the current work, *Michalak and Kitanidis* [2004a] demonstrated the ability of geostatistical inverse modeling to estimate the historical distribution of a contaminant plume using a set of measured concentrations at a subsequent time.

In recent work, *Shlomi and Michalak* [2007] developed a theoretical framework for incorporating flow and transport information for estimating the plume distribution within a deterministically-heterogeneous aquifer contaminated by a single point source with an unknown time-dependent contaminant release history. In that work, a hypothetical heterogeneous aquifer was contaminated, and the resulting plume was sampled at a single time at a small number of wells. These observations were used to infer the full plume distribution using existing geostatistical kriging tools as well as two new proposed methods that incorporate flow and transport information. The authors concluded that, although geostatistical kriging reproduced the available measurements, it was unable to represent the true distribution of the plume. The proposed methods used the same limited concentration information to first estimate the release history of the contaminant into the aquifer, and then use this release and its associated uncertainty to map the plume at the time when measurements were taken. The method assumed a known deterministic transport model and knowledge of the location of the point source. This approach was able to reproduce the true plume shape accurately, and the estimated uncertainty was substantially lower relative to the kriging estimate.

The method proposed here is based on the principle of inverse/forward modeling presented in *Shlomi and Michalak*

[2007]. However, no assumptions are made about the location or nature of the source of contamination, and observations taken at multiple times can be used to constrain the contaminant distribution at any time before, during or after monitoring episodes. The proposed method allows measurements taken at downgradient locations and/or future times to provide additional constraints on the estimated plume distribution. This feature makes the approach applicable to a wide range of problems, such as the estimation of the spatio-temporal evolution of a plume based on measured downgradient breakthrough curves, or the assessment of historical exposure to groundwater contamination.

3. METHODOLOGY

The methodology presented in this work entails the assimilation of concentration measurements taken throughout the monitoring history of the affected site, making use of an available transport model for the aquifer. The proposed method makes use of tools developed for geostatistical contaminant source identification, plume estimation, adjoint state modeling methods, and Kalman filtering / smoothing to provide a framework for assimilating concentration data and knowledge of flow and transport in the affected aquifer to estimate the spatio-temporal evolution of contaminant plumes.

Two approaches are presented. The first involves the sequential integration of concentration data within a Kalman filtering (forward in time) and Kalman smoothing (backward in time) framework, and is outlined in Plate 1. This is the preferred approach in cases where the incremental effect of additional data is to be evaluated or when additional data are made available after initial estimates are made. This first algorithm may also be computationally more efficient for certain transport models. The second approach, illustrated in Plate 2, builds on the inverse/forward modeling approach proposed by *Shlomi and Michalak* [2007], mapping all available measurements to the time at which monitoring began, and mapping this estimated historical contaminant distribution to the time at which the plume is to be estimated. This second approach is more computationally efficient when the plume distribution at multiple times is to be estimated or when measurements are taken at many distinct times (as in the second example presented in this paper). The two approaches are mathematically very similar and the choice between them will be based primarily on implementation considerations, although other minor differences are discussed in Section 5. Both approaches treat the plume distribution as a random function, yielding quantitative uncertainty estimates in addition to descriptions of the plume distributions.

In the following method descriptions, the temporal domain has been discretized as $\mathbf{t} = \{t_1, t_2, \ldots, t_i, \ldots, t_f\}$ where t_i refers to the time for which we wish to obtain an estimate of the plume, and all other indices refer to times at which concentration measurements were taken. As described below, the methods can be applied regardless of whether measurements are also available at time t_i. The discretized plume distributions at different times t_j are termed $\mathbf{z}_{t_j}$, and the available measurements are referred to as $\mathbf{z}^*_{t_j}$. Note that the number of measurements does not have to be the same for all measurement times, and the method can make use of multiple measurement times before and after the estimation time.

3.1. Use Data Sequentially — Kalman Filtering and Smoothing Approach

The first approach is conceptually related to the Kalman filtering approaches discussed in Section 2.2 and involves the sequential assimilation of measurements, starting from the earliest observations and stepping forward through the different measurement times until the time for which the plume is to be estimated (Plate 1). Unlike existing methods, however, this approach also takes advantage of geostatistical inverse modeling tools developed by *Michalak and Kitanidis* [2004a] to assimilate measurements taken after the estimation time.

The method starts by estimating the plume distribution, denoted by the $(m \times 1)$ vector $\mathbf{z}_1$, at the time when monitoring began, t_1, using only the measurements taken at that time, denoted by the $(n_{t_1} \times 1)$ vector $\mathbf{z}^*_1$, in a geostatistical kriging framework (step 1 in Plate 1). The system of linear equations can be expressed as:

$$\begin{bmatrix} \mathbf{SQS}^T + \mathbf{R} & \mathbf{SX} \\ (\mathbf{SX})^T & \mathbf{0} \end{bmatrix} \begin{bmatrix} \mathbf{\Lambda} \\ \mathbf{M} \end{bmatrix} = \begin{bmatrix} \mathbf{SQ} \\ \mathbf{X}^T \end{bmatrix} \quad (1)$$

where $\mathbf{S}$ is an $(n_{t_1} \times 1)$ matrix representing the subsampling of the $\mathbf{z}_1$ plume at locations where measurements are available (this is in effect a sensitivity matrix populated with ones and zeros), $\mathbf{Q}$ is the $(m \times m)$ geostatistical spatial covariance matrix of the discretized plume distribution $\mathbf{z}_1$ at time t_1, $\mathbf{R}$ is the $(n_{t_1} \times n_{t_1})$ model-data mismatch covariance (which represents measurement errors and can sometimes also be used to describe transport model error statistics), $\mathbf{X}$ is an $(m \times p)$ matrix of known base functions defining the geostatistical model of the spatial trend of $\mathbf{z}_1$, and the resulting $\mathbf{\Lambda}$ and $\mathbf{M}$ are used to define the best estimate and uncertainty covariance of $\mathbf{z}_1$, $\widehat{\mathbf{z}}_1$ and $\mathbf{V}_{\widehat{\mathbf{z}}_1}$, respectively:

$$\widehat{\mathbf{z}}_1 = \mathbf{\Lambda}^T \mathbf{z}^*_1 \quad (2)$$
$$\mathbf{V}_{\widehat{\mathbf{z}}_1} = \mathbf{Q} - \mathbf{QS}^T \mathbf{\Lambda} - \mathbf{XM}$$

78 GEOSTATISTICAL DATA ASSIMILATION FOR PLUME ESTIMATION

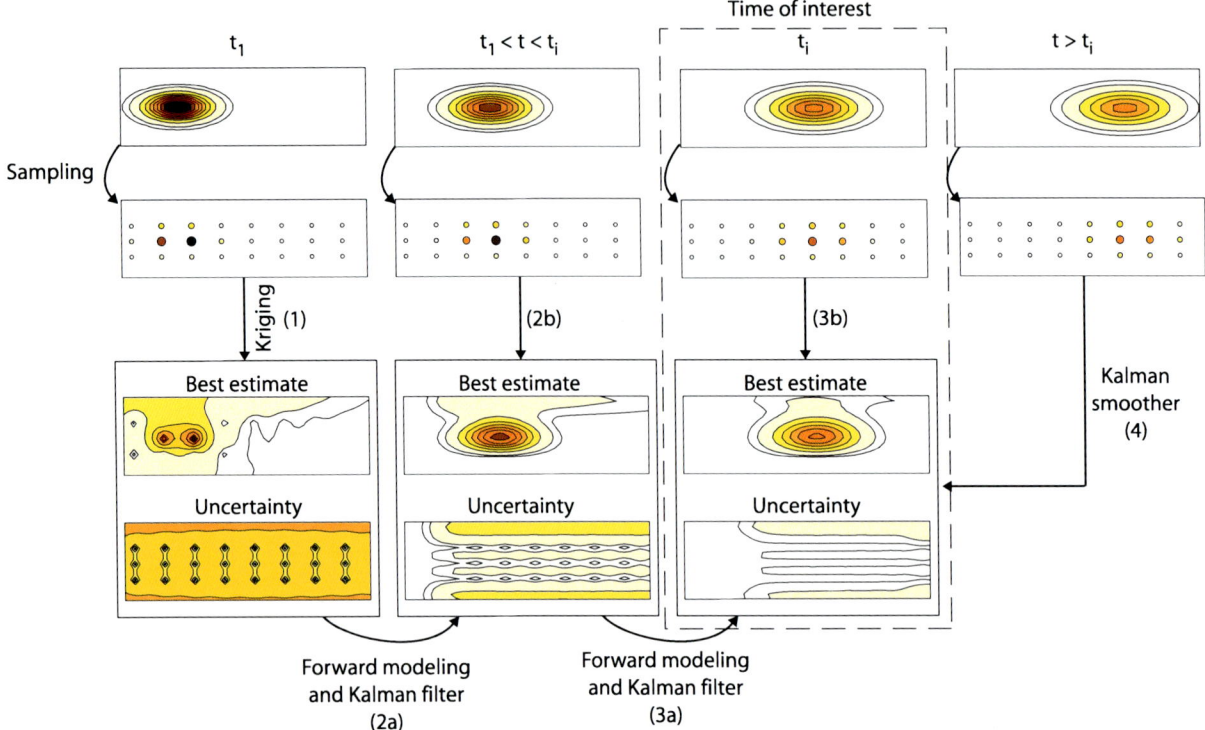

Plate 1. Schematic illustration of Kalman filtering / smoothing approach for sequential data assimilation for plume delineation. The step numbers listed in the text are indicated in parentheses.

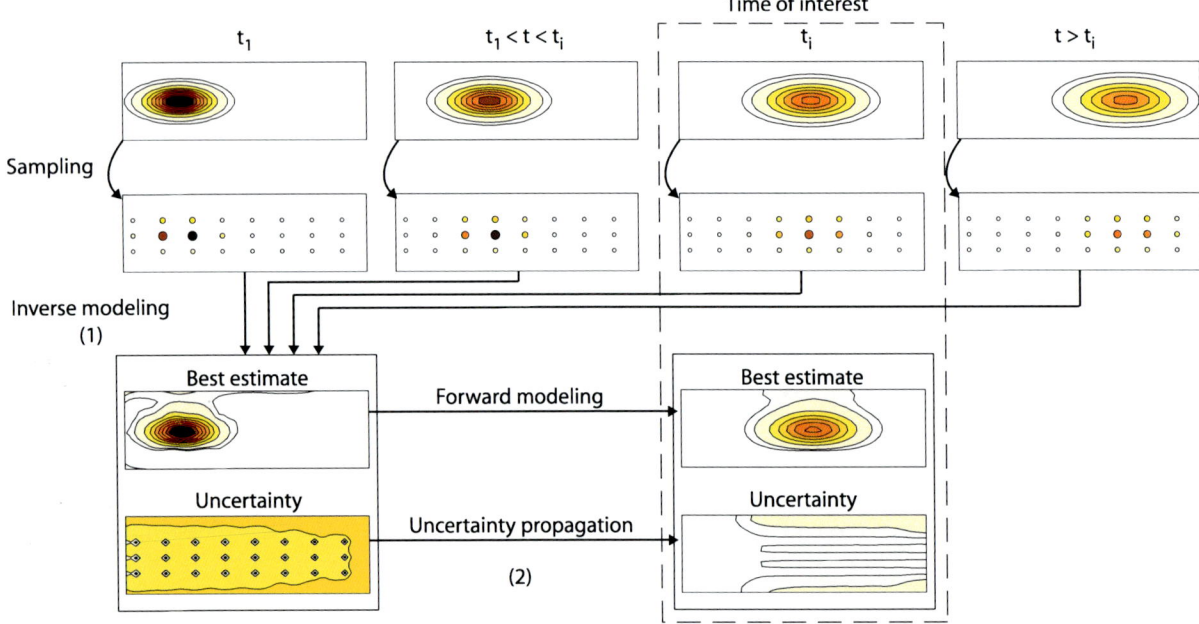

Plate 2. Schematic illustration of inverse/forward modeling for assimilation of multiple datasets for plume delineation. The step numbers listed in the text are indicated in parentheses.

Although written slightly differently, this system is simply a traditional set of kriging equations with either a constant or variable trend (depending on the form of $\mathbf{X}$).

In subsequent steps, the estimate of the earlier plume and its uncertainty are used to define a prior estimate of the plume distribution $\mathbf{z}_2$ at the next sampling time, t_2, making use of the transport information provided by the numerical or analytical model of the affected aquifer (step 2a in Plate 1):

$$\widetilde{\mathbf{z}}_2 = \mathbf{H}_{2,1}\widehat{\mathbf{z}}_1$$
$$\mathbf{V}_{\widetilde{\mathbf{z}}_2} = \mathbf{H}_{2,1}\mathbf{V}_{\widehat{\mathbf{z}}_1}\mathbf{H}_{2,1}^T \quad (3)$$

where $\mathbf{H}_{2,1}$ represents the $(m \times m)$ sensitivity matrix of the discretized plume distribution at time t_2 to that at time t_1, and $\widetilde{\mathbf{z}}_2$ becomes the prior estimate of the distribution $\mathbf{z}_2$. Note that the definition of the sensitivity matrix $\mathbf{H}_{2,1}$ relies on the linearity of the transport model, and the method would need to be modified for nonlinear transport. The matrix $\mathbf{H}_{2,1}$ is square if the plume is estimated at the same locations for both times and can be obtained by a sequence of runs of the model representing transport at the site. This prior estimate is updated in a Kalman filtering step using measurements taken at time t_2 (step 2b in Plate 1). This corresponds to finding the minimum of a least-squared optimization objective function:

$$L_{\mathbf{z}_2} = (\mathbf{z}_2^* - \mathbf{S}\mathbf{z}_2)^T \mathbf{R}^{-1} (\mathbf{z}_2^* - \mathbf{S}\mathbf{z}_2)^T$$
$$+ (\mathbf{z}_2 - \widetilde{\mathbf{z}}_2)^T (\mathbf{V}_{\widetilde{\mathbf{z}}_2})^{-1} (\mathbf{z}_2 - \widetilde{\mathbf{z}}_2) \quad (4)$$

where $\mathbf{S}$ is now an $(n_{t_2} \times m)$ matrix representing the subsampling of the $\mathbf{z}_2$ plume at the n_{t_2} measurement locations, and $\mathbf{R}$ is the $(n_{t_2} \times n_{t_2})$ model-data mismatch covariance of the new set of observations. Note that these can have different dimensions than in equation (1) if the configuration and/or quality of the monitoring network changes over time. This objective function can be represented as the linear system of equations:

$$\left[\mathbf{S}\mathbf{V}_{\widetilde{\mathbf{z}}_2}\mathbf{S}^T + \mathbf{R}\right][\boldsymbol{\Lambda}] = [\mathbf{S}\mathbf{V}_{\widetilde{\mathbf{z}}_2}] \quad (5)$$

which has the solution:

$$\widehat{\mathbf{z}}_2 = \widetilde{\mathbf{z}}_2 + \boldsymbol{\Lambda}^T (\mathbf{z}_2^* - \mathbf{S}\widetilde{\mathbf{z}}_2)$$
$$\mathbf{V}_{\widehat{\mathbf{z}}_2} = \mathbf{V}_{\widetilde{\mathbf{z}}_2} - \mathbf{V}_{\widetilde{\mathbf{z}}_2}\mathbf{S}^T\boldsymbol{\Lambda} \quad (6)$$

where $\boldsymbol{\Lambda}^T$ can be thought of as a typical Kalman gain matrix, and $\widehat{\mathbf{z}}_2$ is the final a posteriori estimate of the distribution $\mathbf{z}_2$. Note that there is no matrix $\mathbf{M}$ of Lagrange multipliers in this set of equations because the a priori distribution is known and defined by $\widetilde{\mathbf{z}}_2$.

This Kalman filtering continues until the time for which the plume distribution is to be estimated, t_i. For the last Kalman filtering step, the prior estimate is obtained from the most recent earlier distribution $\widehat{\mathbf{z}}_{i-1}$ by (step 3a in Plate 1):

$$\widetilde{\mathbf{z}}_i = \mathbf{H}_{i,i-1}\widehat{\mathbf{z}}_{i-1}$$
$$\mathbf{V}_{\widetilde{\mathbf{z}}_i} = \mathbf{H}_{i,i-1}\mathbf{V}_{\widehat{\mathbf{z}}_{i-1}}\mathbf{H}_{i,i-1}^T \quad (7)$$

If data are available at time t_i, the Bayesian updating step takes the form (step 3b in Plate 1):

$$\left[\mathbf{S}\mathbf{V}_{\widetilde{\mathbf{z}}_i}\mathbf{S}^T + \mathbf{R}\right][\boldsymbol{\Lambda}] = [\mathbf{S}\mathbf{V}_{\widetilde{\mathbf{z}}_i}] \quad (8)$$

$$\widehat{\mathbf{z}}_i = \widetilde{\mathbf{z}}_i + \boldsymbol{\Lambda}^T (\mathbf{z}_i^* - \mathbf{S}\widetilde{\mathbf{z}}_i)$$
$$\mathbf{V}_{\widehat{\mathbf{z}}_i} = \mathbf{V}_{\widetilde{\mathbf{z}}_i} - \mathbf{V}_{\widetilde{\mathbf{z}}_i}\mathbf{S}^T\boldsymbol{\Lambda} \quad (9)$$

The estimate at time t_i is then further refined by making use of any available observations taken at a time later than the estimation time, in a Kalman smoother step (step 4 in Plate 1). The objective function for this Bayesian update is:

$$L_{\mathbf{z}_i} = (\mathbf{z}_j^* - \mathbf{H}_{j,i}^*\mathbf{z}_i)^T \mathbf{R}^{-1} (\mathbf{z}_j^* - \mathbf{H}_{j,i}^*\mathbf{z}_i)$$
$$+ (\mathbf{z}_i - \widehat{\mathbf{z}}_i)^T (\mathbf{V}_{\widehat{\mathbf{z}}_i})^{-1} (\mathbf{z}_i - \widehat{\mathbf{z}}_i) \quad (10)$$

where $\mathbf{z}_j^*$ are the n_j measurements taken at any later time(s) $t > t_i$, $\mathbf{H}_{j,i}^*$ is an $(n_j \times m)$ matrix representing the sensitivity of these later measurements to the full discretized distribution at time t_i, and all other terms are defined analogously to the ones in the previous objective functions. Note that $\mathbf{z}_j^*$ can include measurements from more than one time later than t_i. The minimum of this objective function can be expressed as a linear system of equations:

$$\left[\mathbf{H}_{j,i}^*\mathbf{V}_{\widehat{\mathbf{z}}_i}\mathbf{H}_{j,i}^{*T} + \mathbf{R}\right][\boldsymbol{\Lambda}] = \left[\mathbf{H}_{j,i}^*\mathbf{V}_{\widehat{\mathbf{z}}_i}\right] \quad (11)$$

where the final estimate of the plume distribution at time t_i is expressed as:

$$\widehat{\mathbf{z}}_i' = \widehat{\mathbf{z}}_i + \boldsymbol{\Lambda}^T (\mathbf{z}_j^* - \mathbf{H}_{j,i}^*\widehat{\mathbf{z}}_i)$$
$$\mathbf{V}_{\widehat{\mathbf{z}}_i'} = \mathbf{V}_{\widehat{\mathbf{z}}_i} - \mathbf{V}_{\widehat{\mathbf{z}}_i}\mathbf{H}_{j,i}^{*T}\boldsymbol{\Lambda} \quad (12)$$

where the prime denotes the estimate of $\mathbf{z}_i$ after the second Bayesian updating step.

Note that the use of $\mathbf{S}$ in the above equations assumes that the sampled points are a subset of the estimation points, which is not a requirement of the method, but this notation was used here for convenience of illustration. In addition, the estimation points do not necessarily need to be on a grid, although this is convenient for contouring software. This method allows for sequential refinement of estimates as additional data become available, which may present computational savings for some numerical models relative to the second method presented below.

For the special case where the time at which the plume is to be estimated either (i) equals the time at which the latest measurements were taken or (ii) is after the time at which the last measurements were taken, the method is analogous to kriging the earliest measurements and running a Kalman filter that sequentially predicts the plume at the next measurement time and conditions the plume distribution on new observations as they become available (e.g. *Loaiciga* [1989]).

3.2. Use all Data Simultaneously — Inverse/Forward Modeling Approach

The second approach builds on the inverse/forward modeling method recently proposed by *Shlomi and Michalak* [2007]. The original method was applied to the estimation of the timing and intensity of a contaminant release into an aquifer, and this information was used in combination with an available transport model to estimate a plume distribution. Note that *Shlomi and Michalak* [2007] ultimately recommended the use of a second approach (Transport-Enhanced Kriging, TrEK) for this problem, which allowed for separate covariance structures to be defined for the temporal source release history and spatial plume distribution. In the current work, no assumptions are made about the nature of the source of contamination. Therefore, the inverse/forward modeling approach, which focuses on a single covariance structure (the spatial autocorrelation of the plume distribution, in this case), is a more appropriate basis for the proposed method.

For the method presented here, we apply an inverse/forward modeling approach to first estimate the plume distribution at the time when monitoring began (i.e. the time at which the first measurements were taken) using all available measurements from all monitoring episodes. This estimate is then used to recover the plume distribution at a given later time using the available transport model (Plate 2). As in the approach presented in Section 3.1, the approach begins by estimating the plume at the start of monitoring; unlike the first approach, however, all measurements taken at all times are used simultaneously to inform this estimate. The estimate is consistent with all measurements taken at all times.

For the general case where measurements are available before, at, and after the time at which the plume is to be estimated, the objective function is expressed as (step 1 in Plate 2):

$$L_{z_1,\beta} = (z^* - H_1^* z_1)^T R^{-1} (z^* - H_1^* z_1) \\ + (z_1 - X\beta)^T Q^{-1} (z_1 - X\beta) \quad (13)$$

where z^* is the $(n \times 1)$ vector of all available concentration data $(n = n_{t_1} + \ldots + n_{t_f})$, R is the $(n \times n)$ model-data mismatch covariance matrix of all available measurements, Q is the geostatistical spatial covariance of the discretized plume distribution z_1 at time t_1, $X\beta$ is the geostatistical model of the trend, and

$$z^* = \begin{bmatrix} z_1^* & z_2^* & \cdots & z_i^* & \cdots & z_f^* \end{bmatrix}^T \quad (14)$$

$$H_1^* = \begin{bmatrix} S & H_{2,1}^* & \cdots & H_{i,1}^* & \cdots & H_{f,1}^* \end{bmatrix}^T \quad (15)$$

where S is a matrix representing the subsampling of the z_1 plume at locations where measurements are available, $H_{2,1}^*$ represents the $(n_{t_2} \times m)$ sensitivity matrix of each of the z_2^* measurements to the discretized concentration distribution at time t_1, and the other H^* matrices are defined analogously. The sensitivity information is again based on linear transport, and is obtained through the implementation of a numerical or analytical groundwater flow and transport model (e.g. *Michalak and Kitanidis* [2004a]). Equation (13) represents a geostatistical inverse problem, where the distribution z_1 is estimated based on the dual criterion of generating a plume that is consistent with all available measurements z^* and that exhibits spatial autocorrelation structure consistent with Q and $X\beta$.

The solution to this system of equations is obtained by minimizing equation (13) with respect to z_1 and β. The solution can be expressed as a system of linear equations:

$$\begin{bmatrix} H_1^* Q H_1^{*T} + R & H_1^* X \\ (H_1^* X)^T & 0 \end{bmatrix} \begin{bmatrix} \Lambda^T \\ M \end{bmatrix} = \begin{bmatrix} H_1^* Q \\ X^T \end{bmatrix} \quad (16)$$

where Λ and M are used to estimate z_1 and its a posteriori covariance:

$$\begin{aligned} \widehat{z}_1 &= \Lambda^T z^* \\ V_{\widehat{z}_1} &= Q - Q H_1^{*T} \Lambda - XM \end{aligned} \quad (17)$$

The estimate of the plume z_i at time t_i, is then obtained by mapping the estimate $\widehat{z}_1$ and its covariance structure forward in time (step 2 in Plate 2):

$$\hat{z}_i = H_{i,1}\hat{z}_1$$
$$V_{\hat{z}_i} = H_{i,1}V_{\hat{z}_1}H_{i,1}^T \quad (18)$$

where $H_{i,1}$ is the $(m \times m)$ sensitivity matrix of the discretized plume distribution at time t_i to that at time t_1.

For the special case where the time at which the plume is to be estimated either (i) equals the time at which the latest measurements were taken or (ii) is after the time at which the last measurements were taken, the data vector and sensitivity matrix become:

$$\hat{z}_i = H_{i,1}\hat{z}_1$$
$$V_{\hat{z}_i} = H_{i,1}V_{\hat{z}_1}H_{i,1}^T \quad (19)$$

For the special case where no measurements are available for the time at which the plume distribution is to be estimated, the data vector and sensitivity matrix do not include this time step, but the equations remain unchanged. For the case where the time at which the plume is to be estimated precedes any sampling, the method is similar to the work presented in *Michalak and Kitanidis* [2004a] where a spatial array of measurements was used to estimate the previous distribution of a plume. In this case, the data vector and sensitivity matrix become:

$$z^* = \begin{bmatrix} z_i^* & z_{i+1}^* & \cdots & z_f^* \end{bmatrix}^T$$
$$H_i^* = \begin{bmatrix} S & H_{i+1,i}^* & \cdots & H_{f,i}^* \end{bmatrix}^T \quad (20)$$

and the objective function is then written to estimate z_i directly.

In the general case, the method can be applied to estimate the contaminant distribution at any time, integrating transport and concentration information from all available times. As such, this second approach may be more numerically efficient if samples were taken at many different times, or if the plume distribution is to be estimated at multiple times. Note that if the total number of samples is very high, the inversion of the $((n+p) \times (n+p))$ matrix in equation (16) required to solve for Λ and M may become computationally prohibitive, in which case the first approach would be more appropriate.

4. APPLICATIONS

Two sample applications are presented. The first represents the estimation of the spatial distribution of a plume using multiple sets of measurements taken at different times. The second represents the estimation of the temporal evolution of a plume based on measured breakthrough curve information.

4.1. Estimate Plume From Monitoring Network Observations

This example involves the estimation of a contaminant plume distribution in a confined aquifer at a time $t_i = 1600$ days after monitoring of the aquifer began. Measurements are assumed to have taken place at four distinct times, $t_1 = 0$ days, $t_2 = 800$ days, $t_3 = 1600$ days, and $t_4 = 2400$ days. At each time, 12 measurements were taken on a regular grid. A hypothetical example was chosen to illustrate and verify the capabilities of the methods in a setup where the true concentration distributions are known. The Kalman filter and smoother approach (Section 3.1) was implemented for this first example.

The conductivity field for this aquifer, originally used in *Michalak and Kitanidis* [2004a], was generated for a rectangular domain using the numerical spectral approach of *Dykaar and Kitanidis* [1992a,b]. Although a multi-Gaussian representation of the hydraulic conductivity heterogeneity was used here, this was done for simplicity, and is not a requirement for the application of the proposed methods. All four boundaries of this aquifer were defined as constant head, with a 0.034 m/m head gradient in the West to East direction, and a 0.0067 m/m gradient in the North to South direction, inducing flow mainly toward the East with a minor component toward the South. MODFLOW-2000 [*Harbaugh et al.*, 2000] was used to calculate the flow field.

The numerical transport model MT3DMS [*Zheng and Wang*, 1999] was used to simulate the spatio-temporal evolution of the plume and to obtain the sensitivity matrices required to apply the model. Figure 1 shows the solute distribution at the four times when measurements were taken, with time $t = 0$ representing the time of the first sampling event. Data collected at each sampling time at the locations indicated on the plumes in Figure 1 were used to recover the spatial distribution at time $t_i = 1600$ days.

The sensitivity matrices required to implement the Kalman filter / smoother approach were calculated on a grid of 946 points (43×22 cells). This grid also represents the discretization at which the plume was estimated. The transport model was run repeatedly, each time simulating a plume developing from a single unit concentration in one grid cell. This plume was sampled at time $\Delta t = 800$ days, the time intervals between sampling episodes. Note that the method does not require a regular sampling interval in time. By running the model sequentially for each grid cell, the various H matrices were populated, representing the sensitivity of each location in the aquifer to the concentration at a previous time and location. Assuming a steady state flow regime and making use of the linearity of the transport model, the base sensitivity matrix calculated for $\Delta t = 800$ days could be used to calculate the sensitivity over other time intervals ($H_{n\Delta t} = H_{\Delta t}^n$). Separate

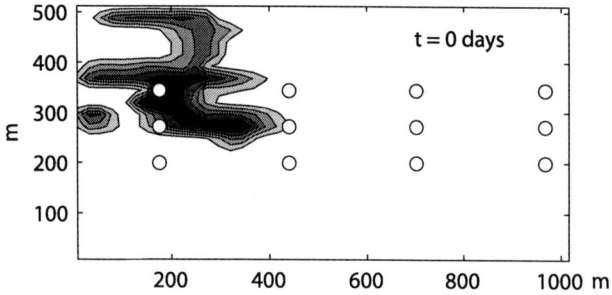
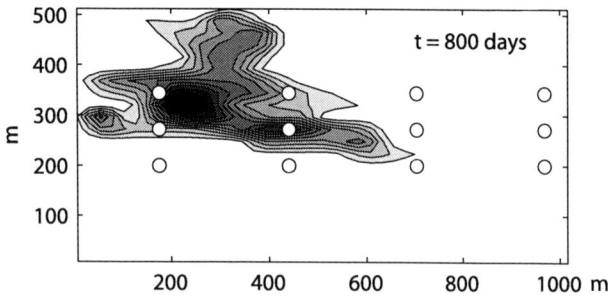
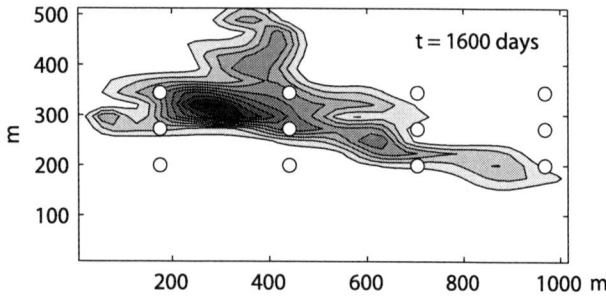
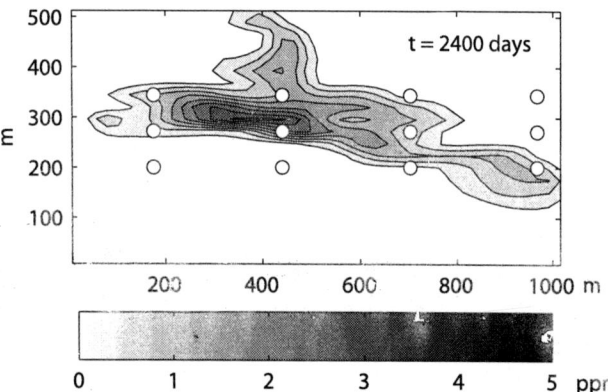

Figure 1. Spatial and temporal evolution of plume used in sample applications. The circles indicate sampling locations for Example 1. Samples were taken only at the times presented in these panels.

runs for each time interval would have been required if the flow were not assumed to be steady state.

The spatial covariance structure of the plume distribution at time $t_1 = 0$ days was estimated using an exhaustive sample from that distribution. In a real situation, this covariance would be estimated using a Restricted Maximum Likelihood approach as discussed in *Kitanidis* [1995] and *Michalak and Kitanidis* [2004a] using only the sampled concentrations, but we opted to use the full plume distribution here to help isolate the behavior of the proposed method. The length (l) and sill (σ^2) parameters for an exponential covariance model were estimated to be:

$$l = 80 \ m, \ \sigma_Q^2 = 1.5 \ ppm^2$$

where

$$Q = \sigma^2 \exp(-h/l) \qquad (21)$$

and h is the separation distance between two points. The plume was assumed to have random, normally distributed measurement error with $\sigma_R = 0.001 \ ppm$, corresponding to an idealized case with no transport model error.

The approach described in Section 3.1 was used to estimate the plume at time $t_i = 1600$ days at $m = 946$ points throughout the domain. Figure 2 shows the best estimate of the plume at the required time, as well as the standard error of estimation, defined as the square root of the diagonal element of $\mathbf{V}_{\hat{z}_i}$.

Figure 3 and 4 show the estimates for the equivalent case obtained by ordinary kriging and Kalman filtering (but without Kalman smoothing), respectively. The ordinary kriging estimate uses only the data ($\mathbf{z}_i^*$) taken at time t_i, because data collected at other times and transport information cannot be directly included in kriging. The spatial covariance parameters used for kriging the $\mathbf{z}_i^*$ measurements for time t_i were:

$$l = 100 \ m, \ \sigma^2 = 1.0 \ ppm^2$$

The Kalman filtering estimate uses measurements taken before (t_1, t_2) and at the time (t_3) at which the plume is to be estimated, but not any subsequent measurements (t_4), yielding results that are similar to those that would be obtained using the approach of *Loaiciga* [1989].

4.2. Estimate Plume Evolution From Downgradient Breakthrough Curves

The second application presents the recovery of the temporal evolution of the spatial distribution of a groundwater plume based on breakthrough curves measured at a small

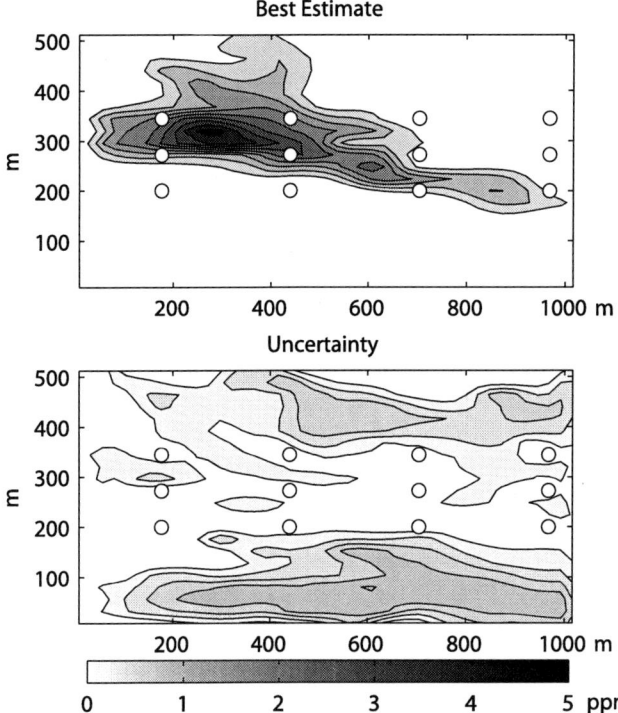

Figure 2. Example 1: Recovered plume distribution for time $t_i = 1600$ days using Kalman filtering / smoothing approach, incorporating information from all four sampling times. The uncertainty is expressed as one standard deviation of the estimation uncertainty. Sampling locations are shown for reference.

number of wells. Unlike the first example presented, existing methods such as Kalman filtering or geostatistical interpolation would not have been practical for this application, because samples are never taken within the area where the plume distribution is to be estimated.

The experimental setup is similar to the first application. The domain is identical, with a net flow from West to East and from North to South. The simulated plume is also identical to that used in the first application, and is presented in Figure 1 for times $t = \{0, 800, 1600, 2400\}$ days. Unlike in the first application, however, the plume is not sampled at these times, but breakthrough curves are instead measured at seven monitoring wells located on the downgradient boundary of the domain at times $t_j = 10j$ days, where $j = 1, \ldots, 400$. The well locations and breakthrough curves are presented in Figure 5.

The breakthrough curves are used in the approach described in Section 3.2 to first estimate the plume at time $t = 0$ days, assuming that the measurements have a small normally distributed random error with $\sigma = 0.001 ppm$. This estimate and its covariance structure are then used to obtain estimates for time $t_i = 1600$ days by applying equation (18). Note that the time of 1600 days was selected to correspond to the time examined in Example 1. Once the plume at time $t = 0$ days is estimated, the plume at any other time can be estimated by applying equation (18) repeatedly for all the times of interest, with different matrices $\mathbf{H}_{i,1}$ corresponding to the sensitivities of the plumes at these various times to the plume distribution at time $t = 0$ days. The resulting plume estimate at time $t = 1600$ days is presented in Figure 6, along with the locations of the breakthrough curve monitoring wells. Conversely to the first application, this represents an estimate based on a large numbers of sampling times at a small number of sampling locations that are not in the vicinity of the plume that is to be estimated.

5. METHOD PERFORMANCE AND APPLICABILITY

The two approaches presented in this work are designed to estimate the spatial distribution of a groundwater contaminant plume through assimilation of concentration measurements taken throughout the monitoring history of a site and knowledge of the flow and transport in the aquifer. The presented approaches yield accurate results in the sense that the actual deviations of the best estimate from the true plume are correctly characterized by the estimation uncertainty.

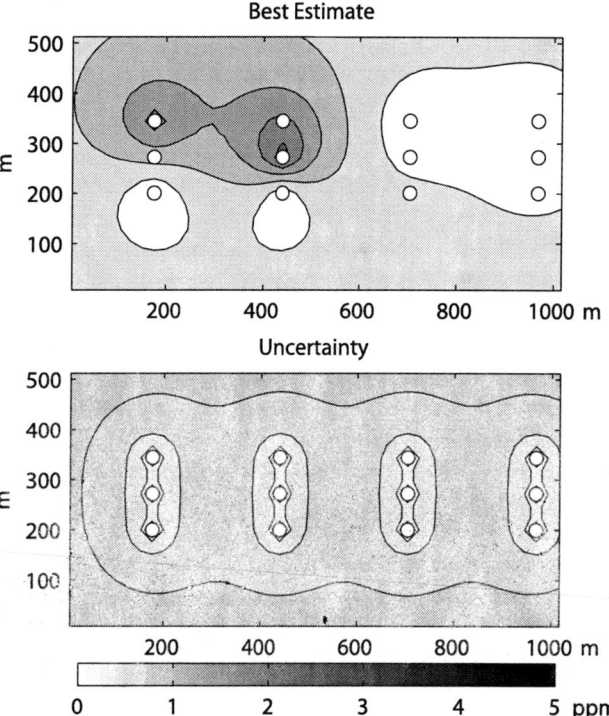

Figure 3. Example 1: Recovered plume distribution for time $t_i = 1600$ days using ordinary kriging, only incorporating information from measurements taken at time $t_i = 1600$ days. The uncertainty is expressed as one standard deviation of the estimation uncertainty.

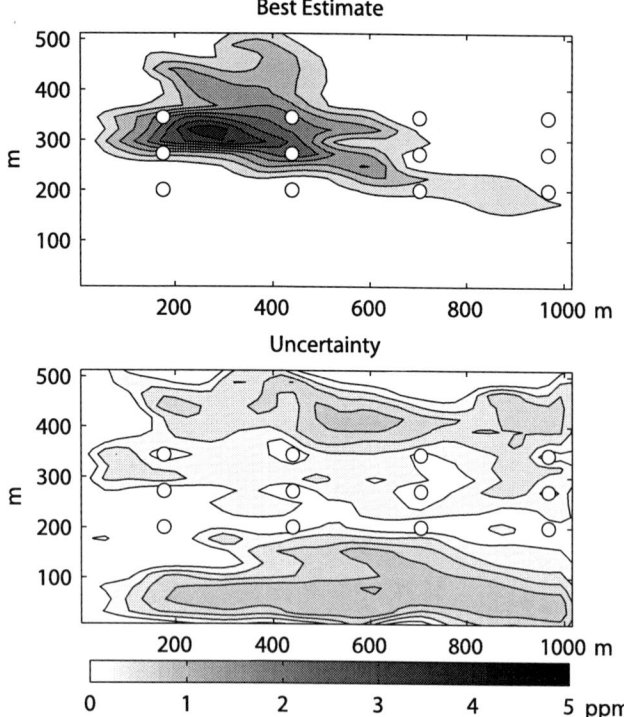

Figure 4. Example 1: Recovered plume distribution for time $t_i = 1600$ days using Kalman filter (but no Kalman smoother step). This estimate incorporates information from times $t = \{0, 800, 1600\}$ days. The uncertainty is expressed as one standard deviation of the estimation uncertainty.

Geostatistical interpolation approaches can only incorporate information from concentration measurements taken at the time when the estimate is sought, or at times close enough to the estimation time such that the plume distribution can be approximated as unchanged. As can be seen from Figure 3, interpolation approaches cannot capture the shape and extent of a groundwater contaminant plume in cases where the monitoring network is sparse, the plume hot spots do not correspond to monitoring locations, and/or the plume is highly heterogeneous.

A Kalman filtering approach can incorporate transport information as well as any measurements taken prior to the time for which the plume is to be estimated. As can be seen by comparing Figure 4 to Figure 2, however, measurements taken at times subsequent to the estimation time provide additional constraints on the plume distribution that cannot be incorporated in a Kalman filtering approach.

The presented Kalman filtering/smoothing and inverse/forward modeling approaches, on the other hand, can assimilate all concentration data. A practical aspect of this advantage is the option to take additional measurements to increase the accuracy of the estimate of a historical plume that was not thoroughly monitored at the time of interest.

In addition, existing plume estimation methods are only directly applicable if the monitoring network samples the plume while it is in the region of the domain that we are interested in. The two proposed approaches, on the other hand, can also make use of breakthrough curves measured at downgradient locations to estimate the spatio-temporal distribution of contaminant plumes. Although not specifically demonstrated through an example here, the data used in Examples 1 and 2 could also have been used concurrently without requiring any modifications to the proposed approaches.

The spatial distribution of the uncertainty is also quite different for the proposed approaches relative to kriging. Whereas for kriging the uncertainty grows uniformly and monotonically away from measurement locations (Figure 3), the uncertainty associated with the proposed approaches is related both to the measurement locations and to the sensitivity of various locations within the aquifer to one another, as expressed through the **H** matrices (Figure 2). As a result, even poorly sampled areas can have relatively low uncertainty if they have a weak hydrologic connection to the locations where contamination was detected.

The two approaches presented here are mathematically very similar, and the selection of a method should be based primarily on computational considerations. From a conceptual perspective, the difference between the two approaches is in estimating the β trend parameters of z_1. Whereas in the Kalman filtering/smoothing approach the estimate of β is based only on the measurements taken at time t_1, in the inverse/forward modeling approach all measurements contribute to this estimate. This dif-

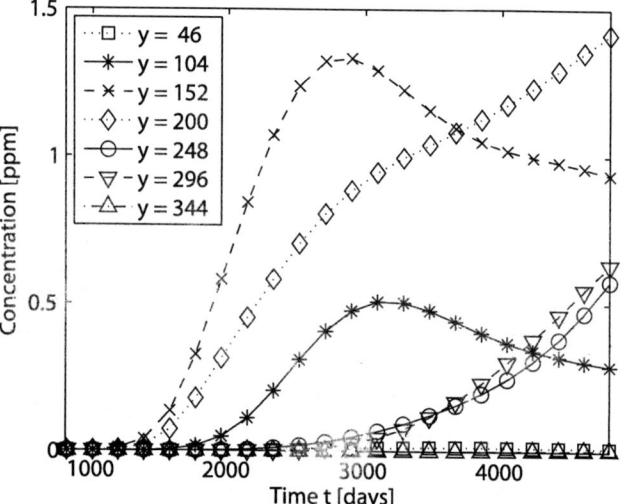

Figure 5. Example 2: Breakthrough curves at observation wells located on the East boundary of the plume domain. The transverse coordinates of the wells are listed in the legend and presented in Figure 6.

ference is only expected to be important if these β parameters are themselves of interest in the analysis.

As mentioned in Section 1, the proposed methods, as presented in the current work, rely on several assumptions. The definition of the sensitivity matrices **H** relies on a linear formulation of transport in the aquifer. In addition, the methods are currently set up for conservative tracers, although some forms of reactive transport, such as linear reaction kinetics, would only require minor modifications to the presented methods. Although the use of a multi-Gaussian representation of the physical heterogeneity of the aquifer is not required (see Section 4.1), the heterogeneity of the plume distribution at the earliest time when measurements are taken is modeled as a second-order stationary process. We have not found this to be a strong limitation in practice even for highly heterogeneous non-Gaussian hydraulic conductivity distributions (results not shown), but this assumption will need to be evaluated explicitly in future applications.

The presented approaches assume that the flow and transport in the aquifer are either known or that transport model errors can be described using an error covariance structure.

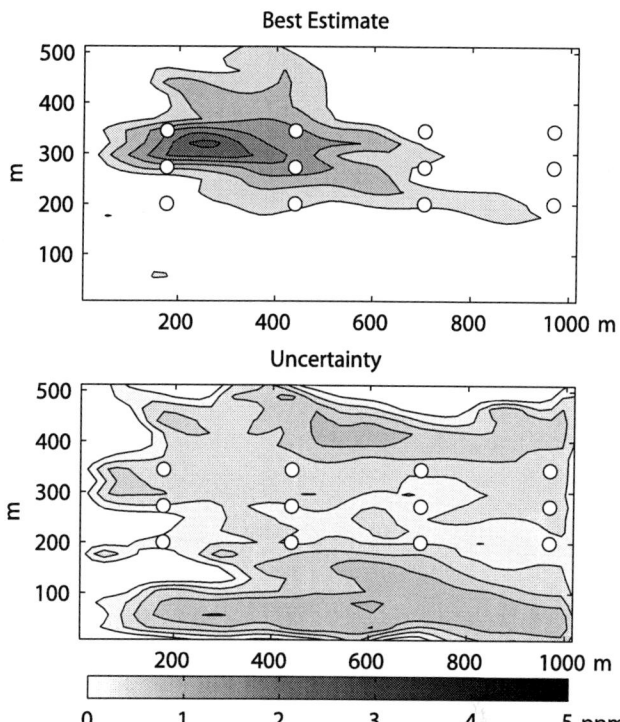

Figure 7. Example 1: Sensitivity analysis with high measurement error ($\sigma^2 = 0.5\ ppm^2$). Recovered plume distribution for time $t_i = 1600$ days using Kalman filtering / smoothing approach. The uncertainty is expressed as one standard deviation of the estimation uncertainty.

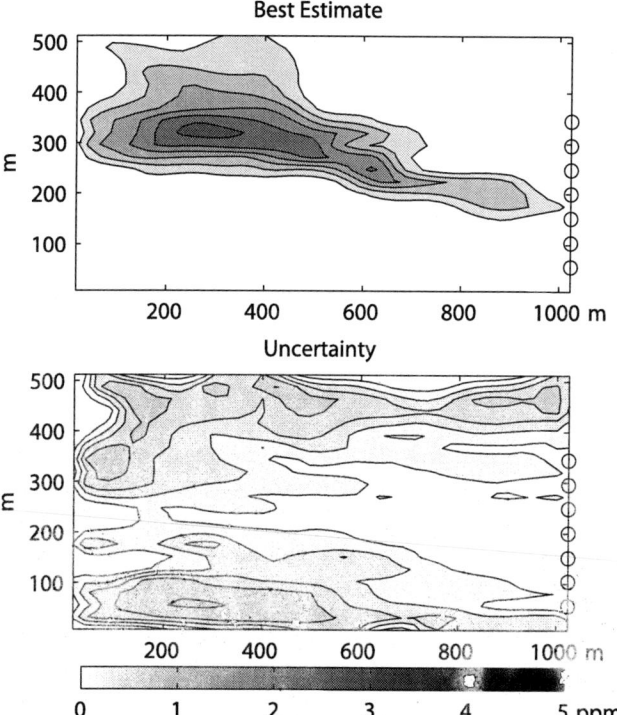

Figure 6. Example 2: Recovered plume distribution at time $t_i = 1600$ days using inverse / forward modeling approach. The uncertainty is expressed as one standard deviation of the estimation uncertainty. Breakthrough curve measurement locations are shown for reference.

The presented examples represent idealized cases with low measurement error and do not explicitly include transport model error. In field cases, the flow and transport in an aquifer are never fully characterized, and this uncertainty will have a substantial impact on the estimated plume distribution. Although a full treatment of transport uncertainty is beyond the scope of this paper, two sensitivity analyses were conducted to assess the impact of errors in the model's ability to reproduce available measurements. In the first sensitivity analysis (Figure 7), the measurement error was increased to a variance of $\sigma^2 = 0.5\ ppm^2$, and random errors with this variance were added to all measurements. If transport errors are present, then measurements cannot be reproduced perfectly, and this error is often parameterized as an additional measurement error. As can be seen in Figure 7, the two main impacts of the increased error are less detail in the best estimate relative to the low error case (Figure 2), and higher uncertainty in the estimate. In the second sensitivity analysis (Figure 8), we defined a spatially correlated model-data mismatch error because transport errors are expected to have spatially-correlated impacts on the measurements (e.g. *Chang and Jin* [2005]).

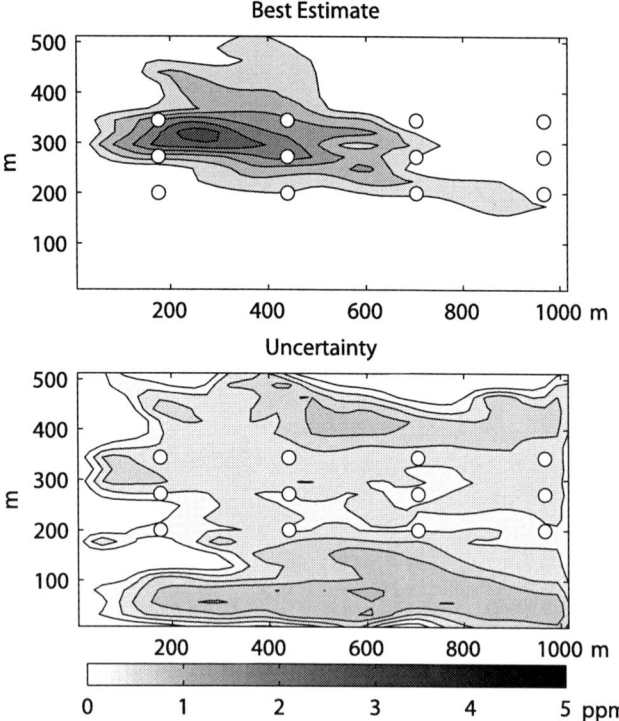

Figure 8. Example 1: Sensitivity analysis with spatially correlated model-data mismatch error ($\sigma^2 = 0.5\ ppm^2$, $l = 100\ m$). Recovered plume distribution for time $t_i = 1600$ days using Kalman filtering / smoothing approach. The uncertainty is expressed as one standard deviation of the estimation uncertainty.

The variance remained $\sigma^2 = 0.5\ ppm^2$, but this error was assumed to have an exponential covariance with a correlation length equal to that of the actual plume at the estimation time ($l = 100m$). Errors with these same characteristics were added to the measurements. In field cases, the correlation length of transport errors could be estimated from tracer experiments by performing a variogram analysis on the difference between actual and predicted tracer concentrations at monitoring wells. In the presented example, the spatial covariance of the error term had little impact on the best estimate and its associated uncertainty. For both sensitivity analyses, the estimates were accurate in the sense that the error of the estimate relative to the actual plume was consistent with the estimated uncertainty. These sensitivity analyses are intended as an illustration of the effect of increased uncertainty on the ability of the methods to obtain accurate estimates. They are not a substitute for the estimation methods described in Section 2.1 that are designed specifically for characterizing the uncertainty associated with subsurface flow and transport parameters.

Both approaches involve the inversion of matrices of size defined by the number of measurements being used to update the estimate of the plume distribution. When a very large number of measurements is available, the inverse/forward modeling approach may become computationally prohibitive because it requires the inversion of a matrix with dimensions slightly larger than the total number of measurements taken at all times. The Kalman filtering/smoothing approach is therefore preferable in these cicumstances. For the case of a finely discretized spatial domain, the matrix multiplications required to obtain the components of the linear system of equations may themselves become computationally expensive, and may need to be implemented as a series of products of smaller arrays. In the case where even such a multi-step approach is prohibitive (e.g. if the transport model has millions of nodes), more sophisticated numerical optimization methods such as ensemble or variational approaches would need to be implemented.

6. CONCLUSIONS

The methods presented in this paper make it possible to assimilate concentration data taken throughout the monitoring history of a site and knowledge of the groundwater flow and transport in the affected aquifer to estimate the distribution of a contaminant plume at any time during or prior to monitoring. The proposed methods produce estimates that are conditioned not only on measurements concurrent to the estimation time and/or measurements taken prior to the estimation time, but can also take full advantage of subsequent concentration data. In addition, knowledge of the source of contamination is not required. These differences allow for more accurate and precise estimates when monitoring continues after the time for which the plume estimate is sought, and make the approaches applicable to a new set of problems. For example, as presented in this work, the breakthrough curve at a few downgradient monitoring wells can be used to estimate the spatio-temporal evolution of a plume.

Finally, although a full statistical analysis of the impact of transport model uncertainty is beyond the scope of the current work, real field sites involve complex uncertainty in the transport parameters. Future work will focus on explicitly characterizing and accounting for this uncertainty. Parameter estimation methods such as the ones described in Section 2.1 could be used to characterize the uncertainty of the flow and transport fields given available measurements, and tools for incorporating this information into the derived methodology could then be developed. Alternately, the flow and transport parameters could be estimated together with the plume distribution (e.g. *McLaughlin et al.*, [1993]), and such tools are the subject of ongoing research.

Acknowledgments. This work was partially supported by National Science Foundation award number 0607002, "Sampling and inversion methods for quantifying effect of incomplete subsurface characterization on uncertainty associated with recovery of contamination history." We thank two anonymous reviewers for providing helpful suggestions that contributed significantly to the final version of this manuscript.

REFERENCES

Alapati, S., and Z. J. Kabala, Recovering the release history of a groundwater contaminant using a non-linear least-squares method, *Hydrol. Processes*, 14, 1003–1016, 2000.

Aral, M. M., J. B. Guan, and M. L. Malia, Identification of contaminant source location and release history in aquifers, *J. Hydrol. Eng.*, 6 (3), 225–234, 2001.

Atmadja, J., and A. C. Bagtzoglou, Pollution source identification in heterogeneous porous media, *Water Resour. Res.*, 37 (8), 2113–2125, 2001a.

Atmadja, J., and A. C. Bagtzoglou, State of the art report on mathematical methods for groundwater pollution source identification, *Environmental Forensics*, 2(3): 205–214, 2001b.

Bagtzoglou, A. C., and J. Atmadja, Marching-jury backward beam equation and quasi-reversibility methods for hydrologic inversion: Application to contaminant plume spatial distribution recovery, *Water Resour. Res.*, 39 (2), 1038, doi:10.1029/2001WR001021, 2003.

Butera, I., and M. G. Tanda, A Geostatistical Approach to Recover the Release History of Groundwater Pollutants, *Water Resour. Res.*, 39(12), 1372, doi:10.1029/2003WR002314, 2003.

Chang, S. Y., and A. Jin, Kalman filtering with regional noise to improve accuracy of contaminant transport models, *J. Env. Eng. ASCE*, 131, 971–982, 2005.

de Marsily, G., J.-P. Delhomme, F. Delay, and A. Buoro, 40 years of inverse problems in hydrogeology, *Earth & Planetary Sci.*, 329, 73–87, 1999.

Diggle, P. J., J. A. Tawn, and R. A. Moyeed, Model-based geostatistics, *J. Royal Statistical Society Series C-Applied Statistics*, 47, 299–326, 1998.

Dykaar, B. B., and P. K. Kitanidis, Determination of the Effective Hydraulic Conductivity for Heterogeneous Porous-Media Using a Numerical Spectral Approach .1. Method, *Water Resour. Res.*, 28, 1155–1166, 1992a.

Dykaar, B. B., and P. K. Kitanidis, Determination of the Effective Hydraulic Conductivity for Heterogeneous Porous-Media Using a Numerical Spectral Approach .2. Results, *Water Resour. Res.*, 28, 1167–1178, 1992b.

Eigbe, U., M. B. Beck, H. S. Wheater, and F. Hirano, Kalman filtering in groundwater flow modelling: problems and prospects, *Stochastic Hydrology And Hydraulics*, 12(1), 15–32, 1998.

Figueira, R., and A. J. S. A. M. G. P. F. Catarino, Use of secondary information in space-time statistics for biomonitoring studies of saline deposition, *Environmetrics*, 12, 203–217, 2001.

Gelb, A., Applied Optimal Estimation, MIT Press, Cambridge, Mass, 1974.

Graham, W. D., and D. B. McLaughlin, A Stochastic-Model Of Solute Transport In Groundwater—Application To The Borden, Ontario, Tracer Test, *Water Resour. Res.*, 27(6), 1345–1359, 1991.

Graham, W., and D. McLaughlin, Stochastic-Analysis Of Nonstationary Subsurface Solute Transport .1. Unconditional Moments, *Water Resour. Res.*, 25(2), 215–232, 1989a.

Graham, W., and D. McLaughlin, Stochastic-Analysis Of Nonstationary Subsurface Solute Transport .2. Conditional Moments, *Water Resour. Res.*, 25(11), 2331–2355, 1989b.

Herrera, G. S., and G. F. Pinder, Space-time optimization of groundwater quality sampling networks, *Water Resour. Res.*, 41(12), W12407, doi:10.1029/2004WR003626, 2005.

Harbaugh, A.W., E. R. Banta, M.C. Hill, and M. G. McDonald, MODFLOW-2000, the U.S. Geological Survey modular groundwater model—User guide to modularization concepts and the Ground-Water Flow Process, U.S. Geological Survey Open-File, 121 pp, USGS, 2000.

Kitanidis, P. K., Quasi-linear geostatistical theory for inversing, Water Resources Research, 31, 2411–2419, 1995.

Kitanidis, P. K., and K. F. Shen, Geostatistical interpolation of chemical concentration, *Advances in Water Resour.*, 19, 369, doi:310.1016/0309-1708(1096)00016-00014, 1996.

Kolovos, A., G. Christakos, D.T. Hristopulos, and M.L. Serre et al., Methods for generating non-separable spatiotemporal covariance models with potential environmental applications, *Advances in Water Resour.*, 27, 815–830, 2004.

Liu, C., and W.P. Ball, Application of Inverse Methods to Contaminant Source Identification from Aquitard Diffusion Profiles at Dover AFB, Delaware, *Water Resour. Res.*, 35 (7), 1975–1985, 1999.

Loaiciga, H. A., An Optimization Approach for Groundwater Quality Monitoring Network Design, *Water Resour. Res.*, 25, 1771–1782, 1989.

McLaughlin, D., An integrated approach to hydrologic data assimilation: interpolation, smoothing, and filtering, *Advances In Water Resour.*, 25(8-12), 1275–1286, 2002.

McLaughlin, D., L. B. Reid, S. G. Li, and J. Hyman, A Stochastic Method For Characterizing Groundwater Contamination, *Ground Water*, 31(2), 237–249, 1993.

McLaughlin, D., and L. R. Townley, A reassessment of the groundwater inverse problem, *Water Resour. Res.*, 32(5), 1131–1161, 1996.

Michalak, A.M. and P.K. Kitanidis, Application of Bayesian inference methods to inverse modeling for contaminant source identification at Gloucester Landfill, Canada, in Computational Methods in Water Resources XIV, Volume 2, edited by S.M. Hassanizadeh, R.J. Schotting, W.G. Gray and G.F. Pinder, p.1259–1266, Elsevier, Amsterdam, The Netherlands, 2002.

Michalak, A. M., and P. K. Kitanidis, A method for enforcing parameter nonnegativity in Bayesian inverse problems with an application to contaminant source identification, *Water Resour. Res.*, 39(2), 1033, doi:10.1029/2002WR001480, 2003

Michalak, A. M., and P. K. Kitanidis, Estimation of historical groundwater contaminant distribution using the adjoint state method applied to geostatistical inverse modeling, *Water Resour. Res.*, 40(8),W08302, doi:10.1029/2004WR003214, 2004a.

Michalak, A. M., and P. K. Kitanidis, Application of geostatistical inverse modeling to contaminant source identification at Dover AFB, Delaware, *J. Hydraulic Res.*, 42 (Special issue), 9–18, 2004b.

Michalak, A. M., and P. K. Kitanidis, A method for the interpolation of nonnegative functions with an application to contaminant load estimation, Stochastic Environmental Research and Risk Assessment, 19(1), 8–23, doi:10.1007/s00477-004-0189-1, 2005.

Neupauer, R.M., B. Borchers, and J.L. Wilson, Comparison of inverse methods for reconstructing the release history of a groundwater contamination source, *Water Resour. Res.*, 36 (9), 2469–2475, 2000.

Saito, H., and P. Goovaerts, Accounting for source location and transport direction into geostatistical prediction of contaminants, *Environmental Science & Technology*, 35, 4823, doi 4810.1021/es010580f, 2001.

Shlomi, S., and A. M. Michalak, A Geostatistical Framework for Incorporating Transport Information in Estimating the Distribution of a Groundwater Contaminant Plume, *Water Resour. Res.*, W03412, doi:10.129/2006WR005121.

Skaggs, T. H., and Z. J. Kabala, Recovering the release history of a groundwater contaminant, *Water Resour. Res.*, 30 (1), 71–79, 1994.

Skaggs, T. H., and Z. J. Kabala, Recovering the history of a groundwater contaminant plume: Method of quasi-reversibility, *Water Resour. Res.*, 31 (11), 2669–2673, 1995.

Skaggs, T. H., and Z. J. Kabala, Limitations in recovering the history of a groundwater contaminant plume, *J. Contaminant Hydrology*, 33, 347–359, 1998.

Snodgrass, M. F., and P. K. Kitanidis, A geostatistical approach to contaminant source identification, *Water Resour. Res.*, 33(4), 537–546, 1997.

Woodbury, A., E. Sudicky, and T. J. Ulrych, Three-dimensional plume source reconstruction using minimum relative entropy inversion, *J. Contaminant Hydrology*, 32, 131–158, 1998.

Woodbury, A. D., and T. J. Ulrych, Minimum relative entropy inversion: Theory and application to recovering the release history of a groundwater contaminant, *Water Resour. Res.*, 32, 2671–2681, 1996.

Zheng, C., and P. P. Wang, MT3DMS, A modular three-dimensional multi-species transport model for simulation of advection, dispersion and chemical reactions of contaminants in groundwater systems; documentation and user's guide, 202 pp, U.S. Army Engineer Research and Development Center, Vicksburg, MS, 1999.

Zimmerman, D. A., G. de Marsily, C. A. Gotway, M. G. Marietta, C. L. Axness, R. L. Beauheim, R. L. Bras, J. Carrera, G. Dagan, P. B. Davies, D. P. Gallegos, A. Galli, J. Gomez-Hernandez, P. Grindrod, A. L. Gutjahr, P. K. Kitanidis, A. M. Lavenue, D. McLaughlin, S. P. Neuman, B. S. RamaRao, C. Ravenne, and Y. Rubin, A comparison of seven geostatistically based inverse approaches to estimate transmissivities for modeling advective transport by groundwater flow, *Water Resour. Res.*, 34(6), 1373–1413, 1998.

Zou, S., and A. Parr, Optimal Estimation of 2-Dimensional Contaminant Transport, *Ground Water*, 33, 319–325, 1995.

A. M. Michalak, Department of Civil and Environmental Engineering, University of Michigan, 183 EWRE Building, 1351 Beal Ave., Ann Arbor, MI 48109-2125, USA. (anna.michalak@umich.edu)

A Bayesian Approach for Combining Thermal and Hydraulic Data

Allan D. Woodbury

Department of Civil Engineering, University of Manitoba, Winnipeg, Manitoba Canada

Incorporating temperatures into a modeling effort can take many forms, and both temperatures and hydrologic data can be combined qualitatively and quantitatively. In the latter category, the least formal would be in calibration, followed by parameter estimation and finally by full-inversion. This paper discusses information-based (specifically Bayesian) approaches of incorporating hydraulic parameters and potentials like temperature and hydraulic head together in a formal procedure. This paper reviews the generalized inverse problem for groundwater and heat; discusses Bayesian solutions to inverse problems; empirical and hierarchical Bayes, upscaling and cokriging and Bayesian interpolation. Along these lines, a list of suggested references is provided, along with suitable mentioning of benchmark papers, monographs and textbooks on the subject.

The technique described in this paper revolves around shallow, low-temperature groundwater flow systems; and that entails steady 2-D fluid and heat flow. The methodology utilizes a perturbation technique to linearize and then couple the governing equations. For the perturbation approach to work, fluid properties must be decoupled from the temperature field. Once this is done, and through the finite element method, a block-linear system of data, kernel, and model parameters is developed.

Two end-members and one set of joint inverse examples are presented. The two end-members are pure heat conduction (an application of Bayesian inversion to Paleoclimate reconstructions), and a pure-groundwater problem which is an example application to the Edwards Aquifer in Texas. Lastly, generic examples of combinations of transmissivity, hydraulic head and temperatures are presented.

1. INTRODUCTION

This paper is about a Bayesian approach to the solution of the joint-groundwater and thermal inverse problem. The reader should note that inasmuch as it is an overview, some details are omitted for the sake of the "big picture". Not all of the relevant literature is cited. Also, this paper

Subsurface Hydrology: Data Integration for Properties and Processes
Geophysical Monograph Series 171
Copyright 2007 by the American Geophysical Union.
10.1029/171GM09

does assume some knowledge on the part of the reader but hopefully it will fit in nicely to some of the other papers within this monograph. Finally, this paper presents the author's view. It is unabashedly Bayesian and may not be agreed to by all.

It may be no longer necessary to justify the use of temperatures in hydrogeological studies, but there are a number of compelling reasons for doing this and perhaps it is worthwhile to repeat these here. A recent paper by Anderson (2005) reviews the essentials of heat and groundwater transport and also contains numerous references. A great textbook chapter can be found in Domenico and Schwartz (1990), and a favorite is a textbook, "Thermal Geophysics" by Jessop (1990). An excellent monograph is that of Beck et al. (1989) "Hydrogeological Regimes and Their Subsurface Thermal Effects". Pioneering work in thermal and groundwater effects can be found in Garven and Freeze (1984), Smith and Chapman (1983) and Hunt et al. (1996). These references are by no means exhaustive and serve only to indicate the importance of thermal and groundwater interactions.

Hydrological site characterization remains a significant challenge in hydrogeology. Numerical models have been widely employed to simulate the responses of groundwater systems under various stresses. One well recognized difficulty for groundwater model applications (and indeed any modeling application) is to obtain sufficient and reliable hydrogeological parameters. Aquifers are often highly heterogeneous and the large spatial variability of a groundwater system controls the distributions of hydraulic heads, contaminants, temperature and other potentials of interest. In practice, there are usually insufficient measurements of hydraulic parameters for a comprehensive site characterization. A common remedy is for practitioners to subdivide an aquifer into a relatively small number of constant property zones; perhaps a conceptualization that should be considered at odds with nearly three decades of research in stochastic hydrology.

Difficulties associated with direct measurements of all the hydrologic parameters needed for physically-based mathematical models are well known. Equally well known are the challenges in trying to adjust parameters within preconceived limits until model output at selected points matches observed values. Quite often questions are raised as to the uniqueness and optimality of these models. A major focus of research over almost three decades has been directed towards inversion techniques and/or parameter estimation as a way of both automatic calibration and as a statistical procedure to quantify the reliability of parameter estimates (see reviews by Ginn and Cushman, 1990; and McLaughlin and Townley, 1996, 1997; Kitanidis, 1997). A "true" inverse problem (one that involves functionals and an exploration of infinite model spaces) is ill-posed, and this is characterized by instability-and non-uniqueness (Ulrych and Sacchi, 2005).

Traditionally, inverse techniques in hydrogeology rely on measurements of hydraulic conductivity and hydraulic heads, and they employ the groundwater flow equation for interpretation. Measurements of hydraulic head, hydraulic conductivity (or transmissivity), seepage flux, and the like could be inputs to an inverse algorithm, and fitted hydraulic conductivity (or other parameters) become the output, along with the parameter covariance structure. Relatively few works have gone beyond this approach to introduce additional information such as tracer data (cf., Carrera et al., 1993), or geophysical measurements (Woodbury and Smith, 1988; Rubin et al., 1992; Hyndman et al., 1994; Hyndman and Gorelick, 1996; Copty and Rubin, 1995; Hubbard et al., 1997; Rubin, 2003; Rubin and Hubbard, 2005). Unfortunately, the sophistication of inverse algorithms cannot replace information and data. This recognition is well demonstrated in the pioneering work of Carrera and Neuman (1986) where it is shown that the instability and non-uniqueness of solutions to the inverse problems can only be eliminated by introducing additional measurements and information. The challenge of course, is to find inexpensive and reliable sources of information, and to find ways to combine them.

The conjunctive use of temperature and hydrogeological data for site characterization has the potential to reduce some of the concerns that have been raised. This potential is evident from published works relating thermal-energy transport to the aquifer's hydrogeological features, and from studies that attempted using temperature data for site characterization. Note that the thermal and hydraulic head fields are linked by the fluid specific discharge. This linkage is discussed further below.

Subsurface temperatures are one example of a source of information that is dependent on the same hydraulic parameters that governs groundwater flow. In the same way we utilize chemical tracers, we can utilize downhole temperatures. In addition, the linkage between the two is physical and not empirical. The basic idea behind a joint-hydrological/thermal inversion scheme is to exploit the sensitivity of the thermal field to hydrogeologic parameters. Woodbury et al. (1987) and Woodbury and Smith (1988) used a joint estimation of hydrogeological data and temperature measurements to characterize site hydraulic parameters. Wang et al. (1989) proposed a generalized least-squares approach for solving thegroundwater inverse problem using both hydraulic parameters and thermal parameters (see also Beck et al., 1989). These studies show that within a system with significant permeability, groundwater movement can redistribute heat, greatly disturbing the conductive thermal regime. Subsurface temperature distributions and groundwater movement are interrelated and groundwater temperature can be used to infer hydraulic parameters. Both Wang et al. (1989) and Woodbury and Smith (1988) solved for the full-nonlinear parameter estimation, but were limited in the actual number of unknowns sought. Parameters were viewed as being "effective" values over a large number of grid blocks.

In recent publications (cf. Woodbury and Sudicky, 1992; Woodbury and Rubin, 2000; Ulrych et al., 2001; Jiang et al., 2005, Painter et al., 2006; Woodbury and Ferguson, 2006) a full-Bayesian approach was used to obtain parameter estimates and variances. The "full-Bayesian" approach signifies that both parameter and hyperparameter determination may be involved. It is the viewpoint of this paper that the full-Bayesian techniques mentioned can be successfully adapted to the inverse problem of coupled groundwater and heat flow. In the following sections the basic concepts of heat and fluid transfer in porous media are reviewed, followed by an adaptation of a "full-Bayesian" approach for coupled, but *linear* inversion of hydraulic head and thermal data. Specifically, the following are dealt with in this paper

1. A background including the pertinent governing equations
2. A section on the generalized inverse problem for groundwater and heat
3. Bayesian solutions to inverse problems
4. Bayesian solutions to linear interpolation of log-transmissivity data
5. Empirical Bayes and hyperparameter estimation by ABIC
6. Hierarchical Bayesian solutions
7. Upscaling and cokriging
8. Perturbation solutions and Bayes conditioning of groundwater and temperature data
9. Some comments on the non-linear problem

Two end-member and one set of joint inverse examples are presented. The two end-members are pure heat conduction (an application of Bayesian inversion to Paleoclimate reconstructions), and groundwater problem which is an example application to the Edwards Aquifer in Texas. Lastly, generic examples of combinations of transmissivity, hydraulic head and temperatures are presented.

2. BACKGROUND: GOVERNING EQUATIONS

An active groundwater system redirects heat flow from the Earth's interior. Heat is transferred through porous media by conduction, advection and radiation. Conductive transport occurs both in the solid medium and fluid, and is dominated by the properties of the porous medium and fluid, and the temperature gradient. Advective transfer occurs under the influence of moving groundwater. Within a system with high permeability, groundwater can redistribute heat, disturbing what would ordinarily be a conductive thermal regime. In shallow subsurface environments radiative transfer is negligible and is usually neglected. A review of studies which consider various aspects of coupled fluid flow and heat transfer in regions with 'normal' geothermal gradients is given by Domenico and Schwartz (1990).

Buoyancy effects, viscosity coupling to hydraulic conductivity, and thermal dispersion due to groundwater velocity fluctuations are important physical phenomena to consider. In Figure 1, both density and viscosity are important fluid parameters. But, for the conditions encountered in a shallow, low-temperature groundwater system, the thermal properties of the medium could be considered independent of temperature (see Figure 2 for shallow continental U.S. subsurface temperatures). Note on Figures 2 and then 1, that a change from 0 °C to 20 °C produces about a 20% reduction in viscosity. Considering that the range of temperatures across any one aquifer would be much smaller, this viscosity coupling to hydraulic conductivity could be a minor effect.

The reader should note that in the case of deep regional groundwater flow systems, the density and viscosity of water would vary significantly with temperature. In addition, in some groundwater-surface water investigations, the viscosity-temperature effect can be the dominant factor altering the flow system (Constantz et al., 1994). Viscous dissipation of energy may also be neglected due to low groundwater velocity and an equilibrium state is assumed to be reached instantaneously between the fluid and the solid aquifer matrix. Note that thermal dispersion is often ignored in thermal applications (e.g. Ferguson and Woodbury, 2005b), which has been postulated by some researchers (for example Sauty et al., 1982) as being equivalent to hydrodynamic dispersion in mass transport. However, the physical processes are different and strong arguments can be made for its omission (Bear and Corapcioglu, 1984). At the present time, the effect of thermal dispersion is still an open question, but perhaps of secondary importance.

A two or three-dimensional groundwater advection-conduction heat transport equation can be represented as a series of linked-partial differential equations and constitutive relationships (see Bear, 1972; Bear and Corapcioglu, 1981). The inverse algorithms presented herein are based on the assumptions of a low-temperature, shallow groundwater system and steady-flow behavior (see Section 9.0). Other

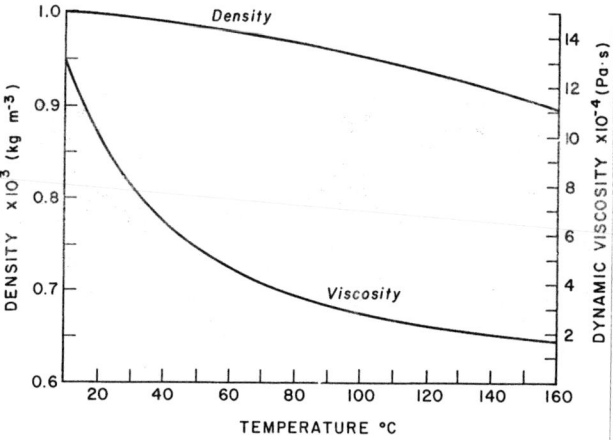

Figure 1. Density and viscosity of water and a function of temperature (see Woodbury and Smith, 1985).

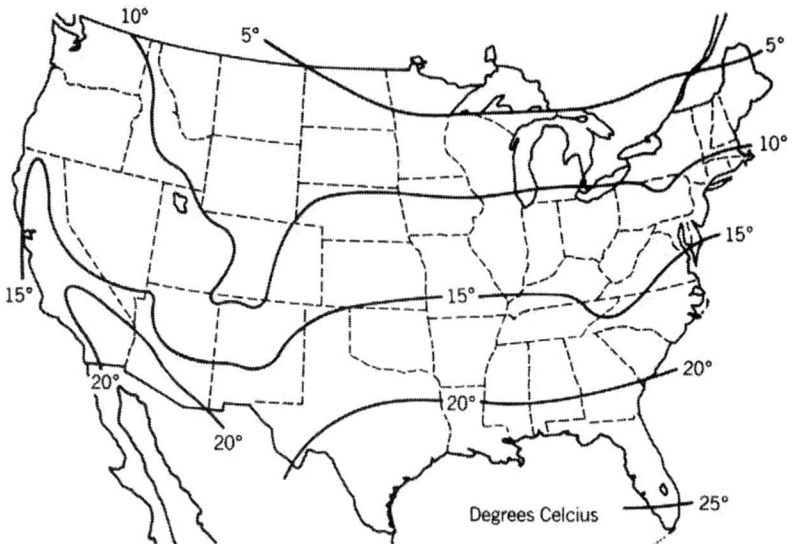

Figure 2. Shallow groundwater temperatures at depths of 10 to 20 m. (Reprinted from Physical and Chemical Hydrogeology by P. A. Domenico and F. W. Schwartz, John Wiley and Sons, Copyright (1990), with permission from Elsevier).

forms of the partial differential equations are certainly possible. Transient expressions involving other process such as deformation within the porous media are developed in Bear and Corapcioglu (1981). For a complete discussion of the equations referred to below the reader is referred to Smith and Chapman (1983).

3. GROUNDWATER AND HEAT: THE INVERSE PROBLEM

A solution of the inverse problem requires that both the forward and inverse problems must be clearly defined. A forward problem is set up by constructing a functional relationship to predict physical data, given a set of input parameters to a physical model. The goal of inverse theory is to use a discrete and finite set of noisy observations to elicit some information model **m** (Ulrych and Sacchi, 2005). For many physical problems the data and the model are related through a linear functional. However, in groundwater hydrology, when a parametric approach and a numerical scheme are used to solve the governing equations, a non-linear functional results; for example:

$$h = \Im[\mathbf{x}, \mathbf{m}] \quad (1)$$

where $\Im$ is a non-linear functional relating h, the potential (e.g., hydraulic head), **x**, a vector of grid coordinates in a numerical scheme and **m**, the actual model parameters, which could consist of hydraulic conductivities, boundary fluxes, sources and sinks, and so on.

In applications of the inverse involving observed data, hydraulic heads have uncertainties associated with their values, resulting from interpolation or measurement errors. In these cases (1) takes the form:

$$h^* = \Im[\mathbf{x}, \mathbf{m}] + v \quad (2)$$

where h^* is the data (interpolated or measured values of hydraulic head), and v is a vector of residuals.

When a numerical scheme is used to solve the functional (1), it can take the form (e.g., steady flow):

$$\mathbf{h} = \mathbf{A}^{-1}\mathbf{b} \quad (3)$$

Where **A** is a matrix, dependent upon a set of parameters, **b** a vector of right hand side terms, and **h** is a vector of computed potentials at nodal grid points. The inverse, then, can be posed as an optimization problem. For example, J below is a functional to be minimized, with respect to other constraints. A generalized L_2 norm can introduced as (after Neuman and Yakowitz, 1979):

$$J = (\mathbf{h} - \mathbf{h}^*)^T \mathbf{V}_h^{-1}(\mathbf{h} - \mathbf{h}^*) + \gamma(\mathbf{m} - \mathbf{m}^*)^T \mathbf{V}_m^{-1}(\mathbf{m} - \mathbf{m}^*) \quad (4)$$

where the above terms are defined as: V_h and V_m are head and model covariance matrices, **m** is a vector of log-conductivities determined by the inverse method, $\mathbf{m}^*$ is a vector of observed or estimated parameters, $\mathbf{h}^*$ is a vector of observed or estimated hydraulic heads, and γ is a scaling factor, which may be unknown. V_h and V_m are defined based on the characteristics of the data set and the numerical mesh. If $\mathbf{h}^*$ and

$\mathbf{m}^*$ are estimated, along with V_h and V_m, then $\gamma = 1$. With an unknown γ we recognize that we may have knowledge about the structure of the covariance matrices but not their magnitudes. Equation (4) can be derived from a maximum likelihood consideration for a Gaussian distribution (Carrera and Neuman, 1986), but one does not have to assume any underlying statistical distributions of $\mathbf{h}$ or $\mathbf{m}$ to apply the norm. The objective function can also be viewed as a weighted sum of L_2 prediction error (heads) and L_2 solution simplicity.

3.1. The Geometry of Non-Uniqueness

Equation (4) represents a constrained non-linear optimization problem and a large number of optimization techniques can be used to solve it. Some methods require that derivatives of the objective function with respect to model parameters be calculated, others do not (see Figure 3; after Sambridge and Mosegaard, 2002). On this Figure, methods to the right hand side offer more thorough explorations of model space but require more computational resources. The derivative or gradient based methods on the left of the figure are usually faster than non-gradient methods but may diverge (fail to find an answer), or converge to a local rather than a global minimum. Cooley's (1977) classic work describes instances of this type of problem. A non-linear inverse problem may have a complex functional surface, and there is no guarantee that any technique will converge to a global minimum. Gradient methods work best if the initial guess is linearly close to the solution. To investigate non-uniqueness it would be desirable to repeat the procedure many times with different initial guesses, and perhaps decluster the results (for example, Vasco et al., 1996). Rath et al. (2006) use a gradient-based approach to solving the linked-groundwater and heat transport inverse problem. Their approach notes that the calculation of the Jacobian matrix is computationally expense and propose an automatic differentiation method.

In inverse problems it is now common to consider both measured values and unknown model parameters as uncertain (see Woodbury and Ulrych, 1998; Ulrych et al., 2001). In a probabilistic approach, we assume that the model can be viewed as a random variable with each model estimate being a realization of a random process. The "true" model is then considered as the expected value of these random variables which, in general, will be dependent and describable in terms of a multivariate pdf. The model estimation problem can now be approached from the viewpoint of probability theory. Given an estimate of the pdf of the model, subject to new information in the form of a sample (data), we "update" the prior pdf with this information. The result of this exercise is still in the form of a pdf but we can obtain an estimate of the model by computing expected values. A detailed discussion

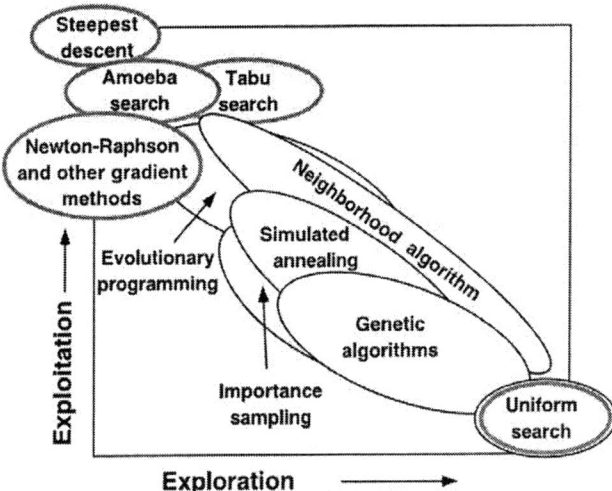

Figure 3. Overview of solution methods from Sambridge and Mosegaard, 2002

of information-based methods (specifically Bayes) is presented in section 4. Additional discussions of the concepts presented below, as well as of the differences between Bayes, maximum likelihood, and minimum relative entropy (MRE) estimators, is provided by Ulrych et al. (2001).

Because (2) is non-linear in its parameters the posterior probability distribution function will *in general* be non-Gaussian, and the maximum likelihood point of a non-Gaussian objective function may not yield the most sensible parameter estimates. Gaussian distributions are symmetric, so the maximum likelihood point always coincides with the mean value. For an arbitrary non-linear surface (for example, multi-modal, skewed) the maximum likelihood point can be quite far from the mean value. Figure 4 shows schematic diagrams of probability surfaces for several non-linear problems (after Mosegaard and Tarantola, 2002). This figure shows an objective function that has several local minima, or points of non-convexity with a well defined global minimum. Plotted on this figure are data d on the y axis, one model parameter on the x axis and a functional relationship given by some form of $d = g(m)$. This relationship could be linear or non-linear. Also shown on Figure 4 is a shaded ellipse representing, say the confidence limits on the model parameter and the observation error. This ellipse is given by the model pdf. A plot of $\sigma(m)$ on the figure represents the resulting posterior pdf of the model parameters as a result of combination of the prior model parameters, data, and functional relationship.

Plots (a) and (b) of Figure 4 show linear or linearizable problems with one solution, (c) shows a finite range of solutions. Plot (d) shows a problem with (possibly) an infinite range of solutions. It is the goal of inverse theory to condition the norm in such away that will yield features like (a), with a well defined solution. As Mosegaard and Tarantola (2002) note what is important is not the intrinsic nature of the non-linearity of the relationship but how linear is the relationship inside the domain of significant probability. Please refer to Mosegaard and Tarantola (2002) for a more complete discussion.

The techniques proposed in this overview utilize an approach that results in Figure (4b), a linearizable problem. Once linearized, the governing equations for fluid flow and heat can be rendered into a form that constitutes a linear inverse problem and we will attack this with Bayesian methods. For an alternate treatment of the coupled heat and groundwater inverse using a non-linear Bayesian maximum aposteriori (MAP) technique, see Rath et al. (2006).

4. BAYESIAN SOLUTION TO INVERSE PROBLEMS

As mentioned in the introduction, it is our goal to reconstruct a vector of hydrogeologic model parameters from observations of hydraulic heads and temperatures. To so called 'Bayesians', inverse problems are problems of inference and this is the philosophy adopted in this work to circumvent the aforementioned concerns about poorly posed problems.

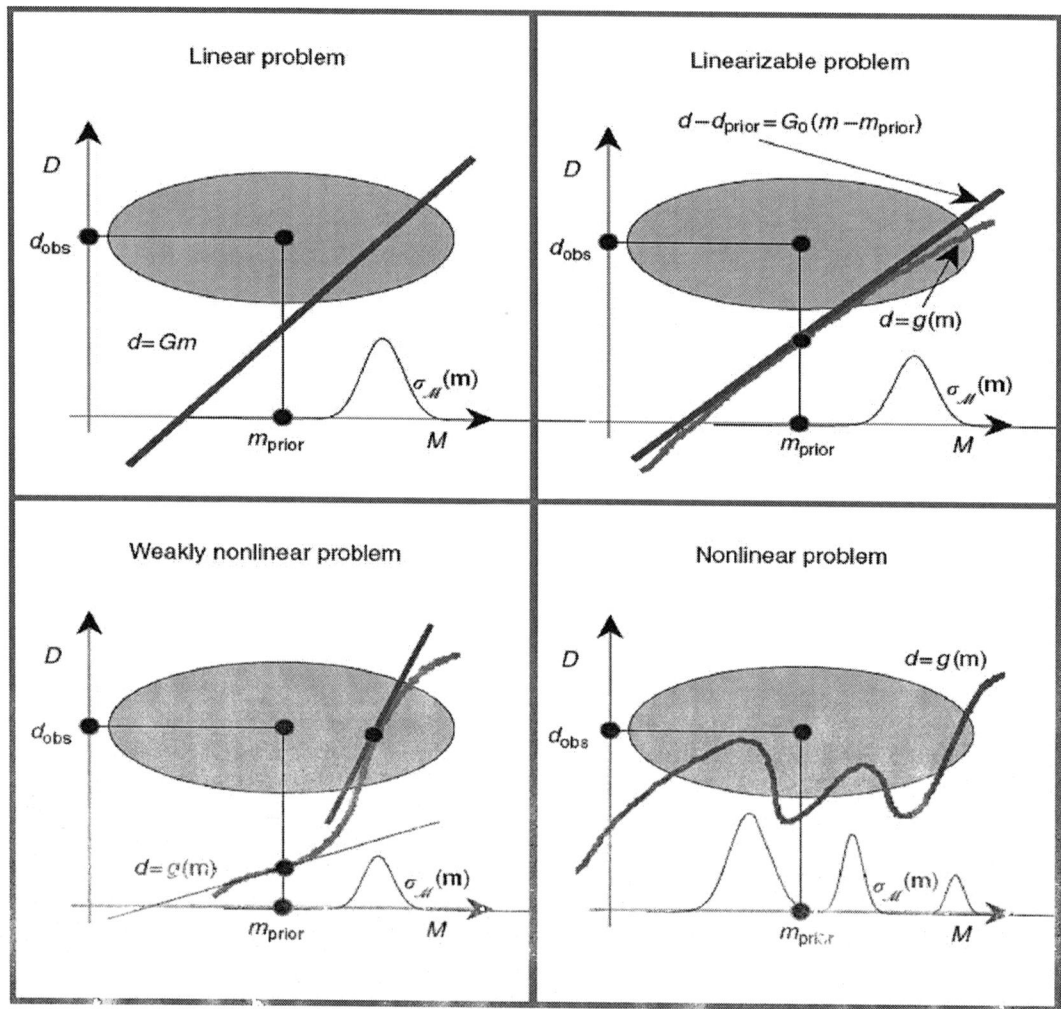

Figure 4. Illustration of the four domains of linearity. Plotted on this figure are data d on the y axis, one model parameter on the x axis and a functional relationship given by some form of $d = g(m)$. This relationship could be linear or non-linear. Also shown is a shaded ellipse representing, say the outer fringes of the ranges of model parameters and the observation errors. This ellipse is given by the model pdf. A plot of $\sigma(m)$ on the figure represents the resulting posterior pdf of the model parameters as a result of combination of the prior model parameters, data, and functional relationship. (Reprinted from Mosegaard, K. and A. Tarantola, 2002, Probabilistic Approaches to Inverse Problems, International Handbook of Earthquake and Engineering Seismology, Vol 81A, 237-265, with permission of Elsevier).

Much has been written on the subject of Bayesian inference and different points of view apply (for review see Ulrych et al., 2001). The reader will note that a "Full-Bayesian" approach signifies that the inference problem will consist of both primary parameter and hyperparameter estimation (Mohammad-Djafari, 1996; Woodbury and Rubin, 2000; Woodbury and Ulrych, 2000); Ulrych et al., 2001).

Bayesian inference supposes that an observer can define a prior probability-density function (pdf) for some random variable **m**. This pdf, $p(\mathbf{m})$, can in principle, be defined on the basis of personal experience or judgment. Bayes' rule (for example; Ulrych et al., 2001) quantifies how the prior pdf can be changed, or updated on the basis of measurements. The updated prior pdf is then referred to as the 'posterior' pdf. However, applications of Bayesian probability theory have been hampered by the precise meaning and interpretation of probabilities and controversy surrounding the appropriate choice of prior pdfs. An orthodox view of probabilities dictates that frequencies measured in an experiment are equated to probabilities and 'prior' information is not allowed. An alternative viewpoint of probability, denoted as the Jaynes-Cox viewpoint by Jowitt (Jowitt, 1979), is one in which probabilities are equated with the degree of plausibility of a proposition and may have no frequency interpretation whatsoever. A necessary component of the Jaynes-Cox view is the 'principle of maximum entropy' (PME). This forces all observers who possess common information to produce consistent results (Woodbury and Ulrych, 1998) and replaces the need for subjective prior information in the Bayesian approach.

Woodbury and Ulrych (1993), Woodbury et al. (1995) and Woodbury (1997; 2004) deal with the estimation of appropriate prior pdf's for hydrogeologic applications. As shown by Woodbury and Ulrych (1993), $p(\mathbf{m})$ may have the form of a multivariate-truncated exponential distribution. This pdf preserves the statistical independence of the parameters. That is, if no correlation is known beforehand the maximum entropy principle does not inject any correlation into the result. In this manner $p(\mathbf{m})$ has the most freedom in assigning realizations of the process. It is important to note that the above approach (PME) of determining $p(\mathbf{m})$ is the one which is the most uncommitted with respect to unknown information.

Simply stated, Bayes' rule is

Posterior $\propto$ Likelihood $\times$ Prior

Consider a vector of observed data $\mathbf{d}^*$. If the conditional pdf of $\mathbf{d}^*$ given **m** and some prior information I, is given by $p(\mathbf{d}^* | \mathbf{m}, I)$, then Bayes' rule states that

$$p(\mathbf{m}|\mathbf{d}^*, I) = \frac{p(\mathbf{d}^*|\mathbf{m}, I) p(\mathbf{m}|I)}{\int p(\mathbf{d}^*|\mathbf{m}, I) p(\mathbf{m}|I) d\mathbf{m}} \quad (5)$$

In the above, $p(\mathbf{m}|I)$ is the prior probability density of the model parameters, given some form of prior information, I, and $p(\mathbf{d}^*|\mathbf{m}, I)$ is the likelihood of observing $\mathbf{d}^*$ given the model parameters and the prior information. This latter term is often referred to as a 'direct' as opposed to a subjective pdf. The term on the left hand side is called the posterior probability (after measurements are taken into account). Finally the term in the denominator is a constant that ensures the posterior is normalized, but is also the actual pdf of observing a set of data, with the uncertainty in the model parameters taken into account.

In the sections below we will outline how the various conditional pdfs and the prior information are defined and show how we can use Bayes' rule to reconstruct a vector of model parameters from heads and temperature data.

4.1. Notes on Empirical and Hierarchical Bayes

Returning to the previous section it is desirable to include in Bayes' theorem some measure of the certainty we have in assigning the prior pdf; term (2) in the numerator of equation (5). In realistic cases more may be known about the form of the underlying pdf of the model (say, Gaussian) than its magnitude σ_Y^2, or other parameters governing the pdf. The parameters that are part of the pdf but which are uncertain are called 'hyperparameters'. It is also most often the case that the variance in the noise, σ_d^2 is unknown. Suppose that the form of the pdf is known, but the values of the 'hyperparameters' such as **s**, σ_d, σ_Y, λ (prior mean values, noise standard deviation, model scale factor, integral scale, respectively) are not. In these cases it may be desirable to generate a series of updated <**m**> values for a wide range of covariances of different magnitudes. It would then remain a problem to choose which candidate solution is 'best' in some sense.

Mohammad-Djafari (1996) also details many strategies with respect to this problem. One interesting technique, as detailed by Jaynes and others (for example Kitanidis, 1986; Loredo, 1990; Rubin and Dagan, 1992; Woodbury and Rubin, 2000) treats the hyperparameters as 'nuisance' parameters that are "removed" from further consideration by integration over these parameters (marginalization). This entire process is referred to as 'hierarchical Bayes', and was the approach adopted by Jiang et al. (2004) for their fluid flow inversion of the Edwards Aquifer.

Recently, Hou and Rubin (2005) derived a multivariate truncated Gaussian pdf, based on the principle of minimum relative entropy (MRE). Then with a likelihood function appropriate for the time-dependent Richard's equation, they developed the posterior non-linear pdf based on Bayes theorem. Observational noise was treated as a hyperparameter and its effect was removed by marginalization. Moments of the resulting pdf were computed by Latin hypercube and Monte Carlo simulations. This above example is an excellent illustration of the combination of hierarchical Bayes and MRE for a non-linear, time dependent problem.

In 'empirical' Bayes, the prior pdf is based on information contained in the input data. In these cases the truth of the prior may be considered irrelevant and is used to constrain the solution to a desired form. In empirical Bayes the hyperparameters themselves are estimated by an external criteria. Examples of both empirical and hierarchical Bayes are given later in this paper.

5. LIKELIHOOD FUNCTION FOR HYDRAULIC HEADS AND TEMPERATURES

Consider a numerical model for the hydraulic head predictions in an aquifer. Equation (3) is now written in terms of a general non-linear model of the type

$$d_i = f_1(\mathbf{x}_i, \mathbf{m}) \qquad (6)$$

for $i = 1 \ldots N$ where N is the number of predicted 'data' points and $\mathbf{x} = (x, y, z)$. Here, $f_1(x)$ depends upon a series of parameters $\mathbf{m}$ which could consist of log-transmissivities, flux conditions and the like.

In the case where head measurements are taken, the associated noise-corrupted case is

$$d_i^* = f_1(\mathbf{x}_i, \mathbf{m}) + \varepsilon_i \qquad (7)$$

Where the data d_i^* consist of a collection of discrete values of hydraulic heads and ε_i is the noise.

The inverse problem consists of trying to reconstruct the parameter vector $\mathbf{m}$, based on the observed data. As mentioned, the inverse problem is viewed in a Bayesian context; that is the inversion is viewed as a problem of inference. In order to solve the inference problem, we will use a Bayesian framework to 'update' a prior probability based on new information in the form of a data sample. To apply Bayes' Theorem we need to determine a pdf for the noise which is consistent with our understanding about its nature. Note, that if one could predict the 'true' data, the difference between d_i and d_i^* is just ε_i. If it is assumed that the noise has a value ε given prior information I, and if the second moment of the noise is known, σ_1, then an application of the maximum entropy principle leads to a Gaussian distribution for ε (Bretthorst, 1988; Kapur, 1989; Rubin, 2005):

$$p(\varepsilon | \sigma_1, I) = \frac{1}{\sqrt{2\pi\sigma_1^2}} \exp\left(-\frac{\varepsilon^2}{2\sigma_1^2}\right) \qquad (8)$$

Here σ_1 is taken as the root mean square (RMS) noise level and (8) is the least informative prior probability density for the noise that is consistent with the given second moment. Note that the central limit theorem leads to the Gaussian form (Jaynes, 1983) even if the second-moment of the noise is not known. In this work (shown later), we treat the noise explicitly as an unknown in Bayes' theorem and then proceed to integrate its effects out.

Having a pdf for the noise and adopting the notation that ε_i is the noise at distance x_i, one can apply the product rule of probability theory (assuming independence) to derive the pdf that one would obtain a set of noise values $\varepsilon_1, \varepsilon_2, \ldots \varepsilon_N$):

$$p(\varepsilon_1, \varepsilon_2, \ldots \varepsilon_N | \sigma_1, I) = \prod_{i=1}^{N}\left[\frac{1}{\sqrt{2\pi\sigma_1^2}} \exp\left(-\frac{\varepsilon_i^2}{2\sigma_1^2}\right)\right] \qquad (9)$$

Kapur (1989) shows that (9) arises naturally in the multivariate case when entropy is maximized with correlations unknown.

Consider another non-linear model for the temperatures, of the type

$$d_j = f_2(\mathbf{x}_j, \mathbf{m})$$

for an additional M points and the associated noise-corrupted case is

$$d_j^* = f_2(\mathbf{x}_j, \mathbf{m}) + \varepsilon_j \qquad (10)$$

Here, the model $f_2(\mathbf{x}_j, \mathbf{m})$ describes the physics of thermal transport and also depends on the same parameters $\mathbf{m}$, as in $f_1(x, \mathbf{m})$; namely the transmissivity, boundary conditions and the like. In this case the noise variance is different than the first N values and is equal to σ_2^2. The data in this case are M observed values of temperature.

In a similar line of reasoning with (9), the noise pdf now becomes

$$p(\varepsilon_1, \varepsilon_2, \ldots \varepsilon_N, \varepsilon_{N+1}, \varepsilon_{N+2}, \ldots \varepsilon_{N+M} | \sigma_1, \sigma_2, I) =$$
$$\prod_{i=1}^{N}\left[\frac{1}{\sqrt{2\pi\sigma_1^2}} \exp\left(-\frac{\varepsilon_i^2}{2\sigma_1^2}\right)\right] \times$$
$$\prod_{j=1}^{M}\left[\frac{1}{\sqrt{2\pi\sigma_2^2}} \exp\left(-\frac{\varepsilon_j^2}{2\sigma_2^2}\right)\right] \qquad (11)$$

Again, if the 'true' model is known, the difference between the data and the model is described by the noise. Taking into account (7) and (10) the pdf that one obtains a set of data $\mathbf{d}^* = (d_1^*, d_2^*, \& d_{N+M}^*)$, given a set of parameters and prior information, is proportional to the likelihood function, L:

$$p(\mathbf{d}^* | \mathbf{m}, \sigma_1, \sigma_2, I) \propto L(\mathbf{m}, \sigma_1, \sigma_2) =$$
$$\prod_{i=1}^{N} \sigma_1^{-1} \exp(-\frac{1}{2\sigma_1^2}[d_i^* - f_1(x_i, \mathbf{m})]^2) \times \quad (12)$$
$$\prod_{j=1}^{M} \sigma_2^{-1} \exp(-\frac{1}{2\sigma_2^2}[d_j^* - f_2(x_j, \mathbf{m})]^2)$$

or,

$$L(\mathbf{m}, \sigma_1, \sigma_2) = \sigma_1^{-N} \sigma_2^{-M} \times$$
$$\exp\left\{-\frac{1}{2\sigma_1^2}\sum_{i=1}^{N}[d_i^* - f_1(x_i, \mathbf{m})]^2 - \frac{1}{2\sigma_2^2}\sum_{j=1}^{M}[d_j^* - f_2(x_j, \mathbf{m})]^2\right\} \quad (13)$$

A non-linear least-squares approach would begin by by minimizing the combined sums in the argument in the exponential of (13). The equivalent maximum likelihood procedure finds the parameter set that maximizes the logarithm of (13). Neither approach incorporates prior information about the model parameters. On the other hand, the Bayesian methodology readily lends itself to the problem of updating prior probabilities based on uncertain field measurements. For example, Kitanidis (1986) and Woodbury and Rubin (2000) outlined the Bayesian approach in which relevant prior information about the model is incorporated. In the current work we adopt a similar approach but following the suggestions of Jaynes and others (for example Kitanidis, 1986; Loredo, 1990; Rubin and Dagan, 1992; Woodbury and Rubin, 2000; Hou and Rubin, 2005) we can treat the two noise variances σ_1^2, σ_2^2 as 'nuisance' parameters that are "removed" from further consideration by integration over these parameters (marginalization). This point is discussed further below in section (8).

6. THE LINEAR INVERSE AND APPLICATION TO INTERPOLATION

In (12) above, the functions f_1 and f_2 are general in form. Here, (and just looking at one term) we investigate a special form for f_1, namely linear in transformation. In matrix-vector form, the data and the model can be related through a linear kernel. Hence,

$$\mathbf{d}^* = \mathbf{Gm} + v \quad (14)$$

where
- $\mathbf{d}^*$ $N \times 1$ vector of observed values, the 'data';
- $\mathbf{G}$ $N \times M$ matrix of coefficients;
- $\mathbf{m}$ $M \times 1$ vector of unknown-actual values of the data, the 'model';
- v $N \times 1$ vector of random observational errors, and

$\mathbf{G}$ is a kernel which transforms data in $\Re^N$, the data space, to $\Re^M$, the model space.

Let us assume the following statistics on the model $\mathbf{m}$ and the noise vector v consistent with the Bayesian framework. The noise v is random with a mean of zero and a covariance $\mathbf{C}_d$:

$$E(v) = 0 \quad (15)$$

$$E(vv^T) = \mathbf{C}_d \quad (16)$$

If v represents measurement error, say independent and identically distributed (iid), then $\mathbf{C}_d$ is a diagonal matrix of variances of the observations, $\sigma_d^2 \mathbf{I}$. The model $\mathbf{m}$ is random and characterized with a prior mean $\mathbf{s}$, and covariance $\mathbf{C}_m$, which physically represents the correlation or spatial variability of the model $\mathbf{m}$.

$$E(\mathbf{m}) = \mathbf{s} \quad (17)$$

$$E[(\mathbf{m} - \mathbf{s})(\mathbf{m} - \mathbf{s})^T] = \mathbf{C}_m \quad (18)$$

$\mathbf{C}_m$ is commonly represented by an exponential correlation structure

$$C_m(k,l) = \sigma_Y^2 \exp\left[-\left(\frac{(x_k - x_l)^2}{\lambda_x^2} + \frac{(y_k - y_l)^2}{\lambda_y^2}\right)^{1/2}\right] \quad (19)$$

where the usual definitions apply, in that σ_Y^2 is the variance of say, $\ln(K)$, λ is the integral scale and k and l refer to two points in question.

If the combination of forward modeling and measurement errors are assumed Gaussian, then the probability of observing a set of data $\mathbf{d}^*$ given the model parameters is (Tarantola, 1987, p 68):

$$p(\mathbf{d}^* | \mathbf{m}, I) = ((2\pi)^N |\mathbf{C}_d|)^{-\frac{1}{2}} \times \exp\left[-\frac{1}{2}(\mathbf{d}^* - \mathbf{Gm})^T \mathbf{C}_d^{-1}(\mathbf{d}^* - \mathbf{Gm})\right] \quad (20)$$

Here N is the length of vector $\mathbf{d}^*$. If the prior distribution of the model is also assumed to be Gaussian then

$$p(\mathbf{m}|I) = ((2\pi)^M |\mathbf{C}_m|)^{-\frac{1}{2}} \exp\left[-\frac{1}{2}(\mathbf{m}-\mathbf{s})^T \mathbf{C}_m^{-1}(\mathbf{m}-\mathbf{s})\right] \quad (21)$$

Here, Tarantola (1987) illustrates the important result that if the forward modeling is linear, i.e., if $\mathbf{d} = \mathbf{Gm}$, and if the above likelihood and prior information are both Gaussian, then the posterior density of $\mathbf{m}$ is Gaussian. The resulting posterior pdf $p(\mathbf{m}|\mathbf{d}^*, I)$ is Gaussian:

$$p(\mathbf{m}|\mathbf{d}^*, I) = ((2\pi)^M |\mathbf{C}_q|)^{-1/2} \times \exp[-\frac{1}{2}(\mathbf{m}-<\mathbf{m}>)^T \mathbf{C}_q^{-1}(\mathbf{m}-<\mathbf{m}>)] \quad (22)$$

The first two moments of this pdf are given by Tarantola (1997, eq 1.93)

$$<\mathbf{m}> = \mathbf{s} + \mathbf{C}_m \mathbf{G}^T (\mathbf{G}\mathbf{C}_m \mathbf{G}^T + \mathbf{C}_d)^{-1}(\mathbf{d}^* - \mathbf{Gs}) \quad (23)$$

$$\mathbf{C}_q = \mathbf{C}_m - \mathbf{C}_m \mathbf{G}^T (\mathbf{G}\mathbf{C}_m \mathbf{G}^T + \mathbf{C}_d)^{-1} \mathbf{G}\mathbf{C}_m \quad (24)$$

where $<\mathbf{m}>$ and $\mathbf{C}_q$ are the expected value and covariance of the posterior pdf, respectively. These results are well known.

6.1. Bayesian Resolution

For Gaussian priors and noise, the Bayes' maximum a posterior solution (23) can also be written as (Tarantola, 1987, p73):

$$<\mathbf{m}> = \mathbf{s} + \left[\mathbf{G}^T \mathbf{C}_d^{-1} \mathbf{G} + \mathbf{C}_m^{-1}\right] \mathbf{G}^T \mathbf{C}_d^{-1}(\mathbf{d}^* - \mathbf{Gs}) \quad (25)$$

If we define a resolution operator $\mathbf{m}_{est} = \mathbf{Rm}_{true}$ and subtracting $\mathbf{s}$ from (25) yields

$$<\mathbf{m}> - \mathbf{s} = \left[\mathbf{G}^T (\mathbf{C}_d^{-1} \mathbf{G} + \mathbf{C}_m^{-1}\right] \mathbf{G}^T \mathbf{C}_d^{-1} \mathbf{G}(\mathbf{m}^{true} - \mathbf{s})$$
$$= \mathbf{R}\left[\mathbf{m}^{true} - \mathbf{s}\right]$$

It can be shown that in the above (Tarantola, 1987, p 200):

$$\mathbf{R} = \mathbf{I} - \mathbf{C}_q \mathbf{C}_m^{-1} \quad (26)$$

Notice that if the posterior covariance of $\mathbf{m}$ is zero then $\mathbf{R} = \mathbf{I}$ and we have perfect resolution, regardless of the prior. Note also that as prior information is introduced the resolution is essentially a variance ratio. In other words, resolution can described as: (Tarantola, 1987, p. 63): "First, a parameter is well resolved by the data set if its posterior error bar is much smaller than the prior one. More generally, if its posterior marginal probability density is significantly different from the prior one. If, for example, the prior and posterior densities are identical, the parameter is completely unresolved."

6.2. Linear Interpolation by Bayesian Update

Bayesian updating methods provide an alternate philosophy to kriging for the characterization of input variables of a stochastic mathematical model. In this approach a priori values of statistical parameters (for instance, mean and covariance) are assumed on subjective grounds or by analysis of a data base from a geologically similar area. As measurements become available during site investigations 'updated' estimates of these parameters are generated.

The Bayesian interpolation scheme naturally follows from the above section on Bayes theorem and the reader is referred to Woodbury (1989) for further details. Let us assume that measurements of a stochastic-random variable $\mathbf{m}$, are made at N locations in a discretized flow domain. The problem is to interpolate the N measured values to $M - N$ other points and closely reproduce the data points themselves at the measurement locations. Therefore, the N measurements of the variable are used to estimate M values of $\mathbf{m}$ by linear inversion. Here, the N measurements of the variable form a vector $\mathbf{d}^*$ which is referred to as the observed 'data'. The M values of $\mathbf{m}$ are the 'model'. The data are mathematically formed as linear combinations of the model and random noise. In this case, $\mathbf{G}$ is a kernel which transforms data in $\Re^N$, the data space, to $\Re^M$, the model space, and consists of 1's and 0's. Suppose there are $i = 1, \ldots\ldots N$ measurement points and $j = 1 \ldots\ldots M$ interpolated points, where $N \ll M$. Where the $i'th$ measurement point corresponds to an interpolated point, $G_{ij} = 1$, otherwise G_{ij} is zero.

Applications of the Bayesian updating approach for interpolating spatial data can be found in Kennedy and Woodbury (2002, 2005). In the latter study, Bayesian updating was successfully used to generate the heterogeneous log-transmissivity field for a vast area of Manitoba's Carbonate Aquifer from geostatistics of the underlying data. The dataset of transmissivity was compiled from measurements from formal pump tests but the majority was estimated from numerous, but highly uncertain specific capacity tests. Even with such a variable quality dataset, Bayesian updating was successfully used to generate the transmissivity field and subsequent hydraulic conductivity field for the Carbonate Aquifer. Without any further adjustments in this hydraulic conductivity, the model was successfully calibrated. However, for the deeper Sandstone Aquifer, where the data were clustered in an eastern freshwater region, the Bayesian Updating method simply assigned a smooth field equal to its prior mean.

7. EMPIRICAL BAYES AND HYPERPARAMETER ESTIMATION BY ABIC

As mentioned, $\mathbf{C}_m$ is often represented with an exponential autocovariance and can play a crucial role in determining an inverse solution. However, the actual statistical parameters embedded into the prior pdf, such as the mean $\mathbf{s}$, the variance σ_y^2 and the integral scale λ may not be well know, and may be difficult to estimate. The idea behind the empirical Bayes approach is that the prior is based on information contained in the input data. As Ulrych et al. (2001) discuss, in this case, the 'truth' of the prior may be considered irrelevant. It is used to constrain the solution to a form that is a priori known to be the desired one. The empirical Bayes approach has been shown to provide many useful solutions in geophysics. For example, the computation of a high resolution DFT (Sacchi et al., 1998) and the compensation of aperture effects in computing the Radon transform (Sacchi and Ulrych, 1995). The Japanese literature is rich in many successful applications (For example, Mitsuhata et al., 2001).

Note that the term in the denominator of (5) represents the actual pdf of observing a set of data, with the uncertainty in the model parameters taken into account. In the empirical Bayes approach it also depends on any hyperparameters that may be embedded into the prior, for example:

$$\int p(\mathbf{d}^* \mid \mathbf{m}, I) p(\mathbf{m} \mid I) d\mathbf{m} = p(\mathbf{d}^* \mid I) = p(\mathbf{d}^* \mid \sigma_y^2, \sigma_d^2, \lambda) \quad (27)$$

(if three hyperparameters are present). Suppose we have two assumptions related to the prior information, say that $p(\mathbf{d}^*|I_1) \geq p(\mathbf{d}^*|I_2)$; we would naturally select I_1 over I_2 as a more appropriate candidate for the prior information.

The particular approach that is described is based on the work of Akaike (1980) whose contributions have had a huge influence on the field of probability and statistics, and is based on the AIC and the ABIC criteria. A much more advanced presentation can be found in Matsuoka and Ulrych (1986). In essence, the AIC is a criterion based on the Kullback-Leibler information measure. The minimum AIC, $\text{AIC}(k)|_{\min}$, is the optimal compromise between errors in parameter estimation and errors in fitting of the model. For normally distributed errors

$$\text{AIC}(k) = N \log s_k^2 + 2k \quad (28)$$

where s_k^2 is the residual sum of squares or, in our case, the variance of the residuals that are computed as the difference between the actual and the computed potentials and k is the number of free parameters.

The first term in the above expression is related to the sample variance and decreases with the number of parameters. The second is related to the fact that the error of fitting the parameters increase with their number. The minimum of the AIC allows the computation of the appropriate number of parameters, a particularly difficult task in problems such as fitting of time series models. The ABIC is similar to the AIC in form and is computed in terms of the Bayesian likelihood defined in equation (5)

$$\text{ABIC} = -2\ln\left\{p(\mathbf{d}^* \mid I)\right\} + 2N_h \quad (29)$$

Here, N_h is the number of hyperparameters in the minimization and the hyperparameters are evaluated at the minimum value of the ABIC. In this way, $p(\mathbf{d}^*|I)$ is maximized for a given set of hyperparameters.

For the linear inverse problem with Gaussian priors and likelihood, Mitsuhata (2004) showed that:

$$p(\mathbf{d}^* \mid I) = (2\pi)^{N/2} \| \mathbf{C}_{dp} \|^{-1/2} \times \exp\left\{-\frac{1}{2}(\mathbf{d}^* - \mathbf{Gs})^T \mathbf{C}_{dp}^{-1}(\mathbf{d}^* - \mathbf{Gs})\right\} \quad (30)$$

Where $\mathbf{C}_{dp} = \mathbf{G}\mathbf{C}_m\mathbf{G}^T + \mathbf{C}_d$.

For the simulations in this paper, it is assumed that there are two principle hyperparameters of interest. The first is σ_d^2, the noise in the observed data, such that $\mathbf{C}_d = \sigma_d^2 \mathbf{I}$. Second, there is a scale parameter for the correlation matrix; i.e., $\mathbf{C}_c \sigma_y^2 = \mathbf{C}_m$. This means that the form of the prior model covariance is known, but not the hyperparameter σ_y^2. Embedded within the correlation matrix $\mathbf{C}_c$ is the hyperparameter, λ (see equation 19) the integral scale. In Woodbury and Ferguson (2006), determination of this quantity was not carried out, and they chose instead to fix this quantity in any one analysis and then select the model set that minimizes the ABIC from various inversion runs.

Using (30) the ABIC is

$$\text{ABIC} = N\ln(2\pi) + \| \mathbf{C}_{dp} \| + (\mathbf{d}^* - \mathbf{Gs})^T \mathbf{C}_{dp}^{-1}(\mathbf{d}^* - \mathbf{Gs}) + 2N_h \quad (31)$$

assuming two hyperparameters, $2N_h = 4$. In the above, terms two and three both depend on σ_d^2 and σ_y^2. The procedure for determining the ABIC is as follows:
- input a starting value for σ_d^2 and σ_y^2.
- form $\mathbf{C}_{dp}$ and compute its determinant.
- form $(\mathbf{d}^* - \mathbf{Gs})^T \mathbf{C}_{dp}^{-1}(\mathbf{d}^* - \mathbf{Gs})$ and minimize the ABIC for σ_y^2 using a Golden search method holding σ_d^2 fixed.
- compute $<\mathbf{m}>$ and estimate σ_d^2 from the misfit between computed and observed temperatures.
- if σ_y^2 converges to a stipulated tolerance then step out of the loop, otherwise go back to (2) above
- the iterations are complete and then compute the final model, covariances, the ABIC, resolution and so on.

This procedure accounts for the uncertainty in σ_y^2 in any one iteration by including its variablity in terms two and three in (31). Although a determinant of the matrix $\mathbf{C}_{dp}$ is required, this is only of the order of the number of data points and for under-determined inverse problems is usually much smaller than the number of model parameters.

7.1. Application of the Empirical Bayes to Paleoclimate Studies

A challenging issue in physical sciences is the recovery of past ground surface temperature (GST) changes from temperature measurements taken in boreholes. Note that one-dimensional heat conduction in a homogeneous medium can be represented in a discrete form as: $\mathbf{d}^* = \mathbf{Gm} + \varepsilon$ where, $\mathbf{m}$ is an N length vector of ground surface temperatures, $m_{N+1} = v_0$ is a time-invariant average-surface temperature and $m_{N+2} = a$ is the geothermal gradient. The matrix $\mathbf{G}$ contains a sub-matrix $\mathbf{H}$ which are the data kernels from the discretization and columns of 1's and z_i values that are appended and correspond to the $N+1, N+2$ locations in $\mathbf{m}$. The vector $\mathbf{d}^*$ contains the actual downhole temperature measurements at various locations z_i, $i = 1, M$ (see Woodbury and Ferguson, 2006; Kennedy et al., 2000). ε is a vector of random measurement error. An example inversion based on the ABIC solution comes from Woodbury and Ferguson (2006). In this application the ABIC technique is used to reconstruct past climate based on downhole thermal measurements. The reconstructed GST record shows (Figure 5) warming between 1800 and 1949 of approximately 0.6° K, with the maximum rate of warming occurring between 1900 and 1949. During the middle of the twentieth century, there is very minor cooling in the reconstructed GST record (1950's and 1970's) and then a sharp rise at the end of the century to about 1.0° K. These results are consistent with other researchers in the climate change area and with instrumental records in the latter half on the 20th century (Beltrami et al., 2003).

8. HIERARCHICAL BAYESIAN APPROACH TO LINEAR INVERSION

Recall the solution to Bayes' theorem for the case of Gaussian priors and a linear functional transformation, (22) noting that, of course,

$$<\mathbf{m}> = \int p(\mathbf{m}|\mathbf{d}^*, I) \mathbf{m} d\mathbf{m} \quad (32)$$

But consider that we have a set of hyperparameters ($\mathbf{s}$, σ_y^2, λ, $\sigma_d^2 = \mathbf{u}$), then

$$<\mathbf{m}> = \int_{\mathbf{m}} \int_{\mathbf{u}} p(\mathbf{m}|\mathbf{d}^*, I, \mathbf{u}) p(\mathbf{u}) \mathbf{m} d\mathbf{u} d\mathbf{m} \quad (33)$$

Changing the order of integration results in

$$<\mathbf{m}> = \int_{\mathbf{u}} p(\mathbf{u}) \left[\int_{\mathbf{m}} p(\mathbf{m}|\mathbf{d}^*, I, \mathbf{u}) \mathbf{m} d\mathbf{m} \right] d\mathbf{u} \quad (34)$$

However, the term within the bracket is equal to (23) and therefore (32) becomes

$$<\mathbf{m}> = \int_{\mathbf{u}} \mathbf{s} + \mathbf{C}_m \mathbf{G}^T (\mathbf{G}\mathbf{C}_m\mathbf{G}^T + \mathbf{C}_d)^{-1}(\mathbf{d}^* - \mathbf{Gs}) p(\mathbf{u}) d\mathbf{u} \quad (35)$$

If the data errors are iid then

$$<\mathbf{m}> = \int_{\mathbf{u}} \mathbf{s} + \mathbf{C}_m \mathbf{G}^T (\mathbf{G}\mathbf{C}_m\mathbf{G}^T + \sigma_d^2 \mathbf{I})^{-1}(\mathbf{d}^* - \mathbf{Gs}) p(\mathbf{u}) d\mathbf{u} \quad (36)$$

Similarly the covariance, $\mathbf{C}_q$ is

$$\mathbf{C}_q = \int_{\mathbf{u}} \mathbf{C}_m + \mathbf{C}_m \mathbf{G}^T (\mathbf{G}\mathbf{C}_m\mathbf{G}^T + \sigma_d^2 \mathbf{I})^{-1} \mathbf{G}\mathbf{C}_m p(\mathbf{u}) d\mathbf{u} \quad (37)$$

Since $<\mathbf{m}>$ and $\mathbf{C}_q$ are functions of $\mathbf{u}$ (eqn. 23), the integration of (36) and (37) involving the hyperparameters must be carried out through some form of numerical integration.

The above integrations can be accomplished using the Monte Carlo method and the concept of importance sampling. The integrals posed by (36) and (37) are evaluated by generating a series of random model vectors using a multivariate random number generator with $p(\mathbf{u})$ as the pdf. The reader will note that while it is conceptually appealing to evaluate (36) and (37), this approach is restricted to a small number of hyperparameters due to computer storage and computational overheads. The fundamental problem then is how to specify, in a logical and consistent manner, the prior pdfs for these hyperparameters. This subject is discussed more fully in Woodbury and Ulrych (2000). See also Hou and Rubin (2005) for an example of hierarchical Bayes with Monte Carlo simulations.

9. STEADY-STATE GROUNDWATER AND THERMAL INVERSION

The basic methodology that we will be following with respect to groundwater inversion is through a stochastic-linearized approach detailed in Woodbury and Ulrych (2000) and Jiang et al., (2004). The essence of their approach is presented here (see also Hoeksema and Kitanidis, 1984). The hydraulic head ϕ and the aquifer transmissivity T satisfy the following partial differential equation in terms of the log-transmissivity, $Y = \ln(T)$.

$$\frac{\partial Y}{\partial x}\frac{\partial \phi}{\partial x} + \frac{\partial^2 \phi}{\partial x^2} + \frac{\partial Y}{\partial y}\frac{\partial \phi}{\partial y} + \frac{\partial^2 \phi}{\partial y^2} = 0 \quad (38)$$

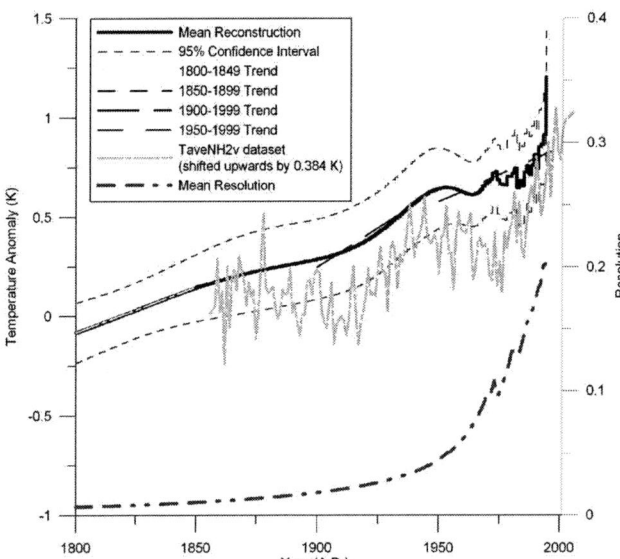

Figure 5. Average of 221 ground surface temperature (GST) reconstructions over Canada. See legend for description of the various lines. See also Woodbury and Ferguson(2006)

The above equation is separated into deterministic and stochastic terms. Letting $\phi = H + h$ and $Y = F + f$, where H is the expected value of the hydraulic head field, h is a zero-mean head perturbation; F is the expected Y value which is not necessarily a constant and f is a zero-mean log-transmissivity perturbation. After some manipulation, the hydraulic head field ϕ becomes the sum of two separate linearly superimposed solutions, one related to the solution to the mean head, H

$$\frac{\partial^2 H}{\partial x^2} + \frac{\partial^2 H}{\partial y^2} + \frac{\partial F}{\partial x}\frac{\partial H}{\partial x} + \frac{\partial F}{\partial y}\frac{\partial H}{\partial y} = 0 \quad (39)$$

and the other to the solution of the perturbation, h

$$\frac{\partial^2 h}{\partial x^2} + \frac{\partial^2 h}{\partial y^2} + \frac{\partial F}{\partial x}\frac{\partial h}{\partial x} + \frac{\partial F}{\partial y}\frac{\partial h}{\partial y} = -\frac{\partial f}{\partial x}\frac{\partial H}{\partial x} - \frac{\partial f}{\partial y}\frac{\partial H}{\partial y} \quad (40)$$

Equation (40) can be solved by the finite element method and takes the form

$$\mathbf{A}\mathbf{h} = \mathbf{B}\mathbf{f} + \mathbf{C}\mathbf{h}_B \quad (41)$$

where $\mathbf{A}$, $\mathbf{B}$, and $\mathbf{C}$ are constant matrices, $\mathbf{h}$ is a vector of nonboundary hydraulic head perturbations, and $\mathbf{h}_B$ is a vector of boundary node head perturbations. These are considered to be known and are typically set to zero. Solving for the hydraulic head perturbations at measurement points, and after some manipulation yields:

$$\mathbf{D}(\mathbf{v} - \mathbf{H} + \mathbf{A}^{-1}\mathbf{B}\mathbf{s}) = \mathbf{D}(\mathbf{A}^{-1}\mathbf{B})\mathbf{m} \quad (42)$$

where $\mathbf{v}$ is a vector of hydraulic heads at discrete points, $\mathbf{H}$ is a vector of conditioned hydraulic heads (solution to [39]), $\mathbf{s}$ is a vector of prior conditional-expected $\ln(T)$ and $\mathbf{m}$ is the model vector, the unknown $\ln(T)$ in the aquifer. The matrix $\mathbf{D}$ is a simple Boolean matrix (consists of 1's and 0's) that filters out the computed values of heads at points other than those corresponding to measurement points. When the left hand side $\mathbf{v}$ is replaced with the actual values of hydraulic heads observed, the system (42) has the same form as $\mathbf{d}^* = \mathbf{G}\mathbf{m}$.

The temperature perturbation is viewed as mainly being caused by groundwater velocity variations, which result from $\ln(T)$ fluctuations. Woodbury (1998) and Jiang and Woodbury (2006) developed the linearized formulation between $\ln(T)$ and the temperature perturbations. According to Darcy's law, the components of the specific discharge are formulated as

$$q_x = -K\frac{\partial \phi}{\partial x}$$

and

$$q_y = -K\frac{\partial \phi}{\partial y}$$

In terms of transmissivity,

$$q_x = -\frac{T}{B}\frac{\partial \phi}{\partial x} = \alpha \exp(Y)\frac{\partial \phi}{\partial x}$$

Here B is the aquifer thickness.
Since $Y = F + f$ and $\alpha = 1/B$. Hence,

$$q_x = \alpha \exp(F)\exp(f)\frac{\partial \phi}{\partial x}$$

If we define

$$u_x = \frac{q_x}{\alpha \exp(F)} = \exp(f)\frac{\partial \phi}{\partial x}$$

However, $\phi = H + h$ and linearizing (neglecting products of perturbation):

$$u_x = \frac{\partial H}{\partial x} + \frac{\partial h}{\partial x} + f\frac{\partial H}{\partial x}$$

For simplicity, we will work with one dimension for illustration. For the steady-state flow of heat in an aquifer,

$$\Lambda \frac{\partial^2 \theta}{\partial x^2} - \rho_f c_f q_x \frac{\partial \theta}{\partial x} = 0$$

$$\frac{\Lambda}{\alpha \exp(F)} \frac{\partial^2 \theta}{\partial x^2} - \rho_f c_f u_x \frac{\partial \theta}{\partial x} = 0$$

$$\Lambda' \frac{\partial^2 \theta}{\partial x^2} - \rho_f c_f u_x \frac{\partial \theta}{\partial x} = 0$$

Let the temperature field be decomposed into the linear superposition of a temperature field produced from a mean solution and a perturbation, $\theta = T + \zeta$ and defining $\rho_f c_f = c$, then after neglecting the products of perturbations and taking expected values,

$$\Lambda' \frac{\partial^2 T}{\partial x^2} - c \frac{\partial H}{\partial x} \frac{\partial T}{\partial x} = 0$$

and

$$\Lambda'' \frac{\partial^2 \zeta}{\partial x^2} - \frac{\partial H}{\partial x} \frac{\partial \zeta}{\partial x} = \frac{\partial h}{\partial x} \frac{\partial T}{\partial x} + f \frac{\partial H}{\partial x} \frac{\partial T}{\partial x}$$

where

$$\Lambda'' = \frac{\Lambda'}{\rho_f c_f}$$

Applying the finite element approach in two dimensions yields

$$\mathbf{M}\zeta = \mathbf{Nh} + \varnothing \mathbf{f} \qquad (43)$$

where $\mathbf{M}$ is a global stiffness matrix, $\mathbf{N}$ is a constant matrix related to conditional-expected temperatures, $\varnothing$ is a constant matrix related to mean head and mean temperature, ζ, $\mathbf{h}$ and $\mathbf{f}$ are vectors of perturbation temperatures, heads and $\ln T$, respectively. In the next section, it will be shown how both groundwater flow and heat transport can be used either collectively, or singularly to invert for log-conductivity.

10. SIMULTANEOUS GROUNDWATER AND THERMAL INVERSION

This section presents a brief condensation of the work of Jiang and Woodbury (2006). The joint use of hydraulic head and temperature measurements in the Bayesian update procedure can be accomplished in a single step. Finite element solutions to the stochastic ground-water and thermal equations can be combined into the following system,

$$\mathbf{D}\left\{\theta - \nu + \mathbf{M}^{-1}(\mathbf{NA}^{-1}\mathbf{B} + \varnothing)\varepsilon\right\} = \mathbf{D}\left\{\mathbf{M}^{-1}(\mathbf{NA}^{-1}\mathbf{B} + \varnothing)\right\}\mathbf{m} \quad (44)$$

where $\mathbf{M}$ is a global stiffness matrix, $\mathbf{N}$ is a constant matrix related to conditional-expected temperature, $\varnothing$ is a constant matrix related to mean head and mean temperature in the aquifer. This equation (44) and that of (42) can be combined into a single block system which is the same form as $\mathbf{d} = \mathbf{Gm}$. See Jiang and Woodbury (2006).

11. EXAMPLE APPLICATIONS

11.1. Bayes Update, Upscaling and Groundwater Inversion

The Bayesian inversion method has been applied to one of the most strategically important aquifers in the United States; the Edwards Aquifer in south-central Texas (Figure 6; see Jiang et al., 2004). For a more complete discussion on all the simulations presented here the reader is referred to Painter et al. (2006). A treatise on the Edwards Aquifer, including more recent data, and the simulations presented here (and considering the karst conduits present), is presented in Lindgren et al. (2004).

The Edwards Aquifer covers an area of about 10,000 km² is virtually the sole source of drinking water for the city of San Antonio. Well-developed secondary porosity and permeability has formed within the aquifer, which is a karstic limestone. Recharge to the aquifer comes mainly from stream losses in the outcrop areas of the Edwards Aquifer. Discharge occurs by pumping wells and at major springs, such as Comal, and these can be at great distances from the recharge areas (240 km).

Inverse simulations of this aquifer, using any method, would be a daunting task. As Painter et al. (2006) note the aquifer is highly heterogeneous, with a variance in ln(K) of 6.4 in the confined areas and 9.7 in the thinner outcrop (recharge) zone. The hydraulic conductivities varying by more than six orders of magnitude through the study area.

Painter et al. (2006) outlined in their paper a comparison of three methods of estimating transmissivity in highly heterogeneous aquifers (see Figure 7). In method one, simple kriging is used to interpolate transmissivity data over the aquifer region. These data consist primarily of single-well specific capacity tests. The reader should note here that simple kriging and the Baysian updating method (should and do) generate similar interpolated fields. The main difference is that kriging will honor the actual observation values while Bayesian updating may not. In method two, an upscaling/cokriging approach removed most of the systematic bias as in indicated in method one. The essence of the upscaling procedure is detailed in Painter et al. (2006) and is also shown on Figure 8. Geostatistics of local scale specific capacity tests are first obtained and then realizations of this field are generated in each cell block of a numerical grid. Grid blocks were successfully removed from simulated conductivity fields and effective transmissivity values were calculated. These simulated values were then analyzed statistically and through cokriging these effective values were conditioned on the local-scale values. Finally, in method three, the Bayesian inversion further conditioned and reduced the mean residuals by more than a factor of ten to about 2.5% of the total head variation in the aquifer.

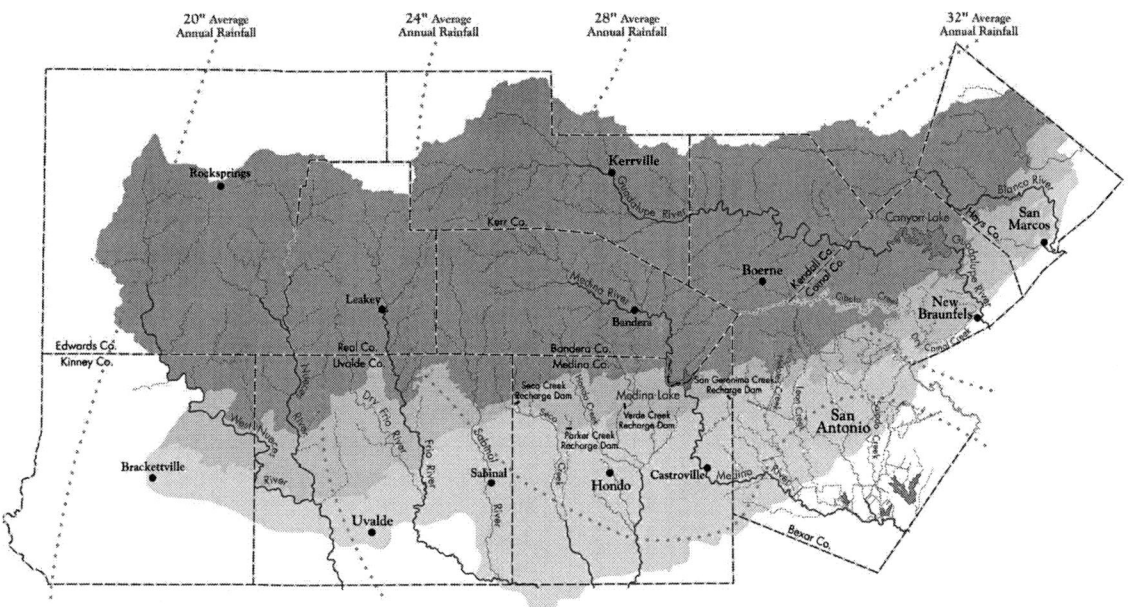

Figure 6. Edwards Aquifer region, courtesy of Edwards Aquifer Authority, 2002. Courtesy of the Edwards Aquifer Authority, 2007.

This agreement demonstrates the utility of the Bayesian methodology on a highly heterogeneous aquifer. Figure 9 from Painter et al. (2006) shows the value for hydraulic conductivity in each grid cell as produced by a upscaling and cokriging, and by Bayesian updating. Locations of major springs are indicated. Note that the two plots have different grey scales. Again, Figure 7 shows that this latter hydraulic conductivity field produced excellent results.

11.2. Example Generic Inversions

Jiang and Woodbury (2006) applied the procedure to a series of test cases, in which the actual values of ln (T), hydraulic head and temperature are generated with known values of stochastic parameters. Samples (50 and 100 points) were randomly taken from the fictitious aquifer. Basic statistics of the sampled data were used to derive pdfs of the hyperparameters (Figure 10).

Jiang and Woodbury (2006) show that joint use of ln (T), head and temperature data in the procedure aids the refinement of ln (T) estimation for s_y (standard deviation in ln (T)) of up to 2.0. Also, conjunctive use of ln (T) and temperature measurements (in absence of head data) is showed to improve ln (T) estimation (even for s_y up to 2) in comparison to the updated ln (T) field conditioned on ln (T) alone. The resolution of reconstructed ln (T) field based on temperature measurements decreases as the variation of the true ln (T) field increases (Figure 11). These results suggest that low-cost temperature measurements are a promising data source

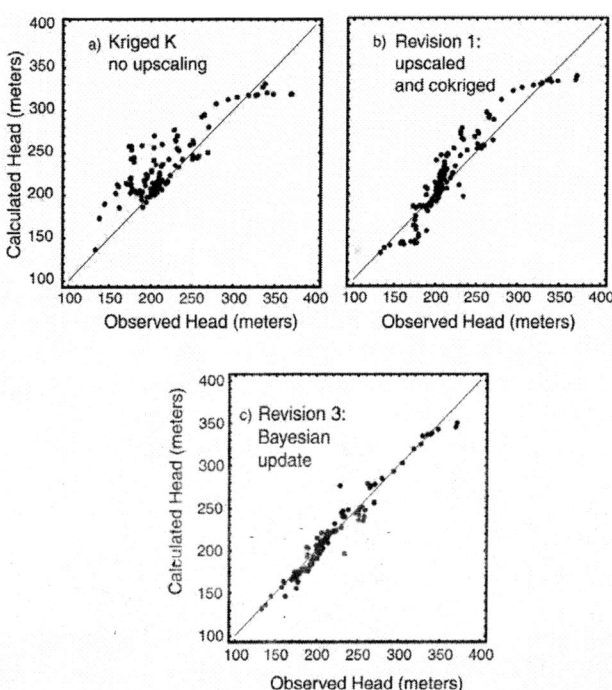

Figure 7. Computed versus observed hydraulic heads for various hydraulic conductivity models of the aquifer. After Painter, S., Woodbury, A. D. and Y. Jiang, 2006, Transmissivity Estimation for Highly Heterogeneous Aquifers: Comparison of Three Methods Applied to the Edwards Aquifer, Hydrogeology Journal, in press 2007, with kind permission of Springer Science and Business Media.

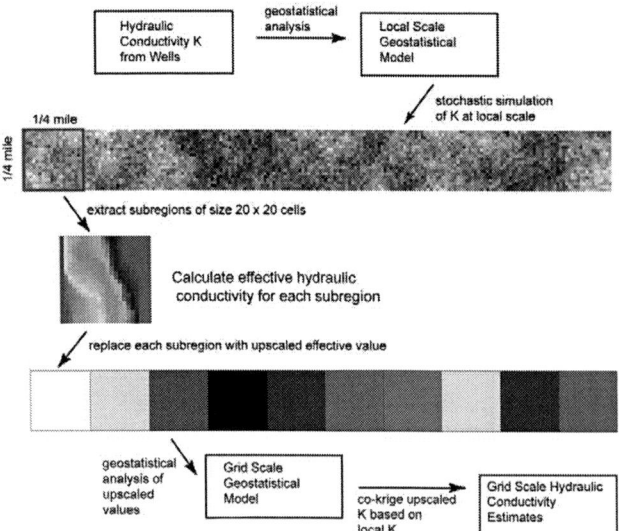

Figure 8. Illustration of the cokriging and upscaling procedure for the Edwards Aquifer. After Painter, S., Woodbury, A. D. and Y. Jiang, 2006, Transmissivity Estimation for Highly Heterogeneous Aquifers: Comparison of Three Methods Applied to the Edwards Aquifer, Hydrogeology Journal, in press 2007, with kind permission of Springer Science and Business Media.

for site characterization, which can be accomplished through the full-Bayesian methodology.

12. DISCUSSION AND CONCLUSIONS

Incorporating temperatures into a modeling effort can take many forms, and both temperatures and hydrologic data can be combined qualitatively and quantitatively. In the latter category, the least formal would be in calibration, followed by parameter estimation and finally by full-inversion. This paper is about a Bayesian approach to the solution of the joint-groundwater and thermal inverse problem. Along these lines, a list of suggested references is provided, along with suitable mentioning of benchmark papers, monographs and textbooks on the subject.

The technique described in this paper revolves around shallow, low-temperature groundwater flow systems; and that entails steady 2-D fluid and heat flow. The methodology utilizes a perturbation technique to linearize and then couple the governing equations. For the perturbation approach to work, fluid properties must be decoupled from the temperature field. Once this is done, and through the finite element method, a block-linear system of data, kernel, and model parameters is developed. The assumptions related to viscosity and density may not be universally applicable.

Two end-member and one set of joint inverse examples are presented. The two end-members are pure heat conduction (an application of Bayesian inversion to Paleoclimate reconstructions), and groundwater problem which is an example application to the Edwards Aquifer in Texas. Lastly, generic examples of combinations of transmissivity, hydraulic head and temperatures are presented.

The work by Jiang et al. (2004) is important in that it shows that the perturbation and linearized approach seems to work well over large ranges of variability in log transmissivity. Jiang et al. (2004) detail these simulations and also outline under what conditions the approach would likely break down. Painter et al. (2006) and Jiang and Woodbury (2006) also confirm and reenforce the idea that it may be possible to effectively "image" aquifers and provide the kinds of details necessary for mass transport calculations. Finally, it shows that it may also be possible to combine various levels and combinations of hydraulic and thermal data. The thermal part of the method has not been tested in a field site though, and a logical extension is to apply the technique in

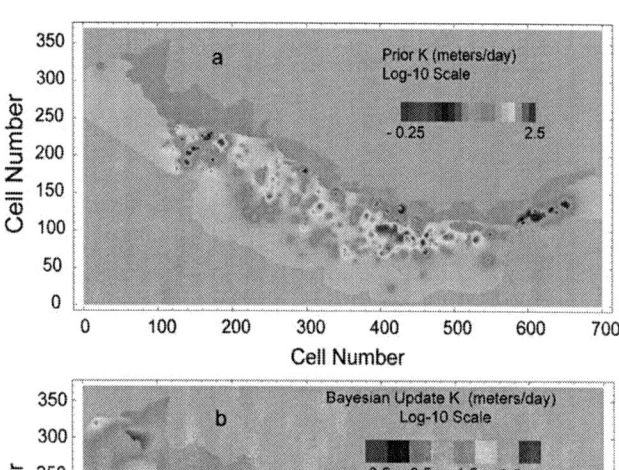

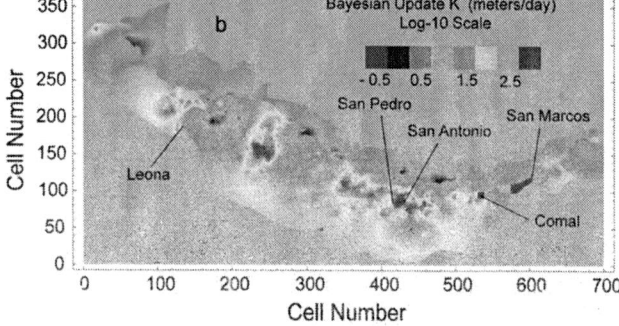

Figure 9. Expected value for hydraulic conductivity in each grid cell as produced by a upscaling and cokriging, and by Bayesian updating. The hydraulic conductivity from upscaling and cokriging was used to set the prior distribution in the Bayesian updating, as described in the text. Locations of major springs are indicated. Note that the two plots have different color scales. After Painter, S., Woodbury, A. D. and Y. Jiang, 2006, Transmissivity Estimation for Highly Heterogeneous Aquifers: Comparison of Three Methods Applied to the Edwards Aquifer, Hydrogeology Journal, in press 2007, with kind permission of Springer Science and Business Media.

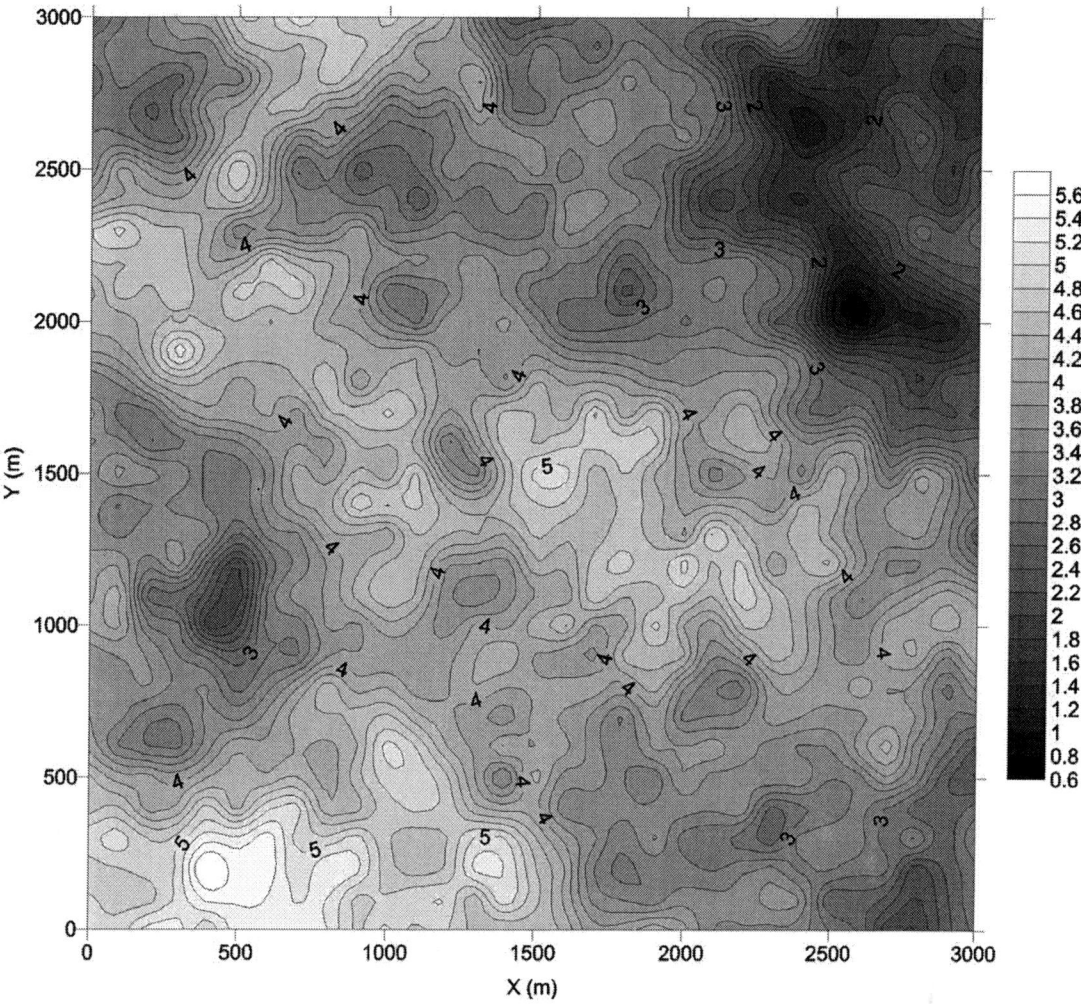

Figure 10. True ln (T) field generated by multivariate Gaussian generator. After Jiang and Woodbury, (2006)

a case-study. Other extensions are possible and those are to peruse 3-D examples, possibly with an upscaling procedure proposed by Painter et al. (2006).

As mentioned by Anderson (2005) "investigators are just starting to explore the full potential for using temperature measurements in a wide variety of hydrogeological settings. The utility of temperature measurements in estimating fluxes in ground water-stream systems is now well established". Clearly, future developments in hydrogeology will see amalgamations of techniques that depend on the acquisition of data from many different sources. Temperatures are one of those "new/old" data sets that will become commonly used, and will be seen as another tool in the practitioner's toolbox. New developments in dataloggers and data acquisition are very encouraging and can allow for long-term collection of many data points. Likely new developments in long-term battery life will allow for an order of magnitude increase in data collection. Miniaturization of devices will continue and absolute measurement accuracy will also improve. But where does this leave us? The techniques and modeling strategies must follow, and developments such as those by Bravo et al. (2002) are encouraging. These authors incorporated transient temperature data in their assessment of inflows in the Wilton, Wisconsin wetlands area. They showed that even when hydraulic head data alone are insufficient to constrain a calibration, addition of temperature and hydraulic information did in fact allow for convergence in a parameter estimation problem. This methodology should offer a wide benefit to others working in groundwater-wetland interactions. Such subjects will become increasing important as more hydrogeologists turn their attention to bio-chemical processes in Riparian and other zones. For example, a recent paper by Conant (2004) shows the utility of temperatures in delineating flows in

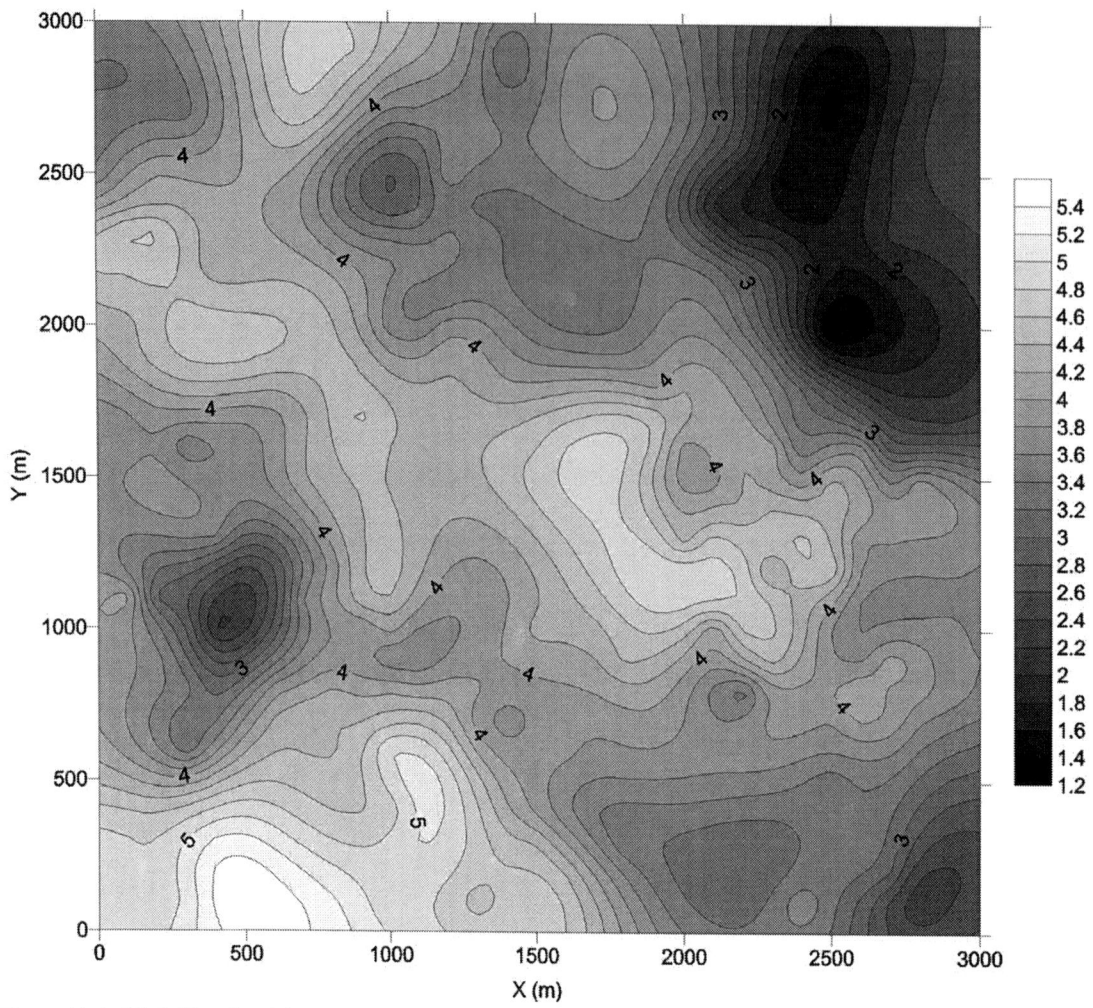

Figure 11. ln(T) field updated from 100 sample points. These include ln(T) measurements with heads and temperatures. Variance of the log-transmissivity field is 1.0. After Jiang et al., (2006)

stream hyporheric zones. Finally, the combination of hierarchical Bayes and MRE as developed by Hou and Rubin (2005) looks most promising to deal with general problems involving time-dependent data.

Acknowledgments. The author wishes to gratefully acknowledge the huge contributions to this work by former students and now valued colleagues, Grant Ferguson and Yefang (Rocky) Jiang. Various papers on inversion were also written with Scott Painter from the Southwest Research Institute and the author wishes to express his sincere appreciation to him, the Institute and to the Edwards Aquifer Authority. Randy Hunt, Mike Fienan and an anonymous reviewer all provided useful feedback on the original manuscript. Finally, the author is deeply indebted for the valuable assistance, suggestions and insight provided by Taduesz J. Ulrych at U.B.C. This research was supported from grants from the Natural Sciences and Engineering Research Council of Canada, Manitoba Hydro, and the Southwest Research Institute.

REFERENCES

Anderson, M. P., 2005, Heat as a ground water tracer,*Ground Water*, 43(6), 951–968.

Bear, J. 1972, Dynamics of fluids in porous media. Elsevier, Amsterdam-New York, 764p.

Bear, J. and M. Y. Corapcioglu, 1981, A Mathematical Model for Consolidation in a Thermoelastic Aquifer Due to Hot Water Injection or Pumping. *Water Resources Research*, 3: 723–736.

Bear, J. and M.Y . Corapcioglu (editors), 1984, Fundamentals of Transport Phenomena in Porous Media, NATO ASI series, Applied Science—No. 82, Martinus Nijhoff Publishers.

Beck, A. E., Garven, G. and L. Stegena (editors) 1989, Hydrogeological regimes and their subsurface thermal effects, *Geophysical Mongraph 47*, IUGG Volume 2, American Geophysical Union.

Beltrami, H., Gosselin, C. and J. C. Mareschal, 2003, Ground surface temperatures in Canada: Spatial and temporal variability, *Geophysical Research Letters*30(10): 1499, doi: 10.1020/2003GL017144.

Bretthorst, G. L., 1988,Bayesian Spectrum Analysis and Parameter Estimation, Lecture Notes in Statistics No. 48,J. Berger, S. Fienberg, J. Gani, K. Krickeberg, and B. Singer (eds), 209p

Bravo, H. R. , Feng, J. and R. J. Hunt, 2002, Using groundwater temperature data to constrain parameter estimation in a groundwater flow model of a wetland system, *Water Resources Research*, 38(8): 1029/2000WR000172.

Carrera, J., and S. P. Neuman 1986, Estimation of aquifer parameters under transient and steady state conditions, 1, Maximum likelihood method incorporating priori information, *Water Resources Research*, 22(2): 199–210.

Conant, B. J., 2004, Delineating and quantifying ground water discharge zones using stream bed temperatures, *Ground Water*, 42(2), 243–257.

Constantz, J., C. L. Thomas, and G. Zellweger, 1994. Influence of diurnal variations in stream temperature on streamflow loss and groundwater recharge, *Water Resources Research*, 30(12): 3253–3264.

Cooley, R. L., 1977. A method of estimation parameters and assessing reliability for models of steady state ground flow, 1, Theory and numerical properties, *Water Resources Research*, 13(2): 318–324.

Copty, N. and Y. Rubin,1995,A stochastic approach to the characterization of lithofacies from surface seismic and well data,*Water Resources Research*,31(7),1673–1686

Domenico P. A. and F. W. Schwartz, 1990, Physical and chemical hydrogeology, John Wiley and Sons, New York, 824p.

Ferguson, G. A. G. and A. D. Woodbury, 2006, Observed thermal pollution and post-development simulations of low-temperature geothermal systems in Winnipeg, Canada, *Hydrogeology Journal*, 14(7), 1206–1215.

Ferguson, G. A. G. and A. D. Woodbury, 2005A, The effects of climate variability on estimates of recharge from temperature profiles, *Ground Water*, 43(6), 837–842.

Ferguson, G. A. G. and A. D. Woodbury, 2005B, Thermal sustainability of groundwater-source cooling in Winnipeg, Manitoba, *Canadian Geotechnical Journal*, 42, 1–12.

Ginn, T. R. and J. H. Cushman,1990,Inverse methods for subsurface flow—A critical review of stochastic techniques,*Stochastic Hydrology and Hydraulics*,4(1),1–26

Hou, Z., and Y. Rubin, 2005, On minimum relative entropy concepts and prior compatibility issues in vadose zone inverse and forward modeling, *Water Resources Research*, 41, W12425, doi:10.1029/2005WR004082.

Hubbard, S. S., Rubin, Y; Majer, E.,1997, Ground-penetrating-radar-assisted saturation and permeability estimation in bimodal systems,*Water Resources Research*,33(5),971–990,

Hunt, R. J., D. P. Krabbenhoft, and M. P. Anderson, 1996, Groundwater inflow measurements in wetland systems, *Water Resources Research*, 32(3): 495–507.

Hyndman, D. W., Harris, J. M. and S. M. Gorelick,1994,Coupled seismic and tracer test inversion for aquifer property characterization*Water Resources Research*,30(7),1965–1977

Hyndman, D. W., S. M. Gorelick, 1996, Estimating lithologic and transport properties in three dimensions using seismic and tracer data: The Kesterson aquifer, *Water Resources Research*, 32(9), 2659–2670, 10.1029/96WR01269

Jaynes, E. T., 1983, Papers on Probability, Statistics and Statistical Physics,D. Redel, Dordrecht-Holland.

Jessop, A. 1990, Thermal Geophysics, Elsevier, Amsterdam, 316p.

Jiang, Y. and A. D. Woodbury, 2006, A full-Bayesian approach to the inverse problem for steady state groundwater flow and heat transport, *Geophysical Journal International*, 167 (2), 1053–1053. doi: 10.1111/ j.1365-246X.2006.03254.

Jiang, Y., Woodbury, A. D., and S. Painter, 2004, Full-Bayesian Inversion of the Edwards Aquifer, *Ground Water*, 2004, 42(5): 724–733

Kapur, J. N,.1989,Maximum Entropy Models In Science and Engineering Wiley, N.Y., 635p

Kennedy, P. L. and A. D. Woodbury, 2002, Geostatistics and Bayesian updating for transmissivity estimation in a multiaquafer system in Manitoba, Canada, *Ground Water*, 40(3), 273–283.

Kennedy, P.A. and A.D. Woodbury, 2005, Sustainability of the bedrock aquifer systems in south-central Manitoba: Implications for large-scale modeling, *Canadian Water Resources Journal*, 30(4), 281–296

Kitanidis, P. K. 1996, On the geostatistical approach to the inverse problem. *Advances in Water Resources*, 19 (6): 333–342.

Kitanidis, P. K., 1997, A reassessment of the groundwater inverse problem—Comment., *Water Resources Research*,33(9),2199–2202

Lindgren, R. J., A. R. Dutton, S. D. Hovorka, S. R. H. Worthington, and S. Painter, 2004, Conceptualization and Simulation of the Edwards Aquifer, San Antonio Region, Texas, USGS Scientific Investigation Report 2004–5277

Loredo, T. J.,1990,From Laplace to supernova SN 1987A: Bayesian Inference, Maximum Entropy And Bayesian Methods,P. F. Fougere,Kluwer Academic Publishers

Matsuoka, T., and T. J. Ulrych, 1986, Information theory measures with application to model identification. *IEEE Trans. Acoust., Speech, Signal Processing*, **ASSP-34**, 511–517.

McLaughlin, D. and L. Townley, 1996. A reassessment of the groundwater inverse problem, *Water Resources Research*, 32(5): 1131–1161.

McLaughlin, D and L. Townley, 1997, A reassessment of the groundwater inverse problem—Reply., *Water Resources Research*,33(9),2203–2204

Mitsuhata, Y.,2004,Adjustment of regularization in ill-posed linear inverse problems by the empirical Bayes approach, *Geop. Prosp.*,52,213–239

Mohammad-Djafari, A. ,1996, A full Bayesian approach for inverse problems,Maximum Entropy And Bayesian Methods, K. M. Hanson and R. N. Silver,Kluwer Academic,135–144

Mosegaard, K. and A. Tarantola, 2002, Probabilistic Approaches to Inverse Problems, International Handbook of Earthquake and Engineering Seismology, Vol 81A, 237–265.

Neuman, S.P. and S. Yakowitz, 1979, A statistical approach to the inverse problem of aquifer hydrology, 1, Theory, *Water Resources Research*, 15(4): 845–860.

Painter, S., Woodbury, A. D. and Y. Jiang, 2006, Transmissivity Estimation for Highly Heterogeneous Aquifers: Comparison of

Three Methods Applied to the Edwards Aquifer, in press, *Hydrogeology Journal*

V. Rath, A. Wolf and H. M. Bucker, 2006, Joint three-dimensional inversion of coupled groundwater flow and heat transfer based on automatic differentiation: sensitivity calculation, verification, and synthetic examples, *Geophys. J. Int.*, 167, 453–466 doi: 10.1111/j.1365-246X.2006.03074.

Rubin, Y. 2003, Applied Stochastic Hydrology, Oxford University press, 391p.

Rubin, Y. and G. Dagan,1992,Conditional estimation of solute travel time in heterogeneous formations—Impact of transmissivity measurements,*Water Resources Research*,28(4),1033–1040

Rubin, Y. and S. S. Hubbard, 2005,*Hydrogeophysics*, Water Science Technology Library, Springer, 540p.

Rubin, Y., Mavko, G. and J. Harris,1992,Mapping permeability in heterogeneous aquifers using hydrologic and seismic data,*Water Resources Research*,28(7),1809–1816

Sacchi, M. D., and T. J. Ulrych, 1995, High-resolution velocity gathers and offset space reconstruction. *Geophysics*, , 1169–1177.

Sacchi, M. D., Ulrych, T. J, and C. Walker, 1998, Interpolation and extrapolation using a high resolution discrete Fourier transform, IEEE Trans. on Signal Processing, 46, 31–38

Sambridge M. and K. Mosegaard, 2202, Monte Carlo methods in geophysical inverse problems, *Rev. Geop*, 40(3), doi:10.1029/2000RG000089

Schmidt, W. L and Gosnold, W. D and J. Enz, 2001, A decade of air-ground temperature exchange from Fargo, North Dakota, *Global and Planetary Change*, 29 (3-4), 311–325.

Smith, L. and D. S. Chapman, 1983, On the thermal effects of groundwater flow: 1. Regional scale systems. *Journal of Geophysical Research*, 88(B): 593–608.

Sauty, J. P., Gringarten, A. C., Fabris, H., Thiery, D., Menjoz, A. and P.A. Landel, 1982, Sensible Energy Storage in Aquifers 2. Field Experiments and Comparison With Theoretical Results, *Water Resources Research*, 18, 253–265.

Tarantola, A. 1987, *Inverse Problem Theory*, Elsevier, 613p.

Ulrych, T. J. and M. Sacchi, 2006, Information-Based Inversion and Processing with Applications, Volume 36 (Handbook of Geophysical Exploration: Seismic Exploration), Elsevier Science, 436p.

Ulrych, T. J., Sacchi, M. and A. D. Woodbury, 2001, A Bayes tour of inversion: A tutorial, *Geophysics*, 66(1): 55–69.

Vasco, D., Peterson, J. E. and E. L. Majer, 1996, Nonuniquenesss in traveltime tomography: ensemble inference and cluster analysis, *Geophysics*, 61(4), 1209–1227.

Wang, K , Shen, P. and A. E. Beck, 1989, A solution to the inverse problem of coupled hydrological and thermal regimes. In Hydrogeological regimes and their subsurface thermal effects, edited by Alan E. Beck, pp.107–118.

Woodbury, A. D, 1989, Bayesian updating revisited., *Mathematical Geology*, 21(3): 285–308.

Woodbury, A. D.,1997,A probabilistic fracture transport model: Application to contaminant transport in a fractured clay deposit,*Can. Geotech. J.*,34,784–798

Woodbury, A. D., 1999. Simultaneous inversion of thermal and hydrologic data: Where do we go from here, EOS Trans. AGU 80(46), F337

Woodbury, A. D.,2004, A FORTRAN program to produce minimum relative entropy distributions,*Computers & Geosciences*,30, 131–138

Woodbury, A. D., and J. L. Smith, 1985, On the thermal effects of three dimensional groundwater flow, *Journal of Geophysical Research*, Vol 90(B1), 759–767.

Woodbury, A. D., and J. L. Smith, 1988, Simultaneous inversion of temperature and hydraulic head data: 2. Application with thermal data, *Water Resources Research*, 24(3): 356–372.

Woodbury, A. D., and T. J. Ulrych,1993,Minimum relative entropy: Forward probabilistic modeling *Water Resources Research*,29(8),2847–2860

Woodbury, A. D. and T. J. Ulrych, 1998, Minimum relative entropy and probabilistic inversion in groundwater hydrology, *Stochastic Hydrology and Hydraulics*, 12: 317–358.

Woodbury, A. D., and E. A. Sudicky,1992, Inversion of the Borden tracer experiment data: Investigation of stochastic moment models, *Water Resources Research*,28(9),2387–2398

Woodbury, A. D. and Y. Rubin, 2000, A full-Bayesian approach to parameter inference from tracer travel-time moments, *Water Resources Research*, 36(1): 159–171.

Woodbury, A. D. and T. J. Ulrych, 2000, A Full-Bayesian approach to the groundwater inverse problem for steady state flow, *Water Resources Research*, 36(8): 2081–2093.

Woodbury, A. D. and G. A. G. Ferguson, 2006, Ground surface paleotemperature reconstruction by information measures and empirical Bayes, *Geophysical Research Letters*, Vol 33, DOI:10.1029/2005GL025243.

Woodbury, A. D., Smith, J. L. and W. S. Dunbar, 1987, Simultaneous inversion of temperature and hydraulic head data: 1. Theory and application using hydraulic head data, *Water Resources Research*, 23(8): 1586–1606.

Woodbury, A. D., F. W. Render and T. J. Ulrych,1995, Practical probabilistic groundwater modeling , *Ground Water*,33(4),532–538

A. D. Woodbury , Department of Civil Engineering, University of Manitoba, Winnipeg Manitoba, Canada R3T 3V5. (woodbur@cc.umanitoba.ca)

Fusion of Active and Passive Hydrologic and Geophysical Tomographic Surveys: The Future of Subsurface Characterization

Tian-Chyi Jim Yeh

Department of Hydrology and Water Resources, The University of Arizona, Tucson, Arizona, USA and Department of Resources Engineering, National Chen Kung University, Tainan, Taiwan, ROC

Cheng Haw Lee and Kuo-Chin Hsu

Department of Resources Engineering, National Cheng Kung University, Tainan, Taiwan, ROC

Yih-Chi Tan

Hydrotech Research Institute, Department of Bioenvironmental Systems Engineering, Disaster Research Center, National Taiwan University, Taipei City, 106, Taiwan, ROC

This chapter first explains the need for high-resolution imaging techniques to characterize the subsurface, and then discusses difficulties of traditional characterization approaches, followed by a presentation of recent advances in hydrologic/ geophysical characterization of the subsurface: information fusion based on active tomographic survey concepts for field scale problems. It finally concludes with examples and propositions regarding how to collect and analyze data intelligently by exploiting natural recurrent events as energy sources for basin-scale passive tomographic surveys. The development of information fusion technologies that integrate traditional point measurements and active/passive hydrogeophysical tomographic surveys, as well as advances in sensor, computing, and information technologies may ultimately advance our capability of characterizing groundwater basins to achieve resolution far beyond the feat of current science and technology.

1. INTRODUCTION

Spatial and temporal variations of subsurface processes are the rule rather than the exception. For instance, inflow (infiltration, recharge, seepage, regional inflows, etc.) and outflow (evaporation, seepage, regional outflows, etc.) are known to be sporadic and highly localized. The variability is controlled in part by the characteristics of basins, which are also heterogeneous at various scales. Currently, we lack the capability to economically obtain three-dimensional (3-D) subsurface information that portrays detailed distributions of water and related properties, as well as the variable spatial and temporal processes. Such 3-D information is necessary to improve our ability to understand and manage groundwater resources that are fundamental to the quality and viability of human life on Earth.

Existing monitoring and characterization technologies can cover only a small fraction of the subsurface, and the resultant information cannot be used to reliably evaluate current and future drought and other water-related conditions. Subsurface sciences need a breakthrough approach or "instrument" to greatly expand and deepen our ability to "see into the Earth." As its key scientific focus, this chapter will present recent successes of data fusion technologies for characterizing and monitoring the subsurface at field scales, and then present a vision and ambition to take on the challenge of developing a system for subsurface imaging at the basin scale. Here, *field scale* refers to areas of tens to hundreds of square meters, and those over hundreds to tens of thousands of square meters are considered to be *basin scale* (e.g., a groundwater basin).

2. DIFFICULTIES OF TRADITIONAL APPROACHES

Quantitative analysis and prediction of subsurface fluid flow and solute transport requires the use of appropriate mathematical models that represent subsurface processes. These models generally rely on partial differential equations (PDE) that express hydrologic, physical, and chemical principles of natural phenomena in the subsurface, extended over space and time. Hereafter, a forward problem (i.e., prediction) refers to solving PDEs for the system states in space and time, with known properties and given initial and boundary conditions. An inverse problem (i.e., characterization, parameter identification or estimation) refers to determining values of the system's properties from information about excitations to the subsurface and observations (monitoring) of responses of state variables to those excitations.

High-resolution prediction demands high-resolution information about the system's properties and initial and boundary conditions. Similarly, high-resolution inverse modeling requires detailed information about excitations to and responses of the system, as well as any pre-existing information on system properties and states. The inherent spatial variation or 3-D heterogeneity of properties at various scales (e.g., pores, lenses, strata, formations, and basins) greatly compounds the difficulties of site characterization and prediction. Traditional in-situ borehole characterization and monitoring methods [i.e., core samples, slug tests, flow meter tests, aquifer tests, multi-level samplers, wells, etc. see Domenico and Schwartz, 1990] are invasive and too costly to emplace in large numbers and significant depths throughout a basin. More critically, the "representativeness" of the properties estimated from these methods has recently been questioned by Butler [1997], Beckie and Harvey [2002], Wu et al. [2005], and others.

Similarly, traditional inverse modeling of groundwater models with distributed parameters based on sparsely observed responses over a large basin (or inverse modeling for short) fails to provide reliable information about the basin characteristics. Difficulties in collecting necessary and sufficient information that makes the inverse problem well posed are the sole cause of the failure. To understand the difficulties, we will consider the governing PDE for groundwater flow in aquifers [Bear, 1972]:

$$\nabla \cdot [K(\mathbf{x})\nabla h(\mathbf{x},t)] = S_s(\mathbf{x})\frac{\partial h(\mathbf{x},t)}{\partial t} \quad (1)$$

where $h(\mathbf{x},t)$ is the hydraulic head which is a function of the position vector, $\mathbf{x}$, and time, t; $K(\mathbf{x})$ is the spatially varying hydraulic conductivity field; $S_s(\mathbf{x})$ is the spatially varying specific storage field of the aquifer. As mentioned previously, a forward model solves the equation with known hydraulic conductivity and specific storage property fields for the hydraulic head in time and space, given initial and boundary conditions. A lack of complete information of the property fields, and initial and boundary conditions, makes the forward problem ill posed; many possible solutions exist, implying that the predictions of groundwater state are uncertain.

For inverse modeling, equation (1) can be rewritten as a corresponding inverse PDE:

$$K(\mathbf{x})\nabla^2 h(\mathbf{x},t) + \nabla K(\mathbf{x}) \cdot \nabla h(\mathbf{x},t) = S_s(\mathbf{x})\frac{\partial h(\mathbf{x},t)}{\partial t} \quad (2)$$

The unknowns in equation (2) are the hydraulic conductivity and the specific storage field, as opposed to the hydraulic head field as in equation (1). Also, notice that the aim of inverse modeling is to correctly determine these hydraulic properties. Prerequisites for a unique solution to equation (2) are: (i) the hydraulic heads everywhere in the solution domain for at least at two time levels, t and t'; and (ii) boundary K values. Then, we have a system of equations for $K(\mathbf{x})$ and $S_s(\mathbf{x})$:

$$[\nabla h(\mathbf{x},t)]\nabla K(\mathbf{x}) + [\nabla^2 h(\mathbf{x},t)]K(\mathbf{x}) = \left[\frac{\partial h(\mathbf{x},t)}{\partial t}\right]S_s(\mathbf{x})$$
$$[\nabla h(\mathbf{x},t')]\nabla K(\mathbf{x}) + [\nabla^2 h(\mathbf{x},t')]K(\mathbf{x}) = \left[\frac{\partial h(\mathbf{x},t')}{\partial t}\right]S_s(\mathbf{x}) \quad (3)$$

According to system (3), the specific storage can be estimated only if the net inflow to a volume of the medium and the head change over time at the volume are known. Therefore, estimation of S_s at a given location, $\mathbf{x}$, requires an observable temporal change in the hydraulic head at the location. These requirements are called necessary and sufficient conditions for the inversion of equation (2) [Yeh and Šimůnek, 2002]. If these conditions are specified, the inverse problem is mathematically well posed; it has a unique solution, and the aquifer can be fully characterized. Otherwise, the problem is ill posed and characterization of the aquifer is uncertain. Note that the above statements implicitly assume that Darcy's law is valid and the scales of the hydraulic head and $K(\mathbf{x})$ and $S_s(\mathbf{x})$ are consistent with the Darcian continuum assumption [Bear, 1972].

Specification of these necessary and sufficient conditions is possible in well-controlled laboratory and field experiments, but unlikely in any field-scale problem. Without fully specifying these conditions, current inverse modeling efforts of basin-scale aquifers have become so called model calibration or history matching exercises that aim at fitting limited observed system responses. History matching, however, does not assure parameter correctness, and it thereby often yields highly subjective aquifer characterizations. Because of this uncertainty in aquifer characterization, as well as our inability to determine temporally and spatially varying boundary conditions (e.g., inflow and outflow) of the aquifers, many grossly misleading predictions of groundwater flow and contaminant migration have been made. Our ability to validate a subsurface model as such has been seriously questioned [see Konikow and Bredehoeft, 1992; Oreskes et al., 1994; Bredehoeft, 2003], as has our ability to predict flow and solute migration in aquifers. Groundwater resources management virtually becomes a matter of political debate without much scientific basis.

Undoubtedly, reducing uncertainty in groundwater resources management is our ultimate goal. It is, however, beyond the scope of this chapter. Instead, we will focus our discussion on the development of a new generation of technologies that can improve or perhaps revolutionize our characterization of aquifer properties over field and basin scales. Development of these technologies is a necessary step toward our final goal of reducing uncertainty in groundwater resources management.

3. DATA FUSION FOR FIELD-SCALE PROBLEMS

3.1. Fusion of the Same Types of Information

Recently, viable alternatives to the traditional in-situ borehole characterization and inverse modeling approaches have emerged, in which data from the traditional characterization and monitoring methods are supplemented with coverage of greater density from indirect, minimally-invasive hydrologic and geophysical tomographic surveys. These tomographic surveys excite the subsurface at different locations and simultaneously monitor responses of the subsurface at a large number of other locations. These surveys thereby yield many pieces of partially "overlapped" information, which are used to constrain interpretation of data collected from each excitation. As a result, the final result is less uncertain. In fact, these tomographic surveys are analogous to CAT scan technology which produces a 3-D picture of an object that is more detailed than a standard X-ray, and which has been widely used in medical sciences to "see" into human bodies non-invasively.

To illustrate the concept and principle of the tomographic survey, consider a composite geologic medium that consists of two layers; each layer has a different hydraulic conductivity value, K_1 and K_2, and the same thickness. Suppose the hydraulic conductivity values of the two layers are the unknowns to be determined. If a steady-state flow experiment is conducted in which water flows in the direction parallel to the layering and if the boundary heads and the total flux are measured, an effective hydraulic conductivity of the composite medium can be determined. It is an arithmetic mean of an infinite number of possible pairs of K_1 and K_2 values (i.e., $K_a = 0.5 \times (K_1 + K_2)$). If the flow experiment is repeated again but allows the flow to enter perpendicular to bedding, the effective hydraulic conductivity becomes the harmonic mean of an infinite number of possible pairs of K_1 and K_2 values (i.e., $K_h = K_1 K_2 / (K_1 + K_2)$). If now we integrate or "fuse" the information from these two experiments (i.e., solve the arithmetic mean and the harmonic mean equations, simultaneously), the number of possible pairs of K_1 and K_2 values becomes only two. This rudimentary example manifests that a tomographic survey—which collects data intelligently and analyzes data smartly—indeed provides additional information for an inverse problem being better posed, and hence reduces the number of possible solutions to the problem. However, both hydrologic and geophysical tomographic surveys are subject to the same noise and measurement errors issues that any traditional hydraulic tests or geophysical surveys confront.

In the following sections, we will discuss the tomographic survey concept applied to hydrologic and geophysical characterization of the subsurface at field scales. These tomographic surveys rely on artificial stimuli (e.g., pumping or injection of water or air, injection of electric current, etc.) which can be well-characterized but have limited area coverage. In hydrology, hydraulic, pneumatic as well as tracer tomography surveys have been developed recently. Likewise, seismic, acoustic, electromagnetic (EM) and other tomography surveys have emerged in geophysics. Our discussion, however, will focus on hydraulic tomography and electrical resistivity tomography only, and then a discussion will follow regarding the strengths and weaknesses of general hydrologic and geophysical tomography.

3.1.1. Hydraulic tomography (HT). Gottlieb and Dietrich [1995]; Renshaw [1996]; Vasco et al. [2000], Yeh and Liu, [2000]; Bohling et al. [2002]; McDermott et al. [2003]; Brauchler et al., 2003; Zhu and Yeh [2005 and 2006]; and others have developed new methods for aquifer characterization, i.e., hydraulic tomography. A simple example of HT involves the installation of at least two wells in an aquifer. Using packers, each well is then partitioned into several intervals along its depth. A sequential aquifer test is subsequently undertaken. During this test, water is injected or withdrawn (a pressure excitation) at a selected interval in a given well, and pressure responses of the subsurface are then monitored at other intervals at this well and the other well(s). This test thus produces a set of pressure excitation/response data of the subsurface. Afterward, the pump is moved to another interval and the test is repeated to collect another set of data. This test is applied to all of the intervals at all of the wells. The data sets from all the tests are then processed by an inverse model to estimate the spatial distribution of hydraulic properties of the aquifer. In other words, a set of pressure excitation/response data in HT is tantamount to an image of subsurface heterogeneity due to light emitting from a given location. Repetition of the test at different intervals merely takes many of these snapshots of the heterogeneity in the aquifer from different angles and directions. Synthesizing all of the snapshots thus maps a 3-D hydraulic property distribution of the tested volume.

Using laboratory sandbox experiments and the HT algorithm by Yeh and Liu [2000], Liu et al. [2002] and Illman et al. [2006] demonstrated that steady-state HT is an effective technique for depicting an aquifer's heterogeneity with a limited number of invasive observations. Recently, Zhu and Yeh [2005] extended the analysis algorithm for steady-state HT to transient HT, and thus both hydraulic conductivity and specific storage fields of aquifers can be estimated. Since great computational resources are required for analyzing data from transient HT, Zhu and Yeh [2006] adapted a temporal moment approach [Harvey and Gorelick, 1995a; Li et. al., 2005] to expedite the analysis.

Although the capabilities of transient HT remains to be fully assessed in the field, results from sand box experiments by Liu et al. [2007] are encouraging. Not only did tomography identify the pattern of the hydraulic conductivity heterogeneity, but also the variation of specific storage values in the sandbox. More importantly, they showed that using the identified spatially varying hydraulic conductivity and specific storage fields, they can predict temporal and spatial evolutions of the drawdown induced by independent hydraulic tests. Likewise, a recent application of HT to a well field at Montalto Uffugo Scalo, Italy, produced an estimated transimssivity field that is deemed consistent with the geology of the site [Straface et al., 2006].

HT can be used to image fracture connectivity in fractured aquifers as well. Figure 1 depicts a synthetic fractured aquifer in which two slanted boreholes intercept two orthogonal fractures. The hydraulic conductivity along the two boreholes was assumed to have been measured prior to a HT survey. Five separate pumping operations were then initiated at specified locations (see Figure 1) to reach five

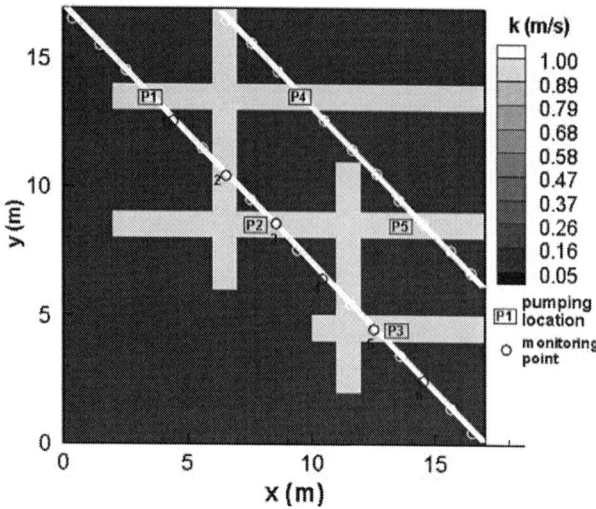

Figure 1. The orthogonal fracture pattern and location of slanted pumping wells used in the numerical experiment. The hydraulic conductivity of the fracture (white) and that of the rock matrix (black) are 1m/m and 0.05 m/m, respectively.

corresponding steady flow fields. During each flow field, pressures along the boreholes were monitored. Using these pressure data and the hydraulic conductivity measurements, the hydraulic conductivity distribution in the entire aquifer (Figure 2) was estimated with the HT algorithm by Zhu and Yeh [2005]. A comparison of Figures 1 and 2 suggests that HT is potentially a promising technology for mapping connectivity of fractures in aquifers.

3.1.2. Electrical resistivity tomography (ERT). Over the past few decades, the dc resistivity survey has been an inexpensive and widely used technique for the investigation of near-surface resistivity anomalies. It recently has become popular for the investigation of subsurface pollution problems [NRC, 2000]. The classic analysis of a resistivity survey relies on analytical formulas that assume a homogeneous earth to derive apparent resistivity. Generally speaking, the electric potential observed at a point in space is influenced by resistivity anomalies over the entire electric potential field created by a survey. In particular, resistivity anomalies near the transmitting and the receiving electrodes have greater influence. But a significant geologic anomaly anywhere within the entire electric current field can also have the same impact. Thus, the apparent resistivity can be highly misleading when derived from a potential measurement using the classical analysis. Similar findings were found in a recent study of traditional analyses of aquifer tests [Wu et al., 2005], which is analogous to the analysis of the apparent resistivity. Indeed, the conventional resistivity survey has been found virtually ineffective for environmental appli-

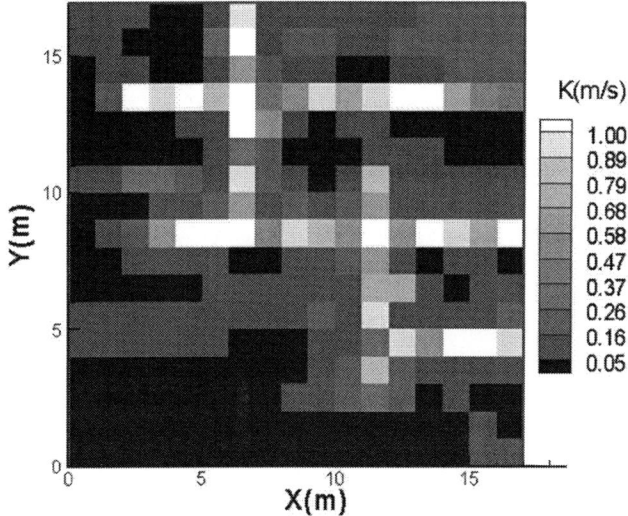

Figure 2. The detected hydraulic conductivity field reflecting fracture pattern, based on the steady hydraulic tomography.

cations, where electrical resistivity anomalies are subtle, complex, and of a multiplicity of scales.

Meanwhile, a contemporary electrical resistivity survey (i.e., ERT) has been designed to collect extensive electric potential data sets in multi-dimensions in a tomograhic survey fashion. The resistivity field is then estimated by inversion of the data sets using a model without the assumption of a homogeneous earth, and using a regularized optimization approach [e.g., Daily et al., 1992; Ellis and Oldenburg, 1994; Li and Oldenburg, 1994; and Zhang et al., 1995].

The general consensus for inverse modeling of resistivity and hydrologic property fields is that prior information about geological structure, and some point measurements of parameters to be estimated, are essential to constrain the solution to the inverse problem [Oldenburg and Li, 1999; Li and Oldenburg 2000; Kitanidis, 1995, McLaughin, and Townley, 1996].

Recently, Yeh et al. [2002] developed a geostatistically-based inverse approach for ERT that includes prior information, i.e. spatial statistics of the resistivity distribution of geologic media and point measurements of resistivity. Applications of this approach to field situations as well as laboratory and numerical experiments have proven its robustness [Yeh et al., 2006]. In particular, Englert et al. [2005] show that, when only scarce potential measurements are available, the geostatistically-based approach yields better estimates than those using the classical regularization method. Accordingly, ERT is an appealing technology for imaging subsurface electrical resistivity anomalies. The resolution of the image nevertheless depends on the design of data collection network. For example, a surface electrode array detects only anomalies near the surface; a down-hole array provides more accurate mapping of the anomalies at great depths. Higher-resolution images can only be obtained if a spatially high-resolution electric potential field is collected using a combination of densely distributed surface and down-hole arrays.

3.1.3. Strengths and weakness of hydraulic and geophysical tomographys. Geophysical tomography (e.g., ERT) generally produces subsurface images at higher resolution than hydraulic or tracer tomography. This is attributed to relative inexpensiveness of geophysical sensors compared to hydrologic sensors. Hence a greater number of geophysical sensors can be deployed to cover a given field site during a tomographic survey to collect more responses and in turn, the survey yields more detailed images. Geophysical sensors can also be easily implemented on the land surface with little invasive operation, whereas hydrologic sensors must be installed in boreholes. Such invasive borehole drilling operations prohibit any dense deployment of hydrologic sensors.

In spite of its shortcomings, hydrologic tomography has its advantages over geophysical tomography for characterization of flow and solute transport processes and properties of geologic formations. Analysis of hydrologic tomography directly yields hydrologic properties. On the other hand, analysis of geophysical surveys yields electrical resistivity or permittivity, which has to be translated into hydrologic properties via some constitutive relation. This relation is often empirical, site specific, scale-dependent, and perhaps ambiguous [Day-Lewis et al., 2005, Moysey et al., 2005, Day-Lewis, and Lane, 2004, etc.] and the translated hydrologic properties, as such, could be misleading. Spatial variability of the relation, as noticed by Yeh et al. [2002], further complicates this translation.

3.2. Fusion of Different Types of Information

Both HT and ERT are typical examples of fusion of the same type of information. They are most appealing because only a small number of invasive operations are needed to obtain a comparable resolution of other conventional characterization methods. However, neither hydrologic nor geophysical tomography alone provides perfect characterization of the subsurface. A tomographic survey merely makes the inverse problem better posed and reduces uncertainty associated with the traditional inverse modeling approaches. Taking advantage of the strength of a particular type of tomographic survey to compensate for the deficiencies of the other becomes a possible means to enhance the resolution of a tomographic survey. This thinking thus promotes fusion of different types of hydrologic information, fusion of hydrologic and geophysical information, and fusion of hydraulic and tracer tomography to enhance our subsurface characterization, as discussed below.

3.2.1. Fusion of different types of hydrologic information. For decades, hydrologists have integrated different types of hydrologic information to obtain better hydrologic characterization of the subsurface. For example, Harvey and Gorelick [1995b] estimated a hydraulic conductivity field using sparse measurements of hydraulic conductivity, heads and solute arrival time. They found that arrival time and head data yielded different estimates. Li and Yeh [1999] estimated the hydraulic conductivity field of variable saturated media conditioned on three types of measurements (i.e., pressure head, solute transport, and solute arrival time). They reported that steady state head measurements are most effective among the three types of measurements, while additional solute concentration data can enhance the estimates based on head measurements alone. Cirpka and Kitanidis [2001] used the first two temporal moments of solute data to estimate the hydraulic conductivity field. They recommended that the use of both head and tracer data could lead to better estimations of the hydraulic conductivity field.

For vadose zone problems, a study by Harter and Yeh [1996] suggested that conditioning the solution transport simulation using pressure head information improves prediction of plume migration. Yeh and Zhang [1996] reported that pressure data can benefit estimation of the saturated hydraulic conductivity field, while moisture content data enhance estimation of the pore-size distribution parameter of the unsaturated hydraulic conductivity curve of the vadose zone. Finally, the use of both pressure and moisture data can result in better characterization of the vadose zone than using either one of them alone.

Clearly, the worth of a type of data rests upon the type of property to be estimated. As an example, information of the hydraulic head gradient and specific discharge is critical to estimating hydraulic conductivity, because these data, along with Darcy's law, define the hydraulic conductivity. By the same token, tracer data are most useful for estimating chemical properties, porosity, and dispersivities. Tracer data alone are, however, less informative about the hydraulic conductivity. The reason is rather straightforward: movement of tracers is governed by the velocity field if the dispersion process is omitted. Velocity is a function of the hydraulic conductivity, but also of the hydraulic gradient and the porosity. Without knowledge of all these controlling factors, estimation of the hydraulic conductivity can be highly uncertain when based on tracer data alone.

On the other hand, propagation of a pressure excitation is a diffusion process which generally smoothes out the effects of heterogeneity (analogous to an electric potential field). The migration of tracers is mainly controlled by advection, which is highly sensitive to variation in hydraulic conductivity. Tracers are thus generally more sensitive to preferential flow paths even at small scales (not identical but similar to high-frequency EM waves, such as ground penetrating radar) than the hydraulic head. Inclusion of tracer data, therefore, can enhance the estimate of the hydraulic conductivity based on the hydraulic head information alone.

3.2.2. Fusion of hydrologic and geophysical information. Near-surface geophysics has become increasingly popular and has played an important role in groundwater investigations over the past few years [NRC, 2000, Rubin and Hubbard, 2005; Vereecken et al., 2006]. While geophysical surveys may not be suitable for mapping hydraulic properties, they are desirable tools for detecting changes in the hydrologic state of geologic media. For instance, Binely et al. [1996] demonstrated that ERT can be used to monitor the breakthrough of chloride tracers in column experiments; Kemna et al. [2002], and Singha and Gorelick [2005] used ERT to monitor the migration of a tracer plume in porous media. Day-Lewis et al. [2003, 2004] used time-lapse radar tomography to monitor tracer migration in fractured rock. Ground penetrating radar (GPR) and self potential measurements were used by Endres et al. [2000], Bevan et al. [2003], Bevan et al. [2005], and Rizzo et al. [2004] to monitor water table responses during aquifer tests; ERT and GPR have been widely used to detect movement of moisture in the vadose zone [e.g., Daily et al., 1992 and Binley et al., 2001]. Caution, however, was raised by Yeh et al. [2002] about using ERT to determine changes in moisture content in the vadose zone due to the inherent variability of the relation between moisture content and resistivity (i.e., parameters of Archie's law). Nonetheless, Liu and Yeh [2004] develop a data fusion approach to overcome this difficulty, which includes in-situ measurements of moisture content, resistivity, and parameters of Archie's law.

Success of these applications suggest that ERT, GPR, and other geophysical surveys may serve as cost-effective tools for obtaining a large number of hydrologic responses of the subsurface over large areas. Spatially dense information of hydrologic responses is a prerequisite for a better hydrologic inversion (section 3.1). To achieve a better hydrologic inversion, it is therefore a logical step to couple geophysical surveys, for the purpose of monitoring states of the subsurface, with hydrologic inversion.

This information fusion idea was demonstrated by Yeh and Šimůnek [2002] for vadose zone monitoring and characterization. Specifically, they used ERT to monitor moisture evolution in the vadose zone during infiltration events.

Electrical potentials from ERT surveys were then analyzed for the moisture content distribution. During the analysis, point measurements of moisture content by neutron probes, core samples, and others were included, as well as their prior knowledge of the spatial statistics of the moisture distribution. Inclusions of point measurements and the spatial statistics not only ensured a correct interpretation of the ERT results in terms of hydrologic and geologic contexts, but also expanded our knowledge about the true distribution of the moisture plume beyond the point measurement locations [e.g., Liu and Yeh, 2004]. As a result, this spatially-extensive moisture information makes a hydrologic inversion better posed, and the estimates of hydrologic properties approach representative values.

Better characterization of geologic media leads to a more accurate prediction of the migration of moisture and in turn, more accurate constraints for the ERT inversion during the monitoring of advancing moisture plumes. Using this iterative information fusion procedure and numerical examples, Yeh and Šimůnek [2002] demonstrated the feasibility of developing a cost-effective monitoring, characterization, and prediction protocol for the vadose zone process.

3.2.3. Fusion of hydraulic and tracer tomography. The potential of fusion of different types of tomography surveys for mapping residual DNAPL distribution was recently studied by Zhu and Yeh [2005]. Figure 3a shows the DNAPL distribution in a synthetic aquifer with four wells, and each well is partitioned into several injection or sampling ports (square and circle, respectively). A hydraulic and partitioning tracer tomography involves injection of water into the aquifer at one of the injection ports to establish a forced gradient flow field. Once a steady flow field is reached, a partitioning tracer is introduced into the aquifer at the same port. Steady flow pressure and the tracer breakthroughs are subsequently collected at the sampling ports of all wells. Afterward, the water and tracer injection operation is moved to another injection port and steady pressure and breakthroughs at all sampling ports are collected again. This operation is repeated until all the selected injection ports are used. Note that a different partitioning tracer is used for each tracer test. After the tests are completed, the pressure data collected during all the injection tests are first used to determine the hydraulic property distribution in the aquifer. This estimated hydraulic property field is subsequently used in the analysis of the partitioning tracer breakthrough data to map the distribution of the DNAPL in the aquifer. This is called hydraulic/partitioning tracer tomography (HPTT).

Figure 3b shows the estimated DNAPL field using conventional direct measurements of DNAPL from the four wells and the kriging method. Using a traditional partitioning

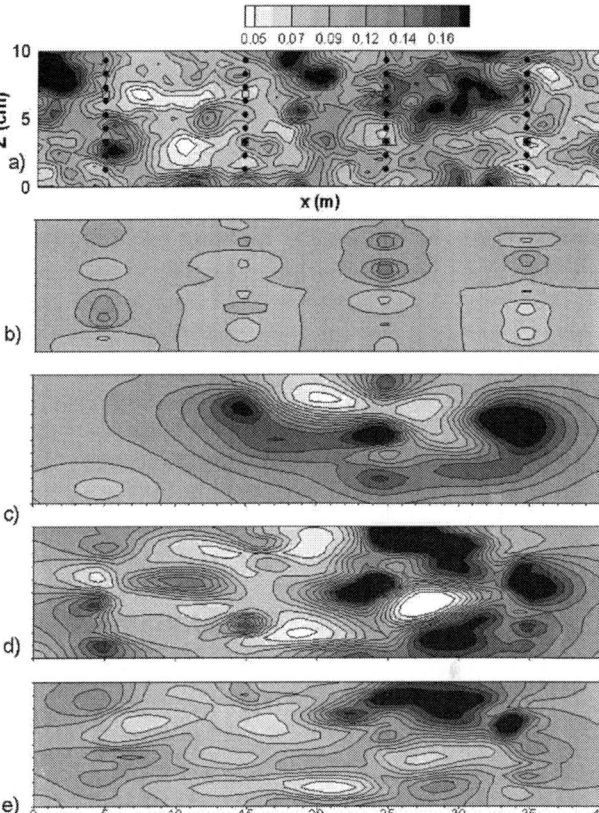

Figure 3. An illustration of the benefit of hydraulic/tracer tomography: a) a synthetic true DNAPL distribution and samples and injection ports of the hydraulic/tracer tomography survey; b) estimated DNAPL distribution based on in-situ borehole samples and geostatistics; c) the estimated distribution using the traditional single injection partitioning tracer test, without taking advantage of hydraulic head information; d) the estimated distribution based on partitioning tracer tomography alone without using the hydraulic head information; e) the estimated field using the hydraulic/tracer tomography.

tracer test (injection of water and the tracer at only one port and monitoring the breakthroughs at the other ports), and analysis of the tracer breakthroughs assuming aquifer homogeneity and without taking advantage of head information lead to an estimated DNAPL distribution shown in Figure 3c. Figure 3d illustrates an estimated DNAPL distribution, using the partitioning tracer tomography (PTT) without any knowledge of the hydraulic heterogeneity of the aquifer or taking advantage of the hydraulic head information. Lastly, the DNAPL distribution resulting from the hydraulic/tracer tomography is plotted in Figure 3e.

Among the approaches used to derive the results shown in Figures 3b, c, d and e, the direct sampling approach (Figure 3b) yields the worst estimate. It detects DNAPL near sampling locations and extrapolates the sample values to its vicinity via the correlation structure, but fails to capture high DNAPL saturation areas between observation wells. A comparison of Figures 3c and d demonstrates the benefit of tracer tomography: tomographic surveys yield many pieces of "partially-overlapped information" such that more detailed DNAPL distribution is identified. A comparison of Figures 3d and e manifests the advantage of fusion of hydraulic and tracer tomography. That is, PTT alone can lead to erroneous estimates of the DNAPL field, which is attributed to the fact the tracer data from one injection test provide only an estimate of the specific discharge (Darcian velocity) field for the given flow scenario. This field is only weakly related to the hydraulic conductivity field unless the hydraulic head field or gradient is specified. While PTT produces many sets of the estimated velocity field, each velocity estimate (in turn, each DNAPL estimate) is independent from one another. Without conditioning each estimate using the available head information during each injection, each DNAPL estimate therefore can be inconsistent with the other. Thus, the final DNAPL estimate deteriorates. A conjunctive use of HT and PTT (i.e., HPTT) is thereby a superior approach for better DNAPL characterization.

4. DATA FUSION FOR BASIN-SCALE PROBLEMS

Undoubtedly, data fusion technologies are still evolving, but results of their current applications are encouraging. Tomographic surveys, in particular, are potentially the future for field-scale as well as basin-scale subsurface characterization. In order to apply these technologies to imaging the subsurface at basin-scale, strong and spatially varying hydrologic and geophysical excitations with wide area coverage and/or significant depth penetration are needed, as are long-term and spatially distributed monitoring of signals on the land surface and in the subsurface. Naturally recurrent stimuli (e. g., lightning, earthquakes, storm events, barometric variations, etc.) with frequent and spatially varying occurrences are ideal energy sources for "illuminating" the subsurface on many occasions and throughout the basin. They thereby can provide the opportunity for an ad hoc, progressive 3-D tomographic survey. Below, we discuss some possibilities and present some numerical examples to illustrate the feasibility of exploiting naturally recurrent stimuli for a passive groundwater basin "CAT scan".

4.1. Fusion of the Same Types of Information

In this category of data fusion for basin-scale characterization, the discussion is focused on innovative approaches that take advantage of river stage fluctuations, cloud-to-ground lightning strikes, and earthquakes.

4.1.1. River-stage tomography. The example given below illustrates the potential of using river stage fluctuations as energy sources for basin-scale tomographic surveys. The influence of stage fluctuation of rivers on the groundwater table and piezometric surfaces has been recognized for decades, as has been the exploitation of the relation between the temporal fluctuation of a river stage and that of the well hydrograph as an alternative to aquifer tests [e.g., Duffy, 1978; Nevulis et al., 1989]. But the conventional analyses of the relation between river stage and well hydrograph again have relied on the assumption of aquifer homogeneity. The potential of using temporal and spatial variations in the stage of a river as an excitation source for basin-scale aquifer characterization was not recognized until the development of HT. Yeh et al. [2004] were the first to propose the use of the river stage during a flood event (and well hydrographs observed at wells adjacent to the river) to map the spatial distribution of properties of underlying aquifers. They suggested that when a flood crest is migrating downstream at any given time, it creates a set of pressure responses at wells at different distances adjacent to the river. When the crest moves to another location, it produces another set of well hydrographs at all the observation wells along the stream. This is analogous to the HT survey. Following this concept, Xiang and Yeh [2005] successfully conducted numerical experiments that ratify this river stage tomography concept for characterizing large-scale aquifers.

4.1.2. Lightning tomography. Cloud-to-ground (CG) lightning strikes are a potential energy source for basin-scale EM tomographic surveys. When lightning EM waves propagate through the subsurface, they will be modified by subsurface heterogeneity at various scales. By measuring these signals at different locations and depths (with distributed smart sensors), and then performing 3-D inverse modeling, we can estimate electrical resistivity and dielectric constant fields of the subsurface, which are indications of geologic structure, hydrologic heterogeneity and chemical distributions in the subsurface. Collecting the signals from lightning strikes at many different locations is equivalent to conducting large-scale EM tomographic surveys, if the amplitude and location of each strike is known.

The exploitation of lightning, here, is different from the conventional magnetotelluric methods (MT). The EM waves for MT arise from lightning (above ~1 Hz) and electric currents flowing in the ionosphere in huge rings around the magnetic poles (below ~1Hz). Because of its low frequency, MT has been used to explore the Earth and geologic basins at great depth, but at low resolution. The suggested lightning tomography takes advantage of the U.S. National Lightning Detection Network (NLDN) that can pin-point the location of each CG strike and provide its peak amplitude with good accuracy [Cummins et al., 1998a and b]. The lightning tomography also takes advantage of the fact that CG lightning produces extremely large EM transients (source powers of 10^9 to 10^{10} watts) [Krider et al., 1976; Krider et al., 1980; Krider, 1992] over a broad frequency range (< 1 to 10^7 Hz). The great power over a broad frequency range implies that it is possible to image the subsurface at various scales over large areas. More importantly, locations of CG strikes vary. These facts facilitate lightning tomographic surveys of groundwater basins.

4.1.3. Earthquake tomography. Earthquakes provide another type of natural stimulus source that may be valuable for both conventional seismic tomography and large-scale hydrologic tomography. Seismologists can use information from earthquakes to generate tomographic images of the subsurface [NRC, 2000]. Effects of earthquakes on groundwater levels or pressures have been investigated in the past as possible precursors for earthquakes. Few however, have explored the relation between groundwater fluctuations due to earthquakes and geologic heterogeneity, and exploited the relation for imaging 3-D hydrologic heterogeneity in a basin. A recent study by Lin et al [2004], using pore-elastic and visco-elastic models and field data during the Chi-Chi earthquakes in Taiwan during 1999, showed that the propagation of groundwater pressure waves induced by earthquakes is indeed influenced by geologic structures and hydrologic heterogeneity. They demonstrated that hydrologic properties of aquifers can be estimated using changes in groundwater levels before and after earthquakes. As a result, an earthquake of sufficient magnitude in a groundwater basin is analogous to an artificially induced excitation to an aquifer during an aquifer test, and the occurrence of successive earthquakes at different locations is similar to sequential excitations at different positions in the aquifer during HT.

Also in this context, it should be noted that storms, storm tides, tidal waves and other types of naturally occurring aquifer loadings [e.g., DeWiest, 1965] take place frequently at different locations in groundwater basins. For each event, groundwater levels respond differently and spatially. Again, information about such disturbances and associated responses of aquifers in a basin is equivalent to a set of data collected during a large-scale hydrologic test. Numerous occurrences of these disturbances originating at different locations, along with corresponding responses of the aquifers, thus constitute naturally occurring HT.

A well hydrograph, of course, can be influenced by a variety of factors (i.e., earth tides, external loadings, barometric pressure variations, precipitation, even by a passing train). Influences of each source generally bear the source signature and characteristics (i.e., frequencies and amplitudes). The different components can be sorted out from a hydrograph if the source characteristics are known. Despite these complications, water level fluctuations caused by pumping, barometric pressure variation, earth tides, etc. have been widely used to estimate aquifer properties in the past [e.g., Nevulis et al., 1989; Ritzi et al., 1991; Desbarats et al., 1999]. The basic principle of these basin-scale tomographic surveys is identical to traditional aquifer model calibrations. But the tomographic surveys collect and analyze data intelligently, and expand our traditional approaches to a new level. In fact, seismologists for decades have been using this rather intuitive concept to pinpoint the earthquake epicenters.

4.2 Fusion of Different Types of Information at Different Scales

Mapping basin-scale hydrogeologic structures is the main objective of exploiting natural stimuli for basin-scale tomography. A basin-scale tomographic survey using only one type of natural stimuli will not be likely to yield a high-resolution map of the structures, as well as meter-scale features in the subsurface. Integrating surveys that use different types of natural stimuli are needed, as is integration of field-scale tomography surveys that utilize different artificial stimuli and point measurements of different types of information. Adaptive fusion of these different types and scales of information thereby can provide the opportunity for an ad hoc, progressive 3-D tomographic survey of a basin. As a result, we can "see" into a basin at a resolution that is beyond current technologies.

5. CONCLUSIONS

Mapping the subsurface using naturally recurring stimuli as basin-scale hydrogeophysical tomographic surveys is an unexplored science and technology. Despite great potential in this new approach, a number of barriers exist. Such barriers, methodological in nature, include a lack of: 1) effective and robust stochastic approaches for fusion of different types of information at various scales, for data screening and discrimination, and for providing the best unbiased estimate and associated uncertainty; 2) efficient computational capability (e.g., data/knowledge-driven adaptive parallel computing technology for processing the massive quantity of information); 3) smart sensor networks, which are driven by results of the stochastic information fusion and data, to collect appropriate types of data at the right time, place, and frequency to minimize the likelihood of information overload. Of course, these are technological challenges. But it is our belief that rapid advances in electronics, sensor technologies, information technology, computer engineering, and smart parallel network computing technologies will ultimately realize this innovative concept and idea. Seeing into groundwater basins at high resolutions will ultimately become possible.

Acknowledgement. The work reported was supported by NSF grant EAR-0229717, NSF IIS-0431079, and a SERDP grant subcontracted through University of Iowa. The first author acknowledges support from Department of Resources Engineering, National Chen Kung University, Tainan, Taiwan, ROC. Our gratitude is also extended to Martha P. L. Whitaker for editing. We are grateful for useful and constructive comments from two anonymous reviewers and Frederick D. Day-Lewis.

REFERENCES

Bear, J., (1972), Dynamics of fluids in porous media. Dover, NY.

Beckie, R., and C. F. Harvey (2002), What does a slug test measure: An investigation of instrument response and the effects of heterogeneity, *Water Resour. Res., 38*(12), 1290, doi:10.1029/2001WR001072.

Bevan, M. J., A. L. Endresb, D. L. Rudolphb, G. Parkinc, (2005), A field scale study of pumping-induced drainage and recovery in an unconfined aquifer, *Journal of Hydrology*, 315, 52–70.

Bevan, M. J., A.L. Endres, D. L. Rudolph, G. Parkin (2003), The noninvasive characterization of pumping induced dewatering using ground penetrating radar. *Journal of Hydrology*, 281, 55–69.

Binley, A., P. Winship, R. Middleton, M. Pokar, J. West (2001), High-resolution characterization of vadose zone dynamics using cross-borehole radar, *Water Resour. Res.*, 37(11), doi:10.1029/2000WR 000089.

Binley, A., S. Henry-Poulter, B. Shaw (1996), Examination of solute transport in an undisturbed soil column using electrical resistance tomography, *Water Resour. Res.*, 32(4), 763–770, 10.1029/95WR 02995.

Bohling, G. C., X. Zhan, J. J. Butler Jr., and L. Zheng (2002), Steady shape analysis of tomographic pumping tests for characterization of aquifer heterogeneities, *Water Resour. Res.*, 38(12), 1324, doi:10.1029/2001WR001176.

Brauchler, R., R. Liedl, and P. Dietrich (2003), A travel time based hydraulic tomographic approach, *Water Resour. Res.*, 39(12), 1370, doi:10.1029/2003WR002262.

Bredehoeft, J. D., From models to performance assessment: the conceptualization problem (2003), *Ground Water* 41.5, p571(7).

Butler, J. J. Jr., (1997), The Design, Performance, and Analysis of Slug Tests, Lewis Pub., 252 pp.

Cirpka, O. A., and P. K. Kitanidis (2001), Sensitivity of temporal moments calculated by the adjoint-state method and joint inversing of head and tracer data, *Adv. Water Resour.* 24(1), 89–103.

Cummins, K. L., E. P. Krider, and M. D. Malone (1998a), The U. S. National Lightning Detection Network and applications of cloud-to-ground lightning data by electric power utilities, IEEE Trans. on EMC, 40(4), 465–480, November.

Cummins, K. L., M. J. Murphy, E. A. Bardo, W. L.Hiscox, R. P. Pyle, and A. E. Pifer (1998b), A combined TOA/MDF technology upgrade of the U.S. National Lightning Detection Network, J. Geophys. Res., 103, 9035–9044.

Daily, W., A. Ramirez, D. LaBrecque and J. Nitao (1992), Electrical resistivity tomography of vadose water movement, *Water Resour. Res.*, 28(5), 1429–1442.

Day-Lewis, F. D., and J. W. Lane, Jr., 2004, Assessing the Resolution-Dependent Utility of Tomograms for Geostatistics, *Geophysical Research Letters*, Vol. 31, L07503, doi:10.1029/2004GL019617, 4p.

Day-Lewis, F. D., J. W. Lane, Jr., J. M. Harris, and S. M. Gorelick, (2003), Time-Lapse Imaging of Saline Tracer Tests Using Cross-Borehole Radar Tomography, *Water Resources Research*, Vol. 39, No. 10, 14 p.,1290, doi:10.1029/2002WR001722.

Day-Lewis, F. D., K. Singha, and A. M. Binley (2005), Applying Petrophysical Models to Radar Traveltime and Electrical Resistivity Tomograms: Resolution-Dependent Limitations, *J. of Geophysical Research*, Vol. 110, B08206, doi:10.1029/2004JB005369, 17p.

Day-Lewis, F. D., J. W. Lane, Jr., and S. M. Gorelick, (2004), Combined Interpretation of Radar, Hydraulic and Tracer Data from a Fractured-Rock Aquifer, *Hydrogeology Journal*, doi: 10.1007/s10040-004-0372-y, Vol. 14, No. 1-2, 1–14.

Desbarats, A. J., D. R. Boyle, M. Stapinsky, and M. J. L. Robin (1999), A dual-porosity model for water level response to atmospheric loading in wells tapping fractured rock aquifers, *Water Resour. Res.*, 35(5), 1495–1506.

DeWiest, R. J. M., (1965), Geohydrology, John Wiley & Sons, Inc., New York.

Domenico, P. A. and F. W. Schwartz (1990), Physical and chemical hydrogeology, Joh Wiley and Son, pp824.

Duffy, C. (1978), Recharge and groundwater conditions in the western region of the Roswell Basin, New Mexico *WRRI. Rept. No-100.*

Ellis, R. G., and S. W. Oldenburg (1994), The pole-pole 3-D dc resistivity inverse problem: a conjugate gradient approach, *Geophysical Journal International* 119,187–194.

Endres, A. L., W. P. Clement, D. L., Rudolph, D. L. (2000), Ground penetrating radar imaging of an aquifer during a pumping test. *Ground Water*, 38, 566–576.

Englert, A., J. Zhu, A. Kemna, J. Vanderborght, H. Vereecken, T.-C. J. Yeh (2005), Potential of electrical resistivity tomography characterizing transport processes in groundwater—synthetic case studies, AGU, H37.

Gottlieb, J., and P. Dietrich (1995), Identification of the permeability distribution in soil by hydraulic tomography, *Inverse Probl.*, 11, 353–360.

Harter, T. and T.-C. J. Yeh (1996) Conditional stochastic analysis of solute transport in heterogeneous, variably saturated soils, *Water Resour. Res.*, 32 (6).

Harvey, C. F. and S. M. Gorelick (1995 b), Mapping hydraulic conductivity: sequential conditioning with measurements of solute arrival time, hydraulic head, and local conductivity, *Water Resour. Res.*, 31(7), 1615–1626.

Harvey, C. F., and S. M. Gorelick (1995 a), Temporal moment-generating equations: Modeling transport and mass transfer in heterogeneous aquifers, *Water Resour. Res.*, 31(8), 1895–1912, 10.1029/95WR01231.

Illman, W. A., X. Liu, and A. Craig (2006), Steady-state hydraulic tomography in a laboratory aquifer with deterministic heterogeneity: Multi-method and multiscale validation of hydraulic conductivity tomograms, submitted manuscript.

Kemna, A., J. Vanderborght, B. Kulessa and H. Vereecken (2002), Imaging and characterization of subsurface solute transport using electrical resistivity tomography (ERT) and equivalent transport models, *J. of Hydrology* 267, 125–146.

Kitanidis, P. K. (1995), Quasi-linear geostatistical theory for inversing, *Water Resour. Res.*, 31(10), 2411–2420.

Konikow, L., and J. D. Bredehoeft (1992), Groundwater models cannot be validated. *Advances in Water Resources*, 15, 75–83.

Krider, E. P. (1992), On the electromagnetic fields, Poynting vector, and peak power radiated by lightning return strokes, *J. Geophys. Res.*, 97 (D14), 15,913–917.

Krider, E. P., R. C. Noggle, A. E. Pifer, and D. L. Vance (1980), Lightning direction-finding systems for forest fire detection, *Bull. Amer. Meteorol. Soc.*, 61 (9), 980–986.

Krider, E. P., R. C. Noggle, and M. A. Uman (1976), A gated, wideband magnetic direction finder for lightning return strokes, *J. Appl. Meteorol.*, 15 (3), 301–306.

Li W., W. Nowak, O. A. Cirpka (2005), Geostatistical inverse modeling of transient pumping tests using temporal moments of drawdown, *Water Resour. Res.*, 41, W08403, doi:10.1029/2004WR003874.

Li, B., and T.-C. J. Yeh (1999), Cokriging estimation of the conductivity field under variably saturated flow conditions, *Water Resour. Res.*, 35(12), 3663–3674.

Li, Y., and D. W. Oldenburg (1994), Inversion of 3D dc-resistivity data using an approximate inverse mapping, *Geophysical Journal International* 116, 527–537.

Li, Y., and D. W. Oldenburg (2000), Incorporating geological dip information into geophysical inversions, *Geophysics*, 65(1), 148–157.

Lin, Y. B., Y-C Tan, T-C J. Yeh, C-W Liu, C-H Chen, (2004), A viscoelastic model for groundwater level changes in the Cho-Shui River alluvial fan after the Chi-Chi earthquake in Taiwan, *Water Resour. Res.* 40 (4): W04213.

Liu, S., T. -C. J. Yeh, and R. Gardiner (2002), Effectiveness of hydraulic tomography: sandbox experiments. *Water Resour. Res.* 38(4): 10.1029/2001WR000338.

Liu, S.-Y, and T.-C.J. Yeh (2004), An integrative approach for monitoring water movement in the vadose zone, *Vadose Zone Journal, Vol.* 3(2), p 681–692.

Liu, X., W. A. Illman, A. J. Craig, J. Zhu and T.-C. J. Yeh (2007), Multi-method and multiscale validation of transient hydraulic tomography, *Water Resour. Res.* to appear.

McDermott CI, M., Sauter, and R. Liedl (2003), New experimental techniques for pneumatic tomographical determination of the flow and transport parameters of highly fractured porous rock samples, *Journal of Hydrology*, 278 (1–4): 51–63.

McLaughlin, D. and L. R. Townley (1996), A reassessment of the groundwater inverse problem, *Water Resour. Res.*, 32(5), 1131–1161.

Moysey, S., Singha K., and Knight, R. (2005), Inferring field-scale rock physics relations through numerical simulation. *Geophysical Research Letters*, Vol. 32, L08304, doi:10.1029/2004GL022152.

National Research Council (NRC) (2000), Seeing into the earth: noninvasive characterization of the shallow subsurface for environmental and engineering application, Board on Earth Sciences and Resources, Water Science and Technology Board, Commission on Geosciencce, Environment, and Resources, National Academy Press, Washington D.C.

Nevulis, R. H., D. R. Davis, and S. Sorooshian (1989), Analysis of Natural Ground-water Level Variations for Hydrogeologic Conceptualization, Hanford Site, Washington, *Water Resour. Res.*, 25(7): 1519–1529.

Oldenburg, D. W. and Y. Li (1999), Estimating depth of investigation in dc resistivity and Ip surveys, *Geophysics*, 64(2), 403–416.

Oreskes, N., K. Shrader-Frechette, and K. Belitz (1994), Verification, validation, and confirmation of numerical models in the earth sciences. Science 263: 641–646.

Renshaw, C. E. (1996), Estimation of fracture zone geometry from steady-state hydraulic head data using iterative sequential cokriging, *Geophysical Research Letters* 23(19): 2685–2688 SEP 15.

Ritzi, R. W., S. Sorooshian, P. A. Hsieh (1991), The estimation of fluid flow properties from the response of water levels in wells to the combined atmospheric and Earth tide forces, *Water Resour. Res.*, 27(5), 883–893, 10.1029/91WR00070.

Rizzo, E., B. Suski, and A. Revil (2004), Self-potential signals associated with pumping tests experiments, *J. Geophys. Res.*, 109, B10203, doi:10.1029/2004JB003049.

Rubin, Y. and S. Hubbard (2005), *Hydrogeophysics*, 523 p., Springer, The Netherlands.

Singha K., S. M. Gorelick (2005), Saline tracer visualized with three-dimensional electrical resistivity tomography: Field-scale spatial moment analysis, *Water Resour. Res.*, 41, W05023, doi:10.1029/2004WR003460.

Straface, S., T.-C. J. Yeh, J. Zhu, and S. Troisi (2006), Sequential Aquifer Tests at a Well Field, Montalto Uffugo Scalo, Italy, *Water Resour. Res.*, in review.

Vasco, D. W, H. Keers, and K. Karasaki (2000), Estimation of reservoir properties using transient pressure data: An asymptotic approach, *Water Resour. Res.*, 36 (12): 3447–3465.

Vereecken, H., A. Binley, G. Cassiani, (2006), Applied Hydrogeophysics, ISBN: 1402049110, Springer, 396 pages.

Wu, C.-M., T.-C. J. Yeh, J. Zhu, T. H. Lee, N.-S. Hsu, C.-H. Chen, and A. F. Sancho (2005), Traditional analysis of aquifer tests: Comparing apples to oranges?, *Water Resour. Res.*, 41, W09402, doi:10.1029/2004WR003717.

Xiang, J., and T.-C. J. Yeh, (2005), Numerical Simulation of River Stage Tomography Numerical Simulation of River Stage Tomography, H41E-0463, *EOS Abstracts*.

Yeh, T.-C. J. and J. Šimůnek (2002), Stochastic fusion of information for characterizing and monitoring the vadose zone, *Vadose Zone Journal,* Vol. 1, p 207–221.

Yeh, T.-C. J., J. Zhu, A. Englert, A. Guzman, and S. Flaherty (2006), A Successive Linear Estimator for Electrical Resistivity Tomography, *Hydrogeophysics,* edited by H. Vereecken.

Yeh, T.-C. J., and S. Liu (2000), Hydraulic tomography: Development of a new aquifer test method, *Water Resour. Res., 36*(8), 2095–2105.

Yeh, T.-C. J., J. Zhang (1996), A geostatistical inverse method for variably saturated flow in the vadose zone, *Water Resour. Res.,* 32(9), 2757–2766, 10.1029/96 WR01497.

Yeh, T.-C. J., S. Liu, R. J. Glass, K. Baker, J.R. Brainard, D. Alumbaugh, and D. LaBrecque (2002), A geostatistically based inverse model for electrical resistivity surveys and its applications to vadose zone hydrology, *Water Resour. Res., 38*(12), 1278, doi: 10.1029/2001WR001204.

Yeh, T.-C. J. Hsu, K., Lee, C., J. Wen, C, Ting (2004). On the Possibility of Using River Stage Tomography to Characterize the Aquifer Properties of the Choshuishi Alluvial Fan, Taiwan, http://adsabs.harvard.edu/abs/2004AGUFM.H11D0329Y.

Zhang, J., R. L. Mackie, and T. Madden (1995), 3-D resistivity forward modeling and inversion using conjugate gradients. *Geophysics,* 60, 1313–1325.

Zhu, J., and T.-C. J. Yeh (2005), Characterization of aquifer heterogeneity using transient hydraulic tomography, *Water Resour. Res.,* 41, W07028, doi:10.1029/2004WR003790.

Zhu, J., and T-C J. Yeh (2006), Analysis of hydraulic tomography using temporal moments of drawdown-recovery data, *Water Resour. Res.,* 42, W02403, doi:10.1029/2005WR004309.

Evaluating Temporal and Spatial Variations in Recharge and Streamflow Using the Integrated Landscape Hydrology Model (ILHM)

David W. Hyndman, Anthony D. Kendall, and Nicklaus R.H. Welty[1]

Department of Geological Sciences, Michigan State University, Michigan, USA

Projections of climate and land use changes suggest that there will be significant alterations to the hydrology of the Upper Midwest. Forecasting those changes at regional scales requires new modeling tools that take advantage of increases in computational power and the latest GIS and remote-sensing datasets. Because of the need to resolve fine-scale processes, fully coupled numerical simulations of regional watersheds are still prohibitive. Although semi-distributed lumped-parameter models are an alternative, they are often not able to accurately forecast across a broad range of hydrologic conditions such as those associated with climate and land use changes.

We have developed a loosely coupled suite of hydrologic codes called the Integrated Landscape Hydrology Model (ILHM), which combines readily available numerical and energy- and mass-balance modeling codes with novel routines. In this paper, the ILHM is used to predict hydrologic fluxes through a 130 km^2 portion of the Muskegon River Watershed in northern-lower Michigan. We combine GIS maps of the land cover, soils, and sediments with a variety of gaged and remotely sensed data for this watershed to simulate evapotranspiration, groundwater recharge, and stream discharge from 1990–2004. These estimates are compared to measured stream discharge data to demonstrate the capability of the ILHM to provide reasonable predictions of groundwater recharge with minimal calibration. The results begin to illustrate critical differences in hydrologic processes due to land cover and climate variability, including a demonstration that approximately 75% of precipitation becomes recharge during leaf-off periods while almost no recharge occurs during the growing season.

INTRODUCTION

Land use and climate changes are expected to alter the spatial and temporal distribution of groundwater recharge over the next century [*Bourari et al.*, 1999; *Houghton et al.*; *IPCC, 2001*]. In the humid midwest, these changes could have far reaching consequences because recharge maintains groundwater supplies that are used as primary drinking water sources, and is critical to stream ecosystem health as groundwater is the main source of streamflow during dry periods. Despite the clear importance of groundwater recharge, its spatial and temporal distribution is generally poorly understood in humid regions. Many hydrologic modeling studies ignore both spatial and temporal variations in recharge rates, either because limited measurements of critical parameters

---

[1] Now at ARCADIS

Subsurface Hydrology: Data Integration for Properties and Processes
Geophysical Monograph Series 171
Copyright 2007 by the American Geophysical Union.
10.1029/171GM11

are available, or because existing modeling methods are not adequate to accurately evaluate these variations at the scales of interest. Integration of available hydrologic and landscape data can help improve estimates of historic recharge rates, and can then provide the basis for evaluating the range of impacts of anthropogenic alterations of the landscape and climate on future hydrological and ecological conditions.

A range of approaches have been developed to estimate recharge rates based on relatively simple analysis of flows and levels in surface water, the unsaturated zone, and the saturated zone, as reviewed by Scanlon et al. [2002]. These methods include analyzing baseflows or tracer concentrations, developing estimates based on changes in groundwater levels (reviewed by Healy and Cook [2002]), and evaluating recharge through the unsaturated zone with lysimeters or well-instrumented field sites. A variety of empirical models have also been developed to estimate recharge across a range of scales (e.g., Bogena et al., [2005]), which can provide estimates with varying degrees of reliability and spatial extent depending on the types, quality, and density of the input data [Scanlon et al., 2002].

Numerical models provide a powerful framework to integrate different data types for recharge estimation. Such models can be broadly categorized as lumped parameter models or process-based models. Lumped-parameter semi-distributed models, such as SWAT [Arnold et al., 1993] and TOPMODEL [Bevan and Kirkby, 1979] have parameters that can be adjusted to fit measurements but can not necessarily be independently measured. As a result, such models tend to have difficulty predicting flow in a new system without independent calibration, or projecting likely changes in a currently modeled system due to changes in factors including climate and land cover.

Process-based codes such as MODFLOW for groundwater flow are based on fully distributed parameters such as hydraulic conductivity, which can be independently measured based on laboratory analyses (e.g., Zhao et al. [2005]) or using field evaluations such as pump or slug tests. Unfortunately most groundwater codes are not designed to estimate recharge rates because they do not incorporate important landscape and unsaturated zone processes that are critical to redistribution of precipitation from the soil surface and the vegetation canopy.

To address this limitation, several codes have previously been developed to link MODFLOW or other groundwater codes to landscape or watershed codes that incorporate aspects of the hydrologic cycle beyond groundwater flow. For example, MODFLOW has been linked with SWAT [Sophocleous et al., 2000] and HSPF [Said et al., 2005], which are both lumped-parameter codes. A new Variably Saturated Flow (VSF) package was developed as a MODFLOW module [Thoms et al., 2006] to add unsaturated zone and overland flow processes to groundwater flow simulations, but the data and computational requirements appear to be too great for large watershed simulations based on our analysis of this code for the watershed presented in this paper.

A variety of process-based models have also been developed to simulate fully coupled surface water, and variably saturated subsurface flow. Such codes, which include SUTRA3D [Voss and Provost, 2002], Mike-SHE [DHI, 1993], WASH123D [Yeh and Huang, 2003], MODHMS [HydroGeoLogic Inc., 2003; Panday and Huyakorn, 2004], and InHM [VanderKwaak, 1999; VanderKwaak and Loague, 2001], provide powerful tools to examine complex interactions between flow and transport across the range of natural conditions observed in the surface and subsurface. Unfortunately, the data requirements and significant computational demands have generally limited the use of these codes to simulate flows through fairly small domains.

In this paper, we present a new Integrated Landscape Hydrology Model (ILHM) to integrate widely available hydrologic and landscape data in a synergistic and computationally efficient manner to assess temporal and spatial changes in important hydrologic processes. Since the focus of this monograph is data integration in hydrology, we begin by describing the watershed that we chose for testing and development of the code along with the available hydrologic and landscape data used in this simulation. This is followed by a detailed description of the model development and results.

METHODS

Study Region: Cedar Creek Watershed

The Cedar Creek Watershed, in southwestern Michigan (Figure 1), was chosen as a site to test the ILHM because it is one of our main field sites in an ongoing ecohydrological monitoring and modeling study. Cedar Creek flows through the lower half of the Muskegon River watershed (7,052 km^2), where urbanization of previously agricultural and forested landscapes is projected to increase runoff volumes and the associated solute transport over the next 35 years based on an empirical model [Tang et al., 2005]. The spatial distribution of land uses within the Cedar Creek Watershed facilitates evaluation of differences in recharge associated with land cover types because the upstream portion of this area is dominated by agriculture while the downstream portion is predominantly forested (Figure 2a). The quaternary geology ranges from medium and coarse-textured glacial tills that drape the northern watershed, to glacial outwash and lacustrine sand and gravel in the central and southern watershed (Figure 2b).

The groundwater source area, which we call a groundwatershed, of Cedar Creek was delineated using a two-layer ground-

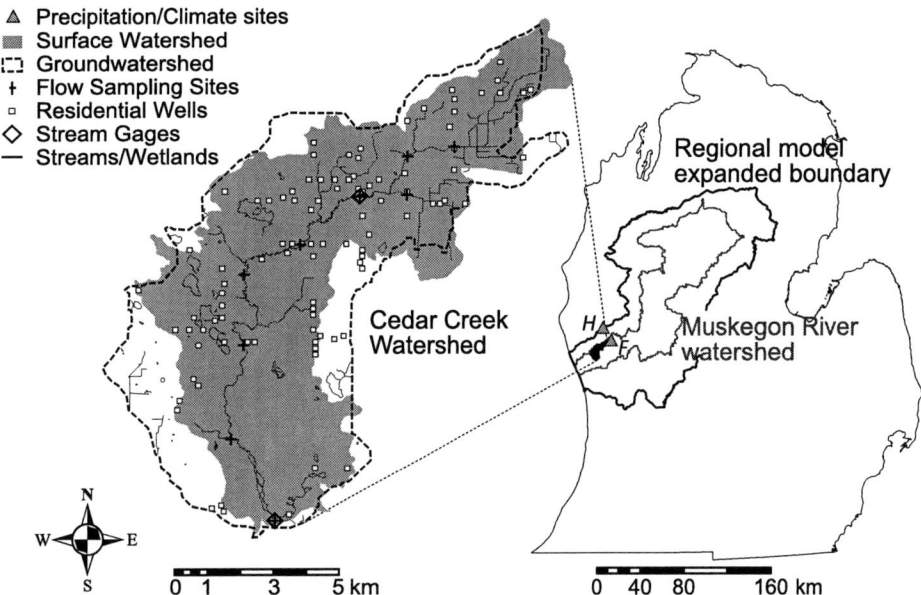

Figure 1. On the left is a map for the Cedar Creek watershed (shaded) along with the groundwater contributing area to Cedar Creek (dashed outline). Also displayed on this map are locations of two stream gages, nine discharge measurement cross sections, and residential drinking water wells located within the watershed. On the right, a map of the lower peninsula of the state of Michigan shows the Cedar Creek watershed within the greater Muskegon River watershed. The boundary of a regional groundwater model of the Muskegon River is shown in bold on the state map. The precipitation and climate gage locations are Hesperia (H), and Fremont (F).

water model of the region encompassing the Muskegon River Watershed (Figure 1). The groundwatershed (~130 km^2) was used in addition to the surface watershed (~100 km^2) for this study because regional modeling of the Grand Traverse Bay Watershed in Michigan by *Boutt et al.* [*2001*] indicated that surface- and groundwatersheds can differ significantly. The regional Muskegon River groundwater model was developed by expanding the watershed boundaries to significant hydrologic features (i.e., the next large stream or lake beyond the surface watershed) to avoid this issue at regional scales (Figure 1). The groundwatershed boundary does fluctuate somewhat with both seasonal and long term climatic variations, but for simplicity in this study we have defined the groundwatershed using the steady-state model.

DATA COLLECTION AND ANALYSIS

Before constructing the groundwater model for the expanded Muskegon River Watershed (MRW) region to define the Cedar Creek groundwatershed, we assembled the available landscape, hydrology, and climate data for the region into a geodatabase. In many parts of the world, the types of data used for this analysis are commonly available as a free download from internet sites. However, supplementary data such as flows and water levels beyond those available from the US Geological Survey will often need to be collected for model calibration or optimization.

Hydrologic Data

Two pressure transducers installed in Cedar Creek recorded stream stage at hourly to sub-hourly intervals [*Wiley and Richards, unpublished data*] from mid to late 2002 through 2004. These surface water levels provide critical information for this study. One transducer was installed in the northern, agricultural portion of the watershed, while the other was installed in forested land near the watershed outlet (See Figure 1). Stream discharge measurements [*Wiley and Richards, unpublished data*] were used to construct rating curves between stage and discharge. The stage discharge relationships were developed between measured streamflows and concurrent water levels from the transducers. For the upper watershed site 19 stage discharge pairs were used, while 23 pairs were used for the lower watershed site.

Groundwater levels for this region were collected from the Michigan Department of Environmental Quality (MDEQ) residential well database, as no monitoring well data are available in this region except at our surface water-groundwater interaction site adjacent to Cedar Creek. Unfortunately, the wells at this site are too close to the stream to provide useful

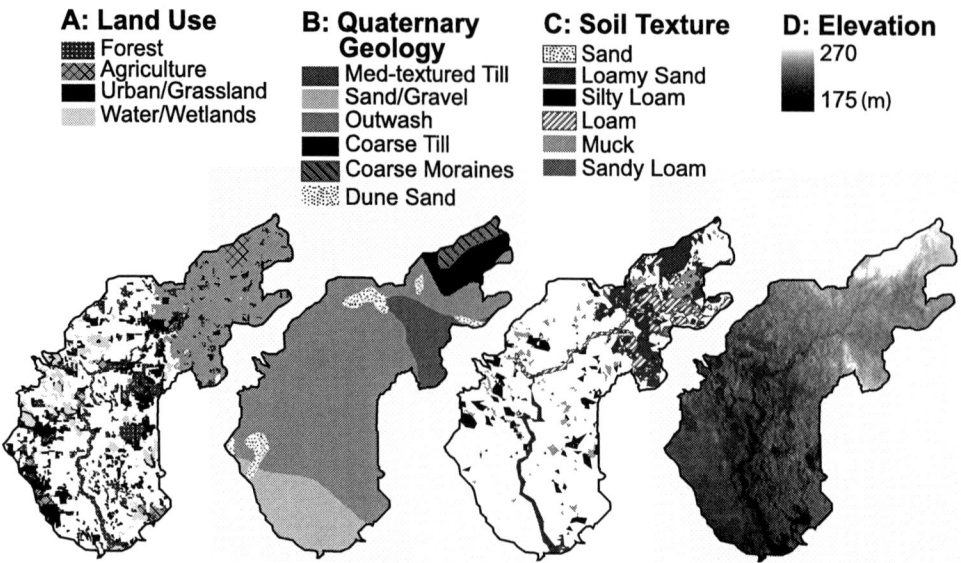

Figure 2. Static GIS datasets that were integrated into the ILHM framework. The datasets are a) land use/cover from IFMAP, b) quaternary geology from *Farrand and Bell* [*1982*], c) SSURGO soil textures, and d) land surface elevation from the NED 26.5 m DEM.

information about groundwater levels for this watershed-scale model testing. Observations were available from 99 wells installed across the watershed during the simulation period in the MDEQ database. For each well, one static water level measurement was taken by the well driller at the time of installation. The static water level measurements were used in a preliminary calibration of an early version of the Cedar Creek groundwater model, but there is a significant amount of error associated with these water level measurements as a variety of methods were used by well drillers to identify the location of the well, the elevation of the ground surface, and the depth to water. Residential wells were not included in the model as extraction wells because most of the extracted water is assumed to return via septic systems and the remainder is assumed to be a very small component of the water budget for this region. There are no known irrigated agricultural areas within the Cedar Creek watershed.

GEOLOGIC, LANDSCAPE, AND REMOTE SENSING DATA

We established a GIS database for the Cedar Creek region with topography, land use, hydrography, and hydrogeology characteristics. These GIS datasets were compiled from the Michigan Geographic Data Library, established by the Michigan Department of Environmental Quality. These datasets are assumed to be static for the purposes of this analysis.

Land surface elevations (see Figure 2d), which were defined based on the National Elevation Dataset 26.5 m digital elevation model (DEM), were used for a range of model inputs. Flow direction and gradients for the overland flow and near surface soil moisture redistribution modules were calculated from the full-resolution DEM and then upscaled. This 4x upscaled DEM was used to set the land surface elevation for the soil water balance model. The drainage network and lake boundaries were defined based on the Michigan Framework GIS dataset, with stream crossing elevations manually extracted from a digitized version of the USGS 1:24000 topography quadrangles. The watershed does not contain any large lakes that are connected to the stream system, thus lakes are not separately considered in the groundwater portion of the integrated model.

The land cover distribution across the Cedar Creek model area (Figure 2) is taken from the Integrated Forest Monitoring Assessment and Prescription (IFMAP) coverage, which is a statewide digital land cover map with 30 m resolution derived from 1997–2000 LANDSAT data [*MDNR, 2001*]. This watershed is dominated by forested/openland/wetland land covers (60%), while 36% of the area is agricultural and the remaining 4% is urban. The IFMAP coverage provides inputs that are used in calculating evapotranspiration and overland flow. Land cover types are associated with transpiration estimates through a variety of terms including stomatal conductance, canopy height, and root depth, and with evaporation estimates through changes in canopy interception, wind speed, and interception of incoming radiation. Overland flow is also associated with land cover through Manning's roughness coefficients. The details of these connections are described in the modeling sub-sections below.

Hydrogeologic zones were parameterized according to a Quaternary Geology coverage of *Farrand and Bell* [*1982*] for the groundwater model along with the Soil Survey Geographic (SSURGO) database (see Figure 2), which was then mapped into saturated and unsaturated zone parameters according to lookup tables based on literature values. The geometry of the aquifer base was interpolated between measured bedrock elevations by de-clustered and polynomial-detrended simple kriging of the elevations of the drift/bedrock contact from oil and gas wells across the entire expanded Muskegon River model domain. Initial hydraulic conductivity values were assigned to the geologic zones based on an optimization of these parameters for the nearby Grand Traverse Bay watershed that has the same geologic zones [*Boutt et al., 2001*].

GIS grids of leaf area index (LAI), the ratio of one-sided green leaf area to ground area [*Myneni et al., 2002*], were also used in calculations of potential evapotranspiration (PET), canopy interception, and solar radiation interception. For this study, we used remotely sensed LAI measurements from NASA's Moderate Resolution Imaging Spectroradiometer (MODIS) eight-day averaged product. Spatially averaged LAI for both forest and agricultural land-use types are plotted for 2003 and 2004 in Figure 3a. In cases where these data were not available (i.e., prior to 2000), we average all LAI grids for each Julian day and apply these multi-year averages to the earlier periods. This will have little effect on our results because we are only comparing simulated and observed flows for mid 2001 through 2004 when all datasets are available. However, it is important to spin up the model using realistic data inputs because we found that it takes between two and three years before the model results are independent of the starting conditions.

CLIMATE DATA

Precipitation data was obtained from the NOAA gage at Hesperia, MI approximately 20 km NNW from the center of the Cedar Creek watershed (see locator map in Figure 1). This gage was chosen because lake effect precipitation is an important meteorological phenomenon in this area, and this gage lies at relatively the same distance from the Lake Michigan shoreline as the Cedar Creek watershed. NOAA data (shown in Figure 3b) included hourly precipitation totals, as well as daily measurements of new snowfall and snow pack depth.

Other climate data, including hourly temperature, relative humidity, wind speed, and incoming solar radiation (Figures 3c–f), were extracted from the Fremont, MI station of the Michigan Automated Weather Network (MAWN) (see Figure 1 for location). This climate network is operated by the Michigan State University Extension, the Michigan Agricultural Experiment Station and the Michigan Department of Agriculture. Since the MAWN data did not exist prior to 1996, from 1990–1995 we used the Julian-day average of the available data.

COMPONENTS OF THE INTEGRATED LANDSCAPE HYDROLOGY MODEL (ILHM)

Figure 4 illustrates our conceptual model of the most important hydrologic processes in the Cedar Creek watershed, and diagrams the linkages between input datasets, ILHM modules, and model outputs. As mentioned earlier, this version of ILHM was developed by linking a novel landscape water balance model with a simple linear -delay unsaturated zone model and MODFLOW-2000 [*Harbaugh et al., 2000*], the most commonly used groundwater flow code. The landscape and near-surface portion of the ILHM combines several existing codes with a set of new modules, in order to speed development and to incorporate the full range of hydrologic processes. The canopy water balance model

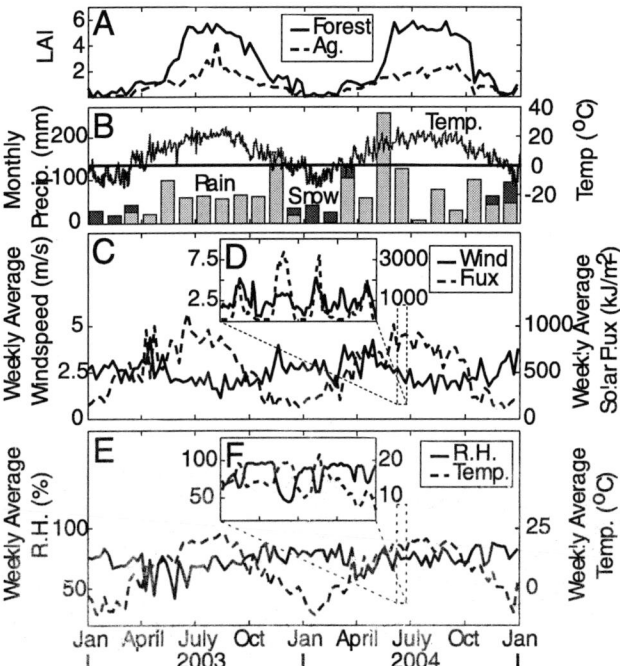

Figure 3. Time series data inputs to ILHM for 2003 and 2004: A) LAI of forest and agricultural land covers; B) monthly rain and snow along with daily average temperature, a horizontal line at 0°C is included as a visual aid; C) weekly averaged windspeed and solar flux; D) inset of hourly windspeed and solar flux data from 6/21/04 through 6/24/04; E) weekly averaged relative humidity (R.H) and temperature; F) inset of hourly R.H. and temperature from 6/21/04 through 6/24/04.

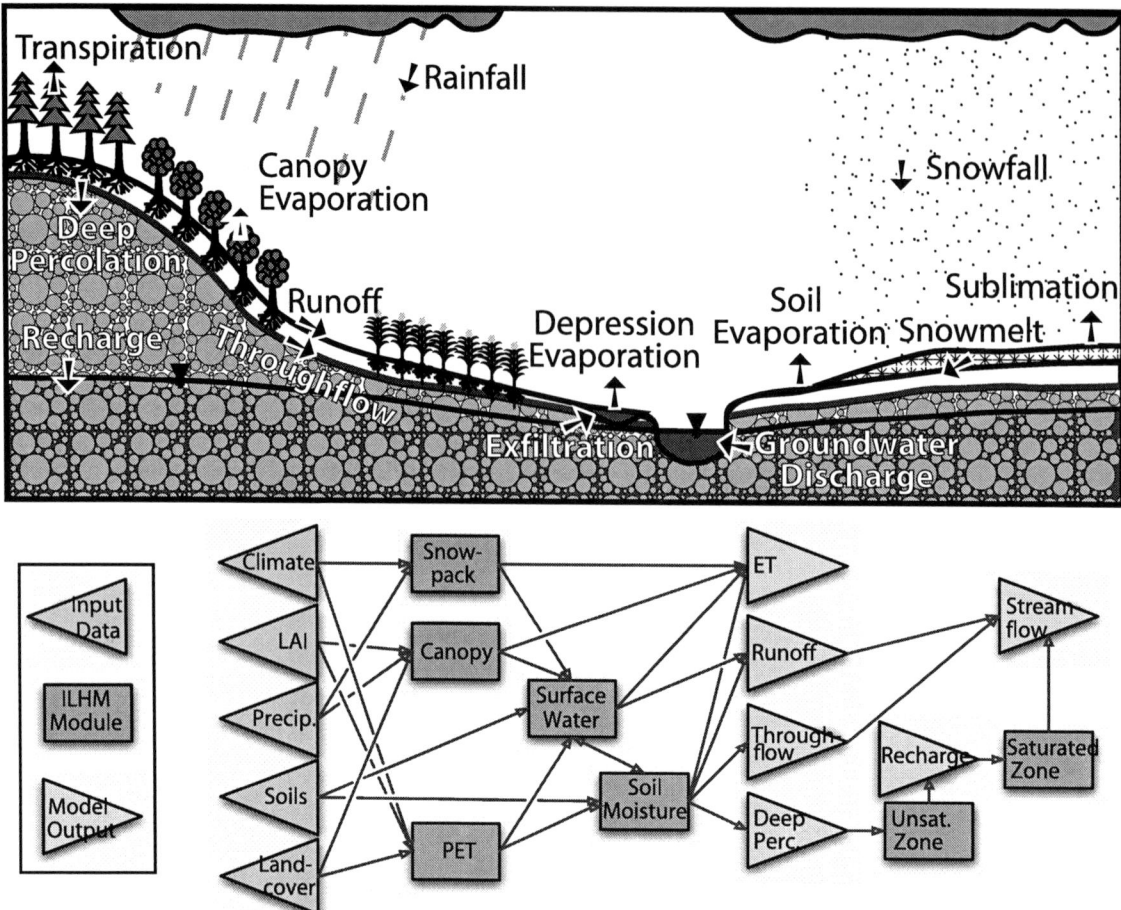

Figure 4. Conceptual model and simplified box diagram of the ILHM. The upper portion illustrates the predominant hydrologic fluxes currently simulated by ILHM. The lower portion of the figure shows the relationships between and among input datasets, models, and model outputs.

is based on equations published in *Chen et al.* [2005]. The surface hydrology model, including infiltration and runoff routing, are modified from the Distributed model for Runoff, Evapotranspiration, and Antecedent soil Moisture (DREAM) model by *Manfreda et al.* [2005]. The snow pack is simulated using the UEB Snow Model by *Tarboton and Luce* [1996]. Soil moisture accounting along with near-surface flows are handled by a set of codes we developed based on common unsaturated zone flow modeling methods.

The ILHM suite calculates each term in the full water balance equation:

$$\Delta S = P - T - E - Pc + Tr - Ex - R \qquad (1)$$

where ΔS is the change in soil moisture storage in the biologically active soil zone, P is watershed available precipitation, T is transpiration, E is evaporation, Pc is deep percolation beneath the biologically active soil zone, Tr is lateral near-surface unsaturated flow called throughflow, and Ex is the exfiltration from each cell, and R is precipitation excess runoff. For Equation 1 and the detailed water balance equations presented in Appendix A, terms are in units of meters per unit time unless otherwise specified.

The landscape portion of our model sequentially calculates the water balance along the paths water takes as it is redistributed from precipitation to various subsurface and surface pathways. Incoming rainfall is first subjected to canopy interception, while snow is routed directly to the snow pack model. Next, canopy thoughfall and snowmelt are applied to the soil surface. These new inputs are then combined with any water stored in surface depressions and allowed to infiltrate into the soil. Any excess water at this point enters surface depression storage up to the available capacity.

Infiltrated moisture is added to the existing surficial soil layer budget, where it can then percolate downward under the influence of hydraulic gradients. Any moisture within the first soil layer is then available for evaporation, along with any transpiration that may occur in any of the biologically-active soil layers. Subsurface lateral throughflow is then calculated, which may cause moisture in down-gradient cells to exceed saturation. At this point, moisture in the lowest biologically-active soil layer may then percolate into the sediments beneath, where it becomes deep percolation. Remaining moisture in excess of saturation is exfiltrated back toward the surface where it also enters depression storage.

Deep percolation is then delayed as a linear function of the thickness of the unsaturated zone, which is estimated based on a steady-state run of the regional MRW groundwater model. The delayed percolation then becomes recharge to the three-dimensional transient groundwater flow model when it crosses the water table. Water stored in surface depressions is then subjected to direct evaporation. If depression storage capacity is exceeded, the excess water becomes surface runoff. Baseflow discharge from the groundwater model is then combined with the surface runoff and throughflow to produce the complete simulated stream hydrograph.

The landscape hydrology components of the model for Cedar Creek are simulated with a 177×153 grid at 106.3 m resolution. As shown in Figure 1, while the watersheds and groundwatersheds overlap for most of the modeling domain, some locations contribute only surface water or groundwater to Cedar Creek. To account for this, the landscape hydrology model is run for the entire domain while the unsaturated zone model only allows groundwater recharge in active cells of the saturated groundwater model, and the stream routing module only includes areas within the surface-watershed of Cedar Creek.

The following sections describe the details of each set of processes simulated in the ILHM, with specific assumptions that were made for the Cedar Creek case. The mathematical descriptions and details about parameters are included in Appendix A.

Precipitation and Snowmelt

Watershed available precipitation (P) is the sum of liquid rainfall and snowmelt. From late December through mid March, precipitation falls predominantly as snow in the Cedar Creek watershed. To model the storage and release of snow we used the UEB Snowmelt Model by *Tarboton and Luce* [1996], which is an explicit energy and water balance model designed to track three state variables: snow water equivalent, energy deficit (i.e., how much energy would be required to return the snow pack and soil layer to the 0 degree C reference condition), and the snow surface age. This model is computationally efficient because it assumes no temperature gradient within the snow pack and the layer of soil with which it interacts. For ease of integration with the rest of the ILHM model suite, we ported the FORTRAN version of this snowmelt code into MATLAB.

The full UEB model requires air temperature, wind speed, relative humidity, and solar insolation. The adjustable parameters in this model component include the density of the snowpack, the thermal conductance of the snow, the liquid water holding capacity of the snowpack, and the depth of soil with which the snow thermally interacts. Preliminary calibration to snow depth data from a single year of record provided the parameter values shown in Table 1.

The current version of the ILHM only runs the UEB model if either the air temperature during a precipitation event is below freezing, or the snow water equivalent of the snowpack is greater than 0.01 mm. Any water remaining in the snowpack below this amount of moisture is then applied to the surface as additional snowmelt. All available snowmelt calculated by the UEB model is then added to any liquid precipitation for each time step to become watershed available precipitation (P).

Evaporation and Transpiration

The evaporation (E) and transpiration (T) terms of the water balance equation first require calculation of potential evaporation and transpiration. All evaporation and transpiration potentials, (canopy, depression, soil, and transpiration) are calculated using the modified Penman-Monteith equation [*Monteith, 1965*] presented by *Chen et al.* [2005] and shown as Equation A1. For each separate potential, the aerodynamic resistance and resistance to vapor transport terms are modified as described in the Appendix.

Evaporation and transpiration rates vary temporally according to land cover types through 8-day LAI scenes, a stomatal conductance coverage, and soil texture. For the Cedar Creek watershed, which is small and has relatively brief surface water residence times, we assume that there is no open water evaporation. Work is ongoing to explicitly model this component of evaporation.

Incoming rainfall is first subjected to interception up to the water holding capacity of the canopy, which is related to LAI in the cell. Water storage in the canopy is simulated as a

Table 1. List of adjustable parameters in the UEB Snowmelt Model and their assumed primary calibration values.

Parameter	Units	Value
Snow Density	kg m^{-3}	200
Liquid Holding Capacity	-	0.15
Thermally Active Soil Depth	m	1.0
Snow Thermal Conductance	m hr^{-1}	0.2

separate storage layer, with losses only from evaporation. We assume that the largely deciduous canopy does not intercept snowfall. The canopy is also assumed to intercept a portion of the incoming solar radiation based on LAI.

The evaporation of moisture in depression storage is calculated after any infiltration and exfiltration (described below) in a given time step have occurred. Total depression storage capacity is determined by land use, soil texture, and slope class; a tabular reference of storage capacities can be found in *Manfreda et al.* [2005]. Depression evaporation occurs at the potential rate until depression storage is depleted. Evaporation from depressions and directly from soil is allowed only from the proportion of soil that is exposed to solar radiation, thus assuming no soil evaporation from the portion shaded by canopy or covered with snow or ice.

Direct soil evaporation is allowed only from the first soil layer, which is a reasonable assumption in this relatively humid region. Soil evaporation occurs at the lesser of calculated potential rate or the soil exfiltration depth (discussed further in the Appendix following Equation A16). We are exploring alternative strategies for calculating evaporation using the model of *Ritchie* [1972]. The total transpiration in each cell T is calculated to be the sum of the root water uptake from each biologically active soil layer. We assume that transpiration only occurs above a dormant threshold temperature, which was chosen to be 40 degrees F for this study. Stomatal conductance is also assumed to be constant for this case, although we plan to incorporate variations due to changes in temperature, carbon dioxide concentrations, and soil moisture as described in *Chen et al.* [2005].

Infiltration, Percolation, Throughflow, and Exfiltration

For this study we assume the biologically active soil can be described by two layers, with a total thickness calculated according to Equation A14 as the depth above which 90% of the root mass lies. The first soil layer, from which evaporation occurs and that controls infiltration capacity, is on the order of several centimeters thick. Infiltration capacity is calculated as the greater of either the soil-texture dependent saturated infiltration capacity, i_{sat}, or the first layer moisture deficit from saturation. We chose the maximum infiltration rate to be $(2 \cdot i_{sat})$ which determines the choice of the first soil layer thickness. This formulation produces similar results to empirical infiltration rate descriptions, and has the advantage that it does not require storm event tracking or single storm event modeling.

Here we assume that i_{sat} can take either its nominal soil-texture dependent value taken from literature values (see Appendix), or $i_{sat} = 0$ if the soil is frozen, which we only allow to occur in agricultural soils based on *Schaetzl and Tomczak* [2001]. The soil is assumed to be frozen if its temperature is below −0.25°C, measured by the MAWN station in Fremont at a depth of 10 cm.

In addition, we assume that no infiltration occurs in cells classified as permanent water features. Although the water features in this watershed often cover only a small portion of each land-use cell, our assumption may nevertheless be fairly realistic. The true physical system process is more accurately described as rapid percolation to a shallow water table followed by equally rapid rise and subsequent relaxation of the water table. The effect is a temporary increase in groundwater discharge that generally does not modify what is more traditionally called stream baseflow. Our stream gage data indicate that the characteristic time of this response may be on the order of twice the surface runoff response, or perhaps 5–7 days. This does not allow for significant losses due to evapotranspiration, thus the combined increase in stream discharge due to percolation to the near-stream water table and direct overland flow would be nearly equal to that expected from an assumption of zero infiltration. Streamflows in our model would thus be expected to peak higher and return to baseflow levels more rapidly than observed.

Throughflow, defined here as lateral subsurface flow within the biologically active soil zone is calculated using a simplified Richards equation model. For a full development of the Richards equation, see [*Hillel*, 1980]. For purposes of computational efficiency, and in order to assure that the subsurface redistribution of moisture occurs in only one dimension, we assume that flow only occurs parallel to the dip of the slope and thus cannot flow uphill on the ~100 meter scale of our model cells. For environments where these assumptions may be invalid, alternate two-dimensional formulations could be substituted for this ILHM module given adequate computational resources. The van Genuchten model was used to calculate all soil-moisture dependent properties [*van Genuchten*, 1980] with parameter values given in the Appendix. The downgradient cell is determined via the D8 flow direction function [*ESRI*, 2003]. However, each cell can have more than one upgradient cell, thus throughflow is the summation of the shallow subsurface flow out of all upgradient cells.

Unsaturated Zone Delay

Deep percolation beneath the biologically-active soil layer is delayed prior to becoming recharged as a linear function of the depth to the water table. The slope of this delay in units of days/meter, was determined from wells installed in the nearby Grand Traverse Bay Watershed, and was fixed at 2.5 for this study. It is important to note that this would not be the same as a solute transport time through the unsaturated zone. Delayed deep percolation then becomes recharge once it reaches the water table. The depth of the water table is assumed temporally invariant for the current version of the unsaturated zone delay module for ease of implementation. By fixing the depth of the water table for this purpose, we do not account for seasonal or trending differences in water travel times through the unsaturated zone. The seasonal differences in the depth of the water table are small (typically <1 meter) relative to average depths to water over most of the model domain. Locations very close to surface water features are an exception, but these areas comprise a small fraction of the total watershed area and thus will not significantly affect the dynamics of modeled recharge.

Groundwater Model

The groundwater model for the expanded MRW was developed using a suite of MATLAB utilities that we developed to create input files from GIS layers for MODFLOW. A regular grid of 1798×1865 cells (106.3 meters on a side) was used so that each cell directly overlies 16 Digital Elevation Model (DEM) cells. The vertical domain of this saturated zone model was then subdivided into two layers with approximately equal saturated thicknesses based on the simulated water table in a single layer model. Automated parameter estimation routines were applied to an early version of the Cedar Creek groundwater and soil balance model to estimate hydraulic conductivity values for aquifer sediments in geologic zones parameterized using a digital map created from *Farrand and Bell* [1982].

The Cedar Creek model is a single layer with dimensions of 184×162 and cell size of 100 meters. In this case, a single vertical layer provided an adequate description of flows through the Cedar Creek watershed, because it does not have significant vertical relief, extensive low-permeability subsurface layers, high-capacity pumping wells, or other features that tend to induce significant vertical head gradients. Groundwater discharge to streams is calculated with the Stream Flow Routing (SFR) package in MODFLOW that routes water via the kinematic wave equation [*Prudic et al.*, 2004].

Runoff and Stream Routing

Surface runoff is routed to the streams using an approach modified from that presented by *Manfreda et al.* [2005]. In this version of the code, we assume that runoff cannot reinfiltrate once it is generated. It is routed overland and through streams according to the D8 flowdirection algorithm in ARC [*ESRI*, 2003] with runoff times given by the velocities in each cell along the flowpath. Runoff is assumed to travel overland at a velocity given by the Kerby time of concentration equation [*Kerby*, 1959]. Once the runoff enters the stream channel its velocity is calculated using Manning's Equation.

For this study, the hydraulic radius, r, for Manning's Equation was determined as a function of discharge using low-flow channel geometry measurements in and around the Cedar Creek watershed along with geometries reported by the USGS for their stream gages. Wetted perimeter was assumed equal to $2 \times depth + width$, while area was simply $depth \times width$. A power law fit to these data produced the empirical relationship for this watershed:

$$r = 0.9046 \cdot Q^{0.283} \quad (2)$$

with a correlation coefficient R^2 of 0.77 (see Figure 5). Q in the above equation is the measured stream discharge in m^3/s.

While streamflow velocity can be dynamically calculated, for simplicity we have assumed a temporally constant v_{stream} for each stream cell. To calculate v_{stream}, we assume that each cell in the Cedar Creek watershed contributes a unit of runoff which is them routed to the gages using ARC's *flowaccumulation* function. To rescale the output to match a typical discharge event in a stream cell (Q_{ij}), we multiply the measured Q at the outlet by the ratio of the *flowaccumulation* value in each cell (i, j) to that of the outlet:

$$Q_{ij} = Q_{outlet} \cdot \left(\frac{flowaccumulation_{ij}}{flowaccumulation_{outlet}} \right) \quad (3)$$

Once v_{stream} has been calculated, the flow time from each grid cell to the outlet(s) can be calculated using ARC's *flowlength* function, which calculates the cost-weighted distance

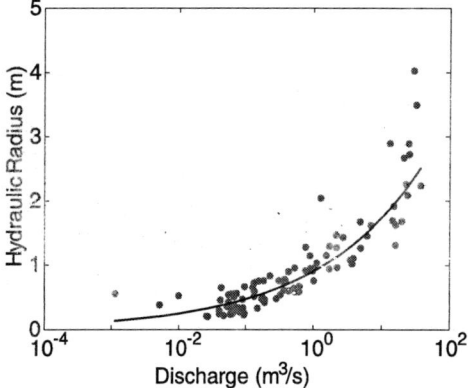

Figure 5. Plot of measured discharge versus hydraulic radius in streams in the greater Muskegon River Watershed and adjacent watersheds. The hydraulic radius calculation assumes rectangular cross-section geometry. The power-law fit to the data produced a correlation coefficient, R^2, of 0.77.

from each cell to the outlet. Here the weighting is the inverse of the velocity in seconds/meter. ARC then multiplies this "cost" by the distance traveled through each cell along the entire flowpath and outputs to the travel time to the outlet. The travel time grid is then used to transform the precipitation excess grid into a runoff hydrograph.

RESULTS

The results of the ILHM simulation for the Cedar Creek watershed are discussed in the context of the broader goals of the code, which are the prediction of temporal and spatial variations in recharge with very little direct calibration of model parameters using readily available remote-sensed and ground-based data sources. All parameters for this prediction were based on literature values (Tables 2 and A1–A3), except for the UEB model parameters (Table 1) as well as two hydraulic conductivity values and one unsaturated zone delay parameter that were calibrated using a very early version of the model (Table 2).

A plot of simulated versus observed heads (Figure 6) across this region shows a reasonable degree of agreement given the measurement uncertainty. Figure 6 shows no trending bias between simulated and observed heads, though a slight high-side bias is present at observed heads lower than approximately 200 meters. We would expect a higher degree of correlation between simulated and observed water levels in regions where pressure transducer data are available from wells, or if a parameter optimization were to be performed for this ILHM simulation.

Detailed evaluation of modeled flows is hampered by the lack of long duration stream gages with stable channels in the basin. All observed stream discharges were collected via established methods, however the rating curves for the two stream gages are currently inadequate to account for temporal adjustments of the channel geometry after flood events. As is commonly the case, we have few high flow measurements, which limits the accuracy of our flows calculated from the rating curve during large floods. In addition, flows in the Lower Cedar gage from January through March appear to suffer from ice-induced over-pressurization not observed at the Upper Cedar gage.

Despite almost no calibration of parameters in the near-surface components of the ILHM code, the model provided

Table 2. List of calibrated parameters for the unsaturated and saturated groundwater models.

Parameter	Units	Value
Outwash conductivity	m d^{-1}	11.0
Till conductivity	m d^{-1}	4.4
Unsaturated zone delay	d m^{-1}	2.5

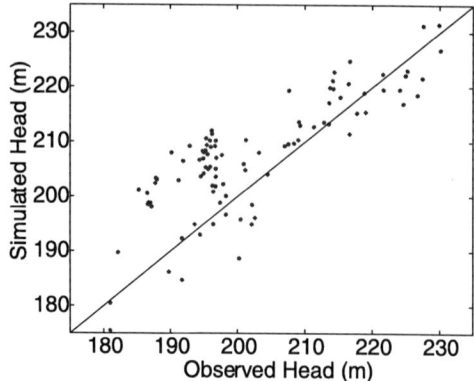

Figure 6. Uncalibrated simulated vs. observed groundwater heads plotted on top of a 1:1 line. Observations are from 96 residential wells installed in the Cedar Creek Watershed between 1990 and 2004 (see locations in Figure 1).

a reasonable prediction of observed flows (Figure 7) for the two available gage sites in this 100 square kilometer watershed (Figure 1) during the fall and winter months. The ILHM also provided reasonable predictions of baseflow for this watershed system during the entire year. Because this prediction is based almost entirely on a set of widely available meteorological inputs and GIS datasets combined with literature parameter values, the code appears suitable for directly simulating streamflows in ungaged basins.

The close agreement between observed and simulated baseflow levels also suggests that the model is providing reasonable predictions of recharge, which provides baseflow in these streams during low flow periods. During May and October 2003, and January–March 2004, the ILHM-simulated total flows typically agreed with gaged values within 10% (when the lower Cedar gage was not affected by ice cover). Baseflow levels in the smaller upper Cedar catchment proved highly sensitive to hydraulic conductivity in the outwash sand/gravel zone (Figure 2a). The conductivity values presented in Table 2 should not be viewed as a fully calibrated parameter set, as optimization of the total streamflow simulation is the subject of ongoing evaluation.

Despite the limitations imposed by some parts the stream gage data, the values from April through December of 2003 are reasonable for quantitative flow comparisons. During lower ET periods from April through early June and October through December, the ILHM-simulated total fluxes are approximately 20% higher than calculated from the flow gages installed at both the Upper and Lower Cedar gage sites. ILHM-simulated total fluxes during higher ET periods from June through October are much greater than observed, also likely due to an incomplete description of ET processes in this version of the ILHM code that is the subject of ongoing development.

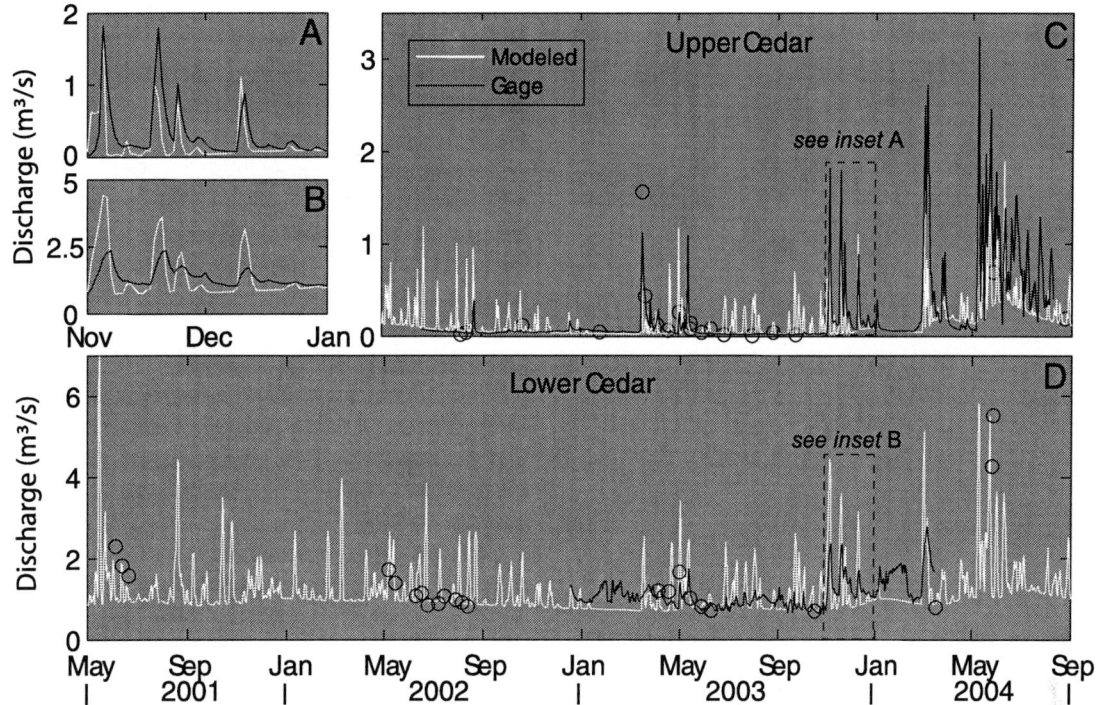

Figure 7. A & C) Upper and B & D) Lower Cedar Creek stream discharges with simulated values shown in white and gage values in black calculated based on stage discharge relationships. Manually measured discharge values are shown as black circles.

A scatter plot of simulated versus observed flows at the two gages on a log-log scale illustrates that most of the moderate to high flows in the system are reasonably described by the model (Figure 8). There is a larger degree of mismatch in the Upper Cedar Creek site, as illustrated by a significant amount of scatter about the 1:1 line. Simulated peak discharge values are similar to those that have been measured, however the simulated discharge peaks are narrower than observed. This temporal offset at near-peak discharge, seen in Figure 7b, appears in Figure 8 as a tendency toward low simulated flows relative to observed values. These narrow simulated peaks are largely related to the simple nature of our stream routing package, and the unidirectional linkage between groundwater and surface water processes in this version of the ILHM code. The assumption of a temporally constant v_{stream} results in average flow velocities lower than those that would be expected, with the effect that simulated peaks would be even sharper if v_{stream} were a function of discharge. However, there are several known flow damping mechanisms, including flow through wetlands and bank exchange, which are not yet represented in this version of the code.

A map of the simulated average annual recharge across the watershed implies that agricultural areas may have higher recharge than forested areas according to this simulation (Figure 9). Several inter-related factors combine to account for the simulated differences. Agricultural areas experience less canopy interception than forested areas due to lower LAI values. Although this increases the infiltration into agricultural soils, a resulting increase in transpiration tends to narrow the difference in recharge between these land use types. Our model may also under-represent soil evaporation in agricultural soils because it does not incorporate solar

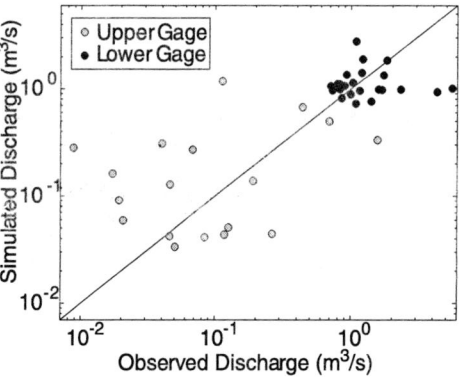

Figure 8. Uncalibrated simulated vs. observed flows for the Upper and Lower Cedar Creek gages plotted on top of a 1:1 line from approximately 20 manual discharge measurements taken at each of the upper and lower Cedar Creek gage locations.

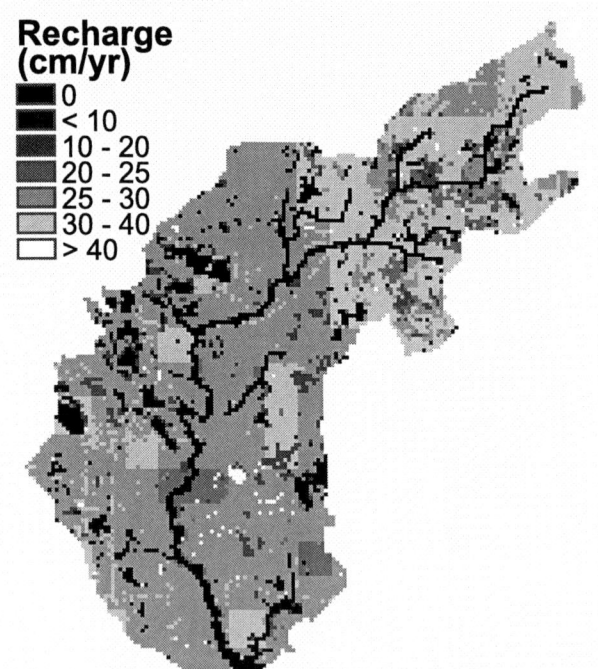

Figure 9. ILHM-simulated average annual recharge for the Cedar Creek watershed. Annual total precipitation averaged approximately 83 cm during the period of simulation.

heating of the shallow soil layer. This simulated difference is the subject of future evaluation across the much larger MRW where more flow data are available.

The blocky nature of the simulated recharge in this map is mainly due to the large (1 km²) LAI cells. The effect of these coarse cells is to decrease forest LAI and increase agricultural LAI in regions with mixed land-uses. We plan to resolve this issue through downscaling the LAI information by assuming that the measured LAI value is a linear combination of the LAI of forest and agricultural land-uses represented by the much higher resolution IFMAP dataset. Thus unique "agricultural LAI" and "forested LAI" values can be approximated at the resolution of the IFMAP data constrained by the total measured LAI from MODIS.

Areas with low hydraulic conductivity soils experience reduced recharge (Figure 9), such as portions of the upper watershed to the south side of the stream where loams and silty loams are common (compare with the soils map in Figure 2c). In contrast, recharge can be greatly enhanced in internally drained regions. In this simulation we deactivated the runoff mechanism in internally drained areas, thus potentially increasing infiltration and shallow subsurface flow. This may be very important in areas with moderate to low-conductivity soils, but the internally drained areas in this watershed tended to also be sandy so the effect is only localized.

The Cedar Creek region experiences very little runoff from upland areas, which is consistent with the simulated map of precipitation excess (Figure 10). Nearly all the simulated precipitation excess in this watershed occurs in cells that are classified as "water" because there is no transpiration or percolation from those cells. The only cells with any significant precipitation excess that are not classified as water are in a region of lower conductivity sediments in the upper watershed. Despite a simple description of runoff processes, we do not expect significant runoff in most of the sediments across this watershed due to high infiltration capacities and saturated hydraulic conductivities. There is also very little simulated subsurface redistribution, or throughflow, throughout most of the watershed due to the highly conductive soils and relatively gentle topography. This process is most active in areas with steep slopes or water tables very near the surface.

The simulation provides evidence for a strong seasonality in recharge rates for the Cedar Creek watershed. The temporal variations in simulated deep percolation are shown in Figure 11. From September through March in the four illustrated years, the model predicts that approximately 70–80% of watershed available precipitation will percolate into the deep aquifer sediments where it eventually recharges groundwater. In contrast, the simulations show virtually no

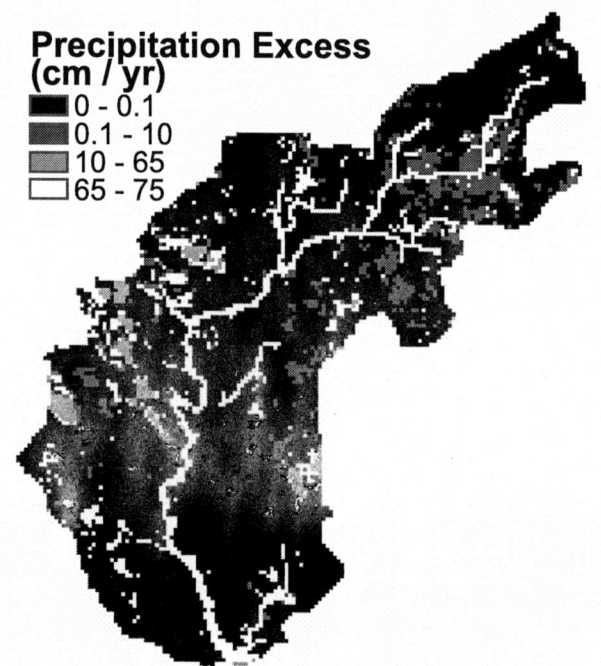

Figure 10. ILHM-simulated average annual precipitation excess runoff for the Cedar Creek watershed. Annual total precipitation averaged approximately 83 cm during the period of simulation.

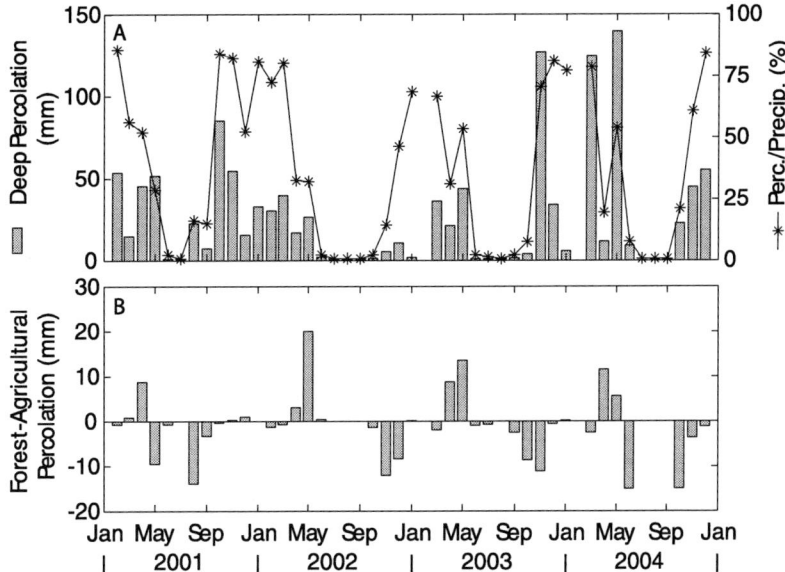

Figure 11. A) Monthly average deep percolation (bars) in the Cedar Creek watershed plotted with the ratio of percolation to total precipitation for each month (stars). B) Monthly difference in deep percolation between forested and agricultural land covers.

deep percolation over the growing season from May through September for the same years, which is consistent with the statistical findings of *Jayawickreme and Hyndman* [2007]. This simulation indicates that agricultural areas have more recharge in the fall months than forested areas, while the opposite occurs with higher relative forest recharge in the spring months. This is reasonable as forests tend to have less extensive frozen soils during snowmelt periods [*Schaetzl and Tomczak, 2001*], and agricultural LAI often begins to decline earlier in the year than in forested areas. Although coniferous forests may transpire year round, they represent only a small percentage of the forested areas in this study region.

Temporal variation in evaporation and transpiration are clearly the causes of most of the simulated variations in deep percolation because these are the primary loss processes. As Figure 12 illustrates, evaporation is generally a much smaller component of water loss in this watershed than transpiration, and this component is larger in forested land relative to agricultural land due to much higher forest canopy interception. Transpiration shows a stronger seasonal trend than evaporation, as it depends more strongly on LAI. Total agricultural transpiration is greater than that of forested areas despite much greater potential transpiration in forested areas. Agricultural areas experience less canopy interception than forests, and thus greater infiltration and higher average soil moisture. As a result, agricultural areas tend to transpire closer to their potential rate than forested areas. Unexpectedly, agricultural transpiration also rises in the spring more quickly than that of forested areas according to these model results due to the similarity in LAI values during early spring and higher stomatal conductance values for agricultural areas relative to forests. As the LAI of forests increases in the late spring, the transpiration in these areas becomes larger than that of agricultural areas, until they reach approximate equality in late June that continues through the rest of the summer. Also during the summer, soil moisture levels reach their lowest point and often approach the permanent wilting point. As deep percolation cannot occur until the field capacity of the soils is reached, most of the water that does infiltrate the soil is transpired. Thus deep percolation is almost non-existent during summer months according to these simulations.

DISCUSSION AND CONCLUSIONS

We present the development and testing of a new suite of loosely coupled process-based codes that we call the Integrated Landscape Hydrology Model (ILHM). This modeling framework has several advantages over existing coupled hydrology codes. It can simulate much larger domains than fully coupled process-based codes, with fewer data requirements. In addition, the ILHM accounts for the processes and mass balance in a more rigorous manner than semi-distributed codes, which tend to lump or oversimplify important watershed processes and use parameters that cannot be independently measured. The ILHM also facilitates model development via direct input of readily available GIS data, in contrast to the impractical level of manual data input

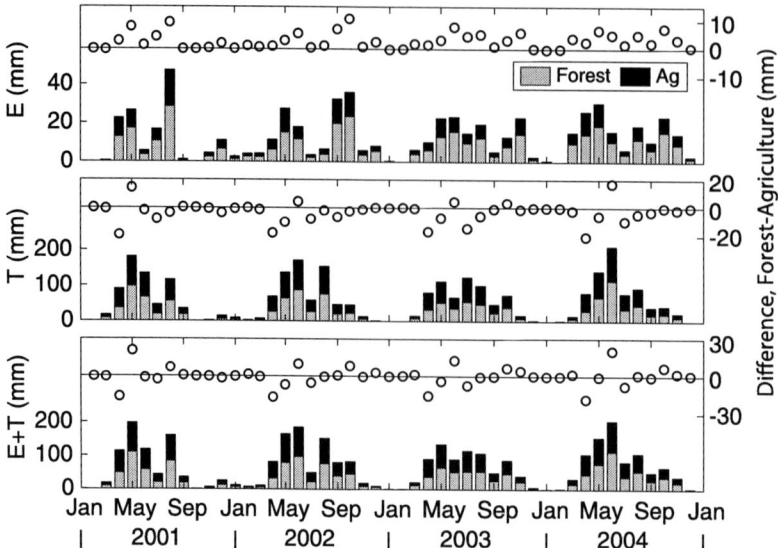

Figure 12. Average monthly evaporation (E), transpiration (T), and evapotranspiration (E+T) plotted for both forested and agricultural landscape as stacked bars along with the difference (Forest-Agriculture) between the two land-use types plotted as circles.

required for large domains from some existing process-based models such as Mike-SHE. Finally, ILHM is well-suited for forecasting purposes because it allows forcing data and component process models to be interchangeable; thus a model developed and calibrated with current data can be rapidly converted to a forecast simulation by adding the appropriate component process code.

This new modeling framework was designed to make development of models for large domains as simple as possible, while maintaining a rigorous fluid mass balance based on the primary processes that drive water movement over the landscape and through the subsurface. The approach is computationally efficient because it allows some processes to be simulated based on full numerical models while others can be described by simpler and thus faster water- and energy-balance approaches. Due to the loose-coupling framework, individual components can also be simulated at a variety of spatial and temporal scales appropriate to the individual processes. This framework also allows more rigorous simulation modules to be used in place of a simpler routine in cases where the additional computational burden provides necessary improvement in the model predictions. Alternatively, in cases where enough data exist to adequately describe a particular process, the data can be used in lieu of that process simulation module.

As currently configured, ILHM is designed to simulate flows through regions with connected surface water and groundwater regimes such as Cedar Creek. This test watershed has a sub-humid and temperate climate, with flow through a glacio-fluvial aquifer, largely covered with deciduous forests, agricultural land, and small percentage of urban cover. Thus the code is expected to provide reasonable predictions for similar environments in the sub-humid Midwest. The general processes are the same in arid and montane regions; however alternate modules would likely provide more accurate simulations in such cases. In particular, some high-relief environments may require a full two-dimensional representation of overland flow, including depth-dependent velocities for sheet or rill flow. Areas with large proportions of urban land uses will require additional modifications, especially when engineered storm water systems have a significant effect on the hydrograph shape after a storm event.

In the Cedar Creek watershed, precipitation excess runoff routing and subsurface moisture redistribution are both largely inactive over most of the modeled domain and timeframe. As a result, the simulation results are similar even if these modules are not active for upland areas. Therefore, further testing in domains where these processes are responsible for a significantly larger percentage of the flow in a river system will be needed for these modules. The current unsaturated zone module is a very simple representation of hydrologic processes, thus we will explore the use of direct solution methods ranging from the Green-Ampt model through the full Richards equations. Additionally, using MODFLOW or any finite difference scheme has the disadvantage of requiring somewhat cumbersome rectangular grids that limit cell refinement at regional scales. However, ILHM can easily be altered to interface with a finite element code capable of representing and accounting for groundwater discharges to streams.

The ILHM was tested in the Cedar Creek watershed because of the need for high-resolution flow simulations that provide the interface between land use change models and ecohydrology models in the near future. This first evaluation of the ILHM modeling framework demonstrated that these codes can reasonably predict groundwater recharge and streamflow through a 130 square kilometer watershed with very little calibration using readily available data. The simulation represented overall basin recharge accurately, but it appears to have slightly overestimated recharge in agricultural areas. This is likely due to an inadequate representation of soil evaporation that will be addressed in a future version of the ILHM. The simulated hydrograph peaks are too narrow and decline more rapidly than is observed because the current surface water/groundwater linkage cannot represent bank storage and release processes, nor can the unidirectional coupling between surface water and groundwater fully represent near-stream processes at our chosen spatial scales. Nevertheless, the recharge and streamflow predictions provide reasonable descriptions of system behavior and will be further refined in future versions of the ILHM applied to much larger domains.

Acknowledgements. We would like to thank Mike Wiley and Paul Richards for supplying us with surface water flow measurements and stream levels from transducers, and Dushmantha Jayawickreme and Cheryl Kendall for their help with data processing and analysis. We are also grateful for the financial support of National Science Foundation grant (EAR-0233648), with supplemental support from both the Great Lakes Fisheries Trust's Muskegon River Initiative and Environmental Protection Agency grants (R830884). Any opinions, findings and conclusions or recommendations expressed in this material are those of the authors and do not necessarily reflect the views of the National Science Foundation or the Environmental Protection Agency.

APPENDIX A: ILHM MODEL DEVELOPMENT

Evaporation and Transpiration

The basis for our potential evaporation and transpiration calculations is the modified Penman-Monteith equation [*Monteith, 1965*] presented by Chen et al. [2005]:

$$PE, PT = \frac{\Delta \cdot F + \rho \cdot c_p \cdot \frac{e_s - e}{r_{ai}}}{\lambda_v \left(\Delta + \gamma \left(1 + \frac{r_{ci}}{r_{ai}} \right) \right)} \quad (A1)$$

Variables appearing in Equation A1 and others that are not explicitly defined in the text are explained in Table A1. In Equation A1, F is the net radiation flux, which is the product of total solar radiation measured at the MAWN gage multiplied by the albedo of each cell. Here we assume that albedo varies seasonally from a leaf-off "brown albedo", a_b, condition to a peak growing season "green albedo", a_g value. When $LAI = 0$, albedo equals a_b and increases linearly to a_g until canopy closure is complete, which we assume occurs at $LAI = 3$. Thus albedo

$$a = \begin{cases} a_g & LAI \geq 3 \\ \frac{LAI}{3}(a_g - a_b) + a_b & LAI < 3 \end{cases} \quad (A2)$$

Values of a_b and a_g are provided in Table A2. ρ (kg m^{-3}) is the density of moist air given by the ideal gas law:

$$\rho = \left(\frac{P_{atm}}{R_d \cdot T} + \frac{e_s}{R_v \cdot T} \right) \quad (A3)$$

The barometric pressure, P_{atm} (Pa) is calculated according to the barometric formula [*Berberan-Santos et al., 1997*]:

$$P_{atm} = P_0 \cdot \exp\left(\frac{-M \cdot g_0 \cdot z}{R_g \cdot T} \right) \quad (A4)$$

where z is the elevation relative to mean sea level (m) given by the DEM.

e_s (Pa) is calculated from the Goff-Gratch equation [*Goff and Gratch, 1946*]:

$$\log_{10} e_s = -7.90298(T_{st}/T - 1) + 5.02808 \log_{10}(T_{st}/T)$$
$$-1.3816 \times 10^{-7} \left(10^{11.344(1-T/T_{st})} - 1 \right) \quad (A5)$$
$$+8.1328 \times 10^{-3} \left(10^{-3.49149(T_{st}/T-1)} - 1 \right) + \log_{10} e_{st}$$

Δ (Pa K^{-1}) is the slope of the saturated vapor pressure-temperature curve, calculated as the numerical derivative of the Goff-Gratch equation; e, the product of e_s and measured fractional relative humidity, is the ambient water vapor pressure; λ_v is the latent heat of vaporization for water (J kg^{-1}) [*Harrison, 1963*]:

$$\lambda_v = 10^3(2500.5 - 2.359T) \quad (A6)$$

and γ is the psychrometric coefficient (Pa K^{-1}) [*Brunt, 1952*]

$$\gamma = 1.61 \cdot c_p \cdot P_{atm}/\lambda_v. \quad (A7)$$

Table A1. Parameters used in calculating PET.

Symbol	Definition	Units	Value	Source
R_d	Gas constant of dry air	J kg^{-1}K^{-1}	287.05	6
R_v	Gas constant of water vapor	J kg^{-1}K^{-1}	461.495	6
P_0	Reference air pressure at mean sea level	Pa	101,325	6
M	Molar mass of dry air	kg	0.029	6
g_0	Gravitational acceleration at mean sea level and 44.5 °N latitude	m s^{-2}	9.80665	7
R_g	Molar gas constant of dry air	J mol^{-1} K^{-1}	8.314	5
c_p	Constant-pressure specific heat of dry air	J kg^{-1}K^{-1}	1006	6
T_{st}	Steam-point temperature	K	373.15	3
e_{st}	e_s at the steam-point temperature	Pa	101,325	3
h_{mv}	Measurement height of relative humidity	m	2	-
h_{m0}	Measurement height of windspeed	m	2	-
f_0	Arbitrary windspeed-at-height offset factor	-	2	-
k	Von-Karman constant	-	0.41	1
α_w	Empirical windspeed-at-height power law coefficient	-	0.143	2
τ	Soil tortuosity	-	2	4
D_v	Molecular diffusion coefficient for water vapor	m^2 s^{-1}	2.5x10^{-5}	4

[1] *Allen et al.* [1998]
[2] *Elliot et al.* [1986]
[3] *Goff and Gratch* [1946]
[4] *Choudhury and Monteith* [1988]
[5] *CODATA* [2002]
[6] *Cengel and Boles* [2001]
[7] *Halliday et al.* [2004]

The canopy resistance to vapor transport, r_{ci}, (m s^{-1}) is calculated as

$$r_{ci} = \frac{1}{g_s \cdot LAI} \quad (A8)$$

where g_s is the stomatal conductance (s m^{-1}). Values for maximum stomatal conductance were taken from *Schulze et al.* [1994]. Unlike *Chen et al.* [2005], the aerodynamic resistance, r_{ai} (m s^{-1}) is calculated based on canopy properties and height-adjusted gaged wind speed [*Allen et al.*, 1998]

$$r_{ai} = \log_e\left(\frac{h_{mw} - h_0}{l_m}\right) \cdot \log_e\left(\frac{h_{mv} - h_0}{l_v}\right) \bigg/ \left(k^2 v_{wa}\right) \quad (A9)$$

Because windspeed was only measured at one height, the effective measurement height is adjusted for canopy height. Here we assume that the height to which windspeed is adjusted, h_{mw} is given by

$$h_{mw} = \max(h_{m0}, f_o \cdot h_c) \quad (A10)$$

where h_c is the canopy height assumed constant for a given land cover (Table A2), and f_o is a factor to move the adjusted wind height some distance above the canopy. The zero displacement height h_0 is assumed to be $2/3 \cdot h_c$ [*Allen et al.*, 1998]. l_m is the roughness length for momentum transport (m) taken as $0.123 \cdot h_c$ [*Allen et al.*, 1998] (same as the zero displacement), and l_v is the roughness length for vapor and heat transport (m) assumed to be $0.1 \cdot l_m$ [*Allen et al.*, 1998]. v_{wa} is the measured wind speed (m s^{-1}) adjusted for measurement height according to the wind profile power law assumption [*Elliot et al.*, 1986].

$$v_{wa} = v_w \left(\frac{h_{mw}}{h_{m0}}\right)^{\alpha_w} \quad (A11)$$

where v_w is the raw measured wind speed (m s^{-1}).

In order to calculate evaporation from leaf surfaces E_c, surface depressions (E_d), and the soil (E_s), Equation A1 is used but the two conductance terms are modified. r_{ci} is set to 0 for evaporation from leaf surfaces and surface depressions, and r_{ai} is calculated using a crop height of 2 cm for surface depression evaporation. To calculate surface soil layer evaporation, r_{ci} is replaced by r_s given by [*Choudhury and Monteith*, 1988]:

$$r_s = \frac{\tau \cdot l_e}{\Phi \cdot D_v} \quad (A12)$$

where l_e is the depth from the surface to the top of the evaporative layer of water (m), here assumed to be half the depth of the top soil layer, and Φ is the total porosity.

Table A2. Land-Use parameters

Symbol	Definition	Units	Land-use Type							Source
			Urban	Ag	Shrub	Forest	Wetlands	Open Water	Bare	
g_s	Stomatal Conductance	s/m	10.0	11.6	7.0	5.0	3.5	0.0	0.0	1
h_c	Canopy Height	m	0.8	1.0	1.0	22.0	1.6	0.01	0.01	2
β	Root Beta	-	0.943	0.961	0.964	0.97	0.96	0.95	0.95	3
a_g	Green Albedo	-	0.2	0.22	0.2	0.16	0.12	0.001	0.001	2
a_c	Brown Albedo	-	0.36	0.4	0.28	0.24	0.4	0.001	0.001	2
μ	Canopy-fraction parameter	-	0.1	0.45	0.35	0.65	0.5	0.0001	0.0001	4

[1] Schulze et al. (1994).
[2] Walko and Tremback (2005).
[3] Jackson et al. (1994).
[4] Manfreda (2005).

The total transpiration in each cell T (m) is the sum of the root water uptake from each biologically active soil layer, T_1 and T_2. Following *Manfreda et al.* [2005], we assume in this version of the ILHM code that the actual transpiration is calculated from the potential value by linearly interpolating between 0 at the permanent wilting point and the potential rate at 75% of saturation according to:

$$T_i = (S_i > \Phi_{-33}) \cdot \min\left[1, 4/3 \cdot S_i/(\Phi_i \cdot l_i)\right] \cdot PT \cdot \frac{l_i}{\sum_i l_i} \quad (A13)$$

where S_i is the soil moisture (m) of the ith soil layer, Φ_{-33} is the permanent wilting point of the soil (Table A3) and l_i is the thickness of the ith soil layer (m). The term $S_i > \Phi_{-33}$ is a logical statement that returns a value of "1" if true and "0" if false. Porosity values, Φ_i are taken as a function of soil type as given by Table A3. Biologically active soil thickness is calculated as the depth above which 90% of the root mass lies using the asymptotic equation [*Gale and Grigal, 1987*]:

$$y = 1 - \beta^d \quad (A14)$$

where y is the cumulative root fraction at depth $d=0.9$ (cm) for this study; β is a land cover-dependent parameter (Table A1). We use Equation A14 to solve for d with a fixed cumulative root fraction $y=0.9$.

$$l = \sum_i l_i = \log_\beta(0.9) \quad (A15)$$

Total evaporation E (m) is the sum of canopy evaporation, E_c, soil evaporation E_s, and surface depression evaporation E_d. Soil evaporation is calculated according to *Chen et al.* [2005] as

$$E_s = \min(PE_s, d_s) \cdot (1 - f_c) \quad (A16)$$

where d_s is the soil-controlled exfiltration depth (m) calculated by

$$d_s = s_e \cdot \Delta t^{-1/2} \quad (A17)$$

and Δt is the model timestep length (s), while s_e is the soil desorptivity (m s$^{-1/2}$), calculated as in *Entekhabi and Eagleson* [1989]:

$$s_e = \left[\frac{8 \cdot \Phi_1 \cdot k_{sat} \cdot \phi_b}{3(1+3 \cdot m)(1+4 \cdot m)}\right]^{1/2} S_0^{(m/2+2)} \quad (A18)$$

where k_{sat} is the saturated hydraulic conductivity of the first soil layer (m s^{-1}), m is the pore size distribution index assumed to be a function of soil texture (Table A3), and $S_0 = \theta_1/(\Phi_1 \cdot l_1)$ is the fractional saturation. As in *Manfreda et al.* [2005], the closed canopy fraction (f_c) is defined by the empirical relationship [*Eagleson, 1982*]

$$f_c = 1 - e^{-\mu \cdot LAI} \quad (A19)$$

where μ is a constant for a given land cover type given by Table A1.

Calculating E_c requires a full canopy water balance model. The canopy water balance is calculated using:

Table A3. Soil properties

Symbol	Definition	Units	Soil Texture Class						Source
			Sand	Loamy Sand	Sandy Loam	Silty Loam	Loam	Muck[†]	
k_{sat}	Saturated Conductivity	×10⁻⁶ m s⁻¹	58.3	17.0	7.19	1.9	3.67	1.19	1
Φ	Total Porosity	-	0.437	0.437	0.453	0.501	0.463	0.398	1
θ_r	Residual Water Content	-	0.020	0.035	0.041	0.015	0.027	0.068	1
i_{sat}	Infiltration Capacity	×10⁻⁶ m s⁻¹	58.3	17.0	7.19	1.9	3.67	1.19	2
Φ_{-33}	Field Capacity	-	0.14	0.15	0.2	0.28	0.25	0.25	2
$\Phi_{-10,000}$	Wilting Point	-	0.07	0.07	0.09	0.10	0.11	0.16	2
α	Van Genuchten parameter	m⁻¹	14.5	12.4	7.5	2.0	3.6	5.9	3
N	Van Genucthen parameter	-	2.68	2.28	1.89	1.41	1.56	1.48	3
ϕ_b	Bubbling Pressure	m	0.073	0.087	0.147	0.208	0.112	0.281	1
m	Pore Size Distribution	-	0.695	0.553	0.378	0.234	0.252	0.319	1

[1] *Stieglitz et al.* [1997]
[2] *Saxton et al.* [1986]
[3] *Carsel and Parrish* [1988]
[†] Assumed same properties as sandy clay loam

$$\Delta S_c = P - Int - E_c \quad (A20)$$

where ΔS_c is canopy water storage (m). Incoming rainfall is first subjected to interception up to the water holding capacity of the canopy given by *Dickinson et al.* [1991]:

$$S_{c\max} = 1\times10^{-4} \cdot LAI (m), \quad (A21)$$

where LAI is the leaf area index (m²/m²). The available interception capacity of the canopy is then given by

$$\delta S = S_{c\max} - S_c^{t-1}. \quad (A22)$$

Additionally, we modify the model of *Chen et al.* [2005] to allow some water to penetrate the canopy at all times based on the assumption that the canopy is not completely closed. Interception at time t is then

$$Int = \min(\delta S, P \cdot f_c) \quad (A23)$$

Canopy evaporation, E_c is then calculated as in *Manfreda et al.* [2005] with

$$E_c = \min\left((S_c / S_{c\max})^{2/3} \cdot PE_c, S_c\right). \quad (A25)$$

Surface depression evaporation, E_d occurs only when water is stored in surface depressions. At each time step, any water stored in surface depressions S_d from the previous timestep is added to throughfall from the canopy (or snowmelt from the UEB model) such that precipitation excess runoff, R_e is given by

$$R_e = \min\left(0, P + S_d^{t-1} - Inf - Int\right), \quad (A26)$$

where infiltration, *Inf*, is calculated as discussed below. S_d is then calculated as

$$S_d = \min(S_{d\max}, R_e + Ex_1) \quad (A27)$$

where Ex_1 is the exfiltration out of the first soil layer and the depression storage capacity. $S_{d\max}$ is assumed to be constant for a given combination of slope, land cover, and soil type (see table in *Manfreda et al.* [2005] using values from *Liu et al.* [2003]). Depression evaporation, E_d is then given by

$$E_d = \min\left[S_d, (1 - F_c) \cdot PE_d\right]. \quad (A28)$$

INFILTRATION, PERCOLATION, THROUGHFLOW, AND EXFILTRATION

The next three terms of the water balance (Equation 1), percolation, Pc; throughflow, Tr; and exfiltration Ex are calculated within the soil water balance model. First, the outputs of the canopy model, snowmelt model, and depression storage model are used to calculate infiltration into the surface soil layer

$$Inf = \min\left(P - Int + S_d^{t-1}, i_{s\max}\right). \quad (A29)$$

The infiltration capacity, $i_{s\max}$ is a function of the moisture content of the surface soil layer and is calculated according to

$$i_{s\max} = \max\left(\Phi_1 \cdot l_1 - S_1^{t-1}, i_{sat}\right) \quad (A30)$$

where l_1 is the thickness of the first soil layer as defined previously, S_1 is the moisture stored within the first soil layer, and i_{sat} is the saturated infiltration capacity, which can vary with time due to the influence of impermeable frozen soils.

Infiltration is applied to the first soil layer, which can then percolate into the second layer. First, the soil moisture storage at the end of each timestep in the first layer is calculated as

$$S_1^{t+} = S_1^{t-1} + Inf - E_s - T_1 + Tr_1 - P_1 + Ex_2 - Ex_1 \quad (A31)$$

where P_1 is the percolation of water from the first soil layer to the second. Note that T_1 requires S_1. To avoid having to solve the coupled equations, T_1 is calculated using an intermediate value of $S_1^t = S_1^{t-1} + I_s - E_s$. Then $S_1 = S_1^{t-} - T_1$, and $S_1^{t+} = S_1 + Tr_1 - P_1 + Ex_2 - Ex_1$. Given S_1, percolation into the second layer, P_1 is given by

$$P_1 = \max\left(S_1 - S_{1\max}, P_{DREAM}\right) \quad (A32)$$

where $S_{1\max} = \Phi_1 \cdot l_1$ and P_{DREAM} is the percolation calculated according to *Manfreda et al.* [2005] given by:

$$P_{DREAM} = \begin{cases} 0 \\ S_{i\max}\left(S_i - \left[\dfrac{\Delta t \cdot k_{sat}(\gamma-1)}{S_{i\max}} + \left(\dfrac{S_i}{S_{i\max}}\right)^{1-\gamma}\right]^{1/(1-\gamma)}\right) \end{cases} \quad (A34)$$

$$S_i(t) \leq \Phi_{-33} \cdot l_i$$
$$S_i(t) > \Phi_{-33} \cdot l_i$$

where $\gamma = (2 + 3m)/m$. P_{DREAM} effectively allows percolation only when soil moisture exceeds the field capacity given by $\Phi_1 \cdot l_1$.

The second-layer soil moisture at the end of the timestep, (S_2), is calculated similarly to Equation A31:

$$S_2^{t+} = S_2^{t-1} + P_1 - P_2 - T_2 + Tr_2 - Ex_2, \quad (A35)$$

where T_2 is calculated from the values of S_2^{t-1} at the previous timestep.

$$Pc = P_2 = P_{DREAM} \quad (A36)$$

is calculated from an intermediate value of $S_1^{t-1} = S_1^{t-1} + P_1 - T_2$.

Throughflow out of a cell is calculated as

$$Tr_{out\,i} = l_i \cdot \Delta x \cdot \left(\dfrac{2}{1/k_{\theta i} + 1/k_{\theta i\,down}}\right) \cdot \left[\dfrac{\Delta z}{\Delta x} + \left(\dfrac{\Delta \psi}{\Delta \theta}\right)_i \cdot \left(\dfrac{\theta_{i\,down} - \theta_i}{\Delta x}\right)\right] \quad (A37)$$

where the effective unsaturated hydraulic conductivity for subsurface flow in layer i is taken as the harmonic average of the unsaturated hydraulic conductivity in the cell, $k_{\theta i}$ (m s^{-1}) and the down slope value $k_{\theta i\,down}$; Δx is the model cell resolution (m); $\frac{\Delta z}{\Delta x}$ is the vertical gradient in the down-slope direction; $\left(\frac{\Delta \psi}{\Delta \theta}\right)_i$ is the slope in meters of the moisture retention curve in layer i, and

$$\theta = \dfrac{S_i(t)/S_{i\max} - \theta_r}{\Phi - \theta_r}$$

where θ_r is the residual volumetric moisture content assumed to be a soil-texture dependent property (see Table A3). The assumption that flow only occurs parallel to the dip of the slope requires that $Tr_{out} \geq 0$. Tr_i is then calculated as

$$Tr_i = Tr_{out\,up} - Tr_{out\,i}. \quad (A38)$$

Finally, Ex_2 is calculated as the soil water in excess of saturation given by

$$Ex_2 = \max\left(0, S_2 + Tr_2 - S_{2\max}\right). \quad (A39)$$

This is then applied to the first layer prior to calculating

$$Ex_1 = \max\left(0, S_1 + Tr_1 + Ex_2 - S_{1\max}\right). \quad (A40)$$

RUNOFF ROUTING

Water exfiltrated from layer one is then applied to the surface depression model, thus R (m) is simply

$$R = R_e - S_d \quad (A41)$$

which is then routed to the streams using an approach modified from that presented by *Manfreda et al.* [2005]. Once generated, runoff cannot infiltrate and is instead routed overland and through streams according to the D8 flowdirection algorithm in ARC [*ESRI, 2003*] with runoff times given by the velocities in each cell along the flowpath. Runoff is assumed to travel overland at a velocity given by the Kerby time of concentration equation [*Kerby, 1959*]

$$t_{cell} = 86.735 \left(\frac{l_{cell} \cdot n}{\sqrt{s}} \right)^{0.467} \quad (A42)$$

where t_{cell} is the time required to completely traverse a model cell (s), l_{cell} is the length of the model cell (m), n is the dimensionless Manning's Roughness coefficient (values from [*McCuen, 2004*]) and s is the fractional slope of the cell in the downslope direction. The velocity (m s^{-1}) is then

$$v_{land} = \frac{l}{t_{cell}}. \quad (A43)$$

Once the runoff enters the stream channel its velocity is calculated using Manning's Equation [*McCuen, 2004*]

$$v_{stream} = \frac{1}{n} r^{2/3} s^{1/2} \quad (A44)$$

where r is the hydraulic radius (m) given by the ratio of the stream cross-sectional area to the wetted perimeter.

REFERENCES

Arnold, J. G., Allen, P. M., and G. Bernhardt (1993), A comprehensive surface-groundwater Flow model, *Journal of Hydrology*, 142, 47–69.

Berberan-Santos, M. N., Bodunov, E. N., and L. Pogliani (1997), On the barometric formula, *American Journal of Physics*, 65(5), 404–412.

Beven, K. J. and M. J. Kirby (1979), A physically-based variable contributing area model of basin hydrology, *Hydrology Science Bulletin*, 24(1), 43–69.

Bouraoui, F., Vachaud, G., Li, L. Z. X., Le Treut, H., and T. Chen (1999), Evaluation of the impact of climate changes on water storage and groundwater recharge at the watershed scale, *Climate Dynamics*, 15(2), 153–161.

Boutt, D. F., Hyndman, D. W., Pijanowski, B. C. and D. T. Long. (2001), Identifying potential land use-derived solute sources to stream baseflow using ground water models and GIS, *Ground Water*, 39, 24–34.

Carsel, R.F. and R.S. Parrish (1988), Developing joint probability distributions of soil water retention characteristics, *Water Resources Research*, 24(5), 755–769.

Cengel, Y.A., and M. Boles (2001). *Thermodynamics: An Engineering Approach*. McGraw-Hill.

Chen, J.M., Chen, X.Y., Ju, W.M., and X.Y. Geng (2005), Distributed hydrological model for mapping evapotranspiration using remote sensing inputs, *Journal of Hydrology*, 305, 15–39.

Choudhury, B.J., and J.L. Monteith (1988), A four-layer model for the heat budget of homogeneous land surfaces, *Quarterly Journal of the Royal Meteorological Society*, 114, 373–398.

CODATA (2002), CODATA Internationally Recommended Values for the Fundamental Physical Constants, *National Institute of Standards and Technology*, http://physics.nist.gov/cuu/Constants/index.html.

Dripps, W. R., Hunt, R. J. and Anderson, M. P. (2006), Estimating recharge rates with analytic element models and parameter estimation. *Ground Water* 44 (1), 47–55.

Eagleson, P. S. (1982), Climate, soil and vegetation, 5, A derived distribution of storm surface runoff, *Water Resources Research*, 18(2), 325–340.

Elliot, D. L., Holladay, C. G., Barchet, W. R., Foote, H. P., and W. F. Sandusky (1986), *Wind Energy Resource Atlas of the United States*, Pacific Northwest Laboratory: Richland, WA.

ESRI (2003), ArcInfo Workstation, Environmental Research Systems Institute: Redlands, CA.

Farrand, W. R., and D. L. Bell (1982), *Quaternary Geology of Southern Michigan*, The University of Michigan, Ann Arbor, MI.

Goff, J. A., and S. Gratch (1946), Low-pressure propreties of water from –160 to 212 F, *Transactions of the American Society of Heating and Ventilating Engineers*, 95–122

Halliday, D., Resnick, R., and J. Walker (2004). *Fundamentals of Physics*, Wiley.

Harbaugh, A. W., E. R. Banta, M. C. Hill, and M. G. McDonald. (2000). *MODFLOW-2000, The U.S. Geological Survey modular ground-water model- user guide to modularization concepts and the ground water flow processes*. Open-File Report 00-92, USGS, Reston, VA.

Healy, R. W., P. G. Cook (2002), Using groundwater levels to estimate recharge, *Hydrogeology Journal*, 10(1), 91.

Houghton, J., Ding, Y., Griggs, D., Noguer, M., van der Linden, P., Dai, X., Maskell, K., Johnson, C. (Eds.), (2001), *IPCC 2001, Climate Change 2001: The Scientific Basis*. Contribution of Working Group I to the Third Assessment Report of the Intergovernmental Panel on Climate Change. Cambridge University Press: Cambridge.

Hillel, D. (1980). *Introduction to Soil Physics*, Academic Press Inc.: San Diego, CA.

Jayawickreme, D. H., and D. W. Hyndman (2007), Evaluating the influence of land cover on seasonal water budgets using Next Generation Radar (NEXRAD) rainfall and streamflow data, *Water Resources Research*, 43, W02408, doi:10.1029/2005WR004460.

Jackson, R. B., Canadell, J., and J. R. Ehleringer (1996), A global analysis of root distributions for terrestrial biomes, *Oecologia*, 108, 389–411.

Kerby, W. S. (1959). Time of concentration for overland flow, *Civil Engineering*, 29(3), 60.

Manfreda, S., Fiorentino, M., and V. Iacobellis (2005). DREAM: a distributed model for runoff, evapotranspiration, and antecedent soil moisture simulation, *Advances in Geosciences*, 2, 31–39.

McCuen, R. H. (2004). *Hydrologic Analysis and Design*, Pearson Prentice Hall: Upper Saddle River New Jersey.

MDNR (2001) *Integrated Forest Monitoring Assessment and Prescription.* Michigan Department of Natural Resources: Lansing, MI.

Prudic, D. E., L. F. Konikow, and E. R. Banta (2004), *A new stream-flow routing (SFR1) package to simulate stream-aquifer interaction with MODFLOW-2000*, Open-File Report 2004-1042, USGS, Reston, VA.

Ritchie, J. T., (1972), Model for Predicting Evaporation from a Row Crop with Incomplete Cover, *Water Resources Research*, 8(5), pp. 1204–1213.

Ross, M., Geurnik, J., Said, A., Aly, A. and P. Tara (2005), Evapotranspiration Conceptualization in the HSPF-Modflow Integrated Models, *Journal of the American Water Resources Association*, 41(5), 1013–1025.

Said, A., D. K. Stevens and G. Sehlke, 2005, Estimating water budget in a regional aquifer using HSPF-MODFLOW integrated model, No. 03151, 41(1), pp. 55–66.

Sanford, W. (2002), Recharge and groundwater models: an overview, *Hydrogeology Journal*, 10(1), 110–120.

Saxton, K. E., Rawls, W.J., Romberger, J. S., and R. I. Papendick (1986), Estimating generalized soil-water characteristics from texture, *Soil Scienec Society of America Journal*, 90, 1031–1036.

Scanlon, B. R., Healy, R. W, Cook, P. G. (2002), Choosing appropriate techniques for quantifying groundwater recharge, *Hydrogeology Journal*, 10(1), 18–39.

Schaetzl, R. J., and D. M. Tomczak (2001), Wintertime temperatures in the fine-textured soils of the Saginaw Valley, Michigan, *The Great Lakes Geographer*, 8(2), 87–99.

Schulze, E. D., Kelliher, F. M., Körner, C., Lloyd, J., and R. Leuning (1994), Relationships among maximum stomatal conductance, ecosystem surface conductance, carbon assimilation rate, and plant nitrogen nutrition: a global ecology scaling exercise. *Annual Reviews of Ecological Systems*, 25, 629–60.

Sophocleous, M. and S. P. Perkins (2000), Methodology and application of combined watershed and ground-water models in Kansas, *Journal of Hydrology*, 236, 185–201.

Stieglitz, M. Rind, D., Famiglietti, J., and C Rosenzweig (1997), An efficient approach to modeling the topographic control of surface hydrology for regional and global climate modeling, *Journal of Climate*, 10, 118–137.

Tang, Z., Engel, B. A., Pijanowski, B. C., and K. J. Lim. (2005), Forecasting land use change and its environmental impact at a watershed scale. *Journal of Environmental Management*, 76. 35–45.

Tarboton, D. G. and C. H. Luce, (1996), "*Utah Energy Balance Snow Accumulation and Melt Model (UEB)*," Computer model technical description and users guide, Utah Water Research Laboratory and USDA Forest Service Intermountain Research Station (http://www.engineering.usu.edu/dtarb/).

Thoms, R. B., Johnson, R.L., and R. W. Healy, 2006, User's guide to the variably saturated flow (VSF) Process for MODFLOW: U.S. Geological Survey Techniques and Methods 6-A18, 58 p.

VanderKwaak, J. E., (1999). *Numerical simulation of flow and chemical transport in integrated surface-subsurface hydrologic systems*, Ph.D. Dissertation, Dept. of Earth Sciences, University of Waterloo: Ontario, Canada.

VanderKwaak, J. E., and K. Loague (2001), Hydrologic-response simulations for the R-5 catchment with a comprehensive physics-based model, *Water Resources Research*, 37(4) 999–1013.

van Genuchten, M. Th. (1980), A closed form equation for predicting the hydraulic conductivity of unsaturated soils, *Soil Science Society of America Journal*, 44, 892–898.

Voss, C. I., and Provost, A. M. (2002), *SUTRA, A model for saturated-unsaturated variable-density ground-water flow with solute or energy transport*, U.S. Geological Survey Water-Resources Investigations Report 02-4231.

Walko, R. L, and C. J. Tremback (2005), *Modifications for the transition from LEAF-2 to LEAF-3*, ATMET Technical Note

Yeh, G. T. and G. B. Huang (2003), *A numerical model to simulate water flow in watershed systems of 1-D stream-river network, 2-D overland regime, and 3-D subsurface media (WASH123D: Version 1.5)*, Technical Report. Dept. of Civil and Environmental Engineering, University of Central Florida: Orlando, Florida.

Integrating Geophysical, Hydrochemical, and Hydrologic Data to Understand the Freshwater Resources on Nantucket Island, Massachusetts

Andee J. Marksamer[1], Mark A. Person[1], Frederick D. Day-Lewis[2], John W. Lane Jr.[2], Denis Cohen[1]; Brandon Dugan[3], Henk Kooi[4], and Mark Willett[5]

In this study we integrate geophysical, hydrologic, and salinity data to understand the present-day and paleo-hydrology of the continental shelf near Nantucket Island, Massachusetts. Time-domain electromagnetic (TDEM) soundings collected across Nantucket and observed salinity profiles from wells indicate that the saltwater/freshwater interface is at least 120 m below sea-level in the northern and central portions of the island, far deeper than predicted (80 m) by modern sea-level conditions. TDEM soundings also indicate that higher salinity conditions exist on the southern end of the island. These findings suggest a relatively high-permeability environment. Paradoxically, a deep, scientific borehole (USGS 6001) on Nantucket Island, sampling Tertiary and Cretaceous aquifers, is over-pressured by about 0.08 MPa (8 m excess head), which is suggestive of a relatively low-permeability environment. We constructed a series of two-dimensional, cross-sectional models of the paleohydrology of the Atlantic continental shelf near Nantucket to understand the flushing history and source of overpressure within this marine environment. We considered two mechanisms for the emplacement of freshwater: (1) meteoric recharge during sea-level low stands; and (2) sub-ice-sheet and glacial-lake recharge during the last glacial maximum (LGM). Results indicate the sub-ice-sheet recharge from the Laurentide Ice Sheet was needed to account for the observed salinity/resistivity conditions and overpressures. Both TDEM soundings and model results indicate that a lateral transition from fresh to saltwater occurs near the southern terminus of the island due to ice sheet recharge. We also conclude that the overpressure beneath Nantucket represents, in part, "fossil pressure" associated with the LGM.

[1] Geological Sciences, Indiana University, 1001 E. 10th Street, Bloomington, Indiana, amarksam@indiana.edu.
[2] U.S. Geological Survey, WRD, Office of Ground Water, Branch of Geophysics, 11 Sherman Place, Unit 5015, Storrs, Connecticut.
[3] Department of Earth Science, Rice University, 6100 Main Street, Houston, Texas.
[4] Vrije Universiteit, De Boelelaan 1085, 1081 HV Amsterdam, The Netherlands.
[5] Wannacomet Water Company, 1 Milestone Rd., Nantucket Massachusetts.

Subsurface Hydrology: Data Integration for Properties and Processes
Geophysical Monograph Series 171
Copyright 2007 by the American Geophysical Union.
10.1029/171GM12

INTRODUCTION

Many coastal aquifer systems on the Atlantic Continental Shelf offshore from New England, USA, have anomalous volumes of freshwater that extend far offshore (>20 km) [*Hathaway et al.*, 1979; *Kohout et al.*, 1977]; these volumes are difficult to explain based on modern sea-level conditions. On Nantucket Island, Massachusetts, for example, monitoring wells PT-12 and PT-13, installed to 90 and 110 meters below sea level (mbsl), respectively, and well USGS 6001, installed to 514 mbsl, contain ground water with salinities of less than 1 part per thousand (ppt) within all permeable units (Fig. 1) [*Folger et al.*,

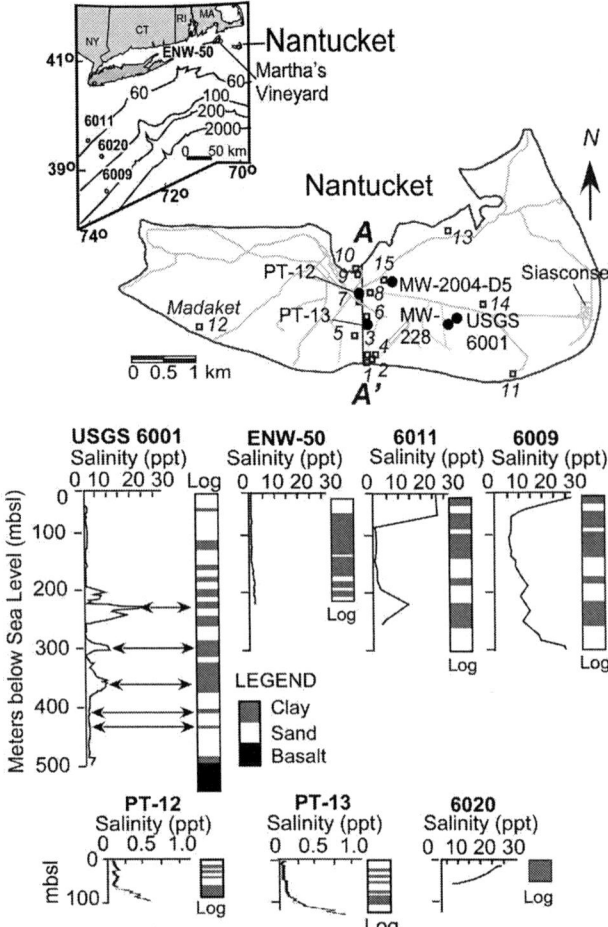

Figure 1. Location of wells on Nantucket Island, Martha's Vineyard Island and the Atlantic Continental Shelf. Numbered squares on the Nantucket map indicate the location of time-domain electromagnetism (TDEM) soundings. Cross section A-A' is associated with the cross section on Figure 6. Salinity profiles are shown of wells on the North Atlantic Continental Shelf. Bathymetry contours are in meters [Kohout et al., 1977; Folger et al., 1978; Hathaway et al., 1979; Hall et al., 1980]. Note the change in scale for wells PT-12 and PT-13.

1978]. The observed freshwater below Nantucket is substantially deeper than would be expected based on the Ghyben-Herzberg principle [Drabbe and Badon-Ghyben, 1889; Herzberg, 1901; McWhorter and Sunada, 1993]. Given water-table elevations of about 2 m at wells PT-12 and PT-13, the Ghyben-Herzberg principle predicts that the freshwater/saltwater interface is approximately 80 m below sea level:

$$Z = \frac{\rho_f}{\rho_s - \rho_f} h \approx 40h, \quad (1)$$

where Z is depth of the freshwater/saltwater interface below sea-level (L), ρ_f is density of freshwater (1000 kg/m^3) (M/L^3), ρ_s is density of saltwater (1025 kg/m^3) (M/L^3), and h is height of the water table above sea level (L). This is substantially shallower than the observed freshwater at 514 mbsl within USGS 6001.

Inspection of long-term water-level records from wells located near Nantucket Island's principle well fields in the center of the island reveal no declining trends over the last two decades; hence it is unlikely that this discrepancy results from mining of ground water by over-pumping. Indeed, the withdrawal rates represent less than 10% of the available recharge on the island. Unexpected freshwater has also been observed on Martha's Vineyard, Massachusetts (to depths of 228 m), and even farther (>100 km) off the New Jersey coast (wells 6009, 6011, 6020) where pore waters with solute concentrations of less than 5 ppt have been noted within Pleistocene, Pliocene, Miocene, and Upper Cretaceous sand units [Kohout et al., 1977; Hathaway et al., 1979; Hall et al., 1980] (Fig. 1).

How and when was freshwater emplaced within the confined aquifers beneath Nantucket Island? During Pleistocene sea-level low stands, hydraulic heads along the Massachusetts coast would have been far too low to drive meteoric water very far offshore [Kooi and Groen, 2001]; moreover, confined aquifers south of Martha's Vineyard and Nantucket outcrop or subcrop below sea level in Nantucket Sound, thus onshore recharge cannot be occurring presently. The fresh and brackish sub-seafloor pore waters that are present along the northeastern U.S. Atlantic coast can be considered key examples of paleo-ground water that was emplaced during Pleistocene sea-level low stands and escaped salinization (e.g., flushing with sea water) during Holocene sea-level rise [Hathaway et al., 1979; Meisler et al., 1984; Kooi and Leijnse, 2000; Person et al., 2003]. Several competing mechanisms have been proposed to explain how paleo ground water is emplaced during glacial low-stand periods. Early studies considered the shore-normal hydraulic gradient associated with the topography of the continental shelf as the primary driving force for freshwater recharge during sea-level low stands [Meisler et al., 1984] (Fig. 2A). More recently, Groen et al. [2002] argued that local flow systems associated with secondary topography of the subaerially exposed and incised shelf are essential to emplace meteoric water far out onto the continental shelf (Fig. 2A), whereas Person et al. [2003] emphasized the role of sub-ice-sheet recharge (Fig. 2B).

Several proglacial lakes formed following the last glacial maximum (LGM) as the Laurentide ice sheet retreated (Fig. 3) [Uchupi et al., 2001]. Seepage from the proglacial lakes during the LGM in Nantucket Sound, Cape Cod Bay, Block Island Sound, and Long Island Sound (Fig. 3) could have provided extensive recharge to the confined aquifers of the continental shelf of New England. When the southern dams

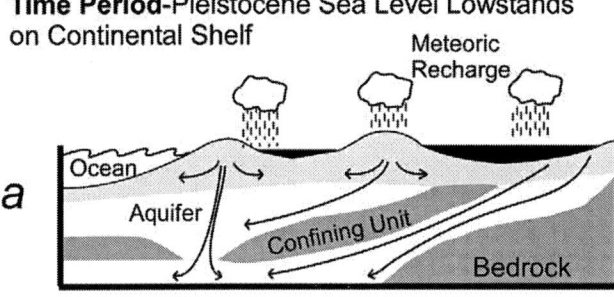
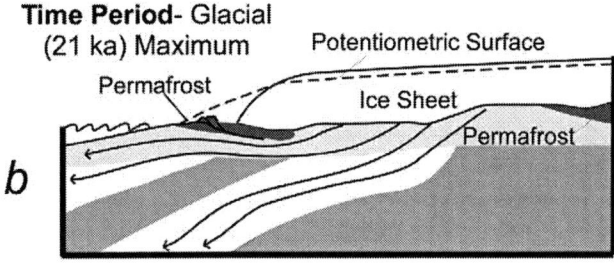
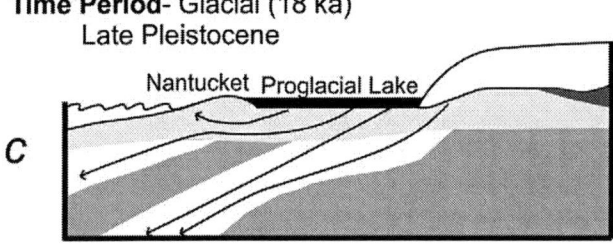

Figure 2. Conceptual models for freshwater plumes: (a) lateral incursion of freshwater during the Pleistocene sea-level low stands [Meisler et al., 1984] and vertical infiltration of meteoric water induced by local flow cells on the Continental Shelf [Groen., 2002]; (b) sub-ice-sheet recharge from the Laurentide ice sheet [Person et al., 2003]; (c) infiltration beneath pro-glacial lakes.

of the proglacial lakes failed, rapid sedimentation on the continental shelf would have also changed the land-surface morphology, created rapid sediment loading, and perhaps driven infiltration of freshwater [Uchupi et al., 2001] (Fig. 2C). Local recharge could be facilitated by gaps in confining units due either to non-deposition or erosion.

There is evidence that flushing with freshwater has occurred on the continental shelf within the deep Cretaceous and Tertiary units on Nantucket during the recent past (USGS 6001, Fig. 1). All permeable Cretaceous and Tertiary units within USGS 6001 contain freshwater; however, the pore waters within the thick low-permeability units show high salinities of up to 12–13 parts per thousand (ppt) and display vertical solute diffusion profiles. Interestingly, the thin confining units contain low-salinity pore water, which indicates that vertical diffusion of the solutes has already run its course (Fig. 1–see salinity profile in well USGS 6001); hence

the Tertiary and Cretaceous sand units once contained high salinity waters and were flushed with fresh ground water, but diffusive transport of saltwater within the thick, low-permeability units has not yet reached equilibrium. Simple vertical diffusion models [Person et al., 2003] suggest that the flushing of the Tertiary and Cretaceous aquifers occurred as recently as 21 ka, about the time of the LGM.

In addition to anomalous amount of freshwater observed below Nantucket, ground-water elevations measured monthly as part of the National Water Information System for the U.S. Geologic Survey (USGS) [http://nwis.waterdata.usgs.gov/nwis/gwlevels] have been used to identify elevated hydraulic heads within the deep Tertiary and Cretaceous units at well USGS 6001 on Nantucket. Hydraulic heads within USGS 6001 are overpressured by up to 8m above sea level (0.08 MPa) and 4m above the local ground-water table as shown in USGS 228 (Fig. 4). Smaller overpressure (approximately 0.3 m) was also observed within a 228-m deep well (ENW-60) on Martha's Vineyard [Kohout et al., 1988]. Even based on the water-table elevations of 7.5 m above sea level in the deep USGS 6001, the Ghyben-Herzberg principle still under predicts the amount of freshwater below Nantucket (predicted 300 m versus observed 514 m).

Overpressure observed in the Tertiary and Cretaceous deposits and the anomalous quantities of freshwater within those units represents a paradox. High fluid pressures are

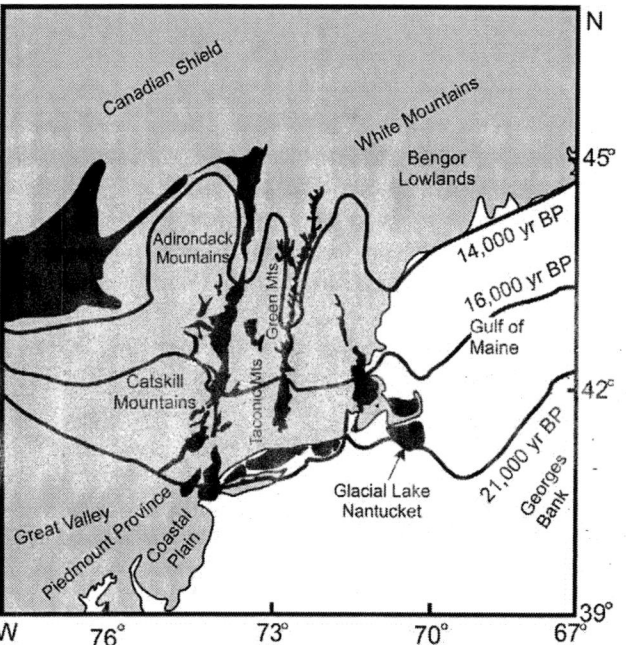

Figure 3. Distribution of proglacial lakes and position of maximum ice sheet extent during the Late Quaternary on the Atlantic Continental Shelf [after Uchupi et al., 2001].

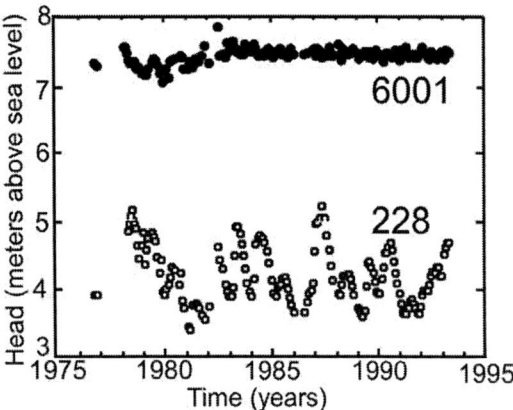

Figure 4. Head data comparing the local water table elevation in well 228 and Cretaceous aquifers in well 6001 beneath Nantucket Island between 1976–1993. The data indicated that the deep aquifers are anomalously pressured and disconnected from the water table flow system. The locations of the wells are shown on Figure 1.

typically suggestive of a low-permeability environment whereas the large quantities of freshwater beneath Nantucket are suggestive of a high-permeability system. However, the elevated heads in the Tertiary and Cretaceous aquifers beneath Nantucket may have occurred in the recent past as a result of rapid sedimentation following glaciation. Near-lithostatic overpressures have been inferred offshore from New Jersey where sedimentation rates were high (~ 1 mm/yr) and confining units are thick [*Dugan and Flemings*, 2000]. Fluid-flow models by these authors demonstrated how overpressures offshore from New Jersey focus fluid migration along permeable sand layers; Dugan and Flemings [2000] argued that Late Pleistocene sedimentation was the sole mechanism responsible for high heads inferred from porosity data and laboratory experiments; however, it is also possible that the high heads are relic features created by extremely high hydraulic heads along the continental shelf due to glacial loading.

To date most studies that have developed quantitative models of ground-water flow beneath ice sheets have utilized static representations of ice-sheet geometry and neglected the effects of hydromechanical loading [*Boulton et al.*, 1995; *Piotrowski*, 1997; *Person et al.*, 2003]; however, if the ice sheet overrides low-permeability sediments, or if rapid sedimentation occurs following glaciation, then hydromechanical effects need to be simulated to account for the effect of loading on ground-water flow [*Lerche et al.*, 1997; *Bekele et al.*, 2003; *Lemieux et al.*, 2006; *Person et al.*, 2007]. Here we use variable-density, cross-sectional models of fluid flow, solute transport, and sediment loading to predict fluid pressures and solute distribution. The hydrostratigraphy represented in this model is based on detailed stratigraphic correlations between available wells on Nantucket Island and Martha's Vineyard, as well as a series of offshore borings. These simulations are compared with time-domain electromagnetic (TDEM) soundings collected from across Nantucket and salinity profiles and hydraulic head pressures from wells on the island. This study helps to unravel the complex paleo-hydrologic history associated with the LGM and subsequent sea level rise while also providing critical information about the distribution of the potable freshwater resources available to Nantucket Island.

STUDY AREA

Basement rocks beneath Nantucket Island consist of Triassic basalts overlain by 460 m of coastal plain sediments of Cretaceous, Tertiary, and Quaternary age [*Folger et al.*, 1978]. Cretaceous sediments on Nantucket range from 118 to 457 mbsl (Fig. 1). The lower portion of that section contains soft, unconsolidated layers of sand and clay, whereas the upper section is mostly clay with interbedded layers of sand [*Folger et al.*, 1978]. Tertiary glauconitic green sand is regionally located above the Cretaceous package; these Tertiary sands have been penetrated by several wells across the island at 80–90 mbsl. Paleontologic studies have linked glauconitic green sand to a similar green sand unit observed on nearby Martha's Vineyard at approximately 30 mbsl [*Hall et al.*, 1980]. Folger et al. [1978] inferred 85 m of Pleistocene sediment overlying the green Tertiary sand within boring USGS 6001. Within the 85 m of Pleistocene sediment, Folger et al. [1978] identified mostly medium-to-coarse grained sand within several layers containing shell fragments up to 7 cm across at depths between 28 and 51 mbsl. The Wannacomet Water Company on Nantucket produces groundwater from these shelly layers. Folger et al. [1978] also describe two zones of either glacial till or weathered soil at 25 and 53 mbsl. Wisconsin-aged glacio-lacustrine deposits of up to 160-m thickness have been identified in Nantucket Sound just north of Nantucket Island [*O'Hara and Oldale*, 1987]; these deposits consist of mostly silt and clay. Similar glacio-lacustrine deposits were observed throughout the western region of Cape Cod [*Masterson et al.*, 1997]. It is likely that these deposits were part of Glacial Lake Nantucket, and therefore, likely extend south below Nantucket to the location of the terminal moraine. Highlands reaching 33 m above sea level in the central eastern portion of Nantucket Island are the remnants of the glacial moraine. Ice contact deposits are also found in this area [*Oldale*, 1985].

Nantucket's demand for potable water has increased significantly over the last two decades due to tourism and development. One new well field has been brought on line (State

Forest) in the past decade and another is planned (North Pasture). The majority of Nantucket's municipal potable water is supplied by well fields (PW-12 and PW-13) located within the central but narrowest portion of the island; this situation highlights the vital need to understand the distribution of freshwater across the island. Water-table elevations vary across the island from up to 3.5 m above sea level in the widest part of the island to less than 2 m at the narrowest part of the island where municipal wells tap into the island aquifers. The water table declines near the coastline where it is even with modern sea level. Numerical modeling, based on projected pumping increases, indicates that saltwater upconing could occur within municipal water wells before 2014 [*Person et al.*, 1998]; however, these models did not account for the presence of the potentially thick confining glacio-lacustrine deposits from former Glacial Lake Nantucket nor the offshore freshwater. Nevertheless, these studies have led to increased interest in monitoring the distribution of freshwater and saltwater across the island. Understanding the regional system and emplacement mechanisms of the aquifers on Nantucket, will enable better design and implementation of a safe, efficient strategy to prevent migration of saltwater into potable water sources.

GEOPHYSICAL DATA

The time-domain electromagnetic (TDEM) method [*Nabighian and Macnae*, 1991] is commonly used for ground-water resource evaluation [*Fitterman and Stewart*, 1986], determination of the depth to the freshwater/saltwater interface [*Fitterman and Deszcz-Pan*, 1997], and mapping the horizontal extent of saline intrusions [*Mills et al.*, 1988]. A TDEM sounding involves the application of a direct current to an ungrounded transmitting coil, which is generally placed at the earth's surface. The transmitter induces eddy currents in subsurface conductors. The current to the transmitting coil is switched off, and the subsequent decay of the secondary fields produced by the eddy currents is recorded at a receiver coil. Inverse modeling of apparent-resistivity data from a single TDEM sounding yields estimates of 1-D (layered) resistivity structure, which provides insight into subsurface lithology and pore-fluid salinity.

Although the TDEM results are useful for understanding subsurface variations in salinity, the method is limited in two important respects. First, the reliability of TDEM results decreases with depth. The maximum depth of investigation in this study is about 120 m based on sensitivity analysis conducted for inversion results. Second, the TDEM soundings may not resolve thin layers or lenses that pinch out laterally, especially at depth

Geophysical Surveys

TDEM soundings were collected on Nantucket Island at 10 sites in May 2002, and at an additional 6 sites in August 2004 (Fig. 1). A Geonics PROTEM 47[1] system was used with a single-turn transmitter coil. For the 2002 data, soundings were conducted using currents of 2.5 to 3.0 amps in a 40-m by 40-m transmitter coil. In 2004, a current of 2.0 amps was used with a 60-m by 60-m transmitter coil. The receiver coil was coplanar with, and located in the center of, the transmitter loop. The TDEM field data were processed using the commercial software package TEMIX. Sounding data were inverted using smoothness-constrained, 2-, 3-, and 4-layer models, depending on the shape of the sounding curves as shown in Fig. 5. Soundings include ultra high frequency (UHF) and very high frequency (VHF) measurements. Inversion results were interpreted for the depths to the freshwater/saltwater interface across the island.

Geophysical Results

TDEM soundings collected along a north-south transect across Nantucket indicate that the freshwater/saltwater interface may be approximately 120 m below the surface in the northern and central portions of the island (Fig. 6). Saltwater was identified at shallower depths in the southern portion of the island (e.g., 35–45 mbsl). The southward shallowing of the saltwater may result from pumping of municipal water supply wells. It is also possible that the freshwater/saltwater interface is shallower along the southern portion of the island because the glacio-lacustrine deposit does not extend that far south, and therefore, the interface may have equilibrated locally with modern sea-level conditions. Based on observations from the Wannacomet Water Company, the northern portion of the island near Nantucket Harbor contains thick low-permeability sediments. These low-permeability sediments may be responsible for preserving the deep freshwater/saltwater interface in that area. Interestingly, results from two TDEM locations (Sites 14 and 15 on Fig. 1) indicate at depths of 80 and 120 m, respectively, a resistivity too high for a sand to be saturated with true seawater. This could be a transition zone or possibly saltwater trapped within a confining unit. It is, however, possible that freshwater exists below these depths as the maximum penetration for the TDEMS soundings is approximately 120 m. The position of the fresh-

---

[1]The use of trade, product, or firm names in this publication is for descriptive purposes only and does not imply endorsement by the U.S. Government.

148 DATA INTEGRATION IN SUBSURFACE HYDROLOGY

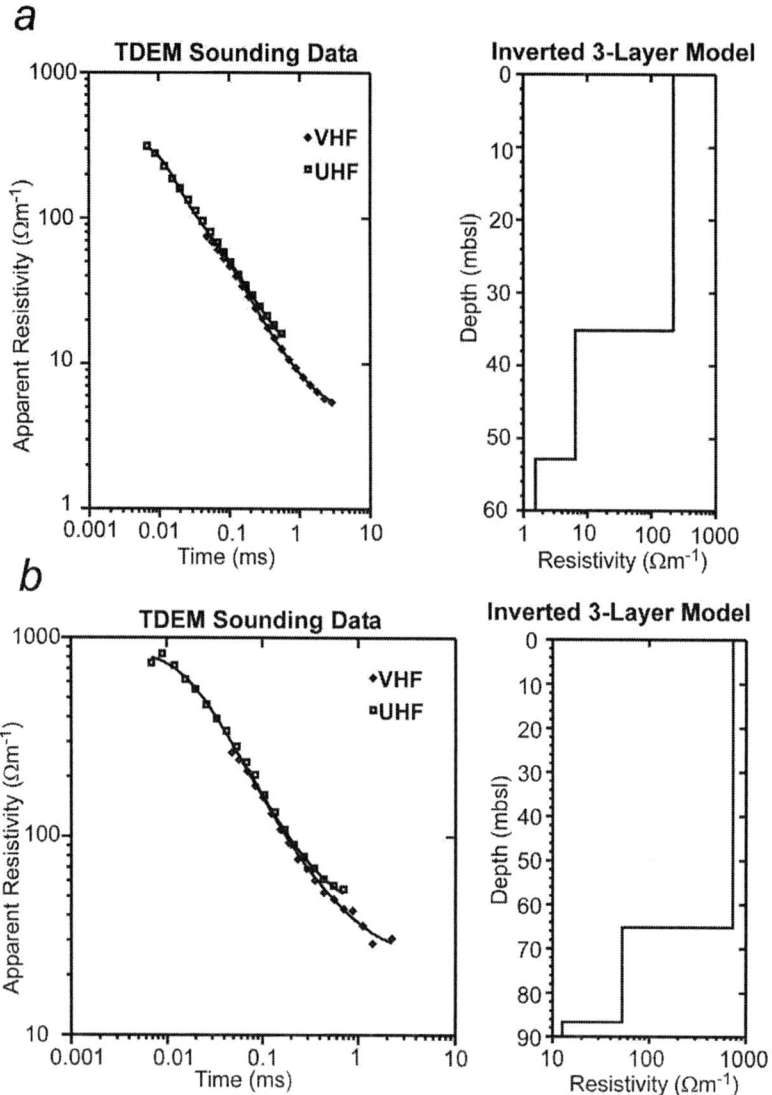

Figure 5. TDEM sounding data and 3-layer inversion results for (a) Site 12 and (b) Site 14 shown on Figure 1. Sounding include ultra high frequency (UHF) and very high frequency (VHF) measurements.

water/saltwater interface based on the Ghyben-Herzberg principle is indicated by the dashed line shown on Fig. 6.

MATHEMATICAL MODELING

A two-dimensional, cross-sectional, finite-element model was developed extending from Nantucket Sound, through Nantucket Island, to the continental slope approximately 250 km southeast of Nantucket (Plate 1a). The purpose of the model was to test how sea-level changes, rapid sediment loading following deglaciation, and increased ground-water recharge from beneath the Laurentide Ice Sheet and from Glacial Lake Nantucket may have influenced the hydraulic heads and the spatial distribution of fresh ground water beneath Nantucket. We begin by presenting results that evaluate whether or not sea-level oscillations alone could have produced the anomalous freshwater and hydraulic heads observed beneath Nantucket. We then consider what impact the Laurentide Ice Sheet and Glacial Lake Nantucket may have had on this system.

The governing ground-water flow equation used to quantify the Pleistocene hydrogeology of the Atlantic Continental Shelf in New England is:

$$\nabla \cdot \left[\rho_f K \mu_r \left(\nabla h + \rho_r \nabla z \right) \right] = S_s \rho_o \left[\frac{\partial h}{\partial t} - \frac{\rho_{ice}}{\rho_f} \frac{\partial \eta}{\partial t} - \frac{\rho_s - \rho_f}{\rho_f} \frac{\partial L}{\partial t} \right], \quad (2)$$

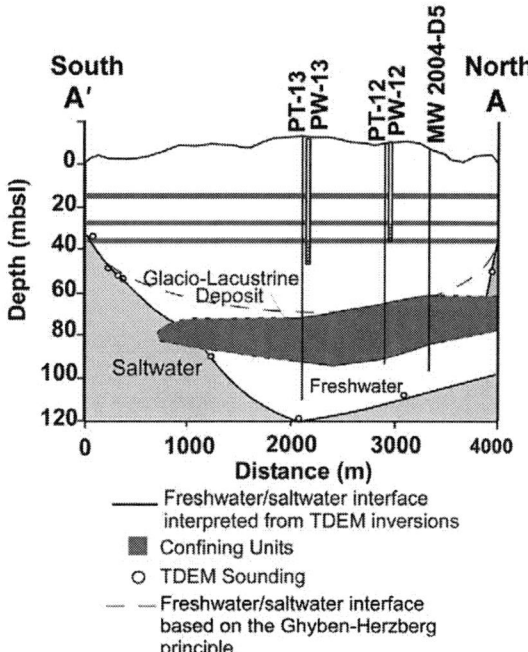

Figure 6. Cross section across Nantucket Island showing the position of the freshwater-saltwater interface estimated by 8 TDEM soundings. The estimated water table elevation and predicted position the freshwater-saltwater interface based using the Ghyben-Herzberg equation (equation 1) is show by a dashed line. The soundings suggest that a lateral transition from fresh to saltwater occurs beneath Nantucket Island from north to south. The location of the cross section is shown on Figure 1.

where t is time (T), ∇ is the nabla operator, K is the hydraulic conductivity tensor (L/T), h is equivalent freshwater hydraulic head (L), ρ_{ice} is ice density (M/L³), ρ_s is bulk sediment density (M/L³), L is the elevation of the sediment-water interface (L), ρ_r is the relative fluid density (M/L³), ρ_f is the fluid density (M/L³), η is the elevation of the top of the ice sheet (L), and S_S is the specific storage (L⁻¹). The relative fluid density, viscosity, and hydraulic conductivity are:

$$\rho_r = \frac{\rho_f - \rho_o}{\rho_o}, \quad (3a)$$

$$\mu_r = \frac{\mu_f}{\mu_o}, \text{ and} \quad (3b)$$

$$K = \frac{k\rho_o g}{\mu_f}, \quad (3c)$$

where ρ_o and μ_o are the reference density (M/L³) and viscosity (M/L·T) at standard state (0 MPa, 0.0 mg/l dissolved solids), respectively, k is the permeability (L²), μ_f is the water viscosity (M/L·T) and g is the acceleration due to gravity (L²/T). Equation 2 is capable of representing ground-water flow induced by water-table gradients, spatial variations in subsurface fluid density as well as sediment and ice-sheet loading [*Person et al.*, 2007].

Thermodynamic equations of state are required to compute the density and viscosity of ground water at elevated temperature, pressure, and salinity conditions. We assume isothermal conditions for these cross-sectional model runs; this assumption is warranted given that the basin was relatively thin (<1200 m). We use the equations of state of Kestin et al. [1981] that are capable of representing fluid density and viscosity.

Transient, advective-dispersive solute transport beneath Nantucket Island is represented by:

$$\nabla \cdot [D\nabla C] - \vec{v} \cdot \nabla C = \frac{\partial C}{\partial t}, \quad (4)$$

where C is solute concentration (mass fraction), v is the seepage velocity (x- and z-directions) (L/T), and D is the advection-dispersion tensor (L²/T) for the porous medium. Equation 4 neglects the effects of solute transport due to porosity changes associated with sediment and ice-sheet loading, which are expected to be small.

The Darcy flux used in the solute transport equation was computed using a variable-density formulation:

$$v_x = -\frac{K_x}{\phi}\mu_r \frac{\partial h}{\partial x}, \text{ and} \quad (5)$$

$$v_z = -\frac{K_z}{\phi}\mu_r \left[\frac{\partial h}{\partial z} + \rho_r\right], \quad (6)$$

where K_x and K_z are the principal components of the hydraulic conductivity in the x- and z-directions (L/T). The components of the advection-dispersion tensor are given by:

$$D_{xx} = \alpha_L \frac{v_x^2}{|v|} + \alpha_T \frac{v_z^2}{|v|} + D_s$$

$$D_{zz} = \alpha_T \frac{v_x^2}{|v|} + \alpha_L \frac{v_z^2}{|v|} + D_s \text{ and} \quad (7)$$

$$D_{zx} = D_{xz} = (\alpha_L - \alpha_T)\frac{v_x v_z}{|v|},$$

where α_L is the longitudinal dispersivity (L), α_T is the transverse dispersivity (L), $|v| = \sqrt{v_x^2 + v_z^2}$, and D_s is the solute molecular diffusion coefficient (L²/T). The solute transport and flow equations are formally coupled through the equations of state for fluid density and viscosity; however, the non-linearity is relatively weak and the two equations can be solved separately and sequentially while stepping through time.

Boundary conditions for the ground-water flow equation include a specified-head (water-table) condition at the Earth's surface and no-flow boundaries along the bottom and sides of the solution domain. The specified-head condition varies in time depending on sea-level, topography, and glacial position. Boundary conditions for solute transport equations include a constant concentration of 0.0 or 35 ppt depending on whether the local land-surface elevation is above or below sea level. No solute flux boundary conditions were imposed along the sides and base of the sedimentary basin.

The fluid flow equation is solved by the finite-element method. We solve the solute transport equation using the modified method of characteristics algorithm [*Zheng and Bennet*, 1995]. Three-node triangular elements are used to discretize the solution domain and the resulting set of algebraic equations is solved using Gaussian elimination. The solution domain is discretized using 12,030 triangular elements composed of 6,167 nodes. The elements size in the x-direction from vary between about 300 m in the western portion of the mesh in the vicinity of Nantucket, to 22 km in the eastern portion of the mesh near the continental slope. Element size in the z-direction varies between less than 1 m and approximately 300 m. We refined the grid within confining units to minimize numerical dispersion. The number of nodes in each vertical column ranges between 18 and 92.

The stratigraphy of the cross section is based on boring logs on Nantucket [*Kohout et al.*, 1977], Atlantic Coring Margin Coring project (AMCOR) wells [*Hathaway et al.*, 1979], and Continental Offshore Stratigraphic Tests (COST) wells G-1 and G-2 located in Georges Bank east of Cape Cod [*Scholle and Wenkam*, 1982; *Schlee and Fritsch*, 1982]. The stratigraphy was divided into five types of sediment including Tertiary and Cretaceous sands, shallow outwash sands, silts, clays, and limestone (Plate 1b), which consists of a wide range of porosities and permeabilities. We do not represent the sand and clay as being continuous as we found a lack of lateral correlation between aquifers and confining units in closely spaced wells (e.g., ENW-50 and USGS 6001). Porosity and permeability had to be assigned to each unit. Over the depth range considered (1.2 km) these properties can vary considerably, even within a lithologic unit due to mechanical compaction and diagenetic processes associated with burial; these effects can amount to orders-of-magnitude changes in the permeability of the sediment at depth. Although the relationship between porosity and permeability is complex, some studies have shown that there is a rough log-linear relationship [*Neuzil*, 1994; *Shenhav*, 1971; and *Lucia*, 1995]. We assume the log-linear relationship

$$\log_{10}(k_{max}) = a + b\phi, \qquad (8)$$

where k_{max} is the maximum permeability in a lithologic unit (L^2), ϕ is porosity (L^3/L^3) and a and b are empirical coefficients [Table 1]. Several studies [*Garven*, 1989; and *Corbet and Bethke*, 1992] have assumed a similar log-linear relationship between porosity and permeability in their numerical simulations. To apply this relationship, it is necessary to assume a compaction curve within the sediment. We relate porosity to effective stress and compressibility using a relationship presented by Hubbert and Rubey [1959] and Bethke and Corbet [1988]:

$$\phi = \phi_o e^{(-\beta \sigma e)}, \qquad (9)$$

where ϕ_o is the initial porosity (L^3/L^3), ϕ is the porosity at the maximum effective stress (L^3/L^3), β is the bulk compressibility of the sediment (L^2/N), and σ_e is the effective stress (N/L^2). σ_e can be calculated based on Terzaghi's principle:

$$\sigma_e = \sigma_v - P, \qquad (10)$$

where σ_v is the total vertical stress (N/L^2) or overburden stress and P is the pore pressure (N/L^2). The pore pressure, assuming hydrostatic conditions, is

$$P = (h-z)\rho_o g, \qquad (11)$$

where h is the hydraulic head (L), z is the depth (L), ρ_o is the base density of the fluid (M/L^3) and g is the acceleration

Table 1. Input parameters for porosity, permeability and effective stress relationships

Sediment Type	Compressibility m^2/N	Φ_o [a] m_3/m_3	a [b] $\log(m_2)$	b [b] $\log(m_2)$
Clay	8.0x10^{-8}	0.48	-20	4.0
Silt	8.0x10^{-8}	0.35	-17	4.2
Tertiary and Cretaceous Sands	4.0x10^{-8}	0.40	-15	7.4
Shallow Outwash Sands and Glacial Till	4.0x10^{-8}	0.35	-16	8.4
Limestone	5.0x10^{-8}	0.40	-17	5.0

[a] Initial porosity of the sediment
[b] a and b parameters used in the permeability-porosity relationship in equation 8.0

due to gravity (L/T²). The total vertical stress or overburden stress is estimated by

$$\sigma_v = g \int_z [\rho_s(1-\phi) + \rho_f \phi] dz, \quad (12)$$

where ρ_s is the bulk density of the sediment (M/L³) and ρ_f is the density of the fluid (M/L³). Compressibilities, initial porosities and a and b parameters used in the permeability-porosity relationship are shown on Table 1. Porosity-depth profiles are shown on Plate 1c. A specific storage of 8.0×10^{-5} m⁻¹ was applied to clays, while 8.0×10^{-6} m⁻¹ was used as the specific storage of the other more permeable units. Permeability ranged in the clay units from 8.3×10^{-19} m² to 9.7×10^{-20} m². Cretaceous and Tertiary sand permeability ranged from 9.1×10^{-13} m² to 1.0×10^{-13} m². Shallow outwash sand and glacial till permeability ranged from 8.7×10^{-14} m² to 5.0×10^{-14} m², and silt permeability ranged from 2.9×10^{-16} m² to 1.0×10^{-16} m². Specific storage and permeability parameters are consist with published values each sediment type [*Freeze and Cherry*, 1979].

It has been suggested that confining units beneath Nantucket have undergone deformation as a result of glacio-tectonic activity associated with the advancement of the Laurentide Ice Sheet [*Oldale and O'Hara*, 1984]. Oldale and O'Hara [1984] identified Cretaceous units that have been thrusted laterally by as much as 1.5 km on Nantucket. Boulton and Cabon [1995] described how high pore pressure and low effective stress are commonly associated with glacio-tectonic features such as those observed on Nantucket Island. It is possible that thrust faulting has created permeability conduits through deep confining units beneath Nantucket that are not represented in our model.

Glacial Lake Nantucket catastrophically drained approximately 17,000 years ago [*Uchupi et al.*, 2001]. This would have released large volumes of sediment that could have resulted in high sedimentation rates on the shelf and massive debris flows on the outer shelf and slope [*Uchupi et al.*, 2001]. This period of rapid sedimentation was simulated in our model by allowing a portion of the upper Pleistocene elements in our mesh to grow until it attained modern Pleistocene elevations after the Laurentide Ice Sheet retreated. We updated the concentration and hydraulic heads of the top nodes using the sea level elevation, land surface elevation and ice sheet thickness. Although actual sedimentation rates immediately following the drainage of Glacial Lake Nantucket are unknown, the majority of glacial lakes associated with the Laurentide Ice Sheet drained within 4,000 years of Glacial Lake Nantucket [*Uchupi et al.*, 2001]; therefore, we allowed our mesh to grow to modern topography for 4,000 years from 17,000–13,000 years ago. Sedimentation rates within the simulation during that time ranged from 0 m/yr to 0.055 m/yr depending on where deposition was observed on the cross section (Fig. 7). Sedimentation rates were highest near the continental slope where Pleistocene sediments were the thickest.

Numerical Simulations

The first model is constructed to test the hypothesis that sea-level variations alone could produce the excess heads and anomalous freshwater beneath Nantucket. Pleistocene sea level varied by as much as 120 m with a period of 40 k.y. to 100 k.y. and an average sea level of approximately 40 m below modern sea level [*Shackleton*, 1987; *Raymo et al.*, 1989, 1997; *Vail and Hardenbol*, 1979; *Summerhayes*, 1986; *Haq et al.*, 1987; *Shackleton and Opdyke*, 1973; *Hays et al.*, 1976; *Clark*, 1994; *Imbrie*, 1985; *Peltier*, 1998]. A sea-level boundary condition is applied during the simulation using a sine curve with an amplitude of 60 m, a period of 100,000 years, and set so that average sea level during the Pleistocene would have been 40 m below modern. Sea level during glacial maximum is set to 120 mbsl. A local sea-level curve for New England [*Redfield and Rubin*, 1962; *Oldale and O'Hara*, 1980; *Gutierrez et al.*, 2003] was used from 16 ka to present (Fig. 8). An initial sea water salinity condition was assigned deep Tertiary and Cretaceous sand and clay units while shallower outwash sands and silts were assigned fresh pore waters. An initial hydrostatic head boundary condition (consistent with local topography) was used to initial heads for the flow equation. The simulation was run for a total of 1.8 million years to reconstruct the Pleistocene and Holocene hydrologic conditions (Fig. 8a).

Our second model incorporates the effects of the ice sheet and glacial lake boundary conditions during the last 21 ka.

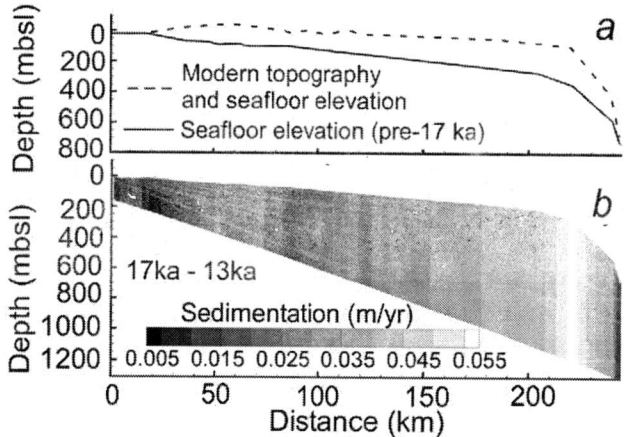

Figure 7. The change in continental shelf elevation between 17 ka and present. (a) Comparison of pre-17 ka topography and modern topography and seafloor elevation of the cross section. (b) Sedimentation rates from 17ka to 13 ka along the cross section.

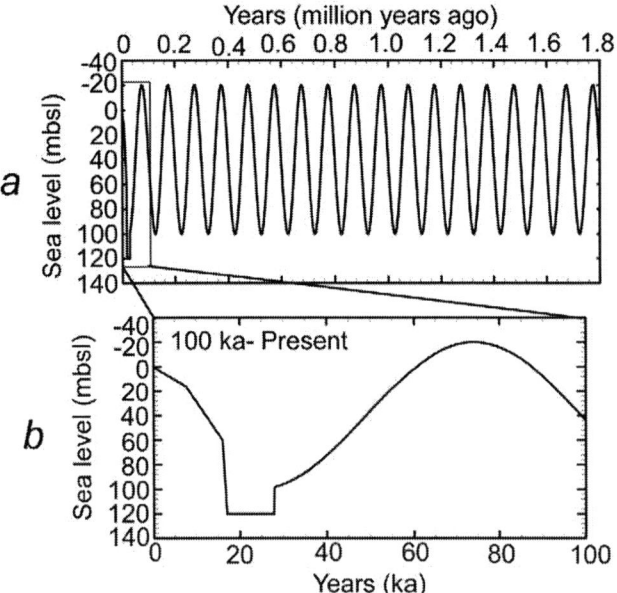

Figure 8. (a) Pleistocene sea-level curve represented using a sine function with an amplitude of 60 m and a period of 100,000 years. (b) Sea-level was set to 120 mbsl for glacial maximum. A local sea-level curve was used for 16 ka to present [Redfield and Rubin, 1962; Oldale and O'Hara, 1980; Gutierrez et al., 2003].

The thickness of the ice sheet was estimated using a polynomial expression presented by van der Veen [1999]:

$$\eta = H\sqrt{\left[\frac{1-x}{L_{ice}}\right]^2} + z_{ls} \qquad (13)$$

where H is the maximum ice-sheet thickness at a particular time step (L); L_{ice} is the ice sheet length at that time step (L); x is the distance from the margin of the basin (L); z_{ls} is the local elevation of the land surface (L); and η is ice-sheet elevation above sea level at distance x from the margin of the basin (L). The advance of the ice sheet is represented such that the ice sheet builds up slowly from 28 ka to the glacial maximum at 20 ka (Fig. 9a). The ice sheet remains at the glacial maximum from 20ka to 18ka. The ice sheet is allowed to retreat at 18 ka. As the glacier retreats, Glacial Lake Nantucket forms and is present until 17 ka, after which it drains (Fig. 9a) [Uchupi et al., 2001]. During the simulation, the ice sheet attains a maximum thickness of 1800 m for our section, which is consistent with maximum ice sheet thicknesses for southern New England [Denton and Hughes, 1981]. The glacier is allowed to extend 70 km along the cross section to the central portion of Nantucket (Fig. 9b). This is the approximate location of the glacial terminal moraines, which represent the maximum extent of the glacier. While the glacier is present, we apply a specified-head boundary condition along the top boundary assuming that the head at the base of the glacier is equal to 90% the local ice sheet elevation due to fluid-ice density differences [Boulton et al. 1995; Person et al. 2003]. A hydraulic head of 10 m above sea level was applied as a boundary condition when Glacial Lake Nantucket was present (Fig. 9b) [Uchupi et al., 2001]. The loading and head caused by the glacier and glacial lake affect the effective stress within the underlying sedimentary layers; the changes in effective stress will change the porosity (Eq. 9) and permeability (Eq. 8). We permitted infiltration to exceed the local melting rate due to the effects of a up-gradient esker systems [Shreve, 1985].

It has been suggested that shallow permafrost conditions existed along the margin of the Laurentide Ice Sheet in New England [Oldale and O'Hara, 1984]. The permafrost would have significantly decreased the permeability of those sediments [Person et al., 2007]. To replicate the permafrost, the permeability of the upper boundary nodes which are above sea level but not located below the ice sheet is decreased to 10^{-20} m^2 during the advance of the ice sheet and during the glacial maximum. This approach represents a simplification of the actual heat transfer involved during freezing and thawing of permafrost [Person et al., 2007]. The ice sheet simulation was run for 1.8 million years using the same

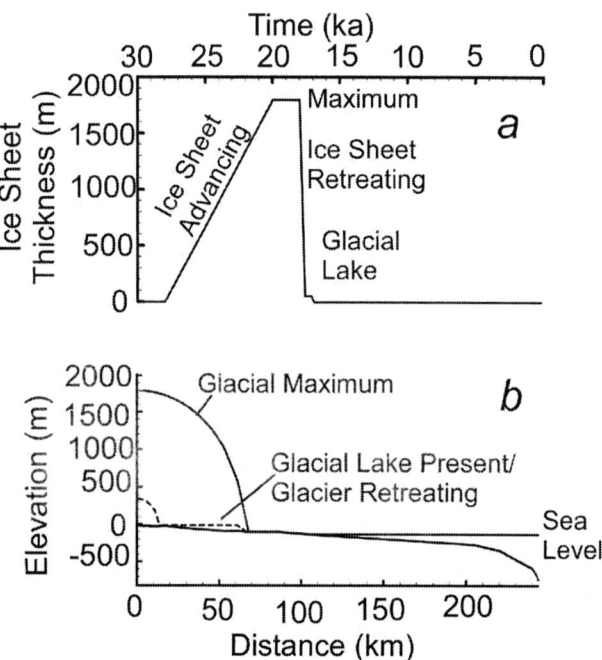

Figure 9. (a) Timing of the advance and retreat of the ice sheet and glacial lake. (b) The elevations of the ice sheet, glacial lake, early Pleistocene topography and sea-level while the ice sheet or glacial lake are present.

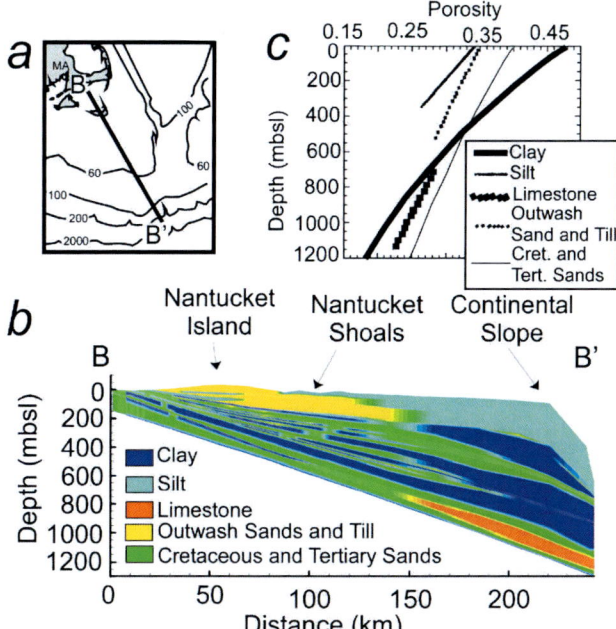

Plate 1. (a) The location of cross section used in the cross-sectional model. (b) Stratigraphy of the cross section. (c) Porosity depth relationship for each stratigraphic unit described in (b).

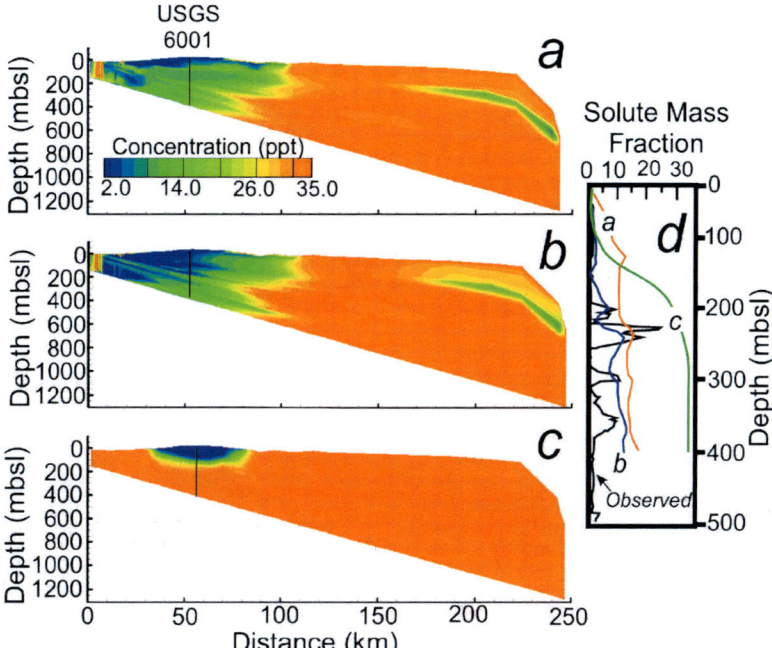

Plate 2. (a) Present-day salinity concentrations from the variable sea-level simulation (b) ice sheet simulation and (c) constant sea-level simulations. (d) Comparison of solute profiles from the variable sea level (a), ice sheet (b), constant sea level and (c) simulation with the observed solute profile from USGS 6001 well. The lateral position of the salinity-depth plot is indicated on Plate 2a-c by the black line segment.

sea-level variations used in the sea-level simulation (Fig. 8). The salinity profile from USGS 6001 on Nantucket and the results of the TDEM experiments were used as ground truth for comparison with our numerical experiments. We ran more than 50 simulations using a range of different permeabilities, porosities, and specific storages representative of continental shelf deposits to better understand what parameters controlled overpressure and salinity distributions on the shelf. A relatively narrow range of values were able to produce both the observed salinity distribution and excess heads. Below we represent one set of parameters that we found was consistent with the observed data. We consider the appropriateness of these parameters in the discussion section.

Simulation Results and Discussion

Solute concentrations using surface boundary conditions that represent only the effects of sea-level fluctuations (referred to herein as the 'variable sea-level' simulation) indicate that freshwater is located only within shallow sediments (<100 m) on Nantucket and at even shallower depths on the southern side of the island (Plate 2a). Permeable units at depths of greater than 200 m contain solute concentrations greater than 10 ppt and are not consistent with the solute concentrations observed within USGS 6001 or interpreted from TDEM soundings. The emplacement of freshwater in deep confined aquifers was only preserved in Nantucket Sound within 20 km of where the confined aquifers outcropped. This conclusion was reached regardless of the permeability/porosity parameters selected from the 50 simulations we completed. These results indicate that meteoric flushing of the Atlantic Continental Shelf during the sea-level lowstands of the Pleistocene cannot be the sole mechanism responsible for the distribution of freshwater beneath Nantucket.

The simulations, which include sub-ice-sheet recharge and glacial lake recharge (referred to herein as the 'ice sheet' simulation), show significant flushing beneath Nantucket compared to the variable sea-level simulation (Plate 2b). Freshwater of solute concentrations of less 1 ppt are predicted within permeable units at depths greater than 300 mbsl. Deep low-permeability clays exhibit solute concentrations with diffusional profiles similar to that shown in USGS 6001. Simulations indicate that a small tongue of slightly saltier (approximately 6 ppt) water invades into the shallow subsurface (<200 m) on the southern end of the island; this is qualitatively consistent with the TDEM results, which similarly indicate that saltwater is found at shallower depths on the southern side of the island. Thus, the higher salinity conditions here may result from paleohydrologic conditions rather than recent pumping of nearby municipal wells. Our models suggest that the transition between fresh to saltwater exists near the southern end of the island.

For comparison, we also present one simulation in which the cross-sectional model is permitted to equilibrate with modern sea-level conditions. Here we run our cross sectional model for 1.8 million years using a modern sea level (referred to herein as the 'constant sea-level' simulation) to predict salinity distribution for the Atlantic continental shelf. This resulted in seawater salinity within pore spaces of sediments everywhere on the continental shelf that is below sea level today. Only beneath Nantucket Island was freshwater found. Results are consistent with the Ghyben-Herzberg approximation where freshwater/saltwater interface beneath Nantucket is located at approximately 120 mbsl. (Plate 2c).

As noted above, the deep Cretaceous and Tertiary aquifers on Nantucket are observed to be overpressured by up to 0.08 MPa (8 m above sea level). The observed overpressures cannot be explained by topographically driven flow, for lack of sub-aerial recharge from the mainland. Computed freshwater heads within deep permeable sediments from the variable sea-level simulation varied between -0.5 and 2.3 m in different sand lithologic units beneath Nantucket Island (Plate 3); these anomalous heads were generated, in part, by Late Pleistocene sedimentation. Results of the ice sheet simulation produce heads that vary from 4.5 to 25.4 m (Plate 3). Negative computed heads are due to: (1) an imposed sea level boundary condition which is significantly lower than modern levels and (2) a continental shelf aquifer system has not fully equilibrated to modern sea-level conditions. Since the USGS 6001 is open to all units below 120 mbsl, observed overpressure at that well represents the average of all permeable units below 120 mbsl. The average freshwater hydraulic heads for these Cretaceous and Tertiary aquifers from the variable sea-level simulation was only 0.56 m. Because shallow water-table elevations across Nantucket are approximately 3 m above sea level, no overpressure would be observed in USGS 6001 in this simulation. However, the average freshwater head from those units in the ice sheet simulation is 15.1 m; this would produce overpressures in USGS 6001 that are 11 m above the local water table, a value that is similar (though not identical) to the observed 4 m of overpressure in that well.

These modeling results indicate that sediment loading associated with the high sedimentation rates of the late Pleistocene was not solely responsible for the overpressure observed beneath Nantucket. Hydraulic heads within sediment below the continental slope are pressured by up to 126 m more within the ice sheet simulation compared to the sea-level only simulation (Plate 3). Given that both simulations were subjected to the same sediment loading, the additional overpressures predicted within the ice sheet run

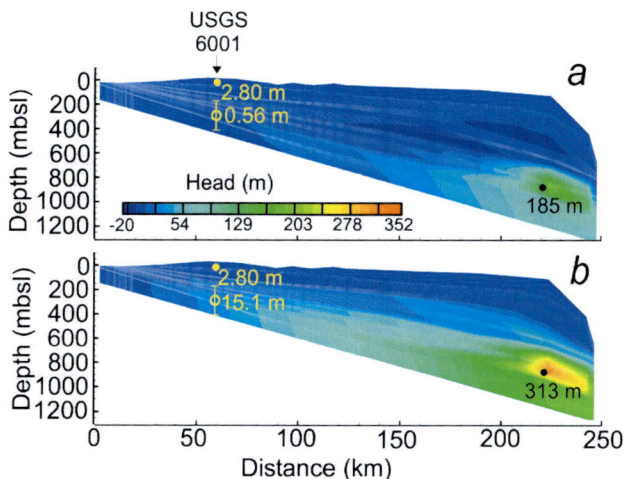

Plate 3. (a) Computed equivalent freshwater hydraulic head from the variable sea-level simulation and (b) ice sheet simulation. Hydraulic heads are reported in meters above sea level. Solid circles represent pressure heads at specific locations. Open circles represent average pressure heads over the length of only the permeable lithologies.

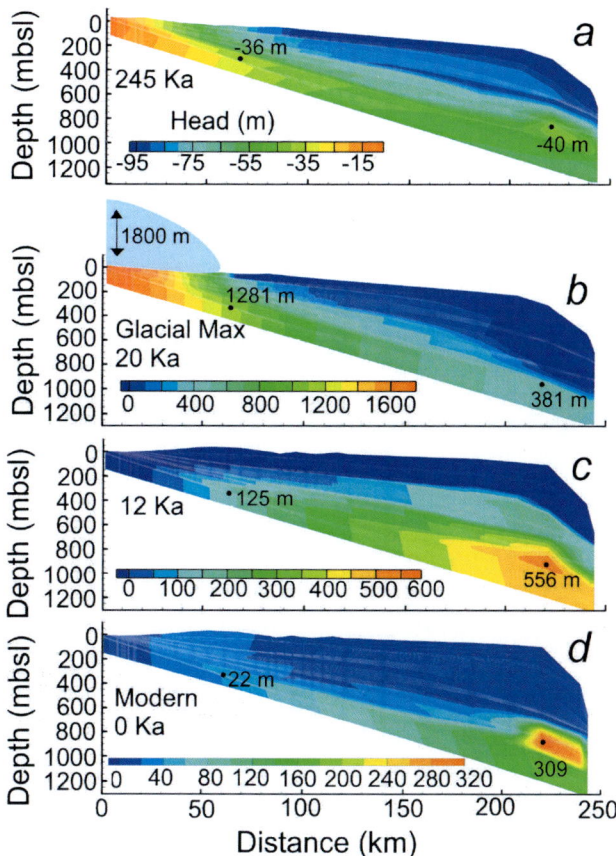

Plate 4. The evolution of hydraulic heads through time in the ice sheet simulation. (a) Hydraulic head at 245 ka. (b) Hydraulic head at glacial maximum 20 ka. (c) Hydraulic head at 12 ka. (d) Hydraulic head at modern 0 ka. The thickness of the sedimentary pile changes between (b) and (c) due to rapid sedimentation associated with the breaching of Glacial Lake Nantucket at 17 ka. Note the change in scale in each plot.

must have originated from the diffusion of high hydraulic heads from the adjacent aquifers during ice-sheet glaciation. The observed excess heads (USGS 6001) thus are inferred to be preserved, fossil hydraulic heads associated with the ice-sheet loading during the LGM. In their sensitivity study of ice sheet loading within sedimentary basins, Person et al. [2007] and Bense and Person [2006] observed this phenomenon.

The preservation of excess head is illustrated by plotting temporal variations in head during the Pleistocene (Plate 4) for our ice-sheet simulation. Hydraulic heads remained low throughout the cross section under normal sea level conditions (Plate 4a). Hydraulic heads within the aquifers below Nantucket became overpressured by up to 1281 m (Plate 4b) while the ice sheet was present. The lateral transfer of this pressure decreased the effective stress within sediments that were experiencing less overburden pressure (i.e. areas not covered by ice). The decrease in effective stress then caused an increase in the effective permeability of the sediment and allowed high pressure to transfer into even deeper sediments below the continental slope. Deep aquifers near the continental slope became pressured by up to 381 m as the ice sheet reached its maximum extent on the continental shelf (Plate 4b). At that point the high fluid pressures diffused into the surrounding clay units. After the ice sheet retreated, the low-permeability clays retained elevated fluid pressure. Fluid pressures subsequently increased due to sediment loading associated with the drainage of Glacial Lake Nantucket and the retreat of the ice sheet. Fluid pressures reached a maximum of 556 m within clay units below the continental slope (Plate 4c). After sedimentation ceased (approximately 12,000 years ago), heads within the confining units on the slope slowly dissipated while some of the elevated heads also got transferred back beneath Nantucket (Plate 4d).

It is also interesting to note that effective stresses below Nantucket decreased to zero while the ice sheet was present. This may have allowed thrust features to form associated with the advance of the ice sheet. Ice sheet thrust features were observed on Nantucket Island and Martha's Vineyard by Oldale and O'Hara [1984]. These thrust faults may have helped transport glacial melt water into deeper aquifers below Nantucket.

We can calculate a response time for the clay to predict the time required for the high hydraulic heads to diffuse out of the clay:

$$\tau = \frac{L_{sed}^2 S_S}{K} \quad (14)$$

where L_{sed} is the length or thickness of the sediment unit (L); S_S is the specific storage of the confining unit (L^{-1}); K is the hydraulic conductivity of the sediment (L/T) and τ is the response time (T). Permeability can be converted to hydraulic conductivity using equation 3c. Based on the thickness of the deep clay unit, approximately 275 m, the Ss of 8.0×10^{-5} m^{-1} and a hydraulic conductivity of 10^{-12} m/s, the time to dissipate the excess head would be approximately 200,000 years. This means that high heads emplaced within the clay units during glacial maximum would not have had enough time to dissipate and equilibrate with modern sea-level conditions and therefore would retain a fossil head. These high head pressures in the deep clay units can laterally transfer pressure through deep permeable sand units to beneath Nantucket. Although the computed heads beneath Nantucket Island are about twice as large, it does show that there is high potential for thick low permeable units on the continental slope to transfer fossil pressures to the near shore environment.

The response time of the clay units in the cross section and their associated flushing ability of the sands are highly dependent on permeability and compressibility of the sediment. Had we raised the permeability or lowered compressibility of the confining units by an order of magnitude, the confining units would have flushed too much and simulations would not reproduce the salinity profiles and observed overpressures beneath Nantucket Island. Although the permeability of the sediment units is reasonable, it may have been possible to obtain similar results using a different cross section where clay units where thicker and the permeability of the clay was higher. Lack of well data in the area of the cross section has made it difficult to predict the stratigraphy and continuity of the units especially at depth. The stratigraphy at depth is based on wells located in Georges Bank, located 200 km east-southeast of Nantucket. Our cross section extends only 1200 mbsl. The sediment column near Georges Bank, located east of this cross section is much deeper (> 5000 mbsl) [Scholle and Wenkam, 1982; Schlee and Fritsch, 1983]. It is possible that had we deepened our cross section to include these units, we might have found that they also contribute to overpressure observed beneath Nantucket.

CONCLUSIONS

Our study demonstrates the utility of integrating geophysical, hydrochemical, and hydrologic data sets with numerical modeling to resolve recharge mechanisms for freshwater emplacement on New England's continental shelf and to better understand the origin of overpressure beneath Nantucket. The geophysical data is a cost-effective method to obtain spatial information regarding the salinity concentrations beneath Nantucket Island that could not be obtained easily by drilling. It provided valuable, qualitative insight for construction of numerical ground-water models; moreover, the geophysical data were used to help evaluate models that might explain the observed modern hydrologic conditions on Nantucket. This work points toward future, fully integrated analyses, in which geophysical data might provide quantitative information for model calibration through regression methods. Integrating the hydrochemical and hydrological data sets with the numerical modeling allowed us to investigate the relationship between hydromechanical loading of the Laurentide Ice Sheet and sediment loading associated with the high sedimentation rate of the late Pleistocene and the hydrologic response of the continental shelf resulting from variable sea-level changes and increased recharge from the LGM.

The TDEM soundings indicate a freshwater/saltwater transition or interface at about 120 mbsl on Nantucket, whereas the solute diffusion profile from USGS 6001 and our numerical modeling indicate a deeper interface. This apparent inconsistency can be explained by either (1) the presence of trapped, saline water in confining units, or (2) confining units that include substantial low-resistivity clay. The approximate maximum depth of investigation for the TDEM surveys is about 120 m; thus any deeper transitions back to freshwater would not be detected by the geophysics. The presence of low-resistivity confining units would also explain sounding results from TDEM sites 14 and 15, where the resistivity was too high for saltwater-saturated sands but too low for freshwater-saturated sands.

Results of the numerical modeling indicate that the high hydraulic heads associated with the Laurentide Ice Sheet and Glacial Lake Nantucket substantially influenced the distribution of freshwater beneath Nantucket and the fluid pressure distribution across the continental shelf. Although sea level varied substantially during the Pleistocene, hydraulic heads experienced during sea-level low stands would not have been high enough to flush salty waters from the deep Cretaceous and Tertiary aquifers below Nantucket. Both TDEM soundings and mathematical model results suggest the transition from fresh to saltwater within confined aquifers of the Atlantic Continental Shelf occurs near the southern terminus of Nantucket Island.

Both the loading associated with the Laurentide Ice Sheet and the high sedimentation rates associated with the late Pleistocene had profound effects on the hydraulic heads on the continental shelf. We believe that we have found, perhaps for the first time, evidence of "fossil pressures" from late Pleistocene glaciation. Large amounts of freshwater were emplaced during the glacial maximum when extremely high hydraulic heads were present on the continental shelf. Long response times of low-permeability clays facilitate the retention of high pressures induced during the glacial maximum. As the ice sheet retreats and sedimentation rates increase on the shelf, fluid pressures continue to increase. After sedimentation rates decrease, high pressures generated in these deep clays could easily have been laterally transferred toward shallower sediment near the coastline. The high heads observed in well USGS 6001 may be the first recorded observations of fossil heads from the LGM. Previous evidence for sub-ice-sheet recharge comes entirely from geochemical and environmental isotopic data [*Boulton et al.*, 1995; *Piotrowski*, 1997; *Siegel and Mandle*, 1984; and *Grasby et al.*, 2000].

The results of this study have important implications for all New England coastal aquifers in that (1) freshwater resources cannot be inferred by modern sea-level conditions, and (2) glacial recharge and high sedimentation rates in the Late Pleistocene may substantially control the modern spatial distribution of freshwater. We found that late Pleistocene recharge rates were at least 10 times greater than present day conditions. This phenomenon may also account for unusually freshwater observed at significant depths offshore from New Jersey and New York [*Hathaway et al.*, 1979; *Kohout et al.*, 1988] as well as the high fluid pressures observed on the continental slope [*Dugan and Flemings*, 2000].

Acknowledgements. The authors would like to thank the two anonymous reviewers for their thorough and constructive comments. We would also like to thank the Wannacomet Water Company and Sarah Oktay and Tony Molis of the University of Massachusetts Boston Field Station for their logistical support. This research was supported by a grant from the National Science Foundation (EAR-0337634) to Mark Person. USGS contributions to this work were funded in part by the USGS Ground Water Resources Program.

REFERENCES

Bekele, E. B., B. J. Rostron, and M. A. Person, Fluid pressure implications of erosion unloading, basin hydrodynamics and glaciation in the Alberta Basin, Western Canada, *Journal of Geochemical Exploration*, 78–79, 143–147, 2003.

Bense, V. F., and M. A. Person, Faults as conduit-barrier systems to fluid flow in siliciclastic sedimentary aquifers, *Water Resources Research*, 42(5), W05421, 2006.

Bethke, C. M., and T. F. Corbet, Linear and nonlinear solutions for one-dimensional compaction flow in sedimentary basins, *Water Resources Research*, 24(3), 461–467, 1988.

Boulton, G. S., P. E. Caban, and K. van Gijssel, Groundwater flow beneath ice sheets; Part I, Large Scale Patterns, *Quaternary Science Reviews*, 14(6), 545–562, 1995.

Boulton, G. S., and P. E. Caban, Groundwater flow beneath ice sheets; Part II, Its impact on glacier tectonic structures and moraine formation, *Quaternary Science Reviews*, 14(6), 563–587, 1995.

Clark, P. U., Unstable behavior of the Laurentide ice sheet over deforming sediment and its implication for climate change, *Quaternary Research, 41(1)*, 19–25, 1994.

Corbet, T. F., and C. M. Bethke, Disequilibrium fluid pressures and groundwater flow in the Western Canada sedimentary basins, *Journal of Geophysical Research, 97(B5)*, 7203–7217, 1992.

Denton, G. H., and T. J. Hughes, The last great ice sheets, John Wiley and Sons, New York, 484, 1981.

Drabbe, J., and W. Badon-Ghyben, Nota in verband met de voorgenomen putboring nabij Amsterdam. In *Tijdschrift van het Koninklijk Instituut van Ingenieurs*. Netherlands, The Hague, 1889.

Dugan, B., and P. B. Flemings, Overpressure and fluid flow in the New Jersey continental slope; implications for slope failure and cold seeps, *Science, 289(5477)*, 288–291, 2000.

Fitterman, D. V. and M. Deszcz-Pan, Geophysical mapping of the freshwater/saltwater interface in Everglades National Park, in Gerould, S., ed.: *U.S. Geological Survey Open-File Report 97-385*, 13–14, 1997.

Fitterman, D. V., and M. T. Stewart, Transient electromagnetic sounding for groundwater, *Geophysics, 51(4)*, 995–1005, 1986.

Folger, D. W., J. C. Hathaway, R. A. Christopher, P. C. Valentine, and C. W. Poag, Stratigraphic test well, Nantucket Island, Massachusetts, *US Geological Survey Circular, C 773*, 28, 1978.

Freeze, R. A., and Cherry, J.A. *Groundwater*, Prentice-Hall, New Jersey, 1979.

Garven, G., A hydrogeologic model for the formation of the giant oil sand deposits of the Western Canada sedimentary basin, *American Journal of Science, 289(2)*, 105–166, 1989.

Grasby, S., K., Osadetz, R., Betcher, and F. Render, Reversal of the regional-scale flow system of the Williston Basin in response to Pleistocene Glaciation, *Geology, 28(7)*, 635–638, 2000.

Groen, K., The effects of transgressions and regressions on coastal and offshore groundwater, Ph. D. Thesis Vrije Universiteit Amsterdam, 2002.

Gutierrez, B. T., E. Uchupi, N. W. Driscoll, and D. G. Aubrey, Relative sea level rise and the development of valley-fill and shallow-water sequences in Nantucket Sound, Massachusetts, *Marine Geology, 193(3–4)*, 295–314, 2003.

Hall R. E., L. J. Poppe, and W. M. Ferrebee, A stratigraphic test well, Martha's Vineyard, Massachusetts, *U. S. Geological Survey Bulletin, B 1488*, 1980.

Haq, B. U., J. Hardenbol, and P. R. Vail, Chronology of fluctuating sea levels since the Triassic, *Science, 235(4793)*, 1156–1167, 1987.

Hathaway, J. C., C. W. Poag, P. C. Valentine, R. E., Miller, D.M., Schultz, F. T. Manheim, F. A. Kohout, M. H. Bothner, and D. A. Sangrey, U.S. Geological Survey core drilling on the Atlantic Shelf, *Science, 206(4418)*, 515–527, 1979.

Hays, J. D., J. Imbrie, and N. J. Shackleton, Variations in the Earth's orbit; pacemaker of the ice ages, *Science, 194 (4270)*, 1121–1132, 1976.

Herzberg, A., Die Wasserversorgung einiger Nordseebader, *Journal für Gasbeleuchtung und Wasserversorgung, 44*, 815–819, 842–844, 1901.

Hubbert, M.K., and W.W. Rubey, Mechanics of fluid-filled porous solids and its application to overthrust faulting, [Part] 1 of Role of fluid pressure in mechanics of overthrust faulting, *Geological Society of America Bulletin, 70(2)*, 115–166, 1959.

Imbrie, J., A theoretical framework for the Pleistocene ice age, *Journal of the Geological Society of London, 142(3)*, 417–432, 1985.

Kestin, J., H. E. Khalifa, and R. Corriea, Tables of the dynamics and kinematic viscosity of aqueous NaCl Solutions in the temperature range of 20–150° and the pressure range of 0.1–35 MP, *Journal of Physical Chemistry Reference Data, 10*, 71–87, 1981.

Kohout, F. A., J. C. Hathaway, D. W. Folger, M. H. Bothner, E. H. Walker, D. F. Delaney, M. H. Frimpter, E. G. A. Weed, and E. C. Rhodehamel, Fresh ground water stored in aquifers under the continental shelf, implications from a deep test, Nantucket Island, Massachusetts, *Water Resources Bulletin, 13(2)*, 373–386, 1977.

Kohout, F. A., H. Meisler, F. W. Meyer, R. H. Johnston, G. W. Leve, and R. L. Wait, Hydrogeology of the Atlantic continental margin, in *The Atlantic continental margin, U.S*, edited by R. E. Sheridan and J. A. Grow, The Geological Society of America, Boulder, CO, 463–480, 1988.

Kooi, H., and J. Groen, Offshore continuation of coastal groundwater systems; predictions using sharp-interface approximations and variable-density flow modeling, *Journal of Hydrology, 246(1–4)*, 19–35, 2001.

Kooi, H., J. Groen, and A. Leijnse, Modes of seawater intrusion during transgressions, *Water Resources Research, 36(12)*, 3581–3589, 2000.

Lemieux, J. M., E. A. Sudicky, W. R. Peltier, and L. Tarasov, Coupling glaciations with groundwater flow models – Surface/subsurface interactions over the Canadian Landscape during the Wisconsin Ian glaciation, IAHR – GW 2006, International Groundwater Symposium, Toulouse, June 12–14, 2006.

Lerche, I., Z. Yu, B. Torudbakken, and R. O. Thomsen, Ice loading effects in sedimentary basins with reference to the Barents Sea, *Marine and Petroleum Geology, 14(3)*, 277–338, 1997.

Lucia, F. J., Rock-fabric/petrophysical classification of carbonate pore space for reservoir characterization, *American Association of Petroleum Geology Bulletin, 79(9)*, 1275–1300, 1995.

Masterson, J. P., B. D. Stone, D. A., Walters, and J. Savoie, Hydrogeologic Framework of Western Cape Cod, Massachusetts, *Hydrologic Investigations Atlas, HA*-741, 1997.

McWhorter, D. B., and D. K. Sunada, *Groundwater Hydrology and Hydraulics*, Water Resources Publications, Highland Ranch, Colorado, 287, 1993.

Meisler, H., P. P. Leahy, and L. L. Knobel, Effect of Eustatic sea level changes on saltwater-freshwater in North Atlantic Coastal Plain, *U.S. Geological Survey Water-Supply Paper 2255*, 1984.

Mills, T., P. Hoekstra, M. Blohm, and L. Evans, Time domain electromagnetic soundings for mapping sea-water intrusion in Monterey County, California, *Ground Water, 26(6)*, 771–782, 1988.

Nabighian, M. N., and J. C. Macnae, Time Domain Electromagnetic Prospecting Methods, in *Electromagnetic Methods In Applied Geophysics Volume 2, Application, Part A*, edited by M. N. Nabighian, Society of Exploration Geophysicists, Tulsa, OK, 427–520, 1991.

Neuzil, C.E., How permeable are clays and shales?, *Water Resources Research, 30(2)*, 145–150, 1994.

O'Hara, C. J., and R. N. Oldale, Maps showing geology, shallow structure and bedform morphology of Nantucket Sound, Massachusetts, *Miscellaneous Fields Studies Map—U. S. Geological Survey Report, MF-1911*, 1987.

Oldale, R. N., and C. J. O'Hara, New radiocarbon dates from the inner continental shelf off southeastern Massachusetts and a local sea level rise curve for the past 12,000 yr, *Geology, 8(2)*, 102–106, 1980.

Oldale, R. N., and C. J. O'Hara, Glaciotectonic origin of the Massachusetts coastal end moraines and a fluctuating late Wisconsinan ice margin, *Geological Society of America Bulletin, 95(1)*, 61–74, 1984.

Oldale, R. N., Geologic map of Nantucket and nearby islands, Massachusetts, *Miscellaneous Investigations Series—U. S. Geological Survey Report, I-1580*, 1985.

Peltier, W. R., Postglacial variations in the level of the sea; implications for climate dynamics and solid-Earth geophysics, *Reviews of Geophysics, 36(4)*, 603–689, 1998.

Person, M., B. Dugan, J. B. Swenson, L. Urbano, C. Stott, J. Taylor, and M. Willett, Pleistocene hydrogeology of the Atlantic continental shelf, New England, *Geological Society of America Bulletin, 115*, 1324–1343, 2003.

Person, M., J. McIntosh, V. Bense, V. H. Remenda, Pleistocene hydrology of North America; The role of ice sheets in reorganizing groundwater flow systems, *Reviews of Geophysics*, in press, 2007.

Person, M., J. Z. Taylor, and S. L. Dingman, Sharp-interface models of salt water intrusion and well head delineation on Nantucket Island, Massachusetts, *Ground Water, 36(5)*, 731–742, 1998.

Piotrowski, J., Subglacial hydrology in north-western Germany during the last glaciations; Groundwater flow, tunnel valleys and hydrological cycles, *Quaternary Science Reviews, 16*, 169–185, 1997.

Raymo, M. E., W. F. Ruddiman, J. Backman, B. M. Clement, and D. G. Martinson, Late Pliocene variation in Northern Hemisphere ice sheets and North Atlantic deep water circulation, *Paleoceanography, 4(4)*, 413–446, 1989.

Raymo, M.E., D.W. Oppo and W. Curry, The mid-Pleistocene climate transition; a deep sea carbon isotopic perspective, *Paleoceanography, 12(4)*, 546–559, 1997.

Redfield, A. C., and M. Rubin, The age of salt marsh peat and its relation to recent changes in sea level in Barnstable, Massachusetts, *Proceedings of the National Academy of Sciences of the United States of America, 48 (10)*, 1728–1735, 1962.

Schlee, J. S., and J. Fritsch, Seismic stratigraphy of the Georges Bank Basin complex offshore New England, *American Association of Petroleum Geologists Memoir, 34*, 223–251, 1982.

Scholle, P. A., and Wenkam, C.R., Geological studies of the COST nos. G-1 and G-2 wells, United States North Atlantic outer continental shelf, *Geological Survey Circular, C-861*, 1982.

Shackleton, N. J., and N. D. Opdyke, Oxygen isotope and paleomagnetic stratigraphy of Equatorial Pacific core V28-238: oxygen isotope temperatures and ice volumes on a 10^5 year and 10^6 year scale, *Quaternary Research, 3(1)*, 39–55, 1973.

Shackleton, N. J, Oxygen isotopes, ice volume, and sea level, *Quaternary Science Reviews, 6 (3-4)*, 183–190, 1987.

Shenhav, H., Lower Cretaceous sandstone reservoirs, Israel; petrography, porosity, permeability, *American Association of Petroleum Geologists Bulletin, 55(12)*, 2194–2224, 1971.

Shreve, R. L., Late Wisconsin ice-surface profile calculated from esker paths and types, Katahdin Esker System, Maine, *Quaternary Research, 23(1)*, 27–37,1985.

Siegel, D. I., and R. J. Mandle, Isotopic evidence for glacial meltwater recharge to the Cambrian-Ordovician Aquifer, north-central United States, *Quaternary Research, 22(3)*, 328–335, 1984.

Summerhayes, C. P., Sea level curves based on seismic stratigraphy; their chronostratigraphic significance, *Palaeogeography, Paleoclimatology, Paleoecology, 57(1)*, 27–42, 1986.

Uchupi, E., N. Driscoll, R. D. Ballard, and S. T. Bolmer, Drainage of late Wisconsin glacial lakes and the morphology and late quaternary stratigraphy of the New Jersey – southern New England continental shelf and slope, *Marine Geology, 172(1-2)*, 117–145, 2001.

Vail, P. R., and J. Hardenbol, Sea level changes during the Tertiary, *Oceanus, 22(3)*, 71–79, 1979.

van der Veen, C. J., *Fundamentals of Glacier Dynamics*, A. A. Balkema, Rotterdam, Netherlands, 1999.

Zheng, C., and G. D. Bennett, *Applied contaminant transport modeling*, Van Rostrand Reinhold, New York, 1995.

Integrating Hydrologic and Geophysical Data to Constrain Coastal Surficial Aquifer Processes at Multiple Spatial and Temporal Scales

Gregory M. Schultz[1], Carolyn Ruppel[2*], and Patrick Fulton[3]

School of Earth and Atmospheric Sciences, Georgia Institute of Technology, Atlanta, Georgia, USA

Since 1997, repeated, coincident geophysical surveys and extensive hydrologic studies in shallow monitoring wells have been used to study static and dynamic processes associated with surface water-groundwater interaction at a range of spatial scales at the estuarine and ocean boundaries of an undeveloped, permeable barrier island in the Georgia part of the U.S. South Atlantic Bight. Because geophysical and hydrologic data measure different parameters, at different resolution and precision, and over vastly different spatial scales, reconciling the coincident data or even combining complementary data sets has required a range of approaches. This study uses geophysical imaging and inversion, hydrogeochemical analyses and well-based groundwater monitoring, and, in some cases, limited vegetation mapping to demonstrate the utility of an integrative, multidisciplinary approach for elucidating groundwater processes at spatial scales (tens to thousands of meters) that are often difficult to capture with traditional hydrologic approaches. The case studies highlight regional aquifer characteristics, varying degrees of lateral saltwater intrusion at estuarine boundaries, complex subsurface salinity gradients at the ocean boundary, and imaging of submarsh groundwater discharge and possible free convection in the pore waters of a clastic marsh. This study also documents the use of geophysical techniques for detecting temporal changes in groundwater salinity regimes under natural (not forced) gradients at intratidal to interannual (1998–2000 Southeastern U.S.A. drought) time scales.

1. INTRODUCTION

A key challenge for hydrologic research is the development of the quantitative and predictive tools needed to understand the response of aquifers to changes in forcing variables over a wide range of spatial and temporal scales. Concurrent with the increased emphasis on measuring spatiotemporal changes in aquifers has been the recognition of the need for a systems approach that links physical, chemical, and biological processes through the application of multidisciplinary characterization methods. In this paper, which is in part based directly on *Schultz and Ruppel* [2000] and *Schultz* [2002], we describe the integration of hydrologic (physical and chemical) and geophysical methods to study coastal surficial aquifer processes at spatial scales of centimeters to kilometers and at temporal scales ranging from subtidal to interannual, as depicted in Figure 1. While a key motivation for our work has been

---

[1] Now at Applied Research Associates, 415 Waterman Road, South Royalton, Vermont.
[2*] Corresponding; Now at U.S. Geological Survey, 384 Woods Hole Rd., Woods Hole, Massachusetts. (cruppel@usgs.gov)
[3] Now at Department of Geosciences, Penn State University, University Park, Pennsylvania.

162 INTEGRATING HYDROLOGIC AND GEOPHYSCAL DATA

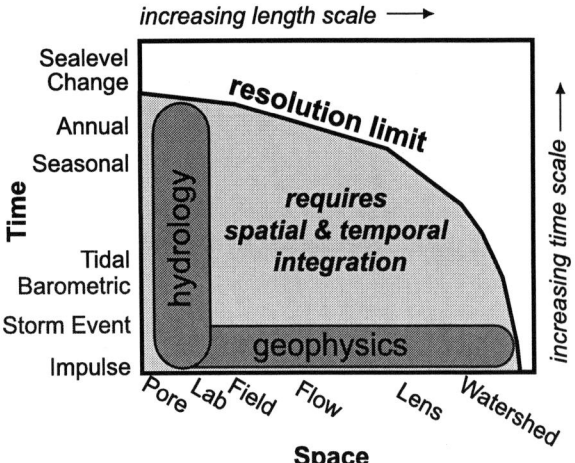

Figure 1. Summary of the spatial and temporal scales relevant to hydrogeophysical studies and the resolving power of geophysical data and groundwater hydrologic data. Geophysical data can integrate over large spatial scales, but only repeated geophysical surveys can integrate over significant time scales. Hydrologic data are often affected by processes occurring over long durations relative to the sampling time of measurements and inherently integrate over multiple time scales, but are usually acquired at discrete points in monitoring wells or piezometers. Only when discrete hydrologic observations are interpolated over larger continuous domains can hydrology alone constrain processes at larger spatial scales. By combining geophysical and hydrologic approaches, the data requirements necessary for resolving system-level processes over various spatial and temporal scales can be sharply reduced.

understanding surface water-groundwater interactions and the connection of complex biogeochemical and ecological processes and physical parameters in the coastal zone, this paper uses the coastal zone data sets merely as examples to explore integration and reconciliation of geophysical and hydrologic data and novel applications of geophysical data particularly for monitoring.

From the beginning of this research in 1997, the studies described here were designed to combine standard hydrologic methods and noninvasive geophysical surveys using a strategy that permits coincident sampling of similar processes and variables at comparable and often nested spatial scales. This philosophical approach contrasts with that of some other studies: Typically, geophysical data are used to constrain hydrologic parameters, but with little groundtruthing of the results. In another common application, geophysical data provide contextual (regional) information needed to enhance interpretation of sparse data acquired in discrete groundwater monitoring wells. Only rarely are geophysical data trusted enough to guide large-scale, process-based, hydrologic interpretations or decisions about instrument-

ing individual sites with monitoring well networks. Both outcomes have emerged from the approach adopted in our studies.

2. OVERLAPPING HYDROLOGIC AND GEOPHYSICAL METHODS

2.1 Hydrologic Methods

The standard hydrologic methods (water level monitoring, groundwater temperature and conductivity measurements, hydrogeochemical analyses, aquifer testing) used for this study were applied in several small (up to 20 wells) monitoring well networks that make up groups of networks consisting of a total of more than 65 wells and piezometers we have installed in this part of the South Atlantic Bight (SAB) between 1997 and 2003. Details of monitoring well and multilevel sampler installation are provided by *Schultz and Ruppel* [2002] and *Snyder et al.* [2004]. All of the ~4.5 m deep monitoring wells on which this study relies are narrow diameter (1.25" to 2") Schedule 40 PVC and were installed using minimally disruptive techniques, either by hand-augering or vibracoring in permeable upland areas and adjacent clastic salt marshes. Screened intervals for upland monitoring wells were placed within the fully saturated zone, and commercial, well-sorted, coarse (20–20 fill) sand was emplaced adjacent to the well screens to act as a filter pack between the fine-grained aquifer sands (d_{10}=120 μm) and the well. Tamped native backfill in the annular space above the screened interval ensures mechanical and hydrologic coupling between the aquifer and the monitoring well and prevents the introduction of non-native materials that could affect the aquifer in the environmentally protected areas in which we conduct our surveys.

To protect against annular flow, special near-surface sealing methods were devised for this high salinity environment (e.g., *Schultz* [2002]). Annular flow, in addition to wellhead protection, was a particular concern in areas that experience periodic tidal inundation, and we used outer PVC casings that stood 1 m or more above the ground surface and that were sunk in cement around the top of the monitoring wells to prevent the wells from being overtopped by tidal waters in these locations.

The vibracoring methods we used after 2001 provided continuous sediment cores for use in laboratory analysis of hydraulic parameters. Extracted sediment samples from vibracores and auger cuttings were archived and formally described using Munsell soil color coding and textural classifications. Geophysical surveys were often used to guide well placement, and additional wells were sometimes added to networks over several years to meet specific scientific objectives. In some cases, we used ancillary, temporary, instrumented piezometers installed in tidal creeks adjacent to monitoring well networks for periods of weeks to months to measure surface water properties and monitor processes (e.g., tidal fluctuations) that affected surficial aquifers in adjacent uplands. The construction, installation, and use of the few multiport sampling wells used for these studies are described by *Snyder et al.* [2004].

2.2 Geophysical Methods

Geophysical methods have been successfully applied in a range of hydrologic investigations of coastal (e.g., *Van Dam and Maeulankamp*, 1967; *Roy and Elliot*, 1980; *Goldman et al.*, 1991; *Frohlich and Urish*, 2002; *Greenwood et al.*, 2006), barrier island (e.g., *Bugg and Lloyd*, 1976; *Stewart*, 1988; *Urish and Frohlich*, 1990; *Ruppel et al.*, 2000), and deltaic aquifers (e.g., *Ebraheem et al.*, 1997; *Collins and Easley*, 1999). Geo-electromagnetic methods (electromagnetic induction, resistivity, and ground penetrating radar) are among the most frequently used in these settings because of their ability to detect variations in pore water conductivity in the near surface. For this study, we acquired data sets using proven, off-the-shelf electrical and electromagnetic (EM) instrumentation along survey lines and in two-dimensional areas that overlap our shallow monitoring well networks. Repeated occupation of survey sites permitted compilations of time-series of geophysical data for quantification of changes in the physics and chemistry of the subsurface.

Noninvasive geophysical techniques provide an effective alternative to in situ sampling and subsequent interpolation of point data collected in monitoring wells. Despite the growing use of hydrogeophysical techniques, the relationship between geophysical and hydrological properties and their covariance remain elusive. For example, geo-electromagnetic methods yield measurements of the volume average bulk conductivity (e.g., EM and resistivity) or dielectric permittivity (e.g., ground penetrating radar) of the subsurface. Yet universal relationships between the measured geophysical parameters and required hydrogeologic properties (porosity, hydraulic conductivity, etc.) are difficult to come by.

2.2.1 Electromagnetic induction (terrain conductivity). We use slingram-type, frequency-domain EM induction instruments to record apparent conductivity to nominal depths of 6 m (Geonics EM31) and 10 to 40 m (Geonics EM34), corresponding to the respective instrument frequency ranges [*McNeill*, 1980]. EM methods have been widely applied in the mapping of saltwater intrusion and Dupuit-Ghyben-Herzberg lens morphology (e.g, *Stewart*, 1988; *Anthony*, 1992; *Ruppel et al.*, 2000). For the studies described in this paper, we also use EM data to quantify changes in the depth to the water table and the groundwater salinity distribution due to dynamic forcing of natural hydraulic gradients (e.g., tidal pumping).

EM data were collected along transects from tens of meters to 3000 m long, with the instrumentation operated in vertical and horizontal dipole modes. The terrain conductivity instruments generate a time-varying EM field in the transmitter coil, which in turn induces very small currents in the earth, giving rise to a secondary EM field. The ratio of the secondary to primary EM fields provides a measure of the apparent conductivity of the subsurface material. At sufficiently low frequencies, certain assumptions (low induction number approximation) allow for relatively simple solutions for mutual coupling between induction loops. In areas with high conductivity, measured values are no longer linearly related to the true apparent conductivities, and more complex data inversions like those included in the commercial EMIX34 [*Stoyer and Butler*, 1994] package or described by *Schultz and Ruppel* [2005] must be implemented.

2.2.2 DC resistivity. DC resistivity surveys in dipole-dipole or Wenner mapping mode were used for near-surface measurements to ~10 m depth. Deeper (to ~30 m depth) constraints on conductivity structure were obtained using Schlumberger vertical electrical soundings (VES). Like EM methods, DC resistivity methods take advantage of differences in conductivity (inverse of apparent resistivity) to map lateral and vertical variations in near-surface hydrologic and lithologic units. In traditional DC resistivity surveys, small currents I (up to 1.2 A for our system) introduced into the ground through a pair of steel electrodes driven approximately 15 cm deep produce a potential difference ΔV between a second pair of electrodes. The results yield a measure of resistivity, which is resistance multiplied by a length scale. For these studies, we conducted both labor-intensive surveys with 4 electrodes that were moved for each measurement and automated multinode surveys with up to 24 electrodes controlled by switching software to generate multiple combinations of current and potential electrodes for the chosen array configuration.

Schlumberger VES surveys yield a single vertical resistivity model characterized by different resistivities and thicknesses for one or more layers overlying a half space. To interpret Schlumberger soundings we used the linear digital filters of *Guptasarma* [1982] and inverted for two- or three-layer conductivity structures. Dipole-dipole and Wenner mapping surveys produce a pseudo-section "image" of the subsurface that must be inverted to yield an accurate representation of subsurface structure. Such mapping surveys can corroborate the results of EM surveys, but lack the penetration depth of EM methods and can experience problems associated with voltage overloads, galvanic coupling, and resistive shielding near the ground surface [*Van Nostrand and Cook*, 1966]. For this study, we present resistivity pseudo-section inversions based on the commercial RESIX software [*Inman*, 1975] and the public domain *ProfileR* software provided by A. Binley (pers. comm., 2003).

2.2.3 Ground penetrating radar (GPR). GPR surveys involve the introduction of radar waves directly into the ground through a transmitting antenna and reception of the returned signal through a receiving antenna. When implemented in bistatic (i.e., fixed source-receiver separation) mode, GPR yields an image of reflectors in the uppermost tens of meters of sediment, with penetration depth dependent on antenna frequency and the dielectric permittivity of the medium. GPR is used to image subsurface geology, to constrain the top of high conductivity (saltwater) layers, and, in some cases, to locate the water table. The greater resolution of radar waves in shallow unsaturated or freshwater saturated sandy sediments renders GPR more suitable than traditional acoustic (seismic) methods for constraining the fine-scale lithologic and stratigraphic structures that define flow pathways for the surficial aquifer.

2.3 Integration of Hydrologic and Geophysical Data

Most hydrologic studies rely on invasive sampling, particularly the installation of monitoring wells or piezometers and direct sampling or invasive monitoring of groundwater properties. Major problems with these techniques include high cost, the difficulty of interpolating between discrete sampling points, and the need to extrapolate sparse data collected at length scales of 10^{-2} to 10^2 m up to ecosystem or watershed scales (10^2 to 10^5 m) [*Sudicky*, 1986; *Millham and Howes*, 1995]. Traditional hydrologic techniques, focused as they are on vertical boreholes, also have limited utility at sites at which hydrogeologic boundaries are arrayed vertically, rather than horizontally, or at which flow fields have strong vertical, not horizontal, components (Figure 2). At the edge of tidally-influenced surficial aquifers in coastal zone settings, vertical flow and vertical hydrogeologic boundaries provide a challenge for studies that rely solely on monitoring wells or piezometers. Once hydrologic data are acquired, an additional challenge is resolving aquifer properties, particularly with respect to horizontal heterogeneity. For example, the success of 3D geostatistical approaches in capturing such heterogeneity is often limited not only by the vertical nature of borehole data and the difficulty of obtaining high data density in the horizontal direction [*Phillips and Wilson*, 1989], but also by the poor capacity of classical geostatistical techniques (e.g., co-kriging) to handle data of different densities in the vertical and horizontal directions. *Dagan* [1986] summarizes these issues most succinctly, demonstrating that hydrologic parameter estimates are inherently dependent on the measurement, computation, and integration scales of a particular groundwater problem.

Hydrogeophysical techniques—the application of geophysical remote sensing to hydrologic problems—provide a powerful approach to characterizing the features of sedimentary aquifer systems. Like standard hydrologic data, geophysical data lack direct information at the range of spatial scales of interest for many problems: Most geophysical methods detect only contrasts in physical properties, not the absolute properties themselves. Geophysical survey techniques also have limited resolving power that depends on the spacing of measurements, the electromagnetic or acoustic frequency of the technique (where applicable), and the characteristics of the site. Geophysical observations are obtained using spatial arrays that average over an area or volume between the sources and sensors, unlike many hydrologic results that average only over the sediments adjacent to a well screen or the area close to the sampling port in a multiport well. Nonuniqueness in the interpretation of geophysical data also presents a formidable challenge. Such nonuniqueness arises from a variety of factors. Most fundamentally, a feature with particular material properties buried at one depth might produce the same geophysical anomaly as a feature with different properties buried at a greater or shallower depth. Furthermore, propagation of electric and electromagnetic energy in the subsurface depends on material properties, the frequency of the source, and other factors, and the amplitude of the energy received is affected nonlinearly by the medium.

The relatively innovative approach of coupling hydrologic and geophysical techniques for aquifer characterization at a variety of spatiotemporal scales can address some of the shortcomings of analyses based exclusively on either hydrologic or geophysical data. At the same time, new difficulties arise when data sets that measure such different parameters, have different resolution, and require different types and degrees of interpretation are combined. Researchers have had success with data fusion techniques that use geophysical data to improve hydrologic results (e.g., *McKenna and Poeter*, 1995; *Chen et al.*, 2004), with applying geostatistical techniques to more robustly connect geophysical observations to hydrogeologic parameters (e.g., *Hyndman and Gorelick*, 1996; *Yeh et al.*, 2002), and with using geophysical data to infer the average degree of spatial variability (correlation structure) in aquifers [*Knight et al.*, 1997; *Hubbard et al.*, 1999]. In the studies presented here, we move away

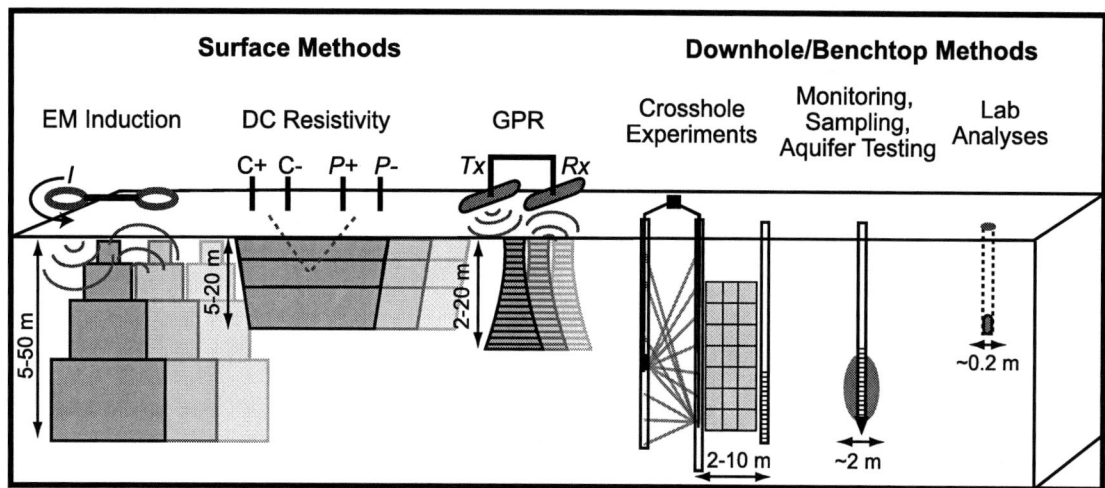

Figure 2. Schematic showing the disparity between and overlap in the relative scales of typical surface, downhole, and laboratory measurements. Geophysical data such as EM, resistivity, or GPR are acquired over one-, two-, or three-dimensional regions to provide constraints on contrasts in bulk physical properties to depths that are dependent on the method and the properties of the geologic media. Hydrologic methods, which range in spatial scale from the laboratory analyses of sediment or pore fluid samples to cross-well tests, average over relatively small spatial scales, but provide a more direct measure of hydrologic processes and properties. Repeated co-located and coincident hydrologic and geophysical data acquisition can add a temporal dimension to data sets for characterization of dynamic hydrologic processes.

from the focus on using geophysical data to constrain the hydrogeologic parameters required as inputs for numerical modeling studies or as descriptors of the scales of aquifer heterogeneity. Instead, we develop interpretations that reconcile geophysical and hydrologic data to elucidate static and dynamic processes in surface water-groundwater interaction and groundwater hydrology.

3. CHARACTERIZATION OF COASTAL AQUIFER STRUCTURE: SPATIAL VARIABILITY

As the transition from land to sea, coastal zones are important for both classical and marine hydrogeology because they contain the regions in which fresh groundwater of largely meteoric origin and saline waters of marine origin meet in the subsurface. For decades, coastal groundwater hydrologists primarily studied the regional freshwater-saltwater interface to assess fresh groundwater reserves and seawater intrusion (e.g., *Bear et al.*, 1999; *Fisher*, 2005) caused by natural or anthropogenic processes. In recent years, much of the coastal zone hydrologic research has shifted to focus on submarine groundwater discharge and groundwater flux to the coastal oceans. Such processes are studied at scales ranging from local field sites (10^1 to 10^2 m) to entire watersheds (10^4 to 10^5 m) and often rely on highly accurate geochemical methods applied to small volume samples.

While the focus of coastal hydrology and hydrogeochemistry has shifted to characterizing shallow flow and transport processes, the cutting-edge is research that attempts truly multidisciplinary, multi-scale approaches that fundamentally link the health of ecological and biological systems (e.g., marsh grass communities, crab and oyster populations) to the physical and chemical factors that affect critical estuarine and shallow nearshore ocean habitats. This is in part the emphasis of some coastal zone Long Term Ecological Research (LTER) programs around the world. Such integrative hydrologic studies require consideration of the interaction between surface (surface water flow, hydrometeorology, ecological zonation) and subsurface processes (groundwater flow and transport, redox zonation, bioirrigation) and the interplay between physical, chemical, and biological systems at scales ranging from that of pores to basins. Already, critical studies at a variety of spatial scales have demonstrated the influence of groundwater and pore water fluxes on nutrient and contaminant fluxes, sediment oxidation potential, and pH (e.g., *Chalmers*, 1982; *Howes et al.*, 1986; *Nuttle and Harvey*, 1988; *Snyder et al.*, 2004) and microbiological processes, estuarine biology, and the distribution of rooted macrophytes [*Hemond and Fifield*, 1982; *Harvey et al.*, 1995; *Thibodeau et al.*, 1998; *Osgood and Zieman*, 1998; *Snyder et al.*, 2004].

3.1. Regional Scale: Lens Aquifer

To provide both local- and regional-scale data, geophysical and hydrologic surveys (Figure 3) were conducted at 5

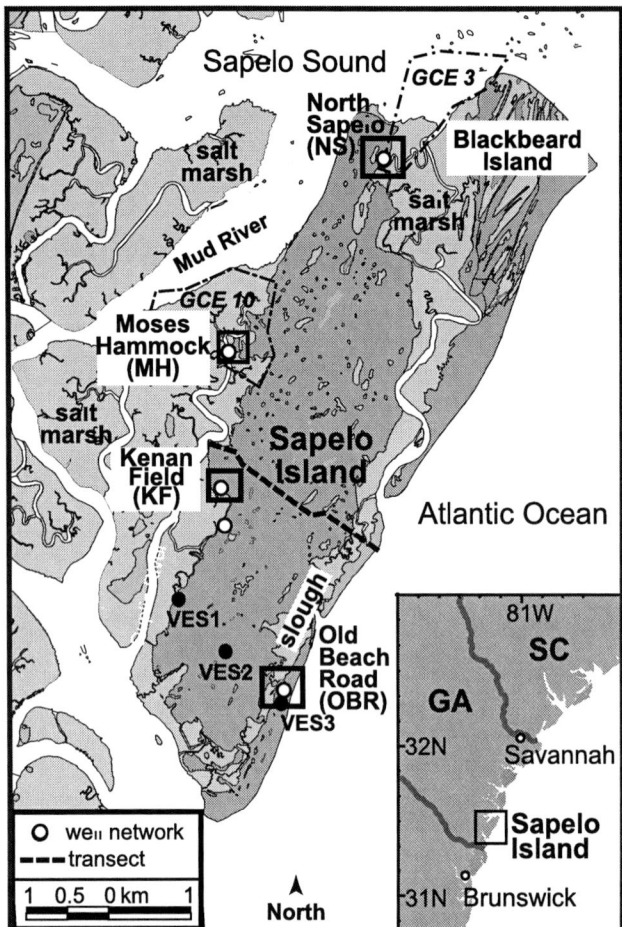

Figure 3. Location maps for the regional and local-scale studies. (a) Map showing the location of Sapelo Island near the center of the South Atlantic Bight. (b) Location of monitoring well network sites and focused geophysical surveys, some of which were completed at designated sites of the Georgia Coastal Ecosystems (GCE) Long Term Ecological Research (LTER) program. Sites discussed here are at the island-estuary (Kenan Field, Moses Hammock or GCE 10), island-marsh (North Sapelo or GCE 3) and island-ocean margins (Old Beach Road). VES, EM, and GPR surveys were also conducted along unpaved east-west trending roads.

field-scale sites (Figure 4) on a sparsely inhabited and largely undeveloped barrier island in the SAB. Sapelo Island, which measures ~16 km long and from 2 to 5 km in width, is one of six low-lying Sea Islands separated from the Georgia mainland by a 4- to 6-km-wide expanse of salt marshes and tidal rivers (Figure 3). The island has a late Pleistocene sedimentary core consisting mostly of well-sorted, clean, fine sands and rimming marsh muds. Holocene beach sand deposits formed by longshore currents and wave action fringe the seaward side of the island. The regional confining layer at the base of the surficial aquifer is a 4- to 30-m-thick blue clay layer whose top lies at an average depth of ~12 to 13 m below ground surface [*Lens*, 1981; *Schultz*, 2002; G. Hebeler, pers. comm., 2002].

The part of the SAB occupied by Sapelo Island is a mixed energy environment with semi-diurnal tides having average amplitude of ~2.4 m and mean spring tide range of 3.4 m [*Chalmers*, 1997]. The horizontal location of the tidal boundary varies by less than 2 m near subvertical tidal creek bluffs to more than 100 m on shallowly-sloping beach faces [*Schultz*, 2002]. The most substantial hydrologic inputs to the island aquifer systems on the Georgia coast include infiltration of rainwater and displaced groundwater that is pumped from the deep Floridan aquifer and discharged at the surface. Sapelo Island is relatively pristine and inhabited by fewer than 100 people, meaning that the impact of deep pumping on eventual infiltration into the surficial aquifer is much more limited than on many other barrier islands in the SAB. Infiltration of precipitation is by far the most important input to the surficial aquifer in our study area, but it is of course dependent on spatial variations in vegetation interception and surface retention.

To characterize the two-dimensional, island-scale distribution of freshwater and saline water beneath Sapelo Island we acquired geophysical data on unpaved roads that run both perpendicular and parallel to the long axis of the island (Figure 3). At selected sites, we also conducted resistivity sounding surveys to constrain the one-dimensional (vertical) conductivity structure.

3.1.1. DC resistivity results. To constrain the bulk vertical electrical conductivity structure at discrete points along a cross-island transect, Schlumberger VES data were acquired at sites (Figure 3) near the island-estuary interface (VES1) on the landward side of Sapelo Island, in center of the island (VES2), and at the island-ocean interface (VES3). The morphology of the raw Schlumberger curves is consistent with high resistivity material near the surface and dramatically lower resistivity at depth. Based on this observation and the assumption that the data should reveal information about the layering of unsaturated, freshwater-saturated, and saline-saturated sediments, we assume a three-layer model and vary the conductivities and thicknesses of the layers in a forward model to obtain a match to the observations. Note that the interpretation of electric sounding data is non-unique (equivalence), and a layer of intermediate conductivity sandwiched between an overlying low conductivity and underlying high conductivity layer can often be difficult to resolve (suppression).

Results of the VES interpretations are detailed in Plate 1. For the survey closest to the island-estuary margin (VES1), forward modeling yielded a best-fit resistivity structure consistent with a low conductivity ($\sigma_1=0.6$ mS m^{-1}; unsaturated?) layer of 2.0 m thickness over a 6.5-m-thick transitional layer ($\sigma_2=2.2$ mS m^{-1}) and a higher conductivity ($\sigma_3=18-25$ mS m^{-1}; saline saturated?) halfspace. The thickness of the unsaturated surface layer is confirmed by static (non-tidally influenced) water levels measured in shallow monitoring wells located elsewhere (e.g., Kenan Field, see below) on the estuary side of the island. The center of the island (VES2) has a gross vertical conductivity structure similar to that at the island-estuary interface (VES1), but with a much thicker intermediate layer that presumably corresponds to the zone of freshwater saturation. The analysis of the VES2 data yields a 2.8-m-thick unsaturated layer ($\sigma_1=0.55$ mS m^{-1}) with an underlying intermediate conductivity ($\sigma_2=4.7$ mS m^{-1}) layer ~28-m-thick and a halfspace with conductivity of 21 mS m^{-1}. The VES3 survey conducted on the ocean side of the island reveals saline saturated sediments at shallow depths. Forward modeling yields a 1.9-m-thick resistive layer of $\sigma_1=3.77$ mS m^{-1} underlain by a layer with $\sigma_2=110$ mS m^{-1} and a halfspace with $\sigma_3=510$ mS m^{-1}. Overall, the results shown in Plate 1b suggest the presence of only a thin veneer of freshwater beneath the Holocene sediments on the Atlantic side of Sapelo Island.

The different halfspace conductivities determined by the 3 VES surveys reflect several factors. First, there is significant lateral variation in the shallow conductivity structure across

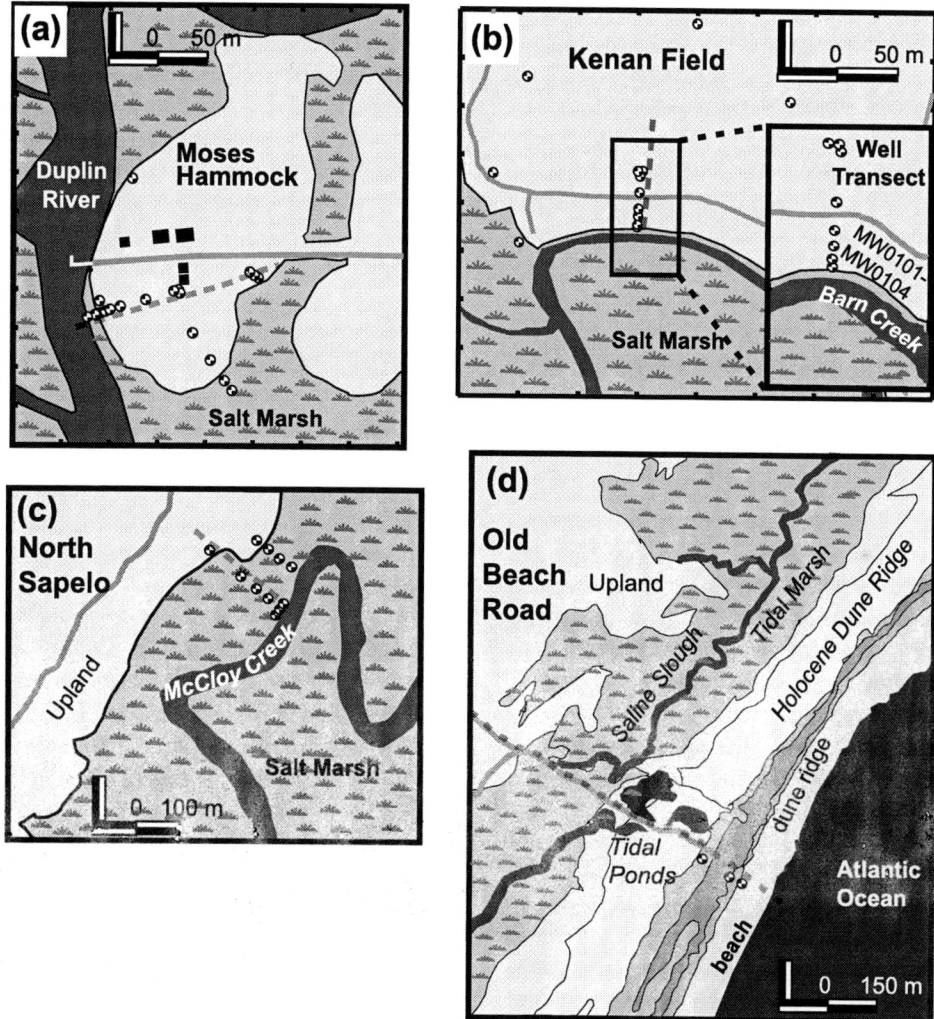

Figure 4. Detailed maps of the well network sites at (a) Moses Hammock (MH), (b) Kenan Field (KF), (c) North Sapelo (NS), and (d) Old Beach Road (OBR). Circles indicate locations of shallow monitoring wells, gray dashed lines show geophysical transects, and black rectangles on (a) denote various outbuildings associated with the seasonal hunt camp. Inset in (b) shows enhanced view of monitoring well transect. Geophysical surveys were primarily conducted coincident with the monitoring well networks, which are oriented perpendicular to the local island boundary.

the island owing to the morphology of the freshwater lens and varying degrees of saline intrusion at the island's boundaries. Second, interpretation of resistivity curves in terms of layered structure may lump together different thicknesses of material in the subsurface in different locations. Finally, we note that the halfspace conductivity (deepest part of the model) is constrained by the data obtained at the largest electrode spacing, and the impact of high conductivity at depth on the raw Schlumberger curves can render the data at these large spacings particularly challenging to interpret.

3.1.2. Terrain conductivity results. In September 1997, EM data were collected along east-west transects at the location of the VES1 survey (island-estuary interface) and VES3 surveys (island-ocean interface). In August 2000, we also conducted a cross-island terrain conductivity survey extending from the island-estuary interface at Kenan Field to the island-ocean interface at Cabretta Island. Local-scale surveys at the island margins were acquired only in horizontal dipole mode, whereas apparent conductivity data for the regional (cross-island) survey were collected in both horizontal and vertical dipole modes and at all possible coil separations of the EM34 instrumentation (10, 20, and 40 m).

The cross-section in Plate 1a shows an inversion of the cross-island EM data with minimally processed 50 MHz GPR data superposed. The EM inversion was generated by application of a regularized tomographic inversion using 2D smoothing constraints [*Schultz and Ruppel, 2005*]. The starting model for the inversion was the vertical conductivity structure constrained by the Schlumberger VES results. However, we note that DC resistivity and terrain conductivity methods do not necessarily produce entirely comparable results. Because DC resistivity relies on galvanic coupling and EM methods use inductive coupling, there are subtle differences in how the methods detect bulk subsurface conductivity structure. During excitation with a quasi-static electric field, electromagnetic contrasts sensed by terrain conductivity methods are theoretically equivalent to those sensed by DC resistivity methods. However, even under the quasi-static assumptions, practical conditions during terrain conductivity surveying may lead to significant differences between terrain conductivity and DC resistivity measurements. Electromagnetic coupling is less problematic in DC resistivity surveying because stationary currents are inherently unable to induce time-varying electromagnetic fields [*Kuras, 2002*]. In areas of high resistivity/low conductivity, magnetically inducing sufficient current to generate magnetic fields that can be detected by the receiver can be challenging, leading to potential complications for EM methods. Although contact impedance may also cause complications for DC resistivity methods, it is generally easier to inject current into resistive ground through galvanic contact rather than through inductive coupling. In conductive environments, direct current signals decay rapidly with depth, and increasing electrode separation does not necessarily yield increased penetration.

The cross-island inversion reveals significant variability in the spatial distribution of apparent conductivity, which should correspond primarily to changes in the conductivity of pore waters. At the ocean interface, the inversion yields conductivities ranging from nearly 1000 mS m^{-1} to 100 mS m^{-1} across the Holocene beach ridge. Lower conductivities (~0.5 to 45 mS m^{-1}) between 900 and 3000 m from the estuary interface (center of the island) are consistent with freshwater saturated sediments.

We also inverted the EM data using EMIX34 to create stitched 1D inversions that yield a sharp boundary between freshwater and saline saturated sediments. The starting model for this inversion was a simplified version of the vertical conductivity structure obtained from the Schlumberger VES surveys, with a single low to intermediate (unsaturated and freshwater saturated) layer overlying a saltwater saturated halfspace with conductivity greater than 900 mS m^{-1}. The apparent conductivity profile shown in Plate 1b is representative of the interface separating low conductivity material in the near-surface from higher conductivities at depth. Predicted profiles produced by EMIX34 forward modeling agree with the actual conductivity measurements to within 13% for vertical dipole data and 19% for horizontal dipole data. The better match obtained with the vertical dipole data probably reflects their lower sensitivity to lateral variations in conductivity structure and their greater penetration depth, which is conducive to better constraining layered conductivity structures.

3.1.3. Composite aquifer morphology. The EM inversions, VES data, and GPR results reveal noticeable thickening of the freshwater aquifer beneath the center of the island, with the thickest part of the aquifer skewed toward the estuary (landward) side of the island. The pseudo-2D EM34 inversions (Plate 1a) imply local thinning of the aquifer between 1100 and 2000 m along the transect in a topographic low occupied by freshwater wetlands. We infer this thinning to be only apparent and attribute it to the presence of higher conductivity (relative to unsaturated sediments) groundwater at the surface in this area, not the upconing of salt water. Such groundwater may be associated with freshwater wetlands or the evaporative residue from these wetlands, which are common at this position on Sapelo Island. In such a pristine environment with such relatively homogeneous sediments, there are few other plausible explanations for this anomaly.

The depths to high conductivity material constrained by EM inversions might not necessarily indicate the base of the freshwater zone and could instead represent a more complicated effect that involves the presence of high-salinity pore waters and the occurrence of clay over a broad range of depths. A prominent GPR reflector inferred from cross-island profiles superposed on the EM inversions in Plate 1a is interpreted as a semi-permeable clay that may confine the base of the lens aquifer to depths shallower than those indicated by EM inversions. With additional data (e.g., induced polarization or IP), it might be possible to distinguish contrasts associated with variations in conductivity from those caused by increased clay content; however, based on our other studies in the area (e.g., *Schultz and Ruppel* [2005]), we assume that increased salinity/conductivity with depth would likely produce a more gradual transition than inferred from the relatively distinct reflector seen in the GPR data. The gentle eastward dipping slope (0.4° to 1.2°) associated with the GPR reflector between 1400 and 2600 m and at depths of 10 to 17 m is consistent with geologic cross-section by *Lens* [1981] and surface seismic wave studies (G. Hebeler, pers. comm., 2001). Although the GPR and seismic data were not topographically corrected, at the scales of interest for this interpretation, the binned elevation data vary by less than 2.5 m along the entire transect. Shoaling of the confining unit is consistent with high GPR attenuation beneath the surficial aquifer on the estuary (west) side of the island and a co-located, near-surface, high conductivity zone in the EM inversion.

The geophysical data largely confirm the predictions of Dupuit-Ghyben-Herzberg (DGH) theory, which describes the lenslike morphology of freshwater aquifers beneath barrier islands. The thickness of the freshwater lens is a function of the density contrast $\alpha = \rho_f/(\rho_s - \rho_f)$ between fresh (ρ_f) and saline (ρ_s) water and the elevation of the freshwater head h above mean sea level (MSL). For a hydrostatic and homogeneous lens aquifer system the depth of the freshwater-saltwater interface below MSL can be estimated from $z = \alpha h$ [*Herzberg*, 1901]. For this study, an important modification to DGH theory is that of *Urish* [1977], who demonstrated that differences in the effective MSL between the ocean and estuary sides of an island typically produce an asymmetric lens skewed toward the landward side of the island, as noted in our EM inversions. The depth z to the freshwater-saltwater interface can be determined at any point x from the estuary side of the island by combining the Dupuit-Ghyben-Herzberg principle with the analytical solution for the shape of the phreatic surface:

$$z = \left[\frac{\alpha \rho_s}{\rho_f} - 1\right]\sqrt{\Delta H^2 - \frac{\Delta H^2(L-x)}{L} + \frac{\varepsilon x(L-x)}{(K + K\alpha)}} - \frac{x\alpha \rho_s \Delta H}{\rho_f L}, \quad (1)$$

where ε denotes infiltration rate, and ΔH represents the difference in freshwater head across an island of width L and hydraulic conductivity K. Equation (1) assumes a sharp interface between fresh and saline groundwater and does not account for aquifer heterogeneity, anisotropy, or the effects of dynamic boundary conditions. In reality, the interface is a transition zone influenced by the morphologic, hydrogeologic, and hydrodynamic properties of the nearshore zone and the aquifer boundary [*Urish and Ozbilgin*, 1989]. An additional factor that has not been widely considered in the literature is the degree to which variations in the salinity of the adjacent surface water body influence density-dependent mixing between fresh and saline water at coastal zone aquifer boundaries.

Superposed on the results shown in Plate 1b is the DGH lens morphology calculated from (1) assuming an effective MSL 0.7 m higher on the ocean side than on the estuary side. Oscillations of the surf zone water level can lead to such an overheight of the water table at the ocean shoreface [*Philip*, 1973]. For this calculation, we assume an average infiltration rate of 2.7×10^{-9} m s^{-1} estimated from monthly-average precipitation rates between 1996 and 2000 taken from NOAA regional data and from published estimates of vertical hydraulic conductivity (1.5×10^{-6} m s^{-1}) [*USDA*, 1959]. The depth to the freshwater-saltwater interface predicted from (1) is consistent with inversions of the cross-island EM data from the August 2000 survey, but not with inversions of local-scale data obtained at the island-estuary interface in September 1997. To fit these older, local-scale data, which were acquired prior to a severe drought that endured from 1998 to 2000, we had to instead assume an infiltration rate that is more than double the estimate obtained from analysis of precipitation data. We infer that variability in the lens thickness between the 1997 local surveys and the 2000 regional survey is more likely attributable to differences in local conditions at the survey locations (edges of island vs. large-scale across island aquifer morphology) and not a regional shrinking of the lens between the September 1997 and August 2000 surveys. On the ocean side of the island, the inversions imply thinning of the freshwater lens to less than a meter beneath the Holocene strip barrier island that is separated from Sapelo Island's core by a complex system of tidal creeks, salt marshes, salt ponds, and sloughs. The freshwater lens in this area may be dissected by the complex distribution of saline surface and groundwater, leaving a self-contained freshwater lens located beneath the ocean-facing dune ridge and separated from the lens beneath the main part of the island.

The exact shape of the lens along the cross-island transect may also be influenced by interaction with the regional confining layer. High-resolution GPR data image the known confining layer at depths of 13 to 18 m, but cannot directly constrain the contrast in hydraulic conductivity between the surficial Pleistocene sediments and the underlying Pliocene unit. If the layer provides a continuous sharp permeability contrast, the freshwater lens will truncate at the layer. Alternatively, if the transition between Pleistocene and Pliocene sediments does not represent a significant change in hydraulic properties, the lens will simply be deflected as it penetrates the deeper, lower permeability unit.

3.2 Local Scale Hydrofacies: Coastal Groundwater Boundaries

The regional scale hydrogeophysical results presented above imply that the detailed morphology of the freshwater-saltwater interface is dependent on local conditions. Previous observational studies [*Ginsberg and Levanon*, 1976; *Ayers and Vacher*, 1986; *Anthony et al.*, 1989] have reported varying degrees of subsurface freshwater-saltwater mixing in different environments, but have provided little explanation of the factors that control the transport of saline groundwater into the edges of freshwater aquifers.

In general, coastal aquifers are subject to tidal boundary conditions regardless of the precise nature of the boundary (e.g., fully exposed beach face, sheltered estuary, or tidal creek-salt marsh complex). Variations in the morphology of the boundary, the degree of hydrodynamic connection between tidal water bodies and the surficial aquifer across the boundary, and the hydrogeologic properties of the nearshore environment can lead to locally different subsurface salinity regimes at the coastal groundwater boundary. In this section, we integrate the results of geophysical surveys with hydrologic data collected in coincident monitoring well networks. This permits us to determine the horizontal and vertical extent of saline intrusion into the permeable uplands and the hydrogeologic conditions that affect flow and transport for the different types of shoreline boundaries represented by the study sites.

Local-scale studies were conducted at the site of monitoring well networks installed at Old Beach Road (OBR), Kenan Field (KF), Moses Hammock (MH), and North Sapelo Island (NS) (see Figures 4a–d). OBR represents an island-ocean interface, while the other 3 sites represent various types of island-estuary interfaces common in barrier island settings. At the KF site, the permeable upland sediments are in direct contact with a tidal creek; at the MH site, a small island is completely surrounded by tidal creek and salt marsh complexes and the permeable upland under study borders salt marsh and a major tidal river; at the NS site, the upland is separated from the tidal creek by a long expanse of mature marsh whose hydrologic characteristics we surveyed as part of our studies.

3.2.1. Island-estuary interface: Upland bordered by tidal creek. The Kenan Field site borders Barn Creek, a tidal creek that is up to 30 m wide at high tide. This width is small relative to the extent of a monitoring well network (~225 m) that stretches inland from the edge of the upland perpendicular to the tidal creek bank. The slope of the creek bank is nearly vertical, and previous studies [*Schultz and Ruppel*, 2002] suggest that clogging of permeable upland sands by finer marsh mud material or iron oxides where the sands border the tidal creek impedes groundwater flow and surface water-groundwater interaction across the tidal creek boundary.

To extend our interpretation of the local hydrogeology at this site, EM data were acquired along the well transect perpendicular to Barn Creek in both horizontal and vertical dipole modes. High apparent conductivities (>100 mS m^{-1}) near the tidal creek imply saltwater intrusion into the creek bank and upland sediments. Because these high conductivity data obtained near the creek bank violate the low induction number approximation, we applied a nonlinear inversion [*Schultz and Ruppel*, 2005] to infer the pore water conductivity structure in the subsurface. The results, shown in Plate 2, reveal the expected deepening of high conductivity groundwaters with distance inland from the tidal creek beneath the local edge of the freshwater lens aquifer with a narrow (< 10 m) transition zone between saline and fresh near-surface groundwaters at this site.

The resistivity inversions shown in Plate 2a are oriented perpendicular to the creek and image the localized region outlined in the EM cross-section. The conductivity increase from ~35 mS m^{-1} at 20 m from the creekbank to ~6000 mS m^{-1} at 5 m from the bank is consistent with the subsurface hydrofacies distribution constrained by the EM surveys. Plate 2a also provides information about the third dimension (parallel to the upland edge) based on inversions from multinode dipole-dipole resistivity surveys. These images confirm lateral intrusion of saline water to only a few meters from the creekbank.

An important component of our studies is integrating hydrologic and geophysical data, despite the different spatial scales resolved by these approaches. At this site, we sampled the conductivity of groundwater in the monitoring well network coincident with the geophysical surveys. Groundwater samples average over a support volume correlated with the length of the screened interval, about 0.7 m in this case. Specific conductance decreased from 42.2 mS at the well positioned 3.5 m inland from the creek bank (MW0101) to 0.183 mS at the well located 10.5 m inland (MW0103) and 0.056 mS at the well most distal from the creek bank (MW0108, 56 m inland).

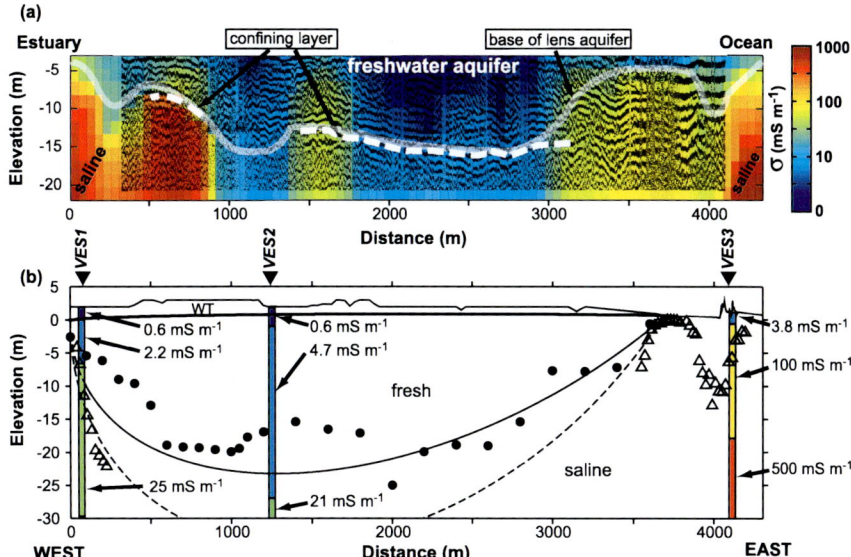

Plate 1. Composite cross-island transects compiled by integrating geophysical and hydrologic data. (a) Two-dimensional model from the inversion of EM34 data extending from the estuary (west) to the ocean (east) are overlaid on processed GPR profile. Lower relative conductivities near the center of the island are indicative of freshwater saturated sediments (dashed-dot curve). A strong coherent reflector in the GPR image (dashed curve) marks the top of a potential confining unit that constrains the form of the freshwater lens aquifer. (b) Depths to the freshwater-saltwater interface predicted by the DGH model and the inversion of EM (points) and VES (columns) data. Inversions of data acquired in September 1997 (open triangles) at the margins of the island aquifer are used to constrain the DGH model that predicts a relatively thick freshwater zone (dashed curve). The DGH model fit (solid black curve) to the inverted interface depths (solid circles) using EM data from the August 2000 cross-island transect yields a shallower zone of freshwater. Parameters used for the DGH model calculations are given in the text.

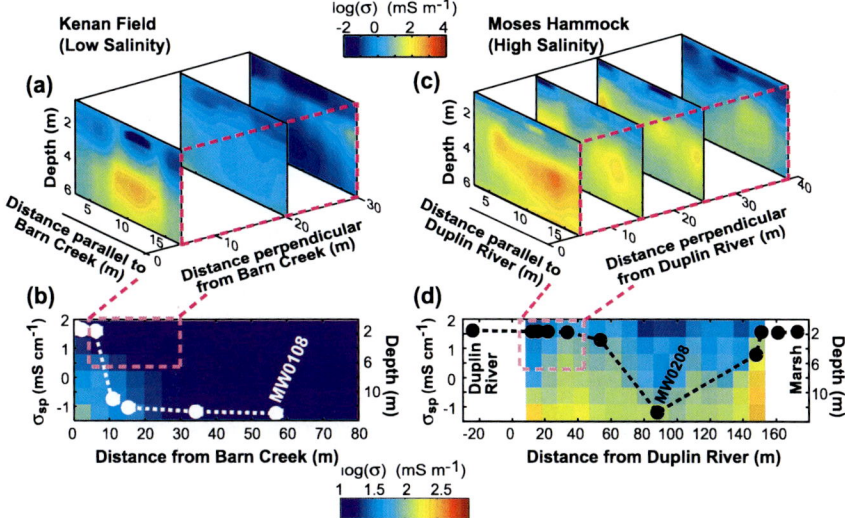

Plate 2. Electrical resistivity data and an inversion of EM data along the main well transects at the Kenan Field and Moses Hammock island-estuary sites. (a) Resistivity inversions produced by public domain code *ProfileR* (A. Binley, pers. comm., 2003) on data from multinode surveys oriented parallel to Barn Creek at Kenan Field. Owing to the geometry of the surveys, the results in the center of these inverted sections and the sections shown in part (c) of this figure should be considered more reliable than conductivities at the edges of the sections. (b) Inversions of EM34 data combined with specific conductivity data from groundwater samples reveal a narrow zone (<10 m) of high conductivity associated with a low degree of saltwater intrusion. (c) Same as (a), but for surveys oriented parallel to the Duplin River. (d) EM34 inversion model and overlaid specific conductivity data corroborate the elevated conductivities extending more than 25 m into the small self-contained freshwater lens aquifer at Moses Hammock.

3.2.2. Island-estuary interface: Upland bordered by tidal creek-marsh complex. The Moses Hammock (MH) site is Georgia Coastal Ecosystems (GCE) LTER Site 10 and is located in the upper reaches of the Duplin River watershed (Figure 4a). The hammock is a ~0.1 km² feature consisting predominantly of Pleistocene sand and is completely surrounded by marsh and tidal creeks. At high tide, the Duplin River occupies over 220 m of the channel on the east side of the hammock, and this width exceeds the ~180 m extent of the monitoring well network that stretches across the hammock. Using water levels monitored over a month-long period, *Schultz and Ruppel* [2002] showed that the relatively gradual slopes of the creek bank and intertidal marsh cause the transfer of tidal energy from the creek to the surficial aquifer to be highly nonlinear, and the geometrical complexity of the upland boundary, the lateral and vertical heterogeneity in hydraulic parameters, and other factors may also play a role in controlling the degree of lateral saline intrusion at this site [*Schultz*, 2002].

The large-scale distribution of saline groundwater at the MH site can be inferred from inversion of EM34 data [*Schultz and Ruppel*, 2005] collected across the hammock (Plate 2d). The cross-section reveals a small (<100 m wide), self-contained freshwater lens confined to distances between 60 and 140 m from the creek bank and shows the extent of saline intrusion at both the Duplin River and marsh boundaries. Elevated conductivities (250–2500 mS m^{-1}) extend ~35 m inland from the Duplin River high tide mark, and another high conductivity zone occupies the 30 m closest to the marsh on the east side of the hammock. Toward the center of the hammock, the inversion results reveal thickening of low conductivity zone (freshwater lens) to ~6–8 m depth.

The inferred large extent of saline water intrusion from the Duplin River side of the hammock is confirmed by multinode DC resistivity surveys carried out both perpendicular and parallel to the shoreline. The inversions shown in Plate 2c reveal greater than 20 m of lateral saltwater intrusion in the upland, significantly larger than the few meters of lateral intrusion at the KF site.

Analyses of groundwater in monitoring wells coincident with the geophysical surveys corroborate inferences about saline and freshwater distribution beneath the upland. Cation and anion concentrations [*Hunter et al.*, 2000; *Snyder*, 2002] confirm the presence of saline waters in the four monitoring wells closest to the Duplin River and in the three monitoring wells adjacent to the salt marsh east of the hammock. Fresh (<5 mS cm^{-1}) groundwater samples were obtained only from monitoring wells near the center of the hammock (e.g., MW0208). Compared to groundwater in MW0101 at KF (3.5 m from the creek bank), groundwater from well MW0201 at MH (5 m from the creek bank) has higher conductivity (33 mS m^{-1}).

3.2.3. Island-estuary interface: Marsh hydrology. To demonstrate the capacity of geophysical methods to constrain hydrologic processes at nested spatial scales in both the vertical and horizontal directions, we present the results of intensive studies carried out across an upland-salt marsh-tidal creek complex at the far northern end of Sapelo Island at a location designated GCE3 by the LTER. Since ~2001, we have conducted hydrogeophysical surveys and installed coincident, specially adapted monitoring well networks in clastic salt marshes stretching from the Okatee watershed in South Carolina [*Sibley*, 2004] to the Satilla watershed near the Georgia-Florida border. The North Sapelo site (NS) was the first to be studied in this way and has yielded important insights into salt marsh hydrology.

The NS site (Figure 4c) is located on the ocean side of Sapelo Island, where the Holocene Blackbeard Island, a small, subsidiary barrier island located ~2 km away across a series of tidal creeks and salt marshes, shelters Sapelo Island from the Atlantic Ocean. The site consists of a sand-dominated upland with a forest of live oak, saw palmetto, cypress, and loblolly pine and an adjacent *Spartina alterniflora* salt marsh with rimming *Juncus romerianus* along the upland edge and *Salicornia* in low-lying, high salinity tidal flats. The marsh is bounded on the east by the tidally-influenced McCloy Creek. *Spartina* grass dominates most of the marshes in this part of the SAB and tolerates low salinity and full ocean salinity, standing water and days without tidal inundation, and both heavily and sparsely burrowed sediments. *Juncus* is an indicator of high marsh, relatively low salinity, and relatively rare tidal inundation. *Salicornia* occurs only in high salinity zones, primarily in salt flats that tend to develop between the *Spartina* that fills most of the marsh between tidal creeks and uplands and the *Juncus* that grows primarily at the edges of uplands. This information is presented here because the ecological zonation of study sites is often an overlooked factor in understanding near-surface hydrologic conditions and salinity gradients, precisely the targets of our investigations.

Since 2001, we have repeated multinode dipole-dipole and Wenner DC resistivity surveys at the NS site. Plate 3a shows the inversion of dipole-dipole data across the upland-marsh interface along one of the two transects we eventually instrumented with monitoring wells. The inversion provides provocative evidence for submarsh freshwater flow up to ~20 m from the upland edge before the tongue of fresher water becomes mixed with saline groundwaters. The finding of submarsh flow in this area largely confirmed the independent hypothesis of *Schultz and Ruppel* [2002] that this process, not seepage, was critical for groundwater-surface water interaction in this part of the SAB. Thermal infrared photographs acquired by the authors in August 2001 also

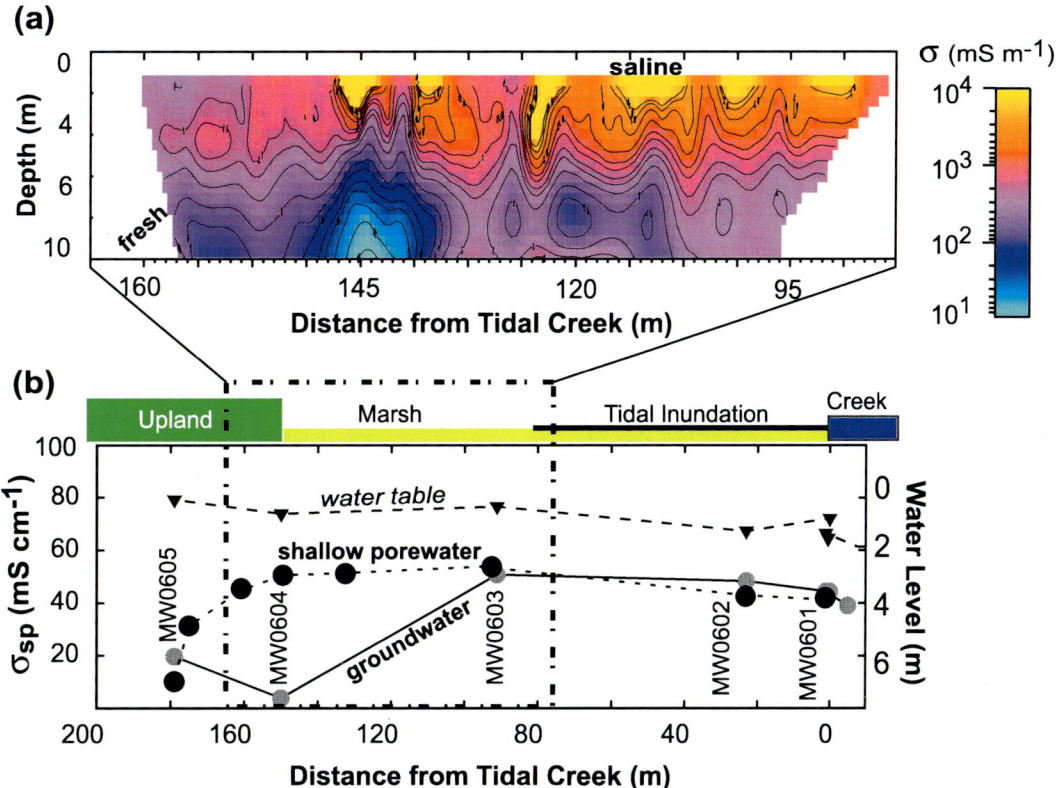

Plate 3. Combination of geophysical and hydrologic results constrain multiple scales of interaction between fresh and saline groundwater at the North Sapelo site. (a) Inversion of DC resistivity across the marsh-upland boundary reveals submarsh freshwater flow from the upland, fingers of saline pore water penetrating downward, and possible evidence for free convection of pore waters. (b) Monitoring well data (gray circles) from deeper than 2 m and porewater analyses at ~10 cm depth (black circles) confirm the presence of less saline groundwater beneath the marsh than inland. Also shown is the water table as measured in the monitoring wells (relative to an arbitrary datum), demonstrating a net hydraulic gradient toward the tidal creek.

revealed no evidence for seepage in the vicinity. Confirming the presence of submarsh freshwater are cation analyses and borehole conductivity logs in coincident monitoring wells that penetrate the marsh to depths of ~4 m. Surprisingly, these data reveal that the marsh well located closest to the edge of the upland (MW0604) has fresher groundwater than that in the upland monitoring well (MW0605).

This example aptly illustrates the meshing of hydrologic and geophysical data at different spatial scales. The geophysical data cannot resolve subtle conductivity differences between the groundwater samples in MW0605 and MW0604. Conversely, the wells, already at much closer lateral spacing than used in most hydrologic studies, do not closely constrain the lateral or vertical extent of the freshwater flow beneath the marsh. Lacking geophysical data, we may never have inferred the true nature of groundwater interactions beneath the marsh from the seemingly anomalous groundwater conductivity detected in MW0604.

The resistivity inversion in Plate 3a also highlights hydrologic conditions and processes at several other spatial scales. First, the data imply an apparent lack of hydraulic connection between the high electrical conductivity material in the shallow marsh and the lower conductivity material universally present deeper in the marsh, even beyond the extent of the freshwater tongue. During installation of the monitoring wells, we obtained continuous core sediment that was analyzed to estimate grain size and hydraulic conductivity using standard formulations such as the Hazen and Beyer formulae. The results revealed that muds up to 2 m thick near the marsh surface had hydraulic conductivity ~2 orders of magnitude lower than that of the underlying fine sands penetrated by MW0604 and MW0602. In addition, pore water conductivities measured at 10 cm depth in the marsh were higher along most of the transect than groundwater conductivity measured in wells (Plate 3b), attesting to the separation of the shallowmost groundwaters from the deeper groundwaters in this marsh. This combination of geophysical data, hydraulic parameters, and hydrogeochemistry implies that tidal inundation of the marsh delivers high salinity surface waters that do not necessarily infiltrate to great depths.

Second, the DC resistivity data, which were collected with redundant, reciprocal pairs of current and potential electrodes as a means of quality control, provide provocative indications of possible free convection in the marsh [*Ruppel and Schultz*, 2003]. The top of the resistivity inversion shows pockets of very high conductivity (low resistivity, ~0.1 Ωm) pore waters between 2 and 5 m across surrounded by zones of lower conductivity (~1 Ωm) pore water. These zones finger downward, while inferred high resistivity, fresher pore waters at greater depth finger upward. The layering of higher density saline pore waters over lower density fresh pore waters has long been recognized as a gravitationally unstable configuration, and the classic Elder problem in hydrology deals with the development of free convection and fingering in such porous systems on timescales of years in models with homogeneous permeability. Also relevant to the interpretation of the resistivity image are laboratory Hele-Shaw observations of fine-scale fingering between layered, gravitationally unstable, miscible fluids having the same viscosities [*Cooper et al.*, 1997].

The resistivity data were acquired at close enough spacing to resolve the fingering features and gross stratification of more saline over fresher waters highlighted here. We caution that, to the best of our knowledge, this is the first instance of geophysical detection of such fingering and convective relationships in the study of salt marsh hydrology and that significant future work is required to confirm our interpretation. More extensive in situ measurement of pore water conductivities, mapping of marsh plants to determine if roots may provide high permeability conduits that facilitate selective, downward infiltration of high salinity tidal waters, and confirmation of the DC resistivity results through repeated surveys are necessary to show unequivocally that hydrogeophysical techniques have for the first time imaged such important hydrologic processes in a salt marsh.

3.2.4. Island-ocean interface. The OBR site is characterized by a wide-intertidal beach and ocean-fringing sand dunes on the east and a complex of behind-dune salt ponds and tidal creek and marsh that make up a slough system (Figure 4d). Such slough systems are common physiographic features in this part of the SAB and generally separate Holocene beach sediments from the Pleistocene core of the main barrier island.

The local groundwater salinity regime at OBR was constrained using both hydrogeochemical analyses of surface water and groundwater and parallel geophysical surveys conducted along a sand footpath running due west from the ocean. Figures 8a–c shows a comparison of raw EM31 data collected during two 1997 surveys separated by 5 hr, an inversion of EM34 data acquired only in horizontal dipole mode, and cation analyses conducted by M. Snyder (pers. comm., 2000). The repeated EM31 surveys shown in Plate 4a generated similar results except in the 50 m nearest the beach and in the region between 350 and 420 m inland. The differences in the observed apparent conductivity are likely due to the influence of tidally induced semi-diurnal fluctuations in both water level and salinity. The lowest conductivities recorded in the upper beach are likely affected by the nonlinear response of the EM31 instrument at high conductivity. The true conductivities, as confirmed by analysis of the EM34 data, are significantly higher than those reported by the raw EM31 data.

The inversion of EM34 data superposed on the GPR cross-section image the larger scale conductivity structure from the

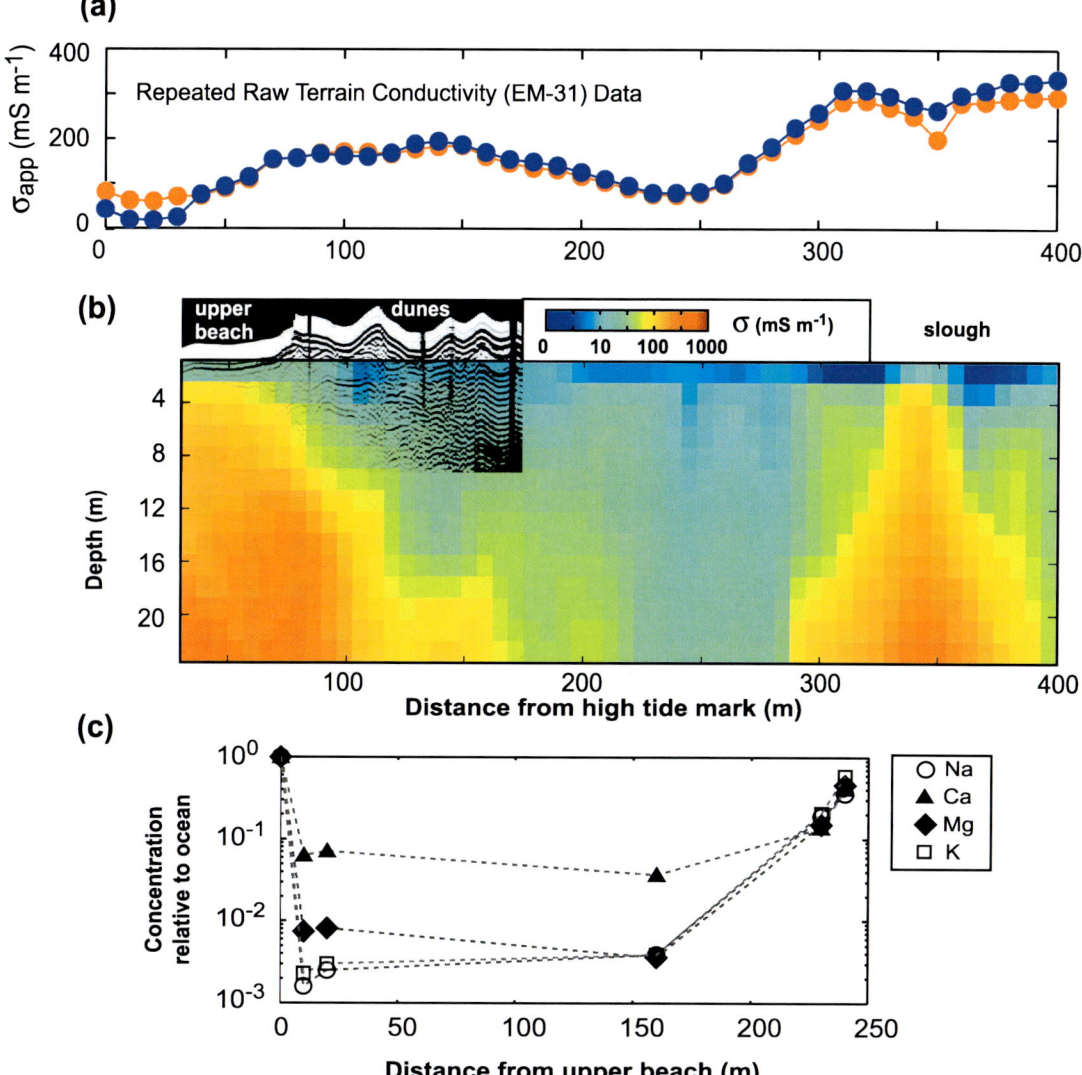

Plate 4. Geophysical survey results and major cation concentrations at the Old Beach Road site at the island-ocean margin. The three parts of the figure are oriented correctly relative to each other in terms of distance and plotted on the same scale so that corresponding points in each data set lie along a single vertical line on the page. (a) Raw EM31 terrain conductivity data acquired 5 hr apart along a survey transect extending from the upper beach across the dune ridge. Differences in the apparent conductivities between the surveys in the 50 m nearest the beach and at the back-barrier slough are indicative of varying tidal water levels and salinities. (b) The conductivity distribution constrained by inversion of EM data, which were here collected in only one coil orientation, reveals the complexity of subsurface salinity gradients at this location. Note the correlation between the inferred base of freshwater beneath the dune ridge and the depth to the attenuated GPR signal in the superposed cross-section. (c) Major cation concentrations along the OBR profile from M. Snyder (pers. comm., 2000).

upper beach, across the dune ridge, and into the topographic low behind the dunes. Like the EM31 data, the inversion of the EM34 data reveals lower conductivity freshwater saturated sediments beneath the dunes and back-barrier areas with a high conductivity region between 280 and 400 m inland. Superposed on the EM inversion are GPR data collected with a 100 MHz antenna. The data reveal eastward-dipping aeolian and marine sedimentary structures and other features [*Schultz*, 2002]. For the purposes of this study, the most important aspect of the GPR survey is the depth at which the radar waves become attenuated (~7 m), which corresponds roughly to the freshwater-saltwater interface as inferred from the EM inversion.

Surface water (ocean and slough) samples and groundwater samples collected from monitoring wells installed in the swales of dunes have major cation concentrations that decrease sharply between 8 and 12 m from the upper beach (M. Snyder, pers. comm., 2000). For example, Na concentrations remain relatively low (~2.3×10^{-2} ppt) across the dune ridge to a distance of ~150 m and then increase to 13 ppt near the back barrier slough. Elevated anion and cation concentrations in groundwater at the western part of the well transect confirm lateral intrusion of saline water from the tidal ponds and slough that intersect the transect. A synthesis of the results of EM inversions, radar attenuation, and ion concentrations suggests that freshwater accumulation in the Holocene dune sediments gives rise to an isolated local freshwater aquifer disconnected from the larger island lens surficial aquifer.

4. CHARACTERIZATION OF COASTAL AQUIFER DYNAMICS: TEMPORAL VARIABILITY

To better understand the dynamic processes controlling flow and transport at the coastal groundwater boundary over various time scales, we conducted repeated geophysical surveys coupled with coincident groundwater sampling. Here we present two examples of this relatively new application of geophysical surveying in the coastal zone. In the first study, EM31 surveys are coupled with water levels and precipitation data to constrain the variability of saline intrusion in response to seasonal and interannual drought conditions. In the second study, we examine the effects of tidal forcing on salt transport over a part of a semidiurnal tidal cycle. In both cases our interpretations of dynamic changes in the groundwater salinity regime assume that the geoelectric response of the system to natural forcings can be directly interpreted in terms of hydrogeology.

4.1. Seasonal Cycle

Seasonal variations in the degree of saline water intrusion were constrained from repeated EM31 surveys collected along the main well transect at MH between September 1998 and April 2000. The self-contained nature of the freshwater lens beneath the hammock makes this an ideal site for focused studies that constrain the impact of variations in the hydrologic cycle on the surficial aquifer. The lens at MH reacts much more quickly to natural or anthropogenic forcing than does the freshwater lens beneath Sapelo Island proper. For scientific studies of only a few years' duration, such smaller lenses serve as convenient, scaled-down microcosm for the lenses beneath larger barrier systems.

Variations in EM31 measurements (Figure 5) constrain the response of the surficial aquifer to seasonally elevated salinities in the Duplin River and decreased recharge due to

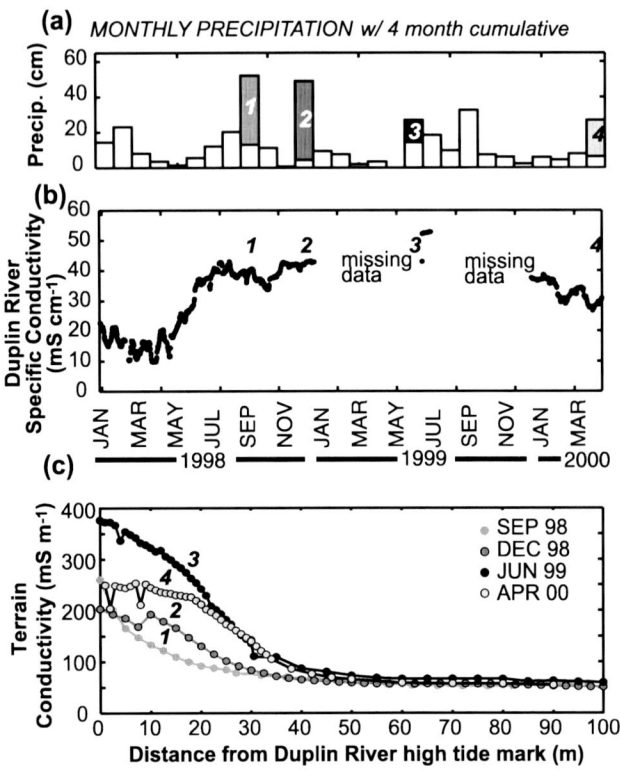

Figure 5. Seasonal variations in terrain conductivity linked to long-term precipitation and salinity data at the island-estuary margin (Moses Hammock site). (a) Cumulative monthly precipitation data between September 1998 and April 2000 highlight the drought that commenced in spring of 1998 following a period of unusually heavy precipitation during the 1997–1998 ENSO event. (b) Specific conductivity observations in the Duplin River correlate with the precipitation record but are also influenced by the long-term variability in salinity of the estuary. (c) EM31 apparent conductivity data acquired at discrete times over the 19-month-long period show a strong correlation with both accumulated precipitation and specific conductivity in the Duplin River, demonstrating the relatively rapid response of this self-contained aquifer to climate and oceanographic variability.

the drought that commenced in mid-1998 following a period of significantly above average rainfall and streamflow in Georgia during the 1997–1998 El Nino event. On average, the lowest raw apparent conductivity values, consistent with the largest freshwater lens and/or the lowest salinity groundwaters making up the saline intrusion, occurred in September 1998, shortly after the start of the drought. The maximum terrain conductivities (>300 mS m^{-1}) occurred in June 1999. In light of the small increase in surface water salinities (~10 mS cm^{-1}) during the December 1998 to June 1999 period, the June 1999 survey almost certainly documents shrinkage of the freshwater lens due to lack of precipitative recharge. Between June 1999 and April 2000, a continued dry period, the EM data reveal little change in the lens at distances greater than 25 m from the edge of the upland. Yet the conductivity measurements decrease markedly at distances less than 25 m, during the same period that the salinity of the adjacent tidal creek also drops sharply (by nearly 25 mS cm^{-1}). This observation implies that the saline intrusion in the 25 m closest to the tidal creek had been replaced by lower salinity waters from the adjacent tidal creek during the winter months of late 1999 and early 2000.

Throughout these surveys, we noted that the point at which the EM measurements returned to background values remained consistent at ~50 m from the tidal creek, regardless of the terrain conductivity changes recorded closer to the creek at this site. *Snyder et al.* [2004] describe redox and other hydrogeochemical data collected in the MH monitoring well and multilevel sampler network and also document no appreciable or systematic change in the hydrogeochemistry at the MH site over the course of 11 months. This observation is consistent with an approximate steady-state for the balance of fresh and saline groundwater. Such hydrogeochemical data provide more detailed information than geophysical measurements, both in terms of the vertical spatial scale over which groundwater samples average and the capacity of the laboratory analyses to yield high resolution estimates of chemical species concentration. On the other hand, the resolution of the geochemical data along the well network is only as fine as the spacing of the wells, meaning that geophysical data are better able to constrain the lateral extent of saltwater intrusion.

4.2. Semi-Diurnal Tidal Cycle

To constrain the dynamics of saline water intrusion at the edge of the upland aquifer over a semi-diurnal tidal cycle, we conducted horizontal dipole EM31 surveys at 1 m spacing along a 50 m transect perpendicular to Barn Creek and coincident with the main well network at the KF site every hour from just before high tide to just after low tide in December 1998. The evolution of terrain conductivity along the transect and changes at a given point during the tidal cycle are shown in Figure 6a. To remove the impact of instrument drift and other systematic changes, each survey was adjusted so that the measurements at the 3 points most distal from the creek along the survey line did not change over the course of a tidal cycle. The data highlight significant changes in terrain conductivity over the course of the tidal cycle, changes that are a function of both water level variations, which can be up to several centimeters, and pore water salinity.

Direct measurements of water levels in the monitoring wells coincident with the EM surveys allowed us to remove the impact of tidally-induced water table fluctuations from the terrain conductivity measurements. We first calculated the terrain conductivity variation that could be attributed to changes in the thicknesses of the unsaturated and freshwater-saturated sediments at each well location during different parts of the tidal cycle. We then interpolated this signal between wells and subtracted this effect from the EM data we measured along the well transect at different times during the tidal cycle. The resulting value represents the residual EM signal attributable to changes in conductivity due to variations in pore water salinity alone. This residual EM signal was then normalized by the maximum conductivity value measured in the entire data set, yielding the results shown in Figure 6b.

The results reveal the largest residual conductivity values at high tide, intermediate values between the tidal extremes, and the lowest values during an intertidal period that followed low tide. To first order, these results are consistent with the most significant intrusion of saline creek water into the edge of the permeable upland aquifer at high tide and probably some outflow of saline groundwater from the creekbank mixing zone at low tide.

The residual conductivity salt signals were groundtruthed by simultaneous conductivity measurements in monitoring wells MW0101–MW0104. Although the porewater conductivity data imply a slightly shorter intrusion distance (~6 m) than the residual terrain conductivity data (~8 m), they follow the same general pattern. The disparity between the independent geophysical and hydrologic measurements is not surprising, given that groundwater analyses average conductivity over the length of the vertical well screen (0.75–1.5 m) while the residual terrain conductivity values represent the integrated conductivity over nominal depths of ~6 m. Furthermore, the lateral spacing of the monitoring wells in which the porewater conductance data were acquired is much greater than the spacing of the EM measurements.

The residual conductivity results also provide insight about the nature of surface water-groundwater exchange at the aquifer boundary. As in the MH dataset, we note the pres-

178 INTEGRATING HYDROLOGIC AND GEOPHYSCAL DATA

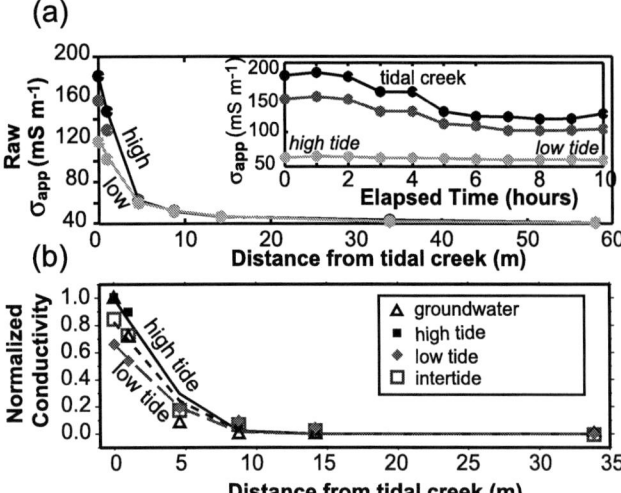

Figure 6. (a) Raw EM31 terrain conductivity measured at high tide (black), intertide (dark gray), and low tide (light gray) as a function of distance along the well network at the KF site. The inset shows EM31 measurements at the edge of the upland (black) and the positions of MW0101 (3.5 m inland; dark gray) and MW0103 (8 m inland; light gray) as a function of time within the tidal cycle. (b) Integrated analysis of terrain conductivity data and water levels measured in coincident monitoring wells over a semi-diurnal tide cycle yield the normalized surface conductivity signal attributed to changes in pore water salinity. The residual conductivity data, groundtruthed by downhole specific conductivity measurements (triangles), reveal a relatively static freshwater-saltwater interface (>8 m). The overall salt signal is adequately modeled by diffusive-dispersive transport ($D=1\times10^{-8}$ m^2 s^{-1}; solid and dashed curves) absent of advective transport.

ence of a so-called static point in this system, here located at ~8 m from the tidal creek. The morphology of the conductivity data curves shown in Figure 6b is consistent with diffusion-dominated transport at the boundary, and we used an analytical solution to the transient diffusive transport equation for constant boundary concentration C_0 of salt to fit the observations and extract an estimated diffusion-dispersion coefficient D:

$$C(x) = C_0 \left[erfc\left(\frac{x}{2\sqrt{Dt}}\right)\right], \quad (2)$$

where x and t denote distance and time, respectively. This analysis assumes that pore water conductivity changes are simply related to the concentration of a species such as chloride.

Setting the concentration C_0 at the boundary (located at MW0101) to the residual conductivity at low, intermediate, and high tides normalized by the high tide value (119 mS m^{-1}), we vary D seeking the best-fit to the residual conductivity structure. Best-fit models shown in Figure 6b are obtained with D of 1.0×10^{-8} m^2 s^{-1} and characteristic time $t=1$ year. We underscore that these results do not indicate the absence of advective processes at the tidal creek boundary, only that, on the time scale and at the resolution of our observations, the system is approximately diffusive. *Snyder* [2002] postulates that the apparently diffusive nature of salt transport at the edge of the MH aquifer, where he completed exhaustive groundwater geochemical studies, may represent the relatively steady-state balance between advective flux of freshwater toward the tidal creek and the net diffusion of chloride into the upland aquifer.

5. DISCUSSION

Spatial heterogeneities appear to exist as a hierarchical system (e.g., *Wheatcraft and Tyler*, 1988; *van de Graaff and Ealey*, 1989; *McLaughlin and Ruan*, 2001). *Schultz* [2002] discuss this point with respect to extensive GPR data collected at various locations on Sapelo Island. The North Sapelo (GCE3) study is an example that demonstrates some of these hierarchies of spatial scales within a single site. At the megascale (up to ~500 m), the results reveal submarsh flow as a possible mechanism for groundwater discharge into the marsh and ultimately the tidal creek and coastal zone. At the macroscale (~5 to 25 m), we infer the impact of lithologic layering on controlling vertical infiltration of tidal waters into the marsh and the interaction of shallow and deeper groundwaters. Mesoscale features (0.1 m to greater than 2 m) include the fine, vertical fingering features that may indicate convection in the marsh and variations in plant heights and distributions that might in part reflect shallow physical or chemical hydrologic conditions.

For temporal data, this study, like those of *Slater and Sandberg* [2000] and *Sandberg and Slater* [2001], demonstrates that even groundwater processes operating under natural, not forced, gradients at time scales as short as a few hours can sometimes be detected by geophysical monitoring. Overall, we found that changes in the distribution of fresh and saline groundwater at the margins of the lens aquifer were most affected by long-term variability such as seasonal changes in tidal creek salinity and recharge. We found that the lens system was generally in equilibrium with conditions in the adjacent tidal river and monthly average rainfall records. Therefore, we inferred that the redistribution of fresh and saline groundwater associated with elevated salinities in the tidal watershed and reduced hydraulic gradients in the aquifer during drought periods occurred relatively rapidly and that the saltwater interface is in a dynamic equilibrium with oceanographic and atmospheric forcing. The response of the aquifer to semi-diurnal tidal forcing and associated transport were constrained by repeated surveys at the KF site.

An important implication of our geophysical results is that the maximum inland extent of the intrusive saline wedge is relatively constant over periods ranging from a tidal cycle (KF) to several weeks or longer (MH). Lens aquifer morphology represents the interplay of many conditions, but particularly local hydraulic parameters and the amount of recharge. The recognition of static points for lateral saline intrusion may point to the dominance of static local hydraulic parameters, such as hydraulic conductivity and time-averaged head differences between the surficial aquifer and adjacent, tidally-influenced creeks, over the more volatile parameter of recharge as the key factor affecting lateral saline intrusion.

The largely static nature of the landward extent of the saline groundwater wedge and the agreement of groundwater cation concentrations and geophysical data with a diffusive model imply that advective exchange is probably not the key mechanism for mixing across the island-estuary interface at our sites. During tidal inundation, the hydraulic head gradient between saline surface water and the water table favors saline water intrusion into the creekbank or beach sediments. Thus a cyclical tidal flushing pattern could theoretically be established, with groundwater discharging during ebb tide and saltwater infiltrating during flood tide [*Hemond et al.*, 1984]. However, it is important to distinguish between localized tidal flushing in the vicinity of the creek bank, beach face, or marsh surface and larger scale diffusive processes that define the extent of saline intrusion on a larger scale. Because seepage velocities are generally less than 1 m d^{-1}, the zone of advective saltwater flushing in nearshore aquifers forced by semi-diurnal tides is less than 0.5 m [*Harvey and Nuttle*, 1995].

Not surprisingly, sustained forcing over periods of months or years has a much greater influence on flow and transport conditions in the surficial aquifer system than semidiurnal tidal forcing from adjacent saline water bodies. The longest time scale resolved in this study was ~18 months, but events at time scales longer (e.g., decadal-scale sea level rise) and shorter (e.g., storm surges) can effect dramatic geologic and hydrologic change in the aquifer system.

6. CONCLUSIONS

Our case studies highlight the potential of combining hydrologic and geophysical data to better understand shallow aquifer processes. Although the integration of these methods has gained wider use throughout the hydrogeophysics community for structural hydrogeologic properties characterization (e.g., *Rubin and Hubbard*, 2005), few process-oriented studies have adopted hydrogeophysical methods from the outset. A major challenge in integration of disparate hydrologic and hydrogeophysical data is transforming geophysical results into meaningful constraints on hydraulic parameters. The development of better joint inversion algorithms should make it possible to refine models using combinations of disparate geophysical and hydrologic data.

Practical considerations for coastal aquifer characterization include the resolution and efficiency of downhole and surface geophysical instrumentation. On one hand, among natural environments coastal aquifers are perhaps uniquely suited to the application of particularly electrical and inductive EM techniques owing to the strong contrast in near-surface conductivity properties and the disparate time scales over which dynamic processes can be detected. On the other hand, these same conditions, coupled with the juxtaposition of high and low saturation conditions (e.g., arid uplands and flooded salt marshes) and high and low gradient processes (e.g., turbulent tidal flow and laminar groundwater flow), can pose difficulties for the use of standard hydrogeophysical techniques and approaches. Geophysical techniques have often evolved independently in terrestrial and marine settings, but coastal zone studies could benefit from the adaptation and application of both classes of methods. At the same time, technical advances in high resolution geophysical methods (e.g., shear wave or radar wave velocity studies) could enhance the detectability of saturation variations, the delineation of subtle changes in lithologic structure, or the impact of transient processes on coastal aquifers.

Although electrical and electromagnetic methods are particularly useful in the coastal zone because contrasts in electrical properties serve as a proxy for changes in salinity, such approaches have a serious limitation in understanding short-term flow patterns. For this application, time-domain methods such as induced polarization (IP) might be more widely applied to constrain the direction of ion mobility in the presence of hydraulic and electrical gradients.

Geophysical data have traditionally been obtained by profiling or sounding techniques, which to first order separate lateral changes from vertical changes in physical properties. However, 3D characterization of heterogeneities is needed to identify the key controls on groundwater flow and transport. Geophysical instruments that automatically acquire data in three orientations (e.g., triaxial EM instrument or 3D multiplexed resistivity arrays, with one set of electrodes in a borehole) could enhance constraints on 3D subsurface heterogeneities. Moreover, just as four-dimensional surveys (including time as a dimension) have revolutionized oil exploration, the addition of a fourth dimension to 3D geophysical surveying, perhaps through the establishment of automated geophysical monitoring networks, could significantly advance the understanding of both the static and dynamic state of shallow aquifers.

The research presented here is relatively unique in providing a time series of particular geophysical data to constrain changes in a coastal aquifer. However, the duration of the study and the nature of the measurements mean that we cannot yet fully constrain aquifer response to climatic, oceanographic, ecological, and anthropogenic processes occurring over time scales of seasons to decades. Ongoing efforts such as those championed by the LTER program, the Consortium of Universities for the Advancement of Hydrologic Science (CUAHSI), and international organizations such as the Global Change Research Program should contribute to the establishment of longer baseline time series. We underscore the importance of long-term and sometimes high-resolution monitoring and analysis, particularly to constrain changes in response to catastrophic events (e.g., storm surges or hurricanes), short-term climate trends (e.g., drought), and decadal events (e.g., sea level rise and urbanization).

Acknowledgments. This research was supported by a subcontract to C.R. from the University of Georgia (UGA) under the auspices of NSF grant OCE9982133 for the Georgia Coastal Ecosystems Long Term Ecological Research (LTER) program from 2000 to 2006. From 1998 to 2000, additional support was provided by a U.S. Geological Survey Georgia Water Resources Institute grant and a Georgia Tech College of Sciences Faculty Development Grant to C.R. and NOAA NERR graduate fellowship NA87OR028 to G.S. We are grateful to teacher participants in the LTER Schoolyard program for assistance with monitoring well installation, to J. Garbisch and the UGA Marine Institute for field logistical support, and to the scores of Georgia Tech students who contributed to the acquisition of geophysical, geochemical, and hydrologic data under adverse conditions between 1997 and 2004. C.R. thanks N. Toksöz for arranging logistical support at MIT during completion of this paper. Comments by L. Slater, F. Day-Lewis, and an anonymous reviewer greatly improved the manuscript.

REFERENCES

Anthony, S. S. (1992), Electromagnetic methods for mapping freshwater lenses on Micronesian atoll islands. *J. Hydrol.*, 137, 99–111.

Anthony, S. S., Peterson, F. L., Mackenzie, F. T. and S. N. Hamlin (1989), Geohydrology of the Laura freshwater lens, Majuro atoll: A hydrogeochemical approach. *Geol. Soc. Am. Bull.*, 101, 1066–1075.

Ayers, J. F. and H. L. Vacher (1986), Hydrogeology of an atoll island: a conceptual model from detailed study of a Micronesian example. *Ground Water*, 24, 185–198.

Bear, J., Cheng A. H.-D., Sorek, S., Ouazar, D. and I. Herrera (1999), *Seawater Intrusion in Coastal Aquifers: Concepts, Methods and Practices*, Kluwer Academic Publishers, Dordrecht, 625 pp.

Bugg S. F. and J. W. Lloyd (1976), A study of fresh water lens configuration in the Cayman Islands using resistivity methods. *Quat. J. Eng. Geol.*, 9, 291–302.

Chalmers, A. G. (1982), Soil dynamics and the productivity of *Spartina alterniflora*. In: *Estuarine Comparisons,* Kennedy, V. S. (Ed.), Academic Press, New York, pp. 231–242.

Chalmers, A. G. (1997), The ecology of Sapelo Island National Estuarine Research Reserve. NOAA Report NA470R0414, 138 pp.

Chen, J., Hubbard, S., Rubin, Y., Murray, C., Roden, E., and E. Majer (2004), Geochemical characterization using geophysical data and Markov Chain Monte Carlo methods: A case study at South Oyster bacterial transport site in Virginia. *Water Resour. Res.*, 40, W12412, doi:1029/2003WR002993.

Collins, W. H. III and D. H. Easley (1999), Fresh-water lens formation in an unconfined barrier-island aquifer. *J. Amer. Water Res. Assoc.*, 35, 1–21.

Cooper, C. A., Glass, R. J., ad S.W. Tyler (1997), Experimental investigation of the stability boundary for double-diffusive finger convection in a Hele-Shaw cell, *Water Resour. Res.*, 33, 517–526.

Dagan, G. (1986), Statistical theory of groundwater flow and transport: Pore to laboratory, laboratory to formation, and formation to regional scale. *Water Resour. Res.*, 22, 120S–134S.

Ebraheem, A. M., Sensosy, M. M., and K. A. Dahab (1997), Geo-electrical and hydrogeochemical studies for delineating groundwater contamination due to salt-water intrusion in the northern part of the Nile delta, Egypt. *Ground Water*, 35, 216–222.

Fisher, A. T. (2005) Marine hydrogeology: recent advances and future accomplishments, *Hydrogeology J.*, 13, 69–97.

Frohlich, R. K., and D. Urish (2002), The use of geoelectrics and test wells for the assessment of groundwater quality of a coastal industrial site. *J. Applied Geophys.*, 50, 261–278.

Ginsberg, A. and A. Levanon, (1976), Determination of saltwater interface by electrical resistivity sounding. *Hydrological Science Bulletin,* 21, 561–568.

Goldman, M., Gilad, D., Ronen, A., and A. Melloul (1991), Mapping of seawater intrusion into the coastal aquifer of Israel by the time domain electromagnetic method. *Geoexploration*, 28, 152–174.

Greenwood, W. J., Kruse, S., and P. Swarzneski (2006), Extending electromagnetic methods to map coastal pore water salinities. *Ground Water*, 44, 292–299.

Guptasarma, D. (1982), Optimization of short digital linear filters for increased accuracy. *Geophys. Prosp.*, 30, 50-1–514.

Harvey, J. W., Chambers R. M., and J. R. Hoelscher (1995), Preferential flow and segregation of porewater solutes in wetland sediment. *Estuaries*, 18, 568–578.

Harvey, J. W., and W. K. Nuttle (1995), Fluxes of water and solute in a coastal wetland sediment. 2. Effect of macropores on solute exchange with surface water. *J. Hydrol.*, 164, 109–125.

Hemond, H. F., Fifield, J. L. (1982). Subsurface flow in salt marsh peat: A model and field study. *Limnology and Oceanography*, 27, 126–136.

Hemond, H. F., Nuttle, W. K., Burke, R. W. and K. D. Stolzenbach (1984), Surface infiltration in salt marshes: theory, measurement, and biogeochemical implications. *Water Resources Research*, 20, 591–600.

Herzberg, B. (1901), Die Wasserversorgung einiger Nordseebader. *J. Gasbeleucht. Wasserversorg*, 44, 815–819.

Howes, B. L., Weiskel, P. K., Goehringer, D. D., and J. M. Teal (1986), Interception of freshwater and nitrogen transport from uplands to coastal waters: the role of salt marshes. In: K. F. Nordstrom and C. T. Roman (Eds.), *Estuarine shores: evolution, environments and human alterations*, Wiley, New York, pp. 289–310.

Hubbard, S., Rubin, Y., and E. L. Majer (1999), Spatial correlation structure estimation using geophysical data, *Water Resour. Res.*, 35, 1809–1825.

Hunter, K. S., Lee, R. Y., Boyd, B., Schultz, G., Joye, S. B., Ruppel, C. (2000), Groundwater geochemistry at the island-estuary interface at perturbed and pristine sites on a Georgia Bight barrier island. *Geol. Soc. Amer. Southeastern Section Annual Meeting*.

Hyndman, D. W. and S. M. Gorelick (1996), Estimating lithologic and transport properties in three dimensions using seismic and tracer data, the Kesterson aquifer. *Water Resour. Res.*, 32, 2659–2670.

Inman, J. R., (1975), Resistivity inversion with ridge regression. *Geophysics*, 40, 798–817.

Knight, R., Tercier, P., and H. Jol (1997), The role of ground penetrating radar and geostatistics in reservoir description. *Leading Edge*, 16, 1576–1583.

Kuras, O. (2002), The capacitative resistivity technique for electrical imaging of the shallow subsurface. Ph.D. Thesis, University of Nottingham, 286 pp.

Lens, L. F. (1981), Quarternary stratigraphy and sedimentary framework of the Sapelo Island area, Coastal Georgia. M.S. Thesis, University of Georgia, 138 pp.

McKenna, S. A., and E. P. Poeter (1995), Field example of data fusion in site characterization. *Water Resour. Res.*, 31, 3229–3240.

McLaughlin, D., and F. Ruan (2001), Macrodispersivity and large-scale variability. *Transport in Porous Media*, 42, 133–154.

McNeill, J. D. (1980), Electromagnetic terrain conductivity measurement at low induction numbers. *Geonics Ltd., Technical Note, TN-6*.

Millham, N. P., and B. L. Howes (1995), A comparison of methods to determine K in a shallow coastal aquifer. *Ground Water*, 33, 49–57.

Nuttle, W. K., and J. W. Harvey (1988), Geomorphological controls on subsurface transport in two salt marshes. In: Kustler, J. A., Brooks, G. (Eds.) *Wetland Hydrology*. Association of State Wetland Managers, New York, 253–261 pp.

Osgood, D. T., and J. C. Zieman (1998). The influence of subsurface hydrology on nutrient supply and smooth cordgrass (*Spartina alterniflora*) production in a developing barrier island marsh. *Estuaries*, 21, 767–783.

Philip, J. R. (1973), Periodic nonlinear diffusion: an integral relation and its physical consequences, *Austral. J. Phys.*, 26, 513–519.

Phillips, F. M., and J. L. Wilson (1989), An approach to estimating hydraulic conductivity spatial correlation scales using geologic characteristics. *Water Res. Res.*, 28, 141–143.

Roy, K. K., and H. M. Elliot (1980), Resistivity and IP survey for delineating saline water and fresh water zones, *Geoexploration*, 18, 145–162.

Rubin, Y., and S. Hubbard (Eds.) (2005), *Hydrogeophysics*, 523 pp., Springer, Water Science and Technology Library, The Netherlands.

Ruppel, C., and G. Schultz (2003), Hydrogeophysics in the coastal zone: spatial and temporal constraints on groundwater processes from coincident wells and geophysical surveys, *EOS, Trans. AGU, 84(46)*, Fall Mtg. Suppl., Abstract H21A03.

Ruppel, C., Schultz, G., and S. Kruse (2000), Anomalous freshwater lens morphology on a strip barrier island. *Ground Water*, 38, 872–881.

Sandberg, S. and L. Slater (2001), Geophysical monitoring through a tidal cycle at Crescent Beach State Park, Maine, *Journal of Environmental & Engineering Geophysics*, 6, 165–174.

Schultz, G. (2002), Hydraulic and geophysical characterization of spatio-temporal variations in coastal aquifer systems, Ph.D. Thesis, Georgia Institute of Technology, 355 pp.

Schultz, G., and C. Ruppel (2000), Geophysical and hydrologic characterization of spatial and temporal variations in the distribution of freshwater and saltwater at the island-estuary interface, Sapelo Island National Estuarine Research Reserve. *NOAA-NERRS Final Report*, 113 pp.

Schultz G., and C. Ruppel (2002), Constraints on hydraulic parameters and implications for groundwater flux across the upland-estuary interface. *J. Hydrol.*, 260, 255–269.

Schultz, G. and C. Ruppel (2005), Inversion of inductive electromagnetic data in high induction number terrains, *Geophysics*, 70, G16–G28.

Sibley, S. (2004), The impact of salt marsh hydrogeology on dissolved uranium, M.S. Thesis, Georgia Institute of Technology, 120 pp.

Slater, L., and S. K. Sandberg (2000), Resistivity and induced polarization monitoring of salt transport under natural hydraulic gradients, *Geophysics*, 65, 408–420.

Snyder, M. (2002). Geochemical trends associated with the seawater-freshwater mixing zone in a surficial coastal aquifer. Sapelo Island, Georgia, M.S. Thesis, Georgia Institute of Technology, Atlanta, 149 pp.

Snyder, M., Taillefert, M., and C. Ruppel (2004). Redox zonation at the saline-influenced boundaries of a permeable surficial aquifer: effects of physical forcing on the biogeochemical cycling of iron and manganese. *J. Hydrol.*, 296, 164–178.

Stewart, M. T. (1988), Electromagnetic mapping of freshwater lenses on small oceanic islands. *Ground Water*, 26, 187–191.

Stoyer, C, and Butler, M., 1994, EMIX34 plus users manual: Interpex Ltd.

Sudicky, E. A. (1986), A natural gradient experiment on solute transport in a sand aquifer: Spatial variability of hydraulic conductivity and its role in the dispersion process, *Water Resour. Res.* 22, 2069–282.

Thibodeau, P. M., Gerdner, L. R., and H. W. Reeves (1998), The role of groundwater flow in controlling the spatial distribution of soil salinity and rooted macrophytes in a southeastern salt marsh, USA. *Mangroves and Salt Marshes, 2*, 1–13.

Urish, D. W. (1977), The fresh water lens in a barrier beach. *Hydraulics in the Coastal Zone, Proc. 25th Annual Hydraulic Division*. Amer. Soc. Civil Eng., Seattle, pp. 161–167.

Urish, D. W., and R. K. Frohlich (1990), Surface electrical resistivity in coastal groundwater exploration. *J. Appl. Geophys., 26*, 267–289.

Urish, D. W, and M. M. Ozbilgin (1989), The coastal ground-water boundary. *Groundwater, 27*, 310–315.

U.S. Department of Agriculture (1959), *Soil Survey of McIntosh County, Georgia.* 33 pp.

Van Dam, J. C., and J. J. Maeulankamp (1967), Some results of the geoelectrical resistivity method in groundwater investigations in the Netherlands. *Geophysical Prospecting*, 92–115.

Van de Graaff, W. J. E. and P. J. Ealey (1989). Geological modeling for simulation studies. *Amer. Assoc. Petr. Geol. Bulletin, 73*, 1436–1444.

Van Nostrand, R., and K. Cook (1966), Interpretation of resistivity data. *U.S. Geol. Survey Prof. Paper 499*, 210 pp.

Wheatcraft, S. W. and S. W. Tyler (1988), An explanation of scale-dependent dispersivity in heterogeneous aquifers using concepts of fractal geometry. *Water Resour. Res., 24*, 566–578.

Yeh, T.-C.J., Liu, S., Glass, R. J., Baker, K., Brainard, J. R., Alumbaugh, D., and D. LaBrecque (2002), A geostatistically based inverse model for electrical resistivity surveys and its applications to vadose zone hydrology. *Water Resour. Res., 38*, 1–13.

Examining Watershed Processes Using Spectral Analysis Methods Including the Scaled-Windowed Fourier Transform

Anthony D. Kendall and David W. Hyndman

Department of Geological Sciences, Michigan State University, East Lansing, Michigan, USA

Important characteristics of watershed processes can be extracted from hydrologic data using spectral methods. We extract quantitative information from precipitation, stream discharge, and groundwater head data from watersheds in northern-lower Michigan using Fourier Transform (FT) methods. By comparing the spectra of these data using similar units, we graphically illustrate the hydrologic processes that link precipitation to stream discharge and groundwater levels including evapotranspiration. We also demonstrate how unit hydrographs can be efficiently and non-parametrically derived using the FT in a manner that allows for a quantitative seasonal comparison of precipitation and the resulting stream discharge response. This analysis clearly illustrates the reduction in summer discharge levels due to canopy interception and evapotranspiration. We also develop a systematic application of the FT we call the Scaled-Windowed Fourier Transform (SWFT), which extracts time-varying spectral content using a similar approach to the wavelet transform. While computationally less efficient than the wavelet transform, the SWFT allows for embedded detrending and tapering. Application of this method clearly illustrates the non-stationarities of spectral content within the three chosen data types, leading to a greater understanding of discharge-generating processes.

INTRODUCTION

Spectral analysis (SA) provides a powerful means of extracting information from hydrologic data. This type of analysis can reveal processes that may be obscured in direct time-series analysis by providing data not just on temporal fluctuations, but also on the spectrum of frequencies within those fluctuations. Furthermore, integrating multiple data types in a quantitative and meaningful way is relatively simple using spectral analysis. Here we apply both existing and novel SA techniques to hydrologic data from two watersheds in northern lower-Michigan. We use these methods to illuminate the processes that link precipitation to stream discharge and groundwater levels.

Spectral analysis has been applied within the hydrologic sciences for decades [e.g., *Bras and Rodriguez-Iturbe*, 1985; *Hameed*, 1984], most commonly as a means of estimating coefficients for linear autoregressive or stochastic models [e.g., *Naff and Gutjahr*, 1983; *Jukic and Denic-Jukic*, 2004; *Zhang and Schilling*, 2004]. During the last decade, SA has been applied to examine fractal and multi-fractal behavior of systems [e.g., *Tessier et al.*, 1996; *Pelletier and Turcotte*, 1997; *Kirchner et al.*, 2000], and to examine linkages between hydrologic and climatic processes [e.g., *Tessier et al.*, 1996; *Pelletier and Turcotte*, 1997; *Coulibaly and Burn*, 2004].

Here we apply SA techniques to explore linkages between hydrologic processes and to provide a deeper understanding of those processes. Previous SA process comparison studies have generally not assured the similarity of measurement units, nor have they (with the exception of more recent wavelet studies including *Gaucherel* [2002] and *Coulibaly and*

Burn [2004]) considered the time-variant nature of process spectra. We demonstrate the quantitative utility of comparing data and spectra with similar physical units for inference of process relationships. Also, using an application of the common Fourier Transform we call the Scaled-Windowed Fourier Transform (SWFT), we illustrate the non-stationary behavior of precipitation, stream discharge, and groundwater head spectra. Comparing Fourier spectra to the SWFT output, we illustrate how ignoring such non-stationarity can result in misinterpretations of spectra and thus of inferred process details. Finally, using a spectral derivation of seasonal unit hydrographs, we demonstrate how simple stream discharge models can be improved by considering the seasonality of watershed processes.

Recognizing that spectral analysis is not a standard tool in the hydrologic science toolbox, *Fleming et al.* [2002] published a practical introduction to SA that focused on applications of the Fourier Transform (FT) including direct frequency domain investigation, spectral filtering, and spectral simulation model validation. Here we apply SA in a similar manner, explaining our assumptions and the common complications, and demonstrating how these methods can be used across the hydrological sciences. Additionally, parts of the methods section are intended to contribute to a set of "best practices" of SA in the hydrologic sciences.

DATASETS AND STUDY AREA

Six data types from two northern lower-Michigan watersheds were used in this study: daily temperature, precipitation, and snowfall, both 15-minute and hourly stream discharge, and bi-hourly groundwater heads. Figure 1 shows the locations of both Michigan watersheds, as well as the locations of our groundwater transducers. The 3711 km² Evart sub-basin of the Muskegon River Watershed (MRW) was selected for this study because there are no actively controlled flow structures on the Muskegon River or its tributaries upstream of this station.

Because the Evart sub-basin lacks a network of monitored groundwater wells, we used water level data from our pressure transducers in the nearly adjacent Grand Traverse Bay Watershed (GTBW), which has a similar hydrogeological setting. Unconsolidated sediments within the Evart sub-basin and the GTBW were deposited by the same set of glacial episodes. These sediments are characterized primarily by coarse to fine sands and gravels with a small percentage of fine-grained material [*Farrand and Bell*, 1982]. The similarity in depositional history and topographic variation between these two basins suggests that GTBW groundwater head fluctuations can be used as a proxy for the nearby upper MRW.

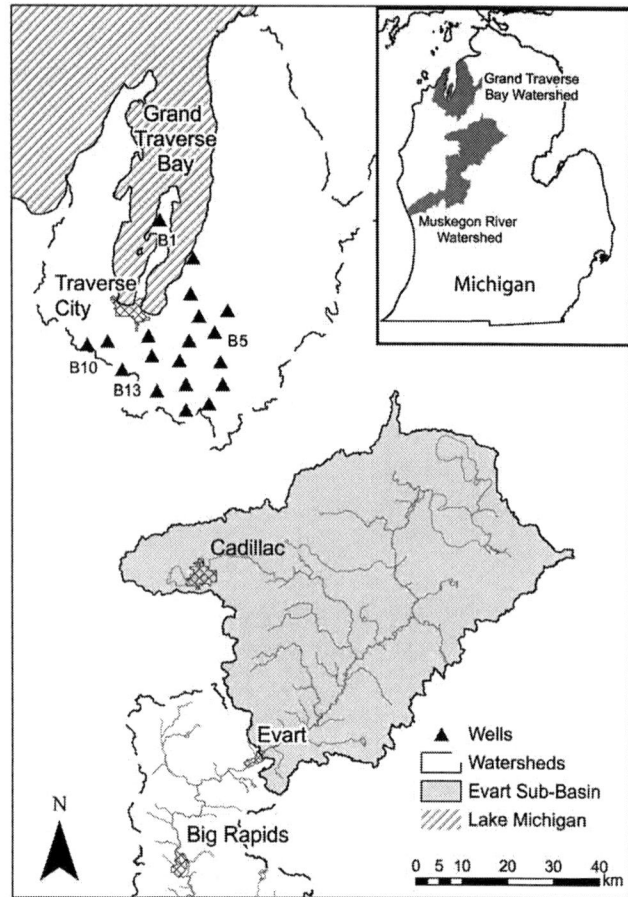

Figure 1. Map of the Muskegon River and Grand Traverse Bay watersheds, with an inset map of Michigan showing their locations. Groundwater wells with transducers are marked with triangles, and the Evart gauge sub-basin is shaded in grey.

Preliminary 15- and 60-minute discharge data from the United States Geological Survey (USGS) stream gauge on the Muskegon River at Evart (Station ID: 4121500) were used in this study [*USGS*, 2005]. Hourly data for this gauge were used from October 1989 until 15-minute discharge data became available in 1997. Combining the hourly data with resampled 15-minute data, the total data record extends from 10/1/1989 through 9/30/2004.

Daily temperature, precipitation, and snowfall records from a US Cooperative Observing Network station in Big Rapids, MI just downstream of Evart on the Muskegon River were available from 1896 to 2001 [*Karl et al.*, 1990]. These data are distributed by the National Climatic Data Center (NCDC) as part of the US Historical Climatology Network dataset [*NCDC*, 2001].

For finer temporal-scale resolution, we processed hourly 4-km NEXRAD data [*Andresen*, 2004] for 2004 and extracted the mean rainfall over the Evart sub-basin at each hourly interval. NEXRAD data show a high degree of correlation to the corresponding gauge values in this region [*Jayawickreme and Hyndman*, 2007]. These radar-based precipitation data are only available for approximately 9 months out of the year in this region (from 4/1/2004 to 11/30/2004) due to errors in snowfall estimates, so they are only suitable for evaluating shorter-period system behavior.

We installed a network of 17 pressure transducers in USGS groundwater wells across the GTBW. Water table elevations were recorded every two hours beginning either in June 2003 (9 transducers) or June 2004 (8 transducers). Data through 7/1/2005 were used in this analysis.

METHODS

Fourier Transform

Fourier's Theorem states that any complex periodic function can be decomposed into a set of periodic basis functions of varying amplitude, period, and phase shift. The most common such decomposition technique is the Fourier Transform (FT), which uses sinusoidal basis functions. *Fleming et al.* [2002] present the basic theory, and a full mathematical treatment of this technique can be found in textbooks [e.g., *Bras and Rodriguez-Iturbe*, 1985; *Percival and Walden*, 1993]. A common implementation of the discrete FT is the discrete Fast Fourier Transform (FFT) popularized by *Cooley and Tukey* [1965]. In this study, we use the FFTW libraries that are integrated into the MATLAB computing environment [*Frigo and Johnson*, 1998]. This particular set of general-radix algorithms does not constrain the user to the requirement in *Cooley and Tukey* [1965] that the number of samples (N) be a power of 2.

The output of the FFT algorithm is a complex array representing the magnitude and phase shifts of the Fourier coefficients. For instance, the FFT of a sinusoid of unit period and amplitude is an array with a single non-zero value corresponding to a period of 1. The FFT of summation of sinusoids would produce non-zero peaks in the spectrum. If instead the time-series input to the FFT were a broad, single-peaked curve such as a gaussian, exponential, or gamma function, spectral amplitudes would be non-zero across a broad range of periods.

Fourier spectra are generally plotted as power vs. frequency, however we find that plots of amplitude vs. period provide a more natural basis for viewing spectra of interrelated physical phenomena. The spectral amplitude, the square-root of spectral power, of a time-domain input corresponds directly to hydrologic flux quantities such as precipitation and stream discharge. Note that the log-log slope of the amplitude spectrum is ½ β, where β is the slope of spectral power in log-log coordinates. A system is typically considered fractal (or multi-fractal) if its power spectrum roughly follows $1/f^\beta$ behavior [*Avnir et al.*, 1998]. This behavior is typical of a wide variety of geophysical systems, and can provide insight into the processes that govern those systems.

If one assures the similarity of the units of each time-series dataset, the amplitudes of multiple spectra can be compared in a physically meaningful manner. To accomplish this, a suitable unit for comparison must first be chosen. For this study, we chose length/time (L/T) units because precipitation is the most common hydrologic forcing mechanism. In order to change the volumetric discharge units (L^3/T) of stream discharge to L/T units, the discharge was divided by the drainage area of the watershed upstream of the gauge. Groundwater head data, measured in L units can be differentiated to yield L/T units. Using the differentiated data, an approximation of the rate of head relaxation at the well location can be obtained by selecting only the negative values. Note, this technique assumes constant lateral inflow. This was then multiplied by an estimate of drainable porosity for the sediments (0.2), to correct for water level differences in porous media vs. open water.

ASSURING PERIODICITY

The FFT algorithm assumes that the data are periodic, namely the starting and ending points of the dataset are identical. Violating this condition results in spurious features in the output spectrum [*Bach and Meigen*, 1999]. However, time-series data from environmental systems rarely satisfy this criterion. There are four primary means of assuring periodicity: taper function multiplication (tapering); trend subtraction; data subset selection (discussed in "Reducing Aliasing and Leakage" below); and filtering which is not considered here, but the interested reader is referred to *Fleming et al.* [2002] for a discussion.

Tapering can be a valid means of forcing periodicity if one considers how it affects the resultant spectra. Tapering refers to the multiplication of a mean-removed signal by a "tapering function" (sometimes referred to as a windowing function) that smoothly tapers from a peak of 1 at its center to 0 at the edges. There are a variety of pre-defined tapering functions [*Blackman and Tukey*, 1958; *Harris*, 1978], and each affects the spectra differently. Tapering has the side effect of reducing the amount of information in the signal, thus limiting its applicability for short data records [*Fleming et al.*, 2002]. The shape of the tapering function, and how gradually it tapers near the edges, controls how much information is lost in the process. There is a tradeoff since the tapering functions that reduce information loss have more severe spectral "side lobes", which distort the transformed spectra by shifting power from primary frequencies to harmonics (integer multiples or factors) of those frequencies. To recover the amplitude of an isolated peak within a spectrum (but not the entire spectrum itself), spectra from each tapering function can be corrected to account for the effect of side lobes. The multiplicative correction factor for each tapering function is given by the ratio of unaltered- to tapered-amplitude [*Harris*, 1978]. In this study, tapering is primarily applied within the SWFT method where distortion is minimal since only a single amplitude is selected from each FT.

Trend removal is often necessary for environmental data series to assure periodicity while minimizing loss or side-lobe distortion from tapering. The simplest method is to linearly remove the trend from the data, which can be effective if the non-periodicity of the data is associated with a nearly linear trend. If there is some roughly sinusoidal long period fluctuation, a more valid means of trend removal may be to subtract a half-period sinusoid from the signal. In this case, the peak and trough of the subtracted function occur at each end of the original signal. A variety of additional trend-removal techniques have also been developed [*Mann*, 2004]. We used the sinusoidal trend removal method in this study due to its simplicity and physical basis.

REDUCING ALIASING AND LEAKAGE

If the FT is blindly applied to data without thought to dominant system processes, sampling rates, or sampling interval, aliasing and leakage may occur. Both aliasing and leakage act to shift spectral power (or amplitude) from "true" frequencies to harmonics of those frequencies, although each acts differently. Aliasing results from under sampling high-frequency fluctuations [*Bras and Rodriguez-Iturbe*, 1985], while leakage or overspill is caused by both the non-periodicity of the system as well as non-periodicity of the processes within that system [*Bach and Meigen*, 1999]. It is important to note that leakage will occur if the endpoints do not match, but the inverse is not always true. Even if the time-series endpoints match, leakage may occur due to variability of the processes that contribute to the sampled time-series.

In theory, aliasing can be avoided by merely increasing the sampling rate until the time series is fully resolved. Specifically, one cannot resolve spectral peaks with frequencies greater than half the sampling rate (the Nyquist frequency) [*Bras and Rodriguez-Iturbe*, 1985]. If a time-series has significant power in frequencies above half the sampling rate, aliased power will be present in the empirical spectrum. In the case of many environmental datasets, aliased peaks may not be significant because, as is shown below, these datasets are typically strongly damped at high frequencies. Our analysis indicates that some data do require sampling frequencies on the order of once per hour, but most of those discussed here are sufficiently sampled with daily sampling rates.

Spectral leakage can be more persistent and troubling than aliasing because periodic processes within a system are rarely sampled over an integer number of cycles. Leakage is commonly reduced by applying a tapering function to the time-series, detrending the time-series, or both [*Fleming et al.*, 2002]. However, when the entire FT spectrum is of interest rather than specific spectral peaks, a more effective means of decreasing leakage in environmental datasets may be to carefully select subsets of the data that correspond to natural breaks in processes, thus assuring near-integer sampling without distortion.

Data subset selection is also important because most environmental processes are non-stationary, thus each occurrence of a process may vary in both period and amplitude. If one includes multiple cycles of a periodic process in the FT, the true peak location, shape, and amplitude can be obscured by differences in system states across cycles. Additionally, selecting a subset of data in which some system processes are inactive can also greatly simplify spectral analyses and reveal the spectra of weaker processes in portions of datasets that would otherwise be obscured by dominant, but intermittent, processes. For example, the diurnal fluctuation of stream discharge is often clearly visible during baseflow conditions but can be obscured by runoff and near-stream groundwater discharge during late spring and early summer. Alternately, if one were only interested in the spectral behavior of runoff, for instance, then selecting data surrounding an isolated moderate-precipitation event during a dry season yields a stream discharge response primarily to direct precipitation and runoff.

UNIT RESPONSE FUNCTIONS

While there are various techniques to derive unit hydrographs from discharge and precipitation data, most are either ill-posed or require assumptions about system behavior [*Yang and Han*, 2006]. But direct FT deconvolution can produce unit hydrographs quickly and deterministically. The total time-series response of a linear system to a forcing input can derived by the convolution of the system unit response and the input time-series as follows [*Smith*, 1997]:

$$q(t) = \int_0^\infty h(\tau) p(t-\tau) d\tau \quad (1)$$

where q is total time-series response (i.e. stream discharge), h is the unit response (for the case of stream discharge, this unit response has the special name of "unit hydrograph"), and p is the input precipitation time series. According to the convolution theorem,

$$Q(k) = H(k)P(k) \qquad (2)$$

where $Q(k)$, $H(k)$, and $P(k)$ are the FTs of $q(t)$, $h(t)$, and $p(t)$, respectively, where k is the spectral frequency. The unit hydrograph time-series is then

$$\mathbf{h}(t) = F^{-1}\left[\frac{Q(k)}{P(k)}\right](t) \qquad (3)$$

where $F^{-1}[..](t)$ denotes the inverse Fourier Transform. Thus the unit-response hydrograph is given by the inverse FT of the ratio of the discharge and precipitation spectra.

Though not strictly necessary, the analysis is simplified if the sampling frequency and units of the precipitation and discharge time-series are identical. The data should be resampled so that the number of samples, n, and the sampling frequency, f, are equivalent. The unit-response function, $\mathbf{h}(t)$, has length n, however, in this case the unit response function is only valid up to the point where it becomes negative, since precipitation can not directly produce a decrease in stream discharge [*Yang and Han*, 2006]. The negative response is therefore the signature of some other watershed process. If baseflow separation is used, this issue can be avoided, though some small uncertainties will remain due to the data themselves, and may result in negative calculated unit responses. Here we chose to not use baseflow separation, because this technique produces a synthetic dataset that is not directly tied to watershed processes and may introduce artifacts of the separation technique that could mask the watershed processes under investigation.

The resultant unit-response hydrograph is not an invariant property of the watershed, as it is sensitive to variations in runoff-generating processes. These processes can be studied by directly comparing different unit-response hydrographs. Differences in response curve timing, peak, and shape can all be used to infer the activity and relative influence of various watershed processes. Applying data subset selection with these process differences in mind can allow for a quantitative sensitivity analysis of system sub-processes.

In this study, we compare seasonally derived unit-response hydrographs for the Evart sub-basin averaged over 10 consecutive years. The derived unit hydrograph will be incorrect if stream discharge is still responding to precipitation inputs that occurred prior to the start of the data period [*Smith*, 1997]. However, applied over entire seasons this error, as well as any error resulting from noisy data, is greatly reduced. Nevertheless, the derived unit-response function remains highly sensitive to edge conditions of the time-series inputs, thus the nominal time period (given by Table 1) of each season was adjusted to remove precipitation events or sudden increases in discharge near the edges of the time-series. The nominal time period for each season does not correspond to starting and ending dates of each season because they were chosen to assure similarity of hydrologic response within a season based on assumptions of process activity, also listed in Table 1. Also note that the fall and winter seasons overlap because the minimum length of the time-series subset must be longer than the watershed response time, which in this case was on the order of 60 days.

To assure periodicity for the FT of each season's data, a Tukey tapering function [*Blackman and Tukey*, 1958] was applied with relatively steep taper (coefficient of 0.1). Tapering was chosen over trend-removal because the magnitude of the discharge response to precipitation was of primary importance. After making these adjustments, some seasons continued to produce non-physical results (characterized primarily by sinusoidal unit-response behavior) and were thus omitted. These omissions are justified on the grounds that the non-physical results reveal only the sensitivities and limitations of the method, and nothing about watershed process.

SCALED-WINDOWED FOURIER TRANSFORM

A key difficulty in applying the FT to environmental datasets is that non-stationarities in the data introduce artifacts in the Fourier spectrum. Here there are two types of

Table 1. List of the nominal time periods used in each seasonal analysis, and of the years omitted from the average unit responses. Data were from 1990–2001.

Season	Nominal Time Periods	Time Period Justification	Years Omitted
Winter	12/1–3/15	Mainly frozen precipitation	1991, 1992
Spring	3/15–5/15	Little evapotranspiration	1990, 1993–99[1]
Summer	6/15–10/1	Maximum evapotranspiration	1990–91, 1993, 1995–96
Fall	10/15–12/25	Mostly liquid precipitation	1991, 1994–95

[1] Only four years were included in the spring average discharge response, 1991, 1992, 2000, 2001.

non-stationarity to consider. The first is non-stationarity of process period, where subsequent cycles of a process within the system have slightly different spectra due to changes in system properties. The second is best described as intermittency, where a process may be active during only a portion of the sampled data. The first results in spreading of spectral power about a central period (if the process is sinusoidal), while the second results in spurious spectral power at harmonics of the primary period.

To avoid these effects and more clearly illuminate important changes in the spectra, we developed a method that we call the Scaled-Windowed Fourier Transform (SWFT). The SWFT is very similar to a sinusoidal wavelet transform (WT), but it differs in a number of respects. The mathematical development of the SWFT presented here is fundamentally different from that commonly presented for the WT, in a way that may increase the utility of this method for hydrologic scientists. In particular we feel that there is great value in demonstrating that the SWFT produces time- and frequency-localized Fourier coefficients, as opposed to the similarly localized WT coefficients that are wavelet-dependent. Additionally, the SWFT is capable of embedded detrending rather than relying on tapering alone to reduce leakage, potentially producing better spectral estimates. Finally as developed, the periods and times queried by the SWFT are more flexible than typical WT schemes, with the tradeoff of decreased computational efficiency.

The SWFT also differs from the traditional windowed FT that only transforms data within a specific subset of the overall time series called a data window (here the window is different from a tapering function). This data window is then slid along the time series to produce a map of spectral power varying in both period and time. Unfortunately, the windowed FT forces a choice between severe aliasing of low-frequency components of the signal or poor resolution of high-frequency non-stationary processes [*Torrence and Compo*, 1997].

By contrast to the windowed FT, the SWFT method scales the width of the window over successive passes along the time-series. At each window position, the data are detrended, multiplied by a tapering function, and Fourier transformed. A single amplitude corresponding to the Fourier coefficient of a single frequency is selected from the complete FT at each window position. The window is then slid along the data, producing a time-varying series of amplitudes for that frequency. When the end of the dataset is reached the window width is rescaled and the process is repeated for the next frequency. Note, hereafter we use the word "period" solely to indicate the spectral period, or the inverse of frequency. We feel that the period of spectral content is generally more applicable to the hydrologic sciences than the frequency.

The SWFT produces the same type of scalogram as would be generated with the WT. These scalograms are well defined mathematically and physically (if proper units are used), and can be examined using standard statistical techniques. Aside from simple examination of the scalogram, comparisons of related scalograms such as the cross-scalogram and the coherence phase map (see *Torrence and Compo* [1997]) can be calculated as well.

Conceptually, the SWFT scalogram is produced via the following:

$$F_{pq} = C \cdot F\left(\left[x(j_{start} : j_{end}) - D\right] \cdot T\right)_k \qquad (4)$$

where p and q are indices defined mathematically below, which correspond to the periods and times at which the SWFT is applied; $F(...)_k$ indicates a single value from the discrete Fourier Transform spectrum; x is the time-series dataset; j_{start} and j_{end} are the beginning and ending indices of the current data window; T is a tapering function (optional); D is a detrending function (optional), each defined over the current data window; and C is a multiplicative correction factor unique to each type of tapering function that is computed as $C = \left|F(\sin(-\infty : +\infty))\right|_k / \left|F(\sin(-\infty : +\infty) \cdot T)\right|_k$, where k is the index that corresponds to the period 2π. Note that a reasonable approximation of C can be obtained using just three or four cycles of the sine function. If a tapering function is not used, $C=T=1$. For our analysis, we chose a Tukey window with a gradual taper (Tukey window coefficient of 0.5) as a tradeoff between frequency resolution and side lobe distortion, resulting in $C \approx 1.33$. Tukey coefficients closer to 1 produce greater side-lobe distortion, while those nearer to 0 reduce side-lobe distortion but increase leakage.

The complex definition of the discrete Fourier Transform (DFT) modified from *Press et al.* [1992] is

$$F_k = \sum_{j=1}^{n} x_j e^{-\frac{2 \cdot i \cdot \pi}{n}(k-1) \cdot (j-1)} \qquad (5)$$

where x is the time-series dataset, and k and j are indices running from $k=[1,...,n]$ and $j=[1,...,n]$ and n is the total number of data points to be transformed. The DFT spectrum F has n points of which the first is the "gain" term, and the next $n/2$ points correspond to the periods $n/f \cdot [1, 1/2, 1/3, ..., 2/n]$.

To compute the SWFT, we would like to extract a single Fourier coefficient corresponding to a certain period, P_p, from the entire DFT spectrum. Also, we need the coefficient not for the entire dataset x, but for a windowed subset x_j with indices $j=[n_{start},...,n_{end}]$ (which relate to times $t = j/f$, where f is the sampling frequency of the dataset) where the total number of points in the window is N_p. To minimize leakage, the window width N_p should be chosen such that an integer number of sinusoidal cycles fit within it, which can expressed as

$$N_P = P_p \cdot f \cdot w. \quad (6)$$

Here w is an integer multiple with possible values [1,2,..., floor(n/ $P_p \cdot f$)] where "floor" is a function that rounds the argument toward zero. In general, higher values of w increase the frequency resolution of the SWFT while decreasing the time resolution. Note that the entire dataset x is used, and the standard DFT Fourier coefficient is produced when w is set equal to $n/P_p \cdot f$.

When Equation 5 is applied to the data window N_p, we note that the indices k in the DFT output then correspond to periods $N_p/f \cdot [1, 1/2, 1/3, ..., 2/N_p]$, and the first point in the DFT output is the DC gain term. The SWFT requires only the value F_k corresponding to P_p. So $P_p = N_p/f \cdot 1/w$, and therefore (replacing k with p) we get

$$F_k = F_p = F_{w+1}. \quad (7)$$

Substituting Equations 6 and 7 into 5, and recalling the definition of N_p, we get:

$$F_p = \sum_{j=n_{start}}^{n_{end}} x_j e^{\frac{-2 \cdot i \cdot \pi \cdot w}{N_p}(j-1)}. \quad (8)$$

The window of width N_p is slid along the dataset producing an array F_{pq} where q corresponds to the center-window time τ_{pq} with indices p and $q=[1,2,..., q(max)_p]$, where $q(max)_p$ is the maximum index of q, defined later in equation 14. The maximum resolution of P is such that the product $Pmax \cdot f = [2,3,...,n]$, since the shortest period allowed is given by the Nyquist critical frequency$(2/f)^{-1}$. However, for computational efficiency when $Pmax$ spans several orders of magnitude, the user can specify any subset P of $Pmax$. For instance, one could choose $P \cdot f = [2, 5, 12, 14, ..., n]$, or any other arbitrary subset of $Pmax$. The index $p=[1,...,length(P)]$ then refers to the periods at which the SWFT will be calculated, where "length" is a function that calculates the size of the 1st dimension of the array P.

Substituting Equation 8 into Equation 4 yields the complete definition of the SWFT:

$$F_{pq} = C \sum_{j=n_{start}}^{n_{end}} (x_j - D_m) \cdot T_m e^{\frac{-2 \cdot i \cdot \pi \cdot w}{N_p}(m-1)} \quad (9)$$

where m is an index $[1,2,...,P_p \cdot f_s \cdot w]$, or $m=(j+1)-n_{start} \cdot n_{start}$ and n_{end} are both functions of p and q:

$$n_{start} = (q-1) \cdot P_p \cdot f \cdot w + 1 = (q-1) \cdot N_p + 1, \text{ and} \quad (10)$$

$$n_{end} = n_{start} + N_p - 1. \quad (11)$$

Because the Fourier coefficients can only be time-localized to a window of width N_p, we include the capability for each data window to overlap the previous one in order to increase the temporal resolution of the transform. As an example, if N_p=10 and no overlap is allowed, the minimum temporal resolution at this value of P_p would be 10. Instead overlap is allowed such that the minimum resolution can be as low as 1. This could either be done with a fixed overlap (i.e., each window overlaps half of the other across all periods), or the overlap can be scaled in any other fashion, such as linearly with P. For this study, we define the overlap value, o_p to be given by a simple linear scaling as

$$o_p = \text{floor}\left[o_{min} + \frac{P_p}{\max(P)}(o_{max} - o_{min})\right] \quad (12)$$

where o_{max} and o_{min} are specified by the user, and max(P) is the maximum value within the array P. The minimum value of $o_{min} = 1$, and the maximum value of $o_{max} = N_p$. With this modification, Equation 10 becomes

$$n_{start} = (q-1) \cdot N_p / o_p + 1 \quad (13)$$

and $q(max)_p$ is given by

$$q(max)_p = \text{floor}\left[(n - N_p) \cdot \frac{o_p}{N_p} + 1\right]. \quad (14)$$

The amplitude or power spectra can be extracted from the full SWFT spectrum in the same way it would be for the standard FT. Here we use the amplitude spectrum that is calculated via

$$A_{pq} = 2 \cdot |F_{pq}| / N_p \quad (15)$$

where the vertical bars indicate the magnitude of the complex value F_{pq} and the factor of 2 arises because the DFT spectrum is symmetric about the vertical axis and thus distributes half of the spectral power at a period to the each of the positive and negative instances of that period. The center-window times τ associated with the arrays F and A are given by

$$\tau_{pq} = f \cdot \left[(q-1)\frac{N_p}{o_p} + \frac{N_p}{2}\right] = f \cdot \left[N_p\left(\frac{q-1}{o_p} + \frac{1}{2}\right)\right]. \quad (16)$$

Note that F, A, and τ are, in general non-rectangular arrays (the exception is when $w=N_p$). MATLAB's cell array capability was used to store these values. For convenience of both visualization and further processing, such as contouring or computing cross-scalograms and phase-coherence maps, these arrays can be interpolated to rectangular grids.

The SWFT array F can either be computed directly via equation 9, or it can be computed via equation 4 using the values of j_{start} and j_{end} from equations 13 and 11. In either case, the tapering and detrending functions T and D are calculated using the data $x(j_{start}:j_{end})$.

The SWFT was developed primarily for flexibly visualizing the non-stationary spectral content of a time-domain signal. Including the ability for o_p to scale with period greatly reduces the computational demand by calculating F_{pq} less frequently for longer periods. Allowing w to vary enables flexibility between time- and frequency-localization, as the needs of the user demand. Though not used here, w could also be a function of p allowing the frequency resolution to also scale with the period. Including explicit tapering and detrending further improves the ability of the SWFT to represent dynamic spectral content when P spans several orders of magnitude.

In order to illustrate the interpretation of the SWFT scalogram, we examine a simple test case using a summation of three separate sinusoids given by: $f(x) = f_1(x) + f_2(x) + f_3(x)$, where $f_1(x) = \sin(4x)$ over $0 \leq x \leq \pi$, and $f_2(x) = \sin(10x)$ over $0 \leq x \leq \pi$, and

$$f_3(x) = \begin{cases} 0 & 0 \leq x \leq 4\pi, 8\pi \leq x \leq 12\pi \\ \sin(x) & 4\pi < x < 8\pi \end{cases}.$$

Figure 2a illustrates the successive superposition of the three sinusoids, the darkest curve plots $f_3(x)$, the mid-tone curve plots $f_3(x) + f_2(x)$, and the lightest curve plots $f(x)$.

The SWFT scalogram (Figure 2b) of the function $f(x)$ reveals the periods, amplitudes, and ranges of activity of each of the simple sinusoids. The method extracts the periods of the three sinusoids and reconstructs the amplitudes accurately, except for the $f_3(x)$. This is simply because only two cycles of the sinusoid were used in $f_3(x)$, and the window width was chosen as twice the period, thus only the very center point should reach an amplitude of 1. The shorter-period sinusoids both have peak amplitudes very near 1, although the middle has amplitudes >1 in some locations because of the interaction between the two longer-period sinusoids. Importantly the longest period sinusoid, which is defined only for $4\pi < x < 8\pi$, only has large amplitudes in this range, thus revealing the time-varying spectral behavior of the input signal.

In cases where the input is something other than a summation of sinusoids, the scalogram output will exhibit large

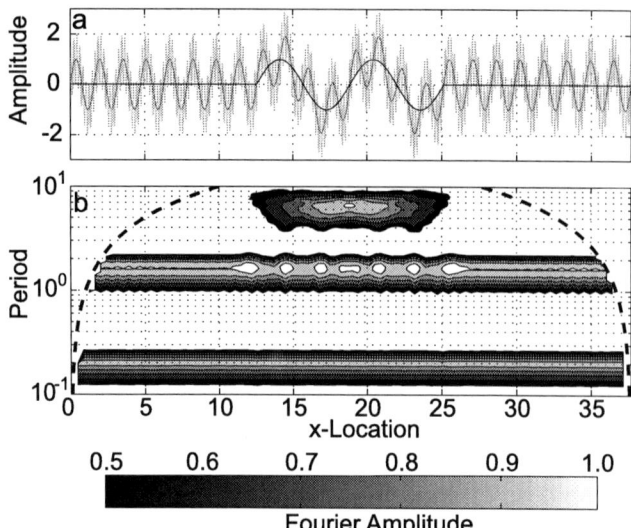

Figure 2. SWFT results for a test function, $f(x)$. a) Curves representing the successive superposition of the three sinusoids with different frequencies; b) the SWFT scalogram. The shaded contours represent amplitudes between 0.5 and 1.0 as shown in the colorbar. The dashed line represents the boundary beyond which the SWFT scalogram is undefined (where half the window width is greater than the available number of points within the dataset).

amplitudes across a range of periods, as expected from Fourier theory. If the broad-spectrum behavior of the input data is relatively time-localized, such as the runoff response to a brief storm event, the scalogram will exhibit lineations corresponding to that event. Thus, temporally continuous but spectrally limited high amplitude regions indicate periodic processes within the input data while temporally limited but spectrally broad lineations indicate broad-spectrum processes.

SCALOGRAM AVERAGING

Averaging the scalogram amplitudes across periods or time gives the period-averaged or time-averaged amplitude spectra, respectively, of the SWFT scalogram [*Torrence and Campo*, 1998]. All of these three averages are computed using a rectangularly interpolated grid, A_{sr} at user-selected periods s and times r, from A_{pq}. Since the period spacing in the SWFT scalogram is not necessarily uniform, we use the period-weighted mean amplitude Ap, calculated using

$$Ap_r = \frac{\sum_{s=1}^{L} P_s A_{sr}}{\sum_{s=1}^{L} P_s}, \quad (17)$$

where L is the dimension of the rectangular grid in the period direction. This provides aggregate information on the temporal variation of spectral amplitude.

The time-averaged amplitude At, given by

$$At_s = \frac{1}{M}\sum_{r=1}^{M} A_{sr} \qquad (18)$$

where M is the dimension of A in the time direction, displays average amplitudes across time at a single period. Referred to as the global SWFT spectrum, this time-averaged spectrum is qualitatively similar to the FT spectrum, but differs in physical meaning.

Bulk changes in the relative influence of short- vs. long-period fluctuations across time can be visualized using the amplitude-weighted average period Pt,

$$Pt_r = \frac{\sum_{s=1}^{L} P_s A_{sr}}{\sum_{s=1}^{L} A_{sr}} \qquad (19)$$

All three of these scalogram averages are demonstrated below.

WATERSHED AVAILABLE PRECIPITATION: SNOWMELT MODELING

Although this study focuses on revealing and exploring watershed process using a purely data-driven SA, one model is required for the analyses. Because precipitation falls as snow during most of the winter months in northern lower-Michigan, a snow storage-and-release model is needed. The data required for a full energy/water balance model were either unavailable or incomplete, thus we implemented a simple heuristic snowmelt model. Because data are available for both fresh snow totals and observed snow depth, the model must simply identify when a decrease in snow pack thickness corresponds to a melt event or densification of the snow pack. This heuristic model tracks snow water equivalent and releases snowmelt based on a three-part rule structure:

1) The density of newly-fallen snow is calculated from daily precipitation and snow fall totals. Since precipitation can be mixed frozen and liquid, a maximum new-snow density cutoff, new_{max}, was determined from the data. Any precipitation in excess of this cutoff is considered equivalent to rainfall and immediately released.

2) The total snowpack density is updated based on the new snow depth and the accumulated water content of the pack.

3) If the average snowpack density exceeds the maximum pack density, $pack_{max}$, it is assumed that some of the snow has melted. Melt is then generated equivalent to the depth of the pack multiplied by the difference between the calculated pack density and the maximum density.

There are three important assumptions in this model. First, we assume that drifting does not affect the observed snow depths at the measurement location (typical of snow models). Second, we assume that the snow pack properties are uniform throughout, which is realistic in the MRW region as total pack thicknesses are typically less than a third of a meter. Finally, this model assumes that the density at which the snow pack releases water remains the same throughout the season.

The maximum densities of the new snowfall and the snowpack are physical quantities that are generally not equal. Maximum snowpack density, $pack_{max}$, is determined by both its composition and water holding capacity, which are functions of the thermal history of the pack and new snowfall conditions. Direct comparison of snow depth and stream discharge in our study area suggests that water is released from the snowpack when the combined snow/water density reaches approximately 0.35 g/cm^3, according to this model, thus this value was chosen for $pack_{max}$. The maximum new snowfall, new_{max} density of 0.23 g/cm^3 (determined as the ratio of new snow water content to new snow depth) was extracted from the data, as this was the relatively abrupt limit above which higher-density new snowfall events were obvious outliers.

To assure that the heuristic model performs acceptably, it was compared to the UEB snow model [*Tarboton and Luce*, 1996] for the winter of 1999/2000. The solid line in Figure 3a is the heuristic snowmelt model, and the dashed line is the output from the UEB snow model. Both models output the combined snowmelt and liquid precipitation, referred to hereafter as watershed-available precipitation. Watershed-available precipitation is that which can be acted upon by physical or biological processes. Note that the models provide very similar results, with the UEB model predicting slightly more melt early in the season and the heuristic model predicting more melt late in the season. The heuristic model predicts approximately 1.5 cm more snowmelt during the season than the UEB model. However, the heuristic model does not allow for sublimation, which entirely accounts for the ~1.5 cm difference at the end of the modeled period. The heuristic model was chosen for our analysis since it performs acceptably, despite minimal data requirements and a simple structure.

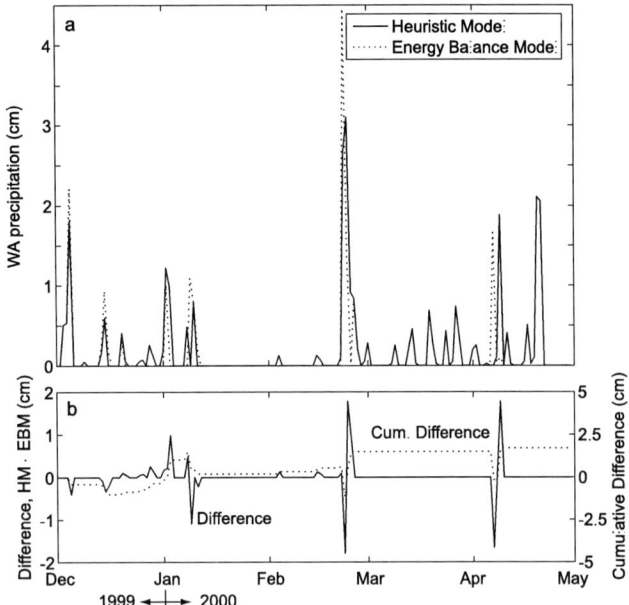

Figure 3. a) Comparison of the watershed available (WA) precipitation simulated by the heuristic snowmelt model and the UEB model; b) the difference between the two (heuristic model (HM)—the UEB energy balance model (EBM)) and the cumulative difference.

RESULTS AND DISCUSSION

Examining Watershed Process: Spectral Comparison

The spectra of stream discharge, watershed-available precipitation, and relaxation of the water table elevation in well B13 (Figure 1) can be directly compared to infer interaction timescales between the data types and to examine details of processes within each type. Well B13 (data not plotted) was chosen because the water table is deep enough (>30 meters) to preclude direct evapotranspiration effects. Four key features of the spectra will be compared to integrate the hydrologic datasets and reveal process details: 1) spectral peaks, 2) log-log linear slopes (i.e., fractal scaling), 3) locations of slope-breaks, and 4) relative spectral amplitudes. Since watershed-available precipitation is the primary forcing function for natural watershed processes in our study region, the spectra of well-head relaxation and stream discharge can be viewed as modifications, or fractal filters [*Kirchner et al.*, 2000], of the watershed-available precipitation spectrum. The same is true to some degree for the well-head relaxation and stream discharge spectra, as groundwater inputs to the Muskegon River account for a majority of its annual discharge [*Jayawickreme and Hyndman*, 2007].

The spectra of these three data types are plotted in Figure 4 along with log-log linear fits and 95% confidence intervals. Also plotted is the fixed period-width binned spectrum to aid visualization. Slopes for selected linear portions of the spectra were calculated using least-squares regression between user selected bounds. These bounds were chosen to match portions of the spectrum that exhibited a linear slope. The slope breakpoints are then calculated at the intercepts of the separate linear fits.

The 95% confidence interval (dotted line) is determined by multiplying the χ-square value for a system with 1 degree of freedom by the average amplitude given by a noise (or scaling) model [*Torrence and Compo*, 1997]. In this case, the noise model was assumed to be given by the linear fits. This enables the flexibility of applying statistical confidence intervals to datasets without assuming *a priori* a particular type of noise. This is useful when working with spectra that exhibit multi-scaling behavior [*Tessier et al.*, 1996; *Dahlstedt and Jensen*, 2004], where a single-scaling noise model, and therefore confidence test, would be inadequate.

All three process spectra have annual-cycle peaks at or near 365 days, although the peak in the head relaxation data is significantly below the 95% CI boundary. This is probably due to the relatively short data record available for the head relaxation. The discharge spectrum has a series of peaks at the harmonics of the 1 day peak that are artifacts, as later discussion will demonstrate. The head relaxation spectrum also has a peak near 117 days, along with weaker peaks near 70 and 50 days (Figure 4d). These are near the integer factors 3, 5, and 7 of the 365 day peak, again suggesting a spectral artifact rather than processes acting at these periods. The watershed-available precipitation spectrum has two additional peaks at ~174 days, and another at ~65 days. These may not indicate sinusoidal processes active at those periods, but may instead be the spectral signature of a non-sinusoidal process characterized primarily by a longer-period oscillation. In particular these two peaks are near the integer factors 2 and 6 of the primary 365 day peak.

The slopes of the spectra and slope breaks (Table 2) provide provocative evidence of linkages between watershed processes. If a quantity, such as stream discharge, is being forced directly by another, such as precipitation, then fractal scaling active in the forcing input should exhibit itself directly in the response variable [*Tessier et al.*, 1996]. However, a non-linear system response behaves like a fractal filter, modifying the input scaling behavior [*Kirchner et al.*, 2000]. There are two examples of this in Figure 4 and Table 2. The first is the segment of the discharge spectrum between one and three hours with a β of 1.5. This is roughly similar to the slope of the NEXRAD hourly precipitation spectrum. The β=0.9 slope in the NEXRAD spectrum is an underestimate of the true spectral slope, given that the hourly NEXRAD spectrum represents just a single year of

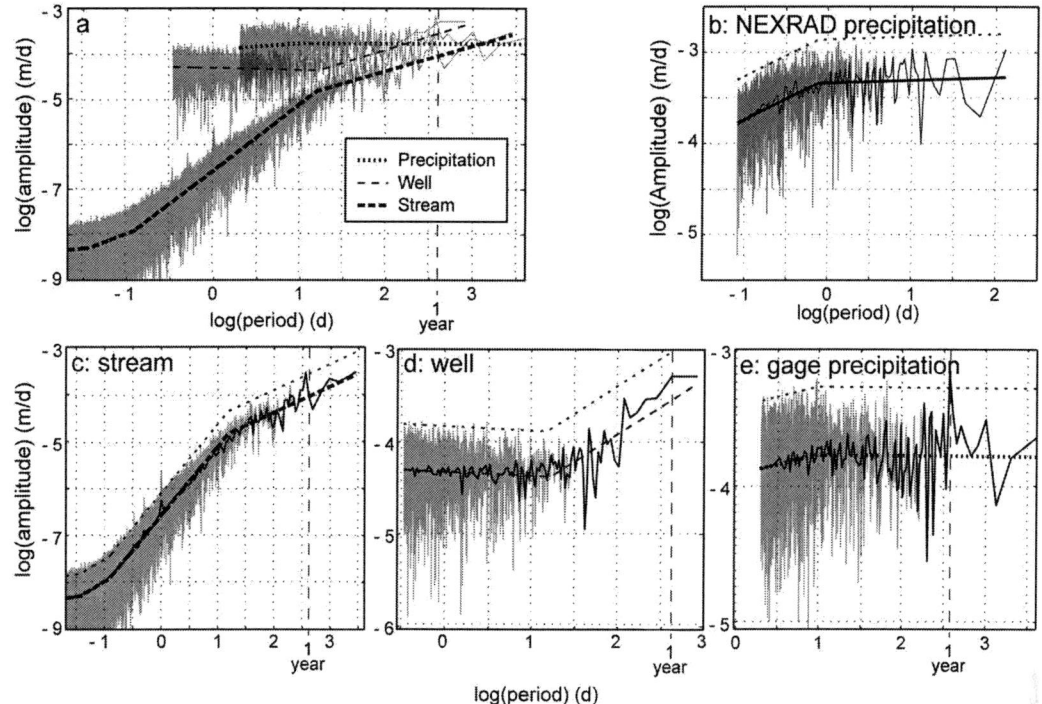

Figure 4. Composite plot of stream discharge, well-head relaxation, and precipitation spectra. Grey lines plot the raw FT spectra, solid lines are the binned-mean amplitudes, and dashed lines give the upper 95% confidence intervals. a) An overlay of the spectra in parts c–e and their linear fits; b) NEXRAD precipitation spectrum for 4/1/2004–11/30/2004, not plotted on part a; c) stream discharge spectra for 10/1/1997–9/30/2004; d) well-head relaxation for well B13 from 7/1/2003–7/1/2005; and e) watershed-available precipitation from 1/1/1990–1/1/2001. Note that, in part c the limited resolution makes it appear as if much of the stream discharge spectrum is above the confidence limit, however fewer than 5% of points exceed the limit in a spectrum with over 130,000 points.

data that undersamples precipitation, and thus suffers from high-frequency aliasing and slope-flattening as described in *Kirchner* [2005]. If, as indicated in these empirical spectra, the "true" spectral slopes of streamflow and precipitation match in this range, we interpret the similarities of slope up to a period of three hours to indicate that direct precipitation is the dominant flow-generation process. Beyond three hours, processes with a different scaling relationship dominate flow.

Process linkages are also apparent between long-period stream discharge and head relaxation spectra. Both spectra have a slope break at approximately 16 days and follow a $\beta=1.1–1.2$ scaling relationship to longer periods. This suggests that variations in groundwater discharge control stream discharge variability at periods longer than approximately 16 days. The exact value of this slope break is approximate because of the gradual transition between linear segments in the discharge spectrum between approximately 10 and 30 days. *Zhang and Schilling* [2004] observed slope breaks in Iowa streams at approximately 30 days. The similarity in slopes between head-relaxation and discharge in Figure 4 suggests that groundwater inputs dominate streamflow in these Midwestern streams for periods longer than 10–30 days while in-stream and near-surface watershed processes appear to dominate at shorter periods.

The spectral slope values also provide useful information. Particularly interesting is the $\beta=2.9$ slope seen at periods between about 3 hours and 16 days in the discharge spectrum. The uniformity of scaling in this portion of the spectrum means that any watershed processes active in this period range also exhibits similar scaling. Fundamentally, this is because linear processes that are additive in the time-domain also add in the spectral domain [*Smith*, 1997]. If the scaling of any hydrologically significant watershed process differed from the others, it would preclude the observed uniform scaling. This uniformity is interesting considering the variety of watershed processes active in this period range, including bank storage and release, precipitation runoff, near-surface "interflow", near-stream saturated groundwater response, and evapotranspiration. Further investigation of this uniformity could be undertaken with a more concentrated set of data designed to explore these processes, or using a detailed process-based hydrological model.

Table 2. Slope break locations and β values for Figure 4.

Discharge		Watershed-available Precip		Head Relaxation		Head Fluctuation	
Break	β	Break[1]	β^1	Break	β	Break[2]	β^2
1.0 hour	0.3						
3.0 hours	1.5	***18.2 hours***	***0.90***			*14 hours*	*0.03*
15.4 days	2.9	9.6 days	0.30	16.5 days	-0.08	*19.3 days*	*1.9*
	1.1		0.01		1.2		*4.4*

[1] These bold italicized entries correspond to results from the NEXRAD dataset, normal entries refer to the watershed-available precipitation from gage data.

[2] These italicized entries correspond to slopes from Well B13, whose binned-mean spectrum is not plotted.

Analysis of the relative amplitudes of the three spectra in Figure 4a provides a hydrologic response time to precipitation. The amplitudes of watershed-available precipitation and discharge converge at a period of approximately 2.5 years. Thus any long-period fluctuation in precipitation will be followed by an equal magnitude fluctuation in stream discharge. Therefore this watershed and its unsaturated and saturated groundwater processes do not appear to control hydrologic fluxes with periodicities greater than 2.5 years. This observation could be used in autoregressive models of basins with short discharge records but more extensive precipitation data. Another interpretation of the similarity in magnitude between the precipitation and discharge spectra is that the combined response time of both surface and groundwater systems is approximately 2.5 years under current conditions, although extended drought periods could certainly affect this value. Therefore, in order to assure insensitivity to initial conditions, a model of this watershed must be "spun up" with realistic meteorological inputs for at least 2.5 years.

Beyond 10–30 days, the well-head relaxation amplitudes are approximately 2–3 times greater than those of stream discharge. This disparity is likely a combination of both physical processes as well as our assumptions. First, the well-head relaxation amplitudes would exceed those of stream discharge because of evapotranspirative losses. Also, the head in an individual well may fluctuate more than the average of all wells in the watershed, particularly if the water table at that well is deep. Thus a more representative comparison would be between stream discharge and the average of spectra from wells distributed across the watershed. However, in this case many of our wells showed signs of periodic anthropogenic disturbance that violate assumptions of our simple differencing technique. Other factors that contribute to the disparity between head relaxation and stream discharge amplitudes may include overestimation of porosity or average recharge rate because of the short data record in the well, as well as differences in annual recharge between the Evart watershed and the location of the well in the Grand Traverse Bay region.

The head relaxation results presented in Table 2 are taken from the spectrum of smoothed-differentiated head fluctuations shown in Figure 4d while the head fluctuation data in Table 2 were taken from the spectrum of actual head fluctuations (not plotted). Unlike head fluctuations, head relaxation (and therefore a decrease in storage) is directly related to groundwater discharge, allowing direct comparison of their spectra. Other studies have reported head fluctuation spectra [*Zhang and Schilling*, 2004; *Lee and Lee*, 2000; *Naff and Gutjahr*, 1983], but because the basic units of head fluctuation and stream discharge differed in these studies, their reported groundwater head and baseflow scaling laws can not directly be related.

EXAMINING INTRA-ANNUAL SPECTRAL VARIABILITY USING THE SWFT

Physical interpretations of Fourier Transform spectra assume that system processes are both non-intermittent and stationary. However, many watershed processes are either inactive for parts of the year (intermittent) or possess different spectral characteristics over successive cycles (non-stationary). Thus, interpreting spectra from many occurrences of a given process is problematic. To overcome this, we use the Scaled-Windowed Fourier Transform (SWFT), which does not require either stationarity or non-intermittency. As Figures 5–10 demonstrate, all three watershed processes examined in this study exhibit both non-stationary and intermit processes.

The SWFT spectrum of stream discharge (Figure 5) displays its time-varying Fourier spectral content, revealing details about how discharge responds to hydrologic inputs and watershed processes. As described previously, the vertical lineations in the SWFT scalogram are caused by the broad spectrum of time-series stream discharge peaks with each such lineation corresponding to a pulse increase in stream discharge. Clearly separated lineations, which primarily occur during the summer months, extend from very short periods and tend to reach maximum amplitudes at periods between 20–40 days. This range of periods corresponds to the width of the time-series discharge peak, and thus to the time-scale of surface and near-surface watershed response to precipitation inputs. As will be shown in the next section, this 20–40 day time scale of surface and near-surface hydrologic response is also similar to the width of the unit hydrographs developed for this watershed.

Spring months typically exhibit increased Fourier amplitudes relative to other seasons across all periods shorter than several hundred days. The spring of 1998 has a very different spectrum than that of 1996 or 1997. Those years had several spring discharge peaks followed by relatively high summer baseflow levels, whereas the single discharge peak of 1998 is followed by low baseflow and weakening of the 365 day amplitude. This summer baseflow portion of the time-series discharge is accompanied by decreased amplitudes across nearly all periods, except near 365 days. Thus, stream discharge during this portion of the year is dominated by long-period fluctuation, most likely from groundwater inputs, as indicated by the amplitude-weighted mean period curve (Pt).

Seasonal differences in spectral character between the summer and spring/fall are not evident in the watershed-available precipitation SWFT scalogram (Figure 6c), but show up very prominently in discharge (Figure 5c). This suggests that although there is spectral power in that period range in watershed-available precipitation, summer evapotranspiration and canopy interception decrease the magnitude of discharge response and thus the spectral amplitudes.

Another important difference between Figures 5c and 6c is that the dominant power in the precipitation spectrum occurs at shorter periods while the reverse is true for stream discharge. This corresponds to the behavior seen in Figure 4c, however the time-averaged amplitude (At) spectrum of watershed-available precipitation differs from the Fourier spectrum in Figure 4e. Such differences are artifacts produced by the violations of the assumptions of stationarity and non-intermittency inherent in the standard FT.

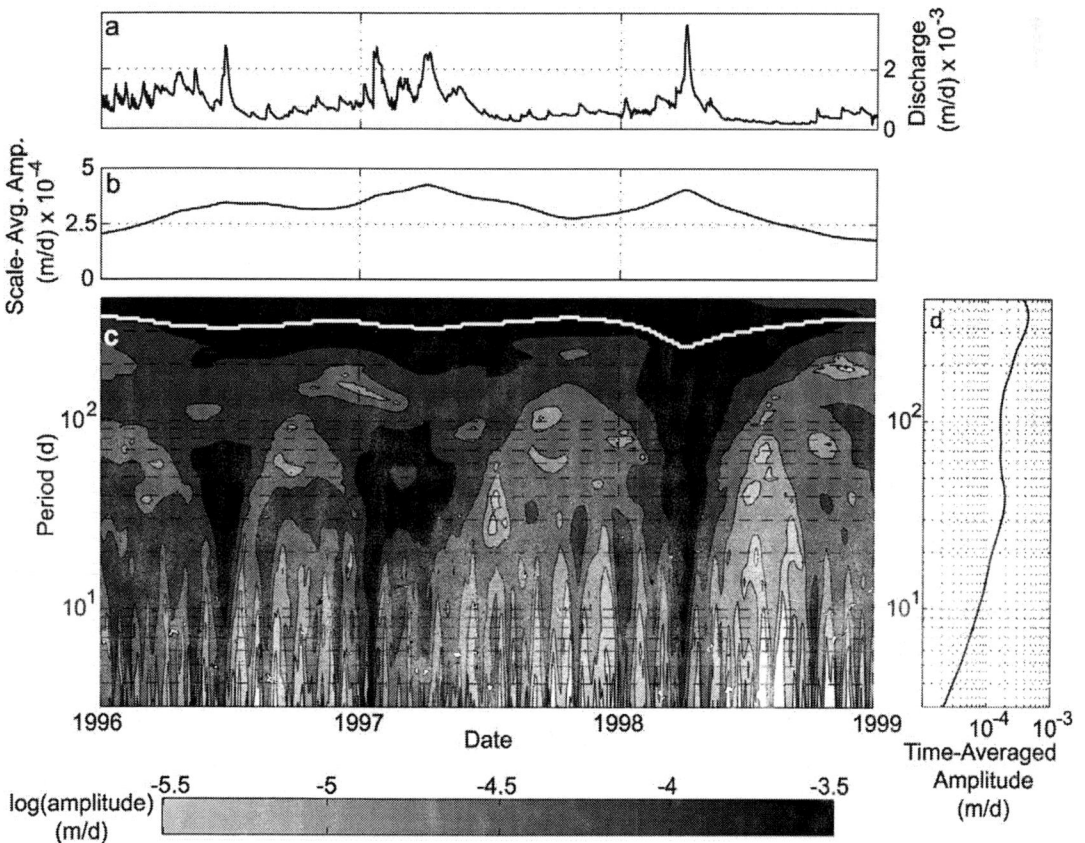

Figure 5. SWFT of stream discharge at the USGS gage in Evart, MI. a) Stream discharge time-series divided by the area of the watershed above this gage; b) Scale-averaged amplitudes, Ap; c) Filled contour scalogram of the amplitude (A) vs. both period (P) and time (τ), along with the location of the amplitude-weighted mean period, Pt (white line); d) Time-averaged amplitude spectrum, At.

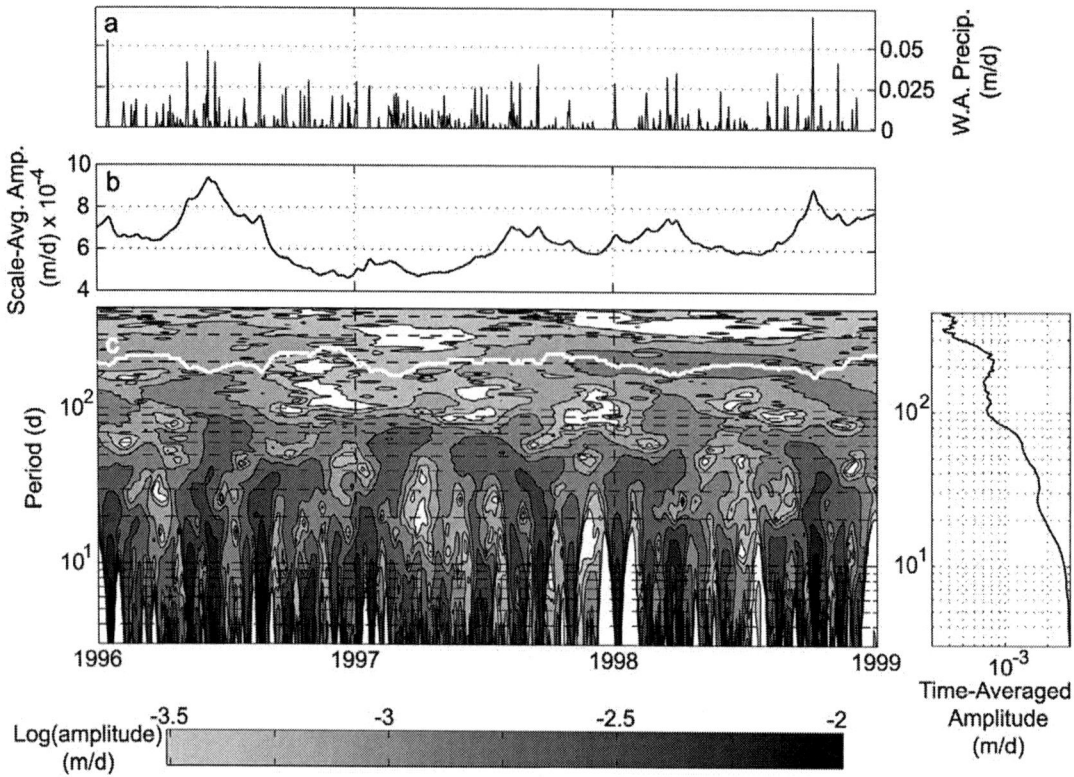

Figure 6. SWFT of watershed-available precipitation at Big Rapids, MI. a) watershed-available precipitation time-series; b) Scale-averaged amplitudes, Ap; c) Filled contour scalogram of the amplitudes (A) vs. both period (P) and time (τ), along with the location of the amplitude-weighted mean period, Pt (white line), (note, white areas have amplitudes < log(amplitude)=-3.5); d) Time-averaged amplitude spectrum, At.

The SWFT discharge scalogram for the water years 10/1/1991–9/30/2004 (Figure 7) displays the spectral content of stream discharge at the Evart gage over periods of 0.3–1000 days. Although the longer-period spectrum is not plotted, 2–3 year and 6–8 year cycles are evident in periods between 100–400 days, perhaps related to climate cycles as seen in *Coulibaly and Burn* [2004]. For 1991–93, long-period fluctuations are most active near a period of 180 days, which then cease during 1994–95 before resuming from 1996–98 at 365 days. Again, this long-period activity switches off for two years and then resumes centered about 180 days. Another interesting set of features in this scalogram are the summer-fall periods of 1998, 2000, and 2002 in which amplitudes are drastically reduced up to periods on the order of 100–200 days.

These observations from Figure 7 enable a deeper understanding of time-averaged spectra such as the FT spectrum or the SWFT time-averaged amplitude (At). The FT spectrum of discharge (Figure 4c) contains a weak spectral peak at 180 days along with a peak at 365 days. Prior to examining the spectrum as a scalogram, we were unable to distinguish between dominant spectral peak harmonics and peaks from independent processes at those periods. The scalogram shows that there are processes generating true peaks at both the 365-day and 180-day periods.

Additionally, we can use the information from the scalogram to indicate when processes that generate particular spectral peaks are most active. A prominent example of this is the 1 day peak in the FT spectrum of discharge. Plausible interpretations of this 1 day peak include diurnal fluctuation in evapotranspiration or streambed conductance during the summer months. However, Figure 7 shows that the dominant 1 day amplitudes occur in the winter and early spring months. A close examination of the time-series reveals two important details: 1) the diurnal fluctuation is strongest during periods where daily maximum temperatures are subfreezing, and 2) discharge peaks during the coldest mid-morning hours of each day. These observations suggest that the diurnal signal may be related to icing effects at the instrument. The 1-day peak seen in Figure 4 is a "true" spectral peak, but it does not

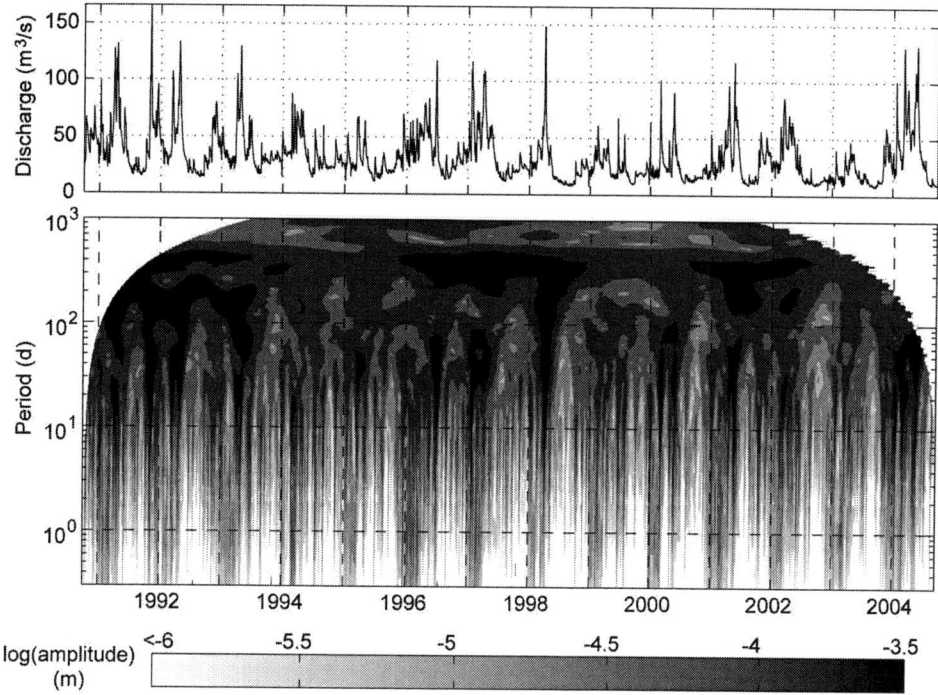

Figure 7. SWFT of stream discharge at Evart from 9/1/1991 to 8/31/2004. a) Stream discharge time-series; b) Filled contour scalogram of the amplitude (A) vs. both period (P) and time (τ). The year labels are centered at January 1. Here the curved white boundaries at longer periods indicate times where the SWFT at those periods is undefined.

indicate cyclic system processes so much as inaccuracies in measurements. Also, there are no significant amplitude peaks at the 0.5-day period, thus confirming that the corresponding peak in Figure 4c was indeed a harmonic. This information could then be used to filter the discharge time-series and remove this apparently erroneous periodic discharge behavior.

The SWFT of head fluctuations from Well B10 (Figure 8) provides another example of the importance of viewing spectra as a function of both period and time. Well B10 was chosen because it exhibited the greatest amplitude of fluctuation in the period 1–30 days of the 17 wells in our study. Figure 8 displays both the head fluctuation data (mean removed) as well as the SWFT scalogram for periods between 0.3 and 100 days. The largest time-series amplitude head fluctuations occur from late October to early April. The scalogram reveals that most of the time series fluctuation is caused by larger Fourier amplitudes of the 1–30 day periods, which are greatest during the winter. Note that the dominant period in this range is not constant throughout the year. During the summer, early fall, and spring, the dominant period is near the 7–10 day range, that then shifts to the shorter 2–7 day range in mid-November. The SWFT scalogram clearly reveals this information that would be difficult to directly extract from the time-series data.

SPECTRALLY DERIVED WATERSHED "UNIT HYDROGRAPH" RESPONSE FUNCTIONS

The watershed annual unit response functions for the portion of the Muskegon River Watershed above Evart gauge were calculated using discharge and Big Rapids precipitation data between 9/1/1999 and 8/31/2000 (chosen arbitrarily). Two different unit response functions are shown in Figure 9a, that of the discharge response to watershed-available precipitation as well as to raw precipitation. Including the snow storage-and-release model creates higher peak discharge responses with a more physically realistic long tail due to groundwater discharge. The higher peak is expected because without a snow model this method treats the precipitation the same in January as it would in July, even though the January precipitation fell as snow and was stored until later, resulting in no significant short-term discharge response.

A convolution of the solid-line unit response function in 9a with watershed-available precipitation according to Equation 1, produces the dashed modeled discharge and residual curves in Figures 10a and 10b. Because the unit-response was truncated as described in the methods section, the convolution is not a perfect reconstruction. The resultant discharge is an overestimate in the summer and fall months, but an underestimate during the spring. This is to be expected as the annually calculated response curve effectively averages the system behavior throughout the year.

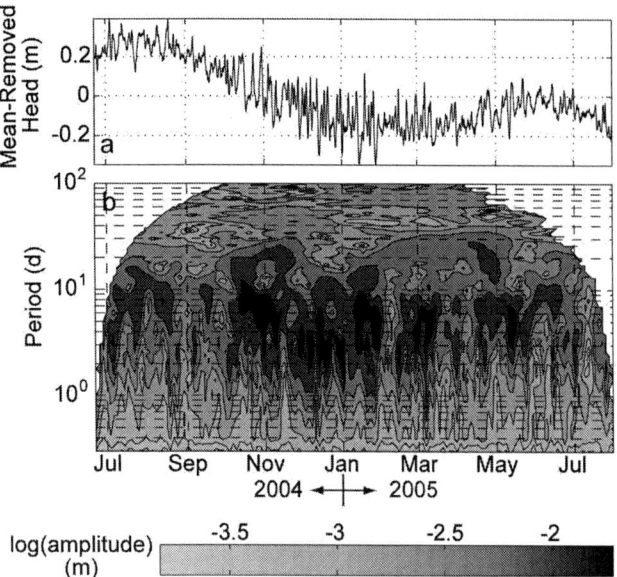

Figure 8. SWFT of groundwater-head fluctuations in Well B10 in the Grand Traverse Bay Watershed. a) Well-head fluctuation time-series; b) Filled contour scalogram of the amplitudes (A) vs. both period (P) and time (τ). Here the curved white boundaries at longer periods indicate times where the SWFT at those periods is undefined.

If the unit response is calculated seasonally rather than annually, the resultant set of unit response curves reveals the seasonal differences among runoff responses in this watershed (Figure 9b). Discharge response during the spring is much higher than either the annual curve or those of the other three seasons, perhaps due to a combination of frozen soils and higher average soil saturation prior to watershed-available precipitation events (which can be either rainfall or snowmelt). Summer discharge response, on the other hand, is highly damped due to canopy interception, lower average soil moisture, and evapotranspiration. The fall and winter responses appear to be very similar, suggesting that there are not large differences in discharge response between these two seasons. This result was somewhat unexpected since frozen soils are generally expected during the winter months due to long periods of sub-freezing temperatures. Significant areas of frozen soils would tend to increase discharge response, as infiltration capacity is greatly reduced. The lack of a response difference suggests that the soils are not homogeneously frozen throughout the winter season, or that this freezing is not important for runoff generation in this watershed during this time period.

The response curves in Figure 9b all converge to near 0 beyond approximately 20 days. Analysis of data from the wells in the GTBW indicates that the delays between peak spring recharge and peak saturated water table response

scales approximately as 2–4 days/meter of unsaturated zone depth. Thus, for depths on the order of 30 meters, the delay between full groundwater response to a precipitation (or snowmelt) event can be as much as 120 days. As the data lengths included in the seasonal response calculations are only on the order of 60–90 days the seasonal unit response curves can not properly represent the groundwater response. The tail in the annually calculated unit response curve of Figure 9a is likely a more reasonable representation of the average groundwater response.

In addition to providing more insight into watershed processes than the annually-derived unit response curve, the seasonally-derived curves produce a much more accurate estimate of stream discharge when convolved with watershed-available precipitation (Figure 10a). The residuals between the two convolutions and stream discharge (Figure 10b) quantitatively demonstrate the improvement in discharge estimation gained by seasonal convolution and consideration of non-stationary system behaviors. Figure 10a illustrates the utility of the seasonally-derived watershed unit response curves for providing a very simple means of forecasting discharge response to precipitation events. Because the seasonal curves average watershed responses across significant variability in watershed state properties (such as soil moisture), the seasonally-derived convolution underestimates the largest peaks in the discharge data by up to 50%, while predicting measured flows to an accuracy of +/- 25% in most other cases. Much of the remaining residual is because the 50-day unit response curves fail to capture

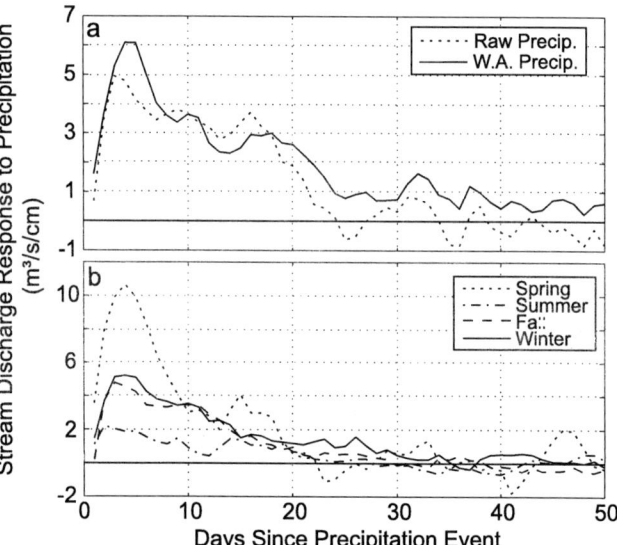

Figure 9. Stream unit response hydrographs. a) Unit response functions obtained using data from 9/1/1999 to 8/31/2000; and b) Seasonal unit response functions using selected years (see Table 1) from 1990–1999.

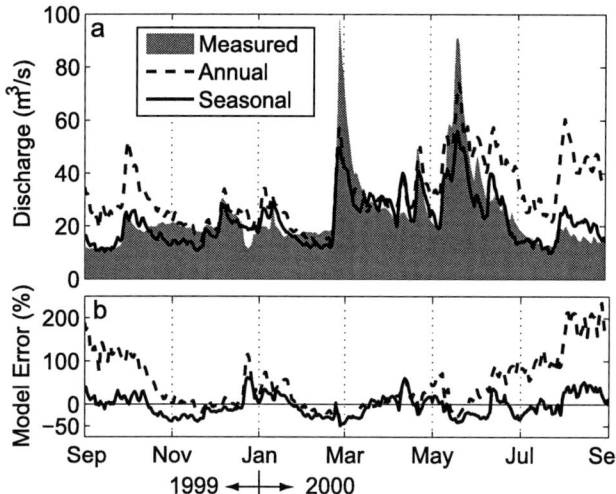

Figure 10. a) Plots of convolved stream discharge from 09/01/1999 to 8/31/2000 using the annually- and seasonally-derived unit hydrographs (initial modeled discharged matched to stream discharge) on top of the measured stream discharge ; b) plot of the percentage difference between each model and the measured discharge (modeled-measured)/(measured)*100.

much of the groundwater response to spring and fall precipitation. An analysis that accounts for the different seasonal responses between wet and dry years, and explicitly incorporates the full groundwater response, might further improve the accuracy of forecasting with this technique.

CONCLUSIONS

We present the application of three spectral analysis techniques, direct spectral comparison, the Scaled Windowed Fourier Transform (SWFT), and the derivation of the unit hydrograph via FT deconvolution. We have discussed and demonstrated how each technique requires consideration of limitations and possible pitfalls in order to be applied successfully, and we elucidated a set of best practices in their application. Most importantly, we have demonstrated that these spectral analysis methods can be used to integrate hydrologic data in order to evaluate watershed processes.

The spectra of related data types were directly compared to infer process linkages between hydrologic inputs, watershed processes, and stream discharge. Similarities in amplitude peaks, log-log linear fractal scaling behaviors, and breaks in scaling slopes among datasets indicate the nature of linkages. For example, fractal scaling in precipitation may be matched in the stream spectrum at periods shorter than approximately 3 hours for the Evart, MI sub-basin of the Muskegon River Watershed. From periods of 3 hours to approximately 10-30 days, stream scaling follows a $\beta=2.9$ slope. This single scaling relationship is notable considering the variety of processes active in this period range, suggesting mathematical similarities among these processes. Beyond 30 days, the scaling apparent in the stream spectrum appears to be controlled by groundwater inputs. But, past a period of approximately 2.5 years, fluctuations in the precipitation spectrum control the stream discharge spectrum. This overall watershed response time should be considered when developing transient predictive simulations of watershed behavior.

We introduced the Scaled Windowed Fourier Transform (SWFT) technique to examine the time-varying content of fundamentally non-stationary hydrologic datasets. The SWFT scalograms revealed both the non-stationarity and intermittency of stream discharge, precipitation, and groundwater head fluctuation spectra. The effect of evapotranspiration and canopy interception is evident in a comparison of the SWFT scalogram of summer precipitation events to the highly damped discharge scalogram for those seasons. Also, the 1-day peak evident in the FT spectrum of stream discharge was shown to be due largely to measurement error rather than diurnal hydrologic processes. Importantly, these are a subset of many possible observations from the rich set of information contained within the SWFT scalogram.

Using direct FT deconvolution, spectral analysis can be also be used to estimate stream unit hydrographs. Because of the simplicity of this method, temporally- and seasonally-varying hydrographs can be quickly derived to better understand non-stationary watershed processes. For the Muskegon River above Evart, MI, the groundwater dominance of the stream discharge spectrum beyond approximately 15–20 days is confirmed by visual inspection of the main unit response peaks. These peaks show a 15–20 day primary stream response period followed by a long-tailed groundwater response that continues out to at least 50 days. Also, the seasonally-derived unit hydrographs quantitatively reveal decreased discharge responses due to evapotranspiration during the summer months and augmented responses during spring snowmelt.

Acknowledgements. We are grateful for the financial support from the National Science Foundation grant (EAR-0233648), and secondary support from the Great Lakes Fisheries Trust. We thank Dr. Jeffrey Andresen and Dushmantha Jayawickreme for providing and processing the NEXRAD precipitation data used in this research. We also thank the hydrogeology research group at Michigan State University for their suggestions and contributions. Finally, we thank Dr. James W. Kirchner and an anonymous reviewer for their comments. Any opinions, findings and conclusions or recommendations expressed in this material are those of the authors and do not necessarily reflect the views of the National Science Foundation.

REFERENCES

Andresen, J. (State of Michigan Climatologist) (2004), Level-III NEXRAD Data. 1996-2004. Archived from the National Weather Service.

Avnir, D., Biham, O., Lidar, D, and O. Malacai (1998), Is the Geometry of Nature Fractal?, *Science, 279,* 39–40.

Bach, M., and T. Meigen (1999), Do's and Don'ts in Fourier Analysis of Steady-State Potentials, *Documenta Ophthalmologica, 99,* 69–82.

Blackman, R. B. and Tukey, J. W. (1958), *The measurement of power spectra, from the point of view of communications engineering.* New York: Dover Publications.

Bras, R. L., and I. Rodriguez-Iturbe (1985). *Random Functions and Hydrology.* New York: Dover Publications.

Cooley, J. W., and J. W. Tukey (1965), An algorithm for the machine calculation of complex fourier series. *Mathematics of Computation, 19,* 297–301.

Coulibaly, P., and D. H. Burn (2004), Wavelet Analysis of Variability in Annual Canadian Streamflows. *Water Resources Research, 40*(3), W03105.

Dahlstedt, K., and H. J. Jensen (2004), Fluctuation Spectrum and Size Scaling of River Flow and Level. *Physica A, 348,* 596–610.

Farrand, W. R., and D. L. Bell. (1982), Quaternary Geology of Southern Michigan. The University of Michigan, Ann Arbor, MI.

Fleming, S. W., A.M Lavenue, A.H. Aly, , and A. Adams (2002), Practical Applications of Spectral Analysis to Hydrologic Time Series. *Hydrological Processes, 16,* 565–574.

Frigo, M., and S. G. Johnson (1998), FFTW: An Adaptive Software Architecture for the FFT, *Proceedings of the International Conference on Acoustics, Speech, and Signal Processing, 3,* 1381–1384.

Gaucherel, C. (2002), Use of Wavelet Transform for Temporal Characterization of Remote Watersheds, *Journal of Hydrology, 269,* 101–121.

Hameed, S. (1984), Fourier Analysis of Nile Flood Levels, *Geophysical Research Letters, 1*(9), 843–845.

Harris, F. J. (1978), On the Use of Windows for Harmonic Analysis with the Discrete Fourier Transform, *Proceedings of the IEEE, 66,* 66–67.

Jayawickreme, D. H., and D. W. Hyndman (2007), Evaluation of the influence of land cover on seasonal water budgets using Next Generation Radar (NEXRAD) rainflow and streamflow data, *Water Resources Research, 43,* W02408, doi: 10.1029/2005WR004460.

Jukic, D., and V. Denic-Jukic (2004), A Frequency Domain Approach to Groundwater Recharge Estimation in Karst, *Journal of Hydrology, 289,* 95–110.

Karl, T. R., Williams, C. N. Jr., Quinlan, F. T., and T.A. Boden, (1990), United States Historical Climatology Network (HCN) Serial Temperature and Precipitation Data, Environmental Science Division, Publication No. 3404, Carbon Dioxide Information and Analysis Center, Oak Ridge National Laboratory, Oak Ridge, TN, 389 pp.

Kirchner, J. W. (2005), Aliasing in $1/f^{alpha}$ Noise Spectra: Origins, Consequences, and Remedies. *Physical Review, 71*(6), 066110(16).

Kirchner, J. W., Feng, X. and C. Neal (2000), Fractal Stream Chemistry and Its Implications for Contaminant Transport in Catchments. *Nature, 403,* 524–526.

Lee, J. and K. Lee (2000), Use of Hydrologic Time Series Data for Identification of Recharge Mechanism in a Fractured Bedrock Aquifer System. *Journal of Hydrology, 229,* 190–201.

Mann, M. (2004), On Smoothing Potentially Non-Stationary Climate Time Series. *Geophysical Research Letters, 31,* L07214.

Naff, R. L., and A. L. Gutjahr (1983), Estimation of groundwater recharge parameters by time series analyses, *Water Resources Research, 19*(6), 1531–1546.

National Climatic Data Center (2001), NESDIS, NOAA, U.S. Department of Commerce, U.S. Daily Surface Data (DS-3200 and DS-3210). National Climatic Data Center, Asheville, NC.

Pelletier, J. D., and D. L. Turcotte (1997), Long-range Persistence in Climatological and Hydrological Time Series: Analysis, Modeling and Application to Drought Hazard Assessment. *Journal of Hydrology,* 198–208.

Percival, D. B. and A. T. Walden (1993), *Spectral Analysis for Physical Applications,* Cambridge: Cambridge University Press.

Press, W. H., Teukolsky, S. A., Vetterling, W. T., and B. P. Flannery (1992), *Numerical Recipes in Fortran 77,* Cambridge: Cambridge University Press.

Smith, S. W. (1997), *The Scientist and Engineer's Guide to Digital Signal Processing,* California Technical Publications.

Tarboton, D. G. and C. H. Luce (1996), *Utah Energy Balance Snow Accumulation and Melt Model (UEB),* computer model technical description and users guide, Utah Water Research laboratory and USDA Forest Service Intermountain Research Station.

Tessier, Y., Lovejoy, S., Hubert, P., and D. Schertzer (1996), Multifractal Analysis and Modeling of Rainfall and River Flows and Scaling, Causal Transfer Functions. *Journal of Geophysical Research, 101*(D21), 26427–26440.

Torrence, C. and G. P. Compo (1998), A Practical Guide to Wavelet Analysis. *Bulletin of the American Meteorological Society, 79,* 61–78.

U.S. Geological Survey (2001), National Water Information System (NWISWeb) data available on the World Wide Web, accessed June 13, 2005.

U.S. Geological Survey (2005). Unofficial Real-Time Data. Direct Communication.

Yang, Z. and D. Han (2006), Derivation of Unit Hydrograph Using a Transfer Function Approach, *Water Resources Research, 42,* W01501.

Zhang, Y. and K. Schilling (2004). Temporal Scaling of Hydraulic Head and River Base Flow and Its Implications for Groundwater Recharge. *Water Resources Research, 40*(3), W03504.

Integrated Multi-Scale Characterization of Ground-Water Flow and Chemical Transport in Fractured Crystalline Rock at the Mirror Lake Site, New Hampshire

Allen M. Shapiro[1], Paul A. Hsieh[2], William C. Burton[1], and Gregory J. Walsh[3]

Estimates of hydraulic conductivity and the effective diffusion coefficient were made in fractured crystalline rock in central New Hampshire over increasingly larger physical dimensions. The hydraulic conductivity of individual fractures ranged over more than six orders of magnitude. Over dimensions of approximately 100 meters, the bulk hydraulic conductivity is controlled by less transmissive fractures; the less transmissive fractures act as "bottlenecks" that impede ground-water flow. Over dimensions of several kilometers, the bulk hydraulic conductivity of the rock was again the same as the network of less transmissive fractures, indicating that there is no interconnected "backbone" of highly transmissive features over kilometers that increases the hydraulic conductivity over larger volumes of rock. In contrast, estimates of chemical diffusion from tracer experiments conducted in rock cores, *in situ* tests over tens of meters, and the interpretation of environmental tracers over kilometers increase as a function of the dimension of the experiment. Estimates of diffusion coefficients from cores were consistent with theoretical interpretations and were less than free-water diffusion coefficients. The wide range of fluid velocities in fractures, however, gives rise to elongated tails in the breakthrough curves of tracer tests conducted over tens of meters. Slow advection from the least transmissive fractures gives the appearance of a diffusive phenomenon. The effective diffusion coefficients resulting from slow advection were greater than free-water diffusion coefficients. The increase in the magnitude of the effective diffusion coefficient with the physical dimension is attributed to the increasing variability in the fluid velocity over larger physical dimensions.

1. INTRODUCTION

Characterizing ground-water flow and chemical transport in fractured rock is regarded as a challenging undertaking. Fractures are not uniformly distributed in formations, and with the wide range of hydraulic properties and the complex connectivity of fractures, convoluted flow paths can exist in fractured rock aquifers from meters to kilometers.

Many issues of societal importance rely on the hydrogeologic characterization of fractured rock aquifers. In issues of water availability, there is intense interest in identifying permeable fractures and other geologic structures that can supply water to individual wells, requiring the characterization of small volumes of rock. At these same sites, however, it is also necessary to understand the regional-scale hydraulic properties of the formation to identify the effect of individual ground-water abstractions on other users and ground-water discharges to surface water drainages.

---

[1]U.S. Geological Survey, Reston, Virginia, USA
[2]U.S. Geological Survey, Menlo Park, California, USA
[3]U.S. Geological Survey, Montpelier, Vermont, USA

Subsurface Hydrology: Data Integration for Properties and Processes
Geophysical Monograph Series 171
This paper is not subject to U.S. copyright. Published in 2007 by the American Geophysical Union.
10.1029/171GM15

Issues that require the understanding of chemical migration in fractured rock also require characterization over increasingly larger physical dimensions. For example, at a site of ground-water contamination, there may be interest in identifying the most permeable fractures and their connectivity in order to design methods of retarding the movement of contaminants downgradient from point sources of contamination. At many industrial sites, contamination in the ground water may have existed for tens of years. Under these circumstances, it is also necessary to assess whether processes and parameters that are appropriate in characterizing chemical migration in fractured rock over dimensions of meters are applicable in the characterization of chemical transport over tens of meters, and even kilometers.

There have been numerous discussions regarding the magnitude of formation properties in heterogeneous subsurface environments as a function of the physical dimension, or scale, over which formation properties are estimated [*Clauser*, 1992; *Gelhar et al.*, 1992; *Neuman*, 1994; *Sánchez-Vila et al.*, 1996; *Neuman and Di Federico*, 2003]. Many of these investigations have relied on data accumulated from multiple sites. The multi-scale measurements that are used in developing hypotheses, however, are rarely drawn from a single site. In addition, sites that have data on hydraulic properties over different physical dimensions do not necessarily have data on chemical transport over increasingly larger dimensions. The general discussions of the relation between the scale of measurements and the associated hydraulic or chemical transport properties that rely on data accumulated from multiple sites rarely points to underlying geologic structures that give rise to the scale dependence of hydraulic and chemical transport properties. For such discussions, data on hydraulic and chemical transport properties estimated over increasingly larger physical dimensions is needed from a single field site, where there is accompanying detailed geologic information.

In this article, the results of multi-scale investigations of chemical transport and hydraulic properties in a fractured crystalline rock site are presented. The investigations were conducted in and around the Mirror Lake watershed in central New Hampshire (Figure 1), which has been a site of detailed multidisciplinary investigations by the U.S. Geological Survey (USGS) in the development of field and interpretive methods of characterizing ground-water flow and chemical transport in fractured rock over dimensions of meters to kilometers [*Hsieh et al.*, 1993; *Shapiro et al.*, 1995; *Shapiro and Hsieh*, 1996a]. This article focuses on the results of hydraulic testing and chemical migration over increasingly larger physical dimensions. The design and interpretation of the hydraulic and chemical migration experiments presented in this article, however, are predicated on the results of geologic and fracture mapping, surface and borehole geophysical logging, and geochemical and isotopic analyses [*Shapiro et al.*, 1999]. The investigations associated with these hydrogeologic disciplines at the Mirror Lake site are not explained in detail in this article; however, the conclusions from these investigations are used extensively in describing the multi-scale results for chemical transport and hydraulic properties. Additional information about the scope of the multi-disciplinary investigations conducted at the Mirror Lake site is given in *Hsieh et al.* [1993], *Shapiro and Hsieh* [1996a], and *Shapiro et al.* [1999], and compilations of articles in *Mallard and Aronson* [1991], *Morganwalp and Aronson* [1996], and *Morganwalp and Buxton* [1999].

2. MIRROR LAKE SITE

The investigations of the chemical transport and hydraulic properties of fractured crystalline rock discussed in this article were conducted in and around the Mirror Lake watershed in Grafton County, New Hampshire (Figure 1). The Mirror Lake watershed is located at the lower end of the Hubbard Brook Valley in the southern part of the White Mountains. The Mirror Lake watershed lies partly in the Hubbard Brook Experimental Forest, which is an ecosystem research facility operated by the U.S. Forest Service [*Likens*, 1985]. The USGS has used the Mirror Lake watershed as a long-term observatory of ground-water flow and the interaction with surface water [*Winter*, 1984]. Multidisciplinary investigations of the crystalline rock underlying the Mirror Lake watershed commenced in 1990 [*Shapiro and Hsieh*, 1991, 1996a; *Shapiro et al.*, 1995, 1999].

In the Mirror Lake area, bedrock is overlain by glacial deposits (drift) that range in thickness from 0 to 50 meters (m). The glacial deposits are mostly till with localized deposits of sand and gravel [*Harte and Winter*, 1996]. The bedrock consists primarily of a coarse-grained, well foliated, ductily deformed, pelitic schist that has been locally intruded by dikes, and pods of granite [*Hsieh et al.*, 1993; *Barton*, 1996; *Burton et al.*, 2000]. Lesser amounts of pegmatite and lamprophyre have also intruded the schist and granite as dikes. Mapping the distribution of rock types at road cuts east of Mirror Lake, and outcrops throughout the Mirror Lake area, indicates that granite and schist have complex, irregular distribution patterns over dimensions of tens of meters [*Hsieh et al.*, 1993; *Barton*, 1996, 1997; *Burton et al.*, 2000].

In the study area, precipitation is the only source of ground-water recharge, and ground water discharges to streams, Mirror Lake, and the Pemigewasset River (Figure 1). The low hydraulic conductivity of both the glacial till

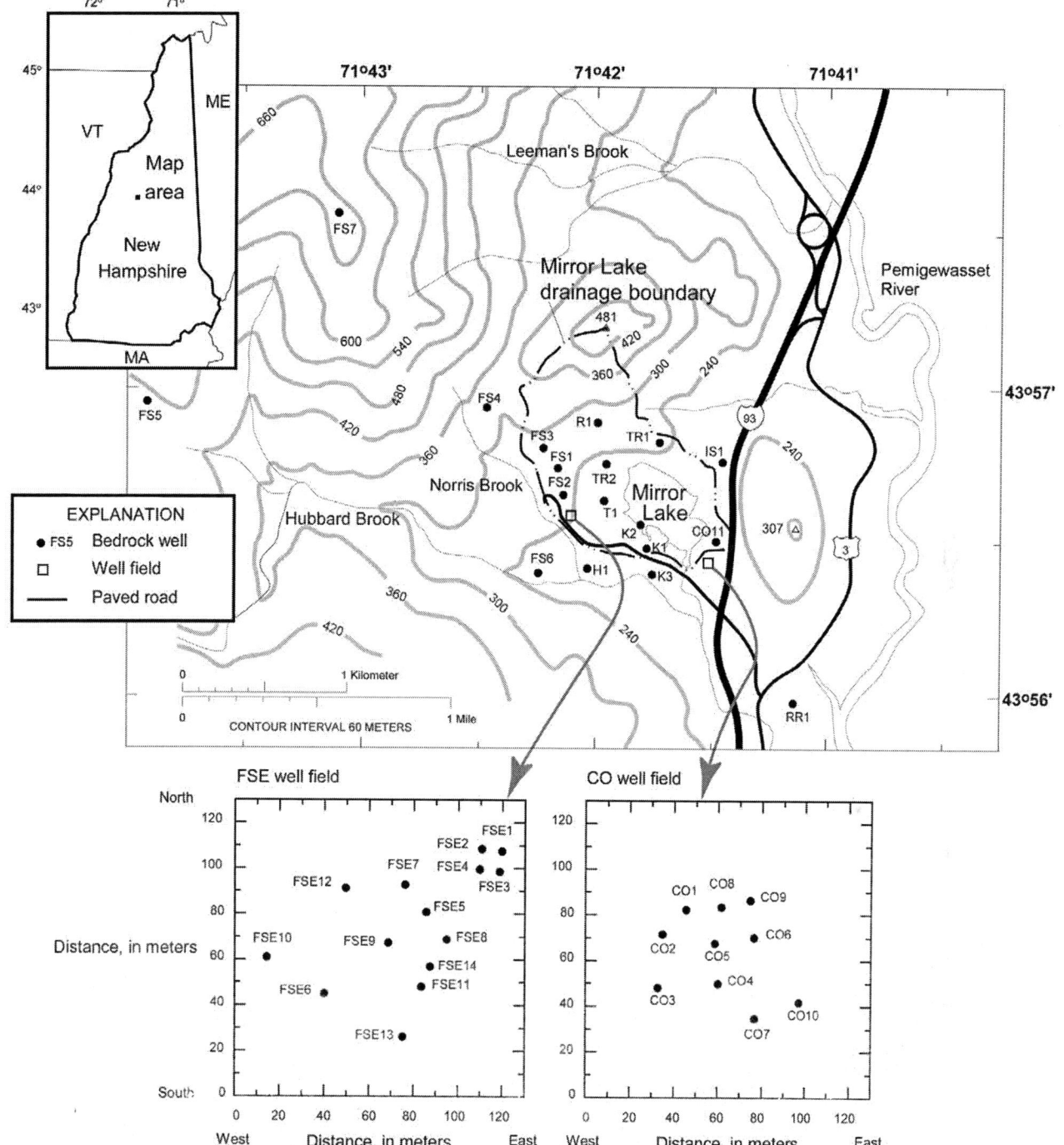

Figure 1. Map of the Mirror Lake area and the location of bedrock boreholes.

and the bedrock results in the water table being close to land surface, even at higher elevations in the Mirror Lake area [*Tiedeman et al.*, 1997, 1998].

Bedrock boreholes, and piezometers and water table wells in the glacial drift were drilled in the Mirror Lake area to investigate ground-water flow and chemical transport over the dimensions of kilometers (Figure 1). Bedrock wells and drift piezometers were also installed in clusters, denoted as the FSE and CO well fields (Figure 1) to investigate ground-water flow and chemical migration over dimensions up to approximately 100 m. The depth of bedrock boreholes ranges from 60 to 300 m, with most wells in the FSE and CO well fields extending approximately 100 m below land surface. After the completion of drilling a bedrock borehole, borehole geophysical logging was conducted to identify the distribution of rock types, the location and orientation of fractures intersecting the borehole, water producing fractures, and other properties of the rock [*Paillet*, 1996; *Johnson and Dunstan*, 1998]. This was followed by the collection of water samples from discrete intervals of the boreholes for chemical and isotopic analyses, and the installation of inflatable packers to isolate the most permeable fractures for long-term monitoring of the hydraulic head [*Hsieh et al.*, 1996].

3. FRACTURES AND GEOLOGIC MAPPING

Fractures are the primary features through which ground water flows in the Mirror Lake area. Quantifying the frequency, orientation, areal extent, and other physical attributes of fractures is important in understanding the chemical transport and hydraulic properties associated with individual fractures and larger volumes of rock containing multiple fractures. Fractures were mapped on road cuts, outcrops, and in bedrock boreholes throughout the Mirror Lake watershed, the Hubbard Brook Valley, and most of the Woodstock 7.5-minute quadrangle [*Barton*, 1996, 1997; *Johnson and Dunstan*, 1998; *Burton et al.*, 2000].

Mapping of fractures and the distribution of rock types was conducted on the north-south trending road cuts of Interstate Highway 93 (I-93) east of Mirror Lake using 4 vertical faces and one subhorizontal natural exposure [*Barton*, 1996, 1997]. Fractures on road cuts having a trace length greater than 1 m were recorded along with their trace length, orientation, aperture, roughness, degree of mineralization, and degree of connectivity with adjacent fractures [*Barton*, 1996, 1997]. On natural exposures and outcrops, joints and joint sets with trace lengths greater than 2 m were recorded in the course of geologic mapping of the Hubbard Brook watershed and the Woodstock quadrangle [*Burton et al.*, 2000]. In bedrock boreholes, fractures were mapped using an acoustic televiewer tool that imaged the borehole wall [*Johnson and Dunstan*, 1998]. A borehole video camera and a high-resolution digital camera were also used to identify fractures and the distribution of rock types over the length of boreholes [*Johnson and Dunstan*, 1998].

Outcrop exposures of the fractured rock in the Mirror Lake area constitute, at most, 3 percent (%) of the land surface [*Burton et al.*, 2000]. The exposures are not uniformly distributed and their size varies from a few square meters to 10's of square meters (m^2), which can result in a bias in data compiled on the distribution of rock types and fractures.

Nevertheless, the mapping of fractures from outcrops over an area encompassing 10's of square kilometers (km^2) provides data on regional fracture orientations [*Burton et al.*, 2000]. Comparing these results with more localized fracture mapping conducted on road cuts and in the boreholes of the FSE and CO well fields provides insight into the spatial persistence of fracture properties and evidence to support the extrapolation of detailed, local-scale fracture mapping to regional scales and larger volumes of rock in the Mirror Lake area.

Mapping fractures in the boreholes of the FSE and CO well fields and the I-93 road cuts revealed two dominant fracture trends. These are subhorizontal fractures and steeply northwest- or southeast-dipping fractures with northeast strikes. The steeply dipping fractures in the boreholes of the CO well field show a predominant northeast orientation of 44 degrees (°) (Figure 2). In comparison, the steeply dipping fractures mapped at the road cut, which is approximately 200 m east of the CO well field, show a predominant orientation of 39°, whereas the fractures mapped in the FSE well field (approximately 1 kilometer to the west) show a peak trend of 57° (Figure 2). The data sets from the road cuts and the CO and FSE well fields also display significant differences. For example, the presence of moderately southeast-dipping fractures in the road cuts are nearly absent in the CO wells, but present in the FSE wells. In addition, a significant percentage of fractures in the boreholes of the FSE well field have orientations of 75° and 96° that are reduced or absent in the data sets from the boreholes of the CO well field and the I-93 road cuts, even though the north-south-trending road cuts are at a favorable orientation to intercept such fractures. Subhorizontal fractures are prominent in all three data sets, but comprise a dominant percentage of the fractures in only the boreholes of the FSE well field. The absence of subvertical fractures in the boreholes of the CO and FSE well fields is likely due to the sampling bias associated with vertical boreholes.

The subhorizontal fractures mapped in boreholes and the road cuts are widespread in New England as sheeting joints, and likely represent the unloading from erosion of overlying rocks and, more recently, stress relief from melting of glacial ice [*Jahns*, 1943]. The density of the subhorizontal fractures mapped in the boreholes of the FSE and CO well fields generally decreases with depth [*Johnson and Dunstan*, 1998]. When considering all mapped fractures in boreholes, there is approximately one fracture every 2 m along boreholes in the Mirror Lake area [*Johnson and Dunstan*, 1998]; however, fracturing is not uniform, and the degree of fracturing depends on the rock type. Almost three-quarters of the mapped fractures occur in granite [*Johnson and Dunstan*, 1998]; however, the granite comprises only a small percentage of the total rock volume in the Mirror Lake area [*Burton et al.*, 2000]. Because granitic intrusions are not necessarily

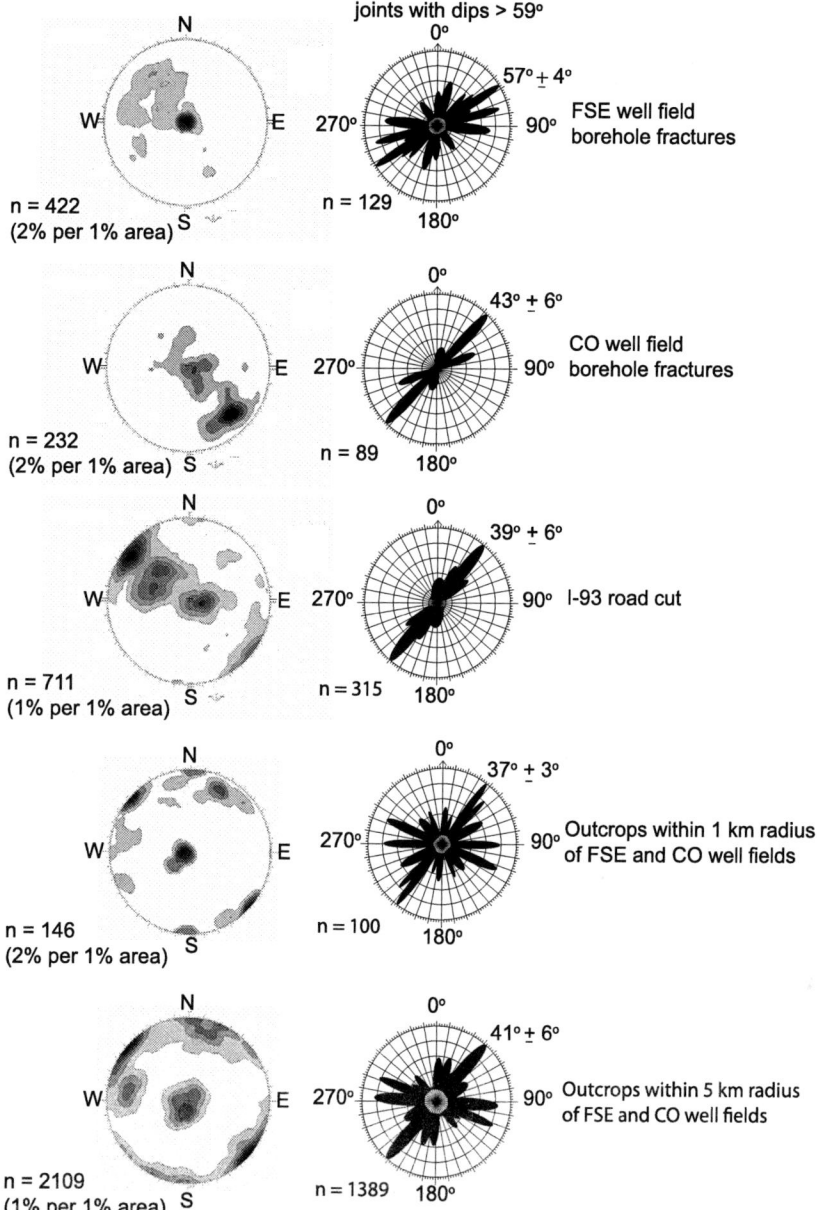

Figure 2. Stereonets and rose diagrams of fractures mapped in boreholes of the FSE and CO well fields, the I-93 road cuts east of Mirror Lake, and outcrops within radii of 1 and 5 kilometers of the FSE and CO well fields.

continuous in the subsurface, the high degree of fracturing in the granite does not necessarily translate to preferential ground-water flow. In general, there is little difference in the range of the transmissivity of fractures in the granite and schist [*Johnson*, 1998]. In addition, lithologic contacts exert little or no control on fracture development or orientation [*Johnson and Dunstan*, 1998; *Burton et al.*, 2000], suggesting a weak link between the bedrock geologic framework, fracture distribution, and hydraulic properties of volumes of the rock containing multiple fractures.

Mapping of fractures on the I-93 road cuts showed trace lengths of fractures that rarely exceed 10 m [*Barton*, 1996]. This condition also seems to be maintained in the boreholes of the FSE and CO well fields, as fractures intersecting one borehole cannot be correlated with fractures intersecting adjacent boreholes. In addition, the fractures mapped on the I-93 road cuts show a relatively low degree of interconnectivity in comparison to the mapping of fractures in other geologic settings [*Barton*, 1996]. A majority of the mapped fractures on the road cuts end without crossing or abutting other fractures.

An analysis of fractures on outcrops over radii of 1 and 5 kilometers (km), respectively, from the FSE and CO well fields shows two dominant fracture sets: subhorizontal sheeting fractures, and subvertical fractures at an orientation of approximately 40° (Figure 2). The similarity of orientations and dip angles of fractures mapped in the boreholes of the FSE and CO well fields, the I-93 road cuts, and outcrops over radii of 1 and 5 km from the FSE and CO well fields suggests a similarity between regional fracture properties and the detailed properties mapped over smaller dimensions.

Only two faults were observed in the Mirror Lake area. One fault was observed at the road cuts east of Mirror Lake and showed significant clay content. A second fault was inferred from geologic mapping and surface geophysical surveys conducted along Hubbard Brook, approximately 1 km west of Mirror Lake [*Powers et al.*, 1999]. In addition, from geologic mapping within the Hubbard Brook watershed, 16 brittle faults, most of which show normal displacements, and 27 Cretaceous-Jurassic mafic dikes were identified [*Burton et al.*, 2000]. The dikes and faults have preferred northeast orientations, similar to the fractures, and *Burton et al.* [2000] suggested that the extensional stress field that controlled dike orientation and faulting also produced the dominant brittle fabrics found regionally. The hydraulic significance of the faults is unclear from their mapped features. Ground-water flow modeling over the dimensions of the Mirror Lake watershed, which is discussed later in this article, was used to infer the hydraulic significance and connectivity of large-scale geologic features in the Mirror Lake area.

4. HYDRAULIC PROPERTIES OF FRACTURED ROCK FROM METERS TO KILOMETERS

Hydraulic properties of a volume of aquifer material can be estimated by performing a controlled hydraulic experiment, where a hydraulic perturbation is introduced by injecting or withdrawing a known volumetric rate of water (the ground-water flux), while simultaneously measuring the associated driving force (the hydraulic gradient). The coefficient of proportionality between the ground-water flux and the driving force is defined as the hydraulic conductivity or transmissivity, depending on the conceptualization of ground-water flow [*Bear*, 1979]. Other formation properties (for example, the specific storage and storativity) can also be estimated from hydraulic experiments [*Bear*, 1979]. In this article, however, only the transmissivity of fractures and the hydraulic conductivity of volumes of rock over increasingly larger physical dimensions will be examined from the interpretation of field experiments and ground-water modeling studies conducted at the Mirror Lake site. Estimates of the specific storage of the bedrock at the Mirror Lake site are discussed in *Hsieh and Shapiro* [1996] and *Hsieh et al.* [1999].

Estimating hydraulic properties from measurements of the ground-water flux and hydraulic gradient can be conducted over increasingly larger physical dimensions. Estimates of hydraulic properties of individual fractures over dimensions of meters can be made by isolating individual fractures in a borehole and inducing a hydraulic perturbation while measuring the associated fluid pressure response. Volumes of rock that include multiple fractures over 10's of meters can also be interrogated by inducing a hydraulic perturbation that influences a much larger volume of the formation. Aquifer tests designed for large volumes of rock, however, need to account for the complex spatially heterogeneous hydraulic properties of fractures and their complex connectivity. Over dimensions of 100's of meters to kilometers, inducing controlled hydraulic perturbations is not feasible, and instead, ambient hydraulic conditions are used to estimate the bulk hydraulic properties of the formation. Over dimensions of 100's of meters to kilometers, the measured hydraulic head and the associated hydraulic gradients are used along with measured or estimated ground water fluxes, such as ground-water discharges into surface-water drainages and ground-water recharge. These approaches to estimating hydraulic properties over increasingly larger dimensions were applied at the Mirror Lake site to investigate the effect of the scale of measurement on the magnitude of hydraulic properties of the rock.

4.1 Single-Hole Tests of Individual Fractures and Closely Spaced Fractures

The transmissivity of an individual fracture or several closely spaced fractures in a borehole was estimated by conducting a single-hole test that hydraulically isolated a discrete interval of a borehole and then either injecting or withdrawing water, while simultaneously measuring the volumetric flow rate and the fluid pressure response in the test interval. The apparatus used to conduct the tests is described in *Shapiro and Hsieh* [1998] and *Shapiro* [2004]. The apparatus consisted of two inflatable packers separated by lengths that ranged from 3 to 5 m. The distance between packers was a function of the fracture locations along the borehole wall. The flexible bladder of each packer is approximately 1-m long, and the position of the packers in the borehole was chosen to coincide with smooth sections of the borehole wall so as to avoid having the packer bladders overlay fractures. Borehole geophysical logging tools that provided information on the location of fractures over the length of the borehole and the ruggedness of the borehole wall were instrumental in identifying locations where hydraulic tests were conducted [*Paillet*, 1996].

The single-hole tests in hydraulically isolated intervals of boreholes were conducted over a short duration (10's of minutes) and are assumed to be representative of the fracture's properties over a physical dimension equivalent to meters in the vicinity of the borehole. Over such dimensions, the hydraulic properties of the fracture are assumed to be homogeneous and interpretations of the hydraulic responses based on assumptions of homogeneity and radially converging or diverging flow were used to estimate the transmissivity of the fractures between the packers in the test interval. Additional information on the methods of conducting the single-hole tests and the assumptions implicit in the interpretations of such tests are given in *Shapiro and Hsieh* [1998].

All bedrock boreholes at the Mirror Lake site were tested by conducting single-hole hydraulic tests where discrete intervals of the boreholes were isolated. The variation of the transmissivity with depth in one borehole from the Mirror Lake area is shown in Figure 3 along with the interpretation

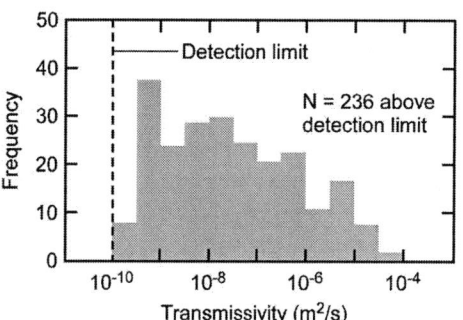

Figure 4. Frequency histogram of transmissivity from discrete-interval tests conducted in bedrock boreholes in the Mirror Lake area.

of the acoustic televiewer log showing fractures on an opened and oriented view of the borehole wall. Fractures over the length of the borehole are not spatially uniform, and multiple sections of the borehole have closely spaced fractures. The intensely fractured sections of the borehole, however, are not necessarily associated with high transmissivity. There are two intensely fractured sections of the borehole shown in Figure 3, at elevations of approximately 150 and 175 m above mean sea level (msl). The transmissivity of the section of the borehole at 150 m above msl is orders of magnitude less than the transmissivity at 175 m above msl. In addition, the transmissivity does not smoothly vary as a function of depth in the borehole; test intervals adjacent to each other can have transmissivities that differ by several orders of magnitude. Similar conditions to those shown in Figure 3 are observed in most of the boreholes in the Mirror Lake area.

A frequency histogram of the transmissivities estimated from the single-hole tests conducted in boreholes in the Mirror Lake area is shown in Figure 4. The transmissivity of the test intervals containing individual fractures or closely spaced fractures ranged from 10^{-10} to 10^{-4} square meters per second (m^2/s). The lower limit of this range is associated with the detection limit of the *in situ* equipment used to conduct these tests. Approximately one-third of the single-hole tests were below the detection limit of the *in situ* equipment. The intervals with transmissivity below the detection limit contained fractures, and thus, the range of the transmissivity of fractures at the Mirror Lake site exceeds the six orders of magnitude shown in frequency histogram of Figure 4.

Selected sections of boreholes that contained no fractures were also tested with the *in situ* equipment. The measured flow rate in these tests was below the detection limit of the apparatus, implying that the transmissivity of the intact rock is also less than 10^{-10} m^2/s. Core samples of unfractured sections of granite and schist in the Mirror Lake area were not tested under laboratory conditions to estimate their transmissivity; however, the permeability of intact crystalline rocks

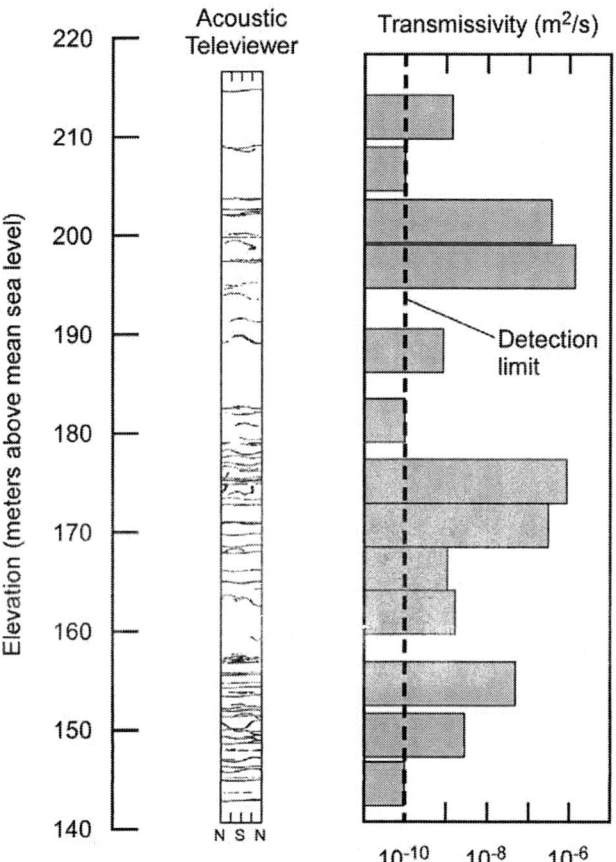

Figure 3. Fractures as interpreted from an acoustic televiewer log and transmissivity as a function of depth in borehole H1 of the Mirror Lake watershed.

from other investigations indicates that the transmissivity of 5-m intervals of unfractured bedrock boreholes could range from 10^{-17} to 10^{-13} m^2/s, depending on the *in situ* stress [*Trimmer et al.*, 1980; *Skoczylas and Henry*, 1995; *Selvadurai et al.*, 2005]. The hydraulic properties of the intact rock would represent the lower limit of the hydraulic properties of the rock.

In this discussion, the transmissivity estimated from single-hole tests is used rather than the hydraulic conductivity. The transmissivity is representative of a two-dimensional flow regime, and represents the ease of fluid movement in the rock over the thickness of the test interval, which includes one or more discrete fractures [*Shapiro and Hsieh*, 1998]. Transmissivity can be converted to hydraulic conductivity by dividing by the thickness of the test interval [*Bear*, 1979]; however, the resulting hydraulic conductivity could be misinterpreted, because the transmissive fractures in the test interval are not uniformly distributed. Increasing or decreasing the length of the test interval without including or excluding transmissive fractures would yield the same transmissivity; however, the hydraulic conductivity would vary as a function of the length of the test interval. In subsequent discussions, the transmissivity of discrete test intervals from single-hole tests is converted to hydraulic conductivity for purposes of comparison with values of hydraulic conductivity estimated from ground-water modeling investigations over larger volumes of rock. In ground-water modeling investigations, hydraulic properties of the rock can vary spatially, but are assumed to be uniform over discrete volume elements of the rock. In this comparison of hydraulic properties over increasingly larger volumes of rock, only order-of-magnitude estimates of the hydraulic conductivity are considered, and thus, variations in the length of the test intervals (between 3 and 5 m) for the single-hole tests do not have an impact on the magnitude of the hydraulic conductivity.

4.2 Cross-Borehole Tests Interrogating 10's of Meters of Rock

Controlled hydraulic tests that are designed to estimate the hydraulic properties of large volumes of rock are usually conducted by withdrawing water from a single borehole over an extended period of time while monitoring the fluid pressure responses in adjacent boreholes. Because fractures are the primary conduits of fluid movement, the design of these cross-borehole tests must account for the fact that fractures of different transmissivity intersect boreholes at different elevations. Boreholes that intersect multiple fractures can act as connections of high hydraulic conductivity between those fractures. To identify the hydraulic properties of the fractured rock in its ambient state over dimensions of 10's of meters, packers are needed to isolate discrete intervals in boreholes in the vicinity where the hydraulic perturbation is introduced.

Cross-borehole hydraulic tests in the FSE well field were conducted by installing packers in boreholes to isolate the most transmissive fractures as identified from the single-hole tests. Figure 5 shows a schematic perspective view of the boreholes in the FSE well field and the location of the highest transmissivity intervals in each borehole as determined from the single-hole tests. The intervals shown in Figure 5 had transmissivity that was greater than 10^{-5} m^2/s. In general, there are usually up to three high transmissivity intervals that intersect each borehole. The connectivity, or lack of connectivity, of the most transmissive fractures will be significant in identifying the hydraulic properties of the rock over 10's of meters.

Figure 6 shows a cross-section through the FSE well field with the location of the packers in the boreholes used in cross-borehole tests; packers were also placed in other boreholes in the FSE well field to isolate high transmissivity intervals. In the test described here, a submersible pump was placed between two packers in the FSE6 borehole and pumped at approximately 10 liters per minute (L/min). Hydraulic responses were monitored in the isolated intervals in each borehole in the FSE well field. The name of each interval is shown in Figure 6 along with the drawdown records for the intervals as a result of withdrawing water from FSE6B.

The drawdown is the time-varying change in the hydraulic head from the ambient hydraulic head at the start of the hydraulic test. The largest drawdown is measured in the pumped interval (FSE6B). Other intervals show less drawdown than the pumped interval, but the spatial distribution of drawdown is not consistent with a homogeneous aquifer. In

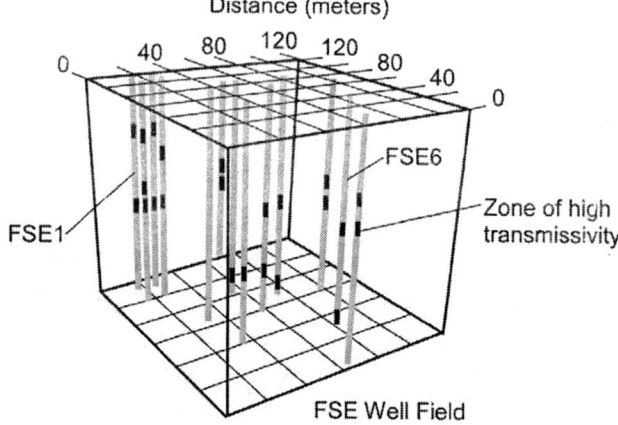

Figure 5. Schematic perspective view of boreholes in the FSE well field and the intervals of highest transmissivity in each borehole as determined from discrete-interval hydraulic tests.

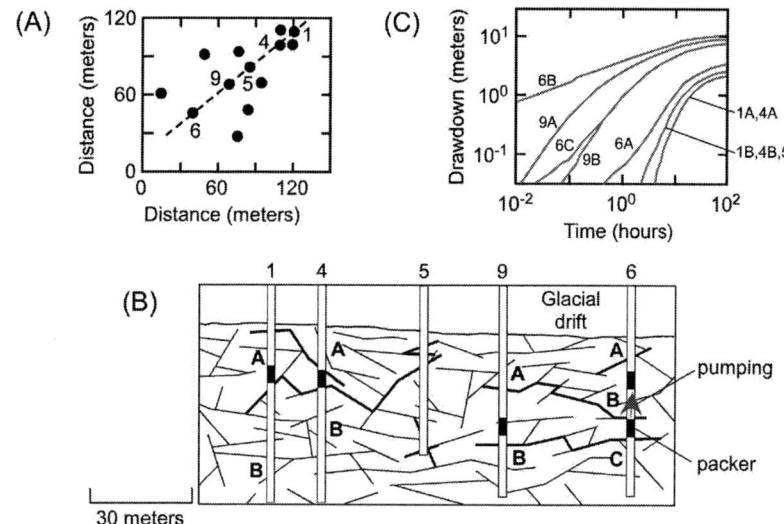

Figure 6. (A) Plan view of FSE well field, (B) a schematic cross-section through the FSE well field showing the location of packers that hydraulically isolated discrete intervals of bedrock boreholes during a cross-borehole hydraulic test, and (C) measured drawdown in hydraulically isolated intervals of bedrock boreholes as a function of time due to pumping in interval 6B.

a homogeneous aquifer, the drawdown should diminish with distance from the pumped interval. The results in Figure 6 show that several intervals have the same drawdown regardless of their distance to the pumped interval.

It is hypothesized that the drawdown responses shown in Figure 6 are dictated by the spatial connectivity of the high transmissivity fractures that intersect each borehole [*Hsieh and Shapiro*, 1996; *Hsieh et al.*, 1999]. For example, it is hypothesized that the high transmissivity fractures intersecting the borehole interval FSE6B also intersects interval FSE9A (Figure 6), resulting in the similarity in the drawdown responses in these two intervals. Furthermore, the borehole interval FSE6C is hypothesized as being hydraulically connected to FSE9B (Figure 6). The drawdown in these intervals is nearly identical because the fluid pressure response migrates rapidly through highly transmissive fractures that are interconnected. Borehole intervals FSE5, FSE4B and FSE1B all have the same measured drawdown, as do borehole intervals FSE4A and FSE1A, respectively (Figure 6).

From the drawdown records shown in Figure 6, a series of highly transmissive, subhorizontal zones in the rock are hypothesized (Figure 7). The orientation of the fractures intersecting the boreholes in the FSE well field, as determined from borehole geophysical logging, indicates that fractures intersecting one borehole do not extend to adjacent boreholes. In addition, there are few fractures measured on road cuts and outcrops in the Mirror Lake area that have trace lengths greater than 10 m [*Barton*, 1996]. Thus, the highly transmissive zones are hypothesized as being composed of multiple intersecting fractures, which may include both the subhorizontal and moderate-to-steeply dipping fractures mapped in the boreholes (Figure 2). Furthermore, the highly transmissive zones that are schematically shown on Figure 7 are assumed to be connected only through a network of fractures in the rock that are less transmissive than the highest transmissivity fractures. The less transmissive fractures impede ground-water flow, which yields the different drawdowns measured in each of the highly transmissive intervals (Figure 6).

Hsieh and Shapiro [1996] and *Hsieh et al.* [1999] performed numerical simulations of ground-water flow in the FSE well field associated with the hydraulic test described above. From those simulations, the hydraulic conductivity of the highly transmissive intervals of the rock was estimated to be 2×10^{-4} meters per second (m/s) and the hydraulic conductivity of the less transmissive fractures was estimated to be 4×10^{-7} m/s. These simulations assumed a deterministic shape of the highly transmissive zones in the rock. *Day-Lewis et al.* [2000] considered the same hydraulic test data in an interpretation that used simulated annealing in generating realizations of the hydraulic conductivity that reproduced the measured hydraulic connections. In addition, *Tiedeman and Hsieh* [2001] conducted a hydraulic test in the FSE well field without packers in the boreholes as a means of comparing the results of hydraulic tests conducted in fractured rock aquifers with and without boreholes that connect highly transmissive fractures.

From the conceptual model of the fractured rock that is shown in Figure 7, a bulk hydraulic conductivity of a volume of the rock that encompasses both highly transmissive fracture intervals and the network of less transmissive fractures is estimated to be 4 x 10^{-7} m/s. The bulk hydraulic conductivity is the coefficient of proportionality between the hydraulic gradient and the volumetric flow rate through a volume of the rock. The bulk hydraulic conductivity of volumes of the rock over dimensions of approximately 100 m is controlled by the less transmissive fractures, because highly transmissive intervals are not interconnected. The lack of connectivity of these zones of high transmissivity is likely a manifestation of the poor connectivity of fractures that was noted in mapping fractures on the I-93 road cuts.

4.3 Ambient Ground-Water Flow Over 100's of Meters to Kilometers

Over dimensions of 100's of meters to kilometers, hydraulic tests conducted by imposing a hydraulic perturbation cannot be conducted at the Mirror Lake site. Instead, estimates of hydraulic properties of the fractured rock over these distances must be determined from ambient hydraulic conditions. A three-dimensional numerical simulation of ground-water flow was conducted over an area that extended beyond the surface-water drainage associated with the Mirror Lake watershed [*Tiedeman et al.*, 1997, 1998]. The numerical simulation used the hydraulic head measured at different depths in bedrock wells and the measured ground-water discharges to streams as a means of estimating the bulk hydraulic conductivity of the bedrock, the ground-water recharge from precipitation, and other hydraulic properties of the ground-water flow system.

The three-dimensional numerical model considered ground-water flow through both the unconsolidated glacial deposits and the fractured bedrock. Over an area of approximately 16 km^2 and 200 m deep in the rock, individual fractures or highly transmissive intervals of interconnected fractures were not specified. Over these dimensions it is hypothesized that the presence or absence of a single fracture or a highly transmissive zone of limited areal extent will not affect the distribution of the hydraulic head in the bedrock or the ground-water discharges to the streams.

Tiedeman et al. [1997] considered different conceptual models of the distribution of hydraulic conductivity in the glacial deposits and bedrock to ascertain which model best reproduced the measured data. The conceptual model that best reproduced the data considered a single bulk hydraulic conductivity of the bedrock. The ground-water flow model reproduced the hydraulic head and stream flow measurements with relatively narrow confidence limits and with unbiased errors. The bulk hydraulic conductivity of the bedrock was estimated to be 3 x 10^{-7} m/s, which is similar in magnitude to the network of less transmissive fractures that connected the highly transmissive, subhorizontal intervals that were identified from cross-borehole testing in the FSE well field. This implies that there are no large-scale, highly transmissive geologic features that need to be considered in estimating the bulk hydraulic properties of the formation over dimensions of 100's of meters to kilometers.

4.4 Hydraulic Properties of Fractured Rock From Meters to Kilometers

The sections above described the methods of estimating the hydraulic conductivity over increasingly larger volumes of fractured rock at the Mirror Lake site. For the purpose of comparison the transmissivity estimated from the single-hole hydraulic tests were converted to hydraulic conductivity by dividing by the length of the test interval, which ranged from 3 to 5 m. A comparison of the estimates of hydraulic conductivity from the single-hole tests, the controlled hydraulic tests conducted in the FSE well field, and the calibration of a ground-water flow over the Mirror Lake watershed is shown in Figure 8 [*Hsieh*, 1998]. In this figure, the logarithm of the estimated hydraulic conductivity is shown for tests conducted over increasingly larger physical dimensions.

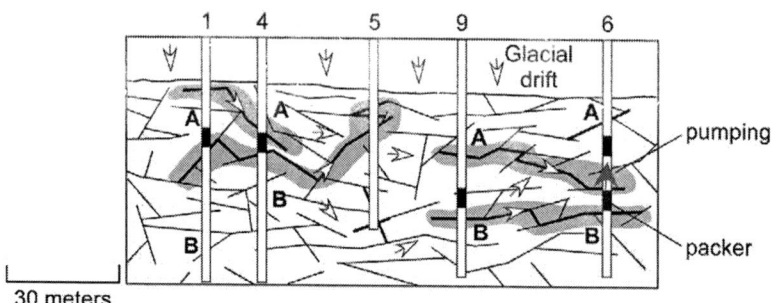

Figure 7. Hypothesized distribution of fractures having high transmissivity (depicted by zones of gray) in the rock underlying the FSE well field; the intervals of high-transmissivity fractures are connected only by less transmissive fractures (modified from *Hsieh and Shapiro*, 1996)

The single-hole tests comprise the largest data set of hydraulic conductivity measurements, as over 200 single-hole tests were conducted in boreholes at the Mirror Lake site. This information is presented by points representing individual tests in Figure 8. The physical dimension associated with the single-hole tests is assumed to be between 1 and 5 m, and there is no significance to the separation of points over these dimensions. The points are intended to show the range of hydraulic conductivity of individual fractures or closely spaced fractures in boreholes, from 10^{-10} to 10^{-4} m/s. The lower limit of this range is associated with the detection limit of the *in situ* testing equipment. The testing indicates that there are fractures in the rock with hydraulic conductivity below the detection limit. The hydraulic conductivity of the intact (unfractured) rock is likely to be several orders of magnitude below the detection limit [*Trimmer et al.*, 1980].

The estimates of the hydraulic conductivity from controlled cross-borehole hydraulic tests conducted in the FSE well field are shown in Figure 8 for physical dimensions that range from 20 to 100 m. There are two estimates of the hydraulic conductivity associated with a dimension of approximately 20 m. These are the two distinct features of the fractured rock that were identified in the cross-borehole hydraulic tests, that is, the highly transmissive, subhorizontal intervals of fractures, and the background network of less transmissive fractures. The magnitude of the hydraulic conductivity of the highly transmissive intervals coincides with the highest hydraulic conductivity estimated from the single-hole tests. The hydraulic conductivity of the background network of less transmissive fractures is approximately equal to the geometric mean of the estimates of the hydraulic conductivity from the single-hole tests.

Over a volume of the rock that encompasses both highly transmissive intervals of fractures and the less conductive background network of fractures (over dimensions of approximately 100 m), the bulk hydraulic conductivity of the fractured rock is equivalent to the hydraulic conductivity of the background network of fractures. The highly transmissive intervals of fractures do not appear to be connected over distances greater than approximately 20 m in the rock.

The hydraulic conductivity of the fractured rock from a calibrated ground-water flow model over the dimensions of the Mirror Lake watershed is of the same order of magnitude as the background network of less transmissive fractures identified from the hydraulic testing in the FSE well field over 10's of meters (Figure 8). This implies that over physical dimensions of 100's of meters to kilometers a network of interconnected highly transmissive fractures that controls ground-water flow is not present. The model results indicate that ground-water flow over 100's of meters to kilometers is again controlled by a background network of less transmissive fractures.

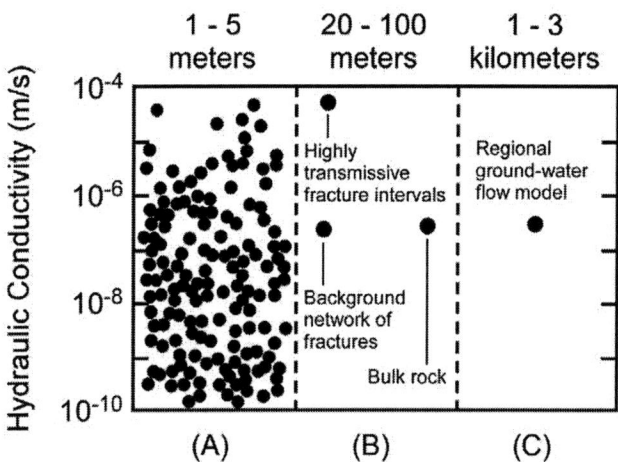

Figure 8. Hydraulic conductivity in the fractured bedrock of the Mirror Lake watershed and its vicinity as estimated over increasingly larger physical dimensions from (A) discrete-interval, single-hole hydraulic tests, (B) cross-hole hydraulic tests, and (C) regional ground-water flow modeling (modified from *Hsieh*, 1998).

Figure 8 indicates that there is no increase in the hydraulic conductivity over larger volumes of rock at the Mirror Lake site. Other authors have speculated that the hydraulic conductivity of heterogeneous aquifers over increasingly larger volumes of the subsurface material may increase as the physical dimensions of the measurement increase [*Clauser*, 1992]. Over increasingly larger volumes of aquifer material it is speculated that large scale heterogeneities will be included in the estimation of aquifer properties. This argument implicitly assumes that the hydraulic properties of the large scale features have not been interrogated over smaller test dimensions, and implies that insufficient sampling at small dimensions is responsible for the failure to detect large scale features, such as major faults or fracture zones, that could control the magnitude of the hydraulic conductivity over regional dimensions.

At the Mirror Lake site, a large number of hydraulic conductivity measurements have been made at small test dimensions (meters), and the results from these tests have likely shown the full range of the hydraulic conductivity of the fractures in the rock. The estimates of the bulk hydraulic conductivity of the fractured rock at the Mirror Lake site over increasingly larger dimensions indicates that the connectivity of the fractures is important in characterizing the bulk hydraulic properties of the rock. If the highest conductivity fractures are not connected over any significant distance, then the hydraulic conductivity of the fractures that connect the more transmissive fractures are the ones that will control the bulk hydraulic properties of the rock at large dimensions. At other fractured rock sites with different geologic controls on fracture properties, the connectivity of fractures may yield different trends in the estimates of the hydraulic connectivity over increasingly larger physical dimensions [*Hsieh*, 1998].

5. CHEMICAL MIGRATION IN FRACTURED ROCK

Because fractures are the features that control the majority of fluid movement in the fractured rock aquifers, such as those at the Mirror Lake site, fractures will also be responsible for the majority of the chemical migration in the subsurface. The complex connectivity of fractures coupled with the large range in the hydraulic properties of fractures can give rise to highly convoluted flow paths, which in turn, will yield complex spatial distributions of chemical constituents in the rock.

Other complexities associated with chemical transport in fractured rock arise because of the complex topology of individual fractures. The asperities on fracture surfaces and points of contact between fracture surfaces can give rise to a complex flow regime within an individual fracture. Regions of an individual fracture could be subject to relatively rapid fluid movement, whereas other areas of the same fracture surface could be subject to relatively slow advection, or stagnant water. *Neretnieks et al.* [1982] referred to this phenomenon as "channeling" within a fracture surface. Chemical constituents will move preferentially through portions of the fracture with the highest volumetric flux; however, diffusion will also occur due to chemical gradients. Chemical constituents could then have long residence times in regions of a fracture subject to slow advection or stagnant water.

There are very few physical examples illustrating the complexities of chemical migration in fractured rock, because at most sites where there are chemical plumes, there are only a sparse number of boreholes at which to conduct chemical sampling. Plate 1 shows a natural analog that illustrates the complexity of chemical transport in fractured rock. Plate 1 shows a photograph of one of the faces of the I-93 road cut east of Mirror Lake, where the distribution of rock types and fractures were mapped [*Barton*, 1996]. In addition, the presence or absence of iron-hydroxide staining on the rock adjacent to fracture surfaces was also noted. The iron-hydroxide staining is hypothesized to be an artifact of the migration of oxygenated water, presumably infiltrating from the surface. As oxygenated water moves through fractures, oxygen diffuses into the primary porosity of the rock, also referred to as the rock matrix. The rock types in the vicinity of Mirror Lake are rich in iron bearing minerals, and thus, it is anticipated that iron will be in solution in the primary porosity of the rock. The interaction between iron and oxygen results in an iron-hydroxide precipitate [*Wood et al.*, 1996]. Not all the fractures on the road cut have an iron-hydroxide staining (Plate 1). At some locations, only a portion of a fracture has the iron-hydroxide staining, indicating that oxygenated water did not migrate over the entire fracture surface. Plate 1 shows that even in areas of interconnected fractures there will not be uniform migration of chemical constituents.

5.1 Diffusion in Fractured Rock

The primary (or matrix) porosity of the rock also plays a role in the migration of chemical constituents in fractured rock aquifers. As a chemical constituent migrates through fractures, a chemical gradient will exist between the fluid in the fractures and the fluid in the rock matrix in contact with the fracture. In rock types such as those at the Mirror Lake site, the matrix porosity can range from less than 1 to more than 3% [*Wood et al.*, 1996]. *Ohlsson and Neretnieks* [1995] noted a similar range for the matrix porosity of igneous rocks from other field sites.

A matrix porosity of 3% may not seem like a huge volume of fluid; however, the void volume associated with fractures in crystalline rock sites is also on the order of 1 to 3%. These estimates of fracture porosity are made from mechanical considerations, where apertures of fractures are assigned and the volume associated with all fractures is summed. A void volume of the matrix porosity of 3% represents an extensive fluid reservoir into which chemicals can diffuse. Because chemical migration into and out of the matrix porosity is controlled by diffusion and the surface area of fractures, chemical constituents that diffuse into the matrix porosity are likely to have extremely long residence times in the formation.

The diffusion coefficient for a nonsorbing constituent in the matrix porosity is described by the equation,

$$D = n_m \gamma D_w = n_m D_m \quad (1)$$

where D is the coefficient of diffusion associated with the matrix porosity of the rock, n_m is the matrix porosity of the rock, γ is the formation factor that is inversely related to the turtuosity of the matrix porosity, D_w is the free-water diffusion coefficient for the constituent under consideration, and D_m is the effective diffusion coefficient in the matrix porosity for the constituent under consideration. In general, D_m will be less than D_w because the tortuosity of the matrix porosity reduces the capacity for diffusion ($\gamma < 1$). The diffusion coefficient, D, can be evaluated from laboratory experiments conducted on cores [*van der Kamp et al.*, 1996; *Novakowski and van der Kamp*, 1996].

Figure 9 shows the hypothetical results for the case of one-dimensional chemical transport in a fracture by advection

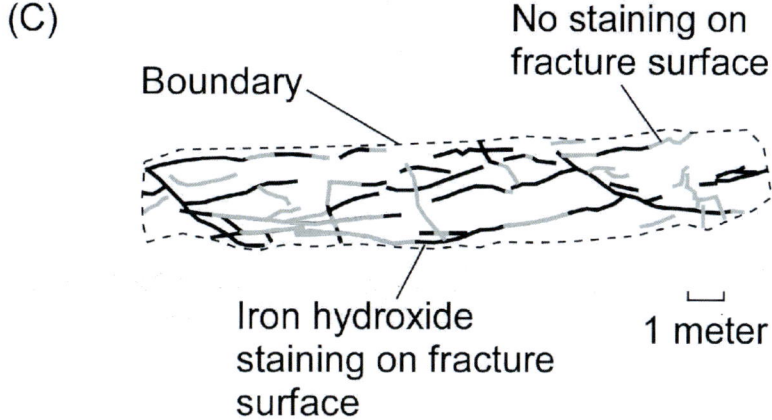

Plate 1. (A) Photograph (courtesy of Christopher C. Barton) of a face of the road cut adjacent to I-93 east of Mirror Lake, (B) a photograph (courtesy of Warren W. Wood) of iron-hydroxide staining in the rock adjacent to a fracture, and (C) the mapping of fractures with iron-hydroxide staining on a apportion of the I-93 road cut (modified from *Barton*, 1996).

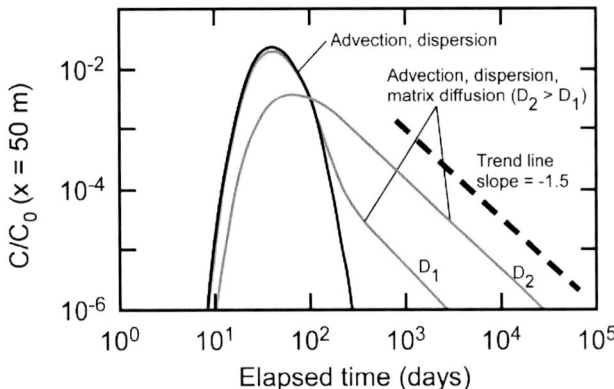

Figure 9. Hypothetical breakthrough curves at 50 m downgradient from a point of a pulse injection into a formation with fractures and rock matrix, where the average fluid velocity is 1 meter per day.

and dispersion with diffusion into the adjacent rock matrix. The time-varying concentration of the tracer at a location 50 m downgradient from the point of a pulse injection is shown in Figure 9, where the logarithm of the concentration is plotted as a function of the logarithm of the elapsed time from the start of the tracer injection. Also shown in Figure 9 is the breakthrough curve 50 m downgradient from the injection point for the case where matrix diffusion is absent, that is, only advection and dispersion are controlling the chemical migration in the fracture.

For log-log plots of concentration versus time, the breakthrough curve for advection and dispersion shows a parabolic shape. When diffusion into the primary porosity is considered, the peak concentration of the breakthrough curve is reduced because of the loss of mass into the matrix porosity. After the peak of the tracer concentration has migrated downgradient in the fracture, concentration gradients are conducive for the tracer to migrate out of the rock matrix and into the fracture, and then be advected and dispersed as it migrates downgradient in the fracture. This phenomena yields elongated tails on breakthrough curves (Figure 9). For larger values of the free-water diffusion coefficient, more mass is diffused into the primary porosity, resulting in a lower peak concentration and an extended breakthrough curve. For diffusion from planar fractures into a porous matrix, the declining limb of the breakthrough curve on log-log plots of concentration versus time is a straight line with a slope of -1.5, regardless of the magnitude of the free-water diffusion coefficient.

The magnitude of chemical diffusion in fractured rock is important in the design of geologic settings for waste isolation, the remediation of contaminated ground water, and the interpretation of the chemical evolution and isotopic signatures of ground water in fractured rock aquifers [*Birgersson and Neretnieks*, 1990; *Maloszewski and Zuber*, 1991; *Parker et al.*, 1994; *Shapiro*, 2001]. A question that naturally arises is whether the results from experiments conducted on cores can be applied in physical settings over larger dimensions. Over dimensions of meters, 10's of meters, and kilometers, diffusion into or out of the matrix porosity is undoubtedly ongoing, but the estimates of the diffusion coefficient obtained from laboratory experiments may not account for all diffusive-type processes that manifest themselves in complex networks of fractures in communication with the rock matrix.

Garnier et al. [1985] conducted controlled tracer experiments in a fracture in a chalk aquifer using wells separated by approximately 10 m. The experiment consisted of using a suite of tracers with different free-water diffusion coefficients. The tracer with the largest free-water diffusion coefficient showed the most pronounced effect of diffusion into the matrix porosity of the chalk (Figure 10). The results of tests conducted by *Garnier et al.* [1985] are plotted as the logarithm of the tracer concentration versus the logarithm of the elapsed time of the test. *Moench* [1995] showed that the magnitude of the free-water diffusion coefficients for the tracers can describe the separation of the breakthrough curves from the tests conducted by *Garnier et al.* [1985], similar to the hypothetical breakthrough curves shown in Figure 9. In addition, the declining limbs of the breakthrough curves on log-log plots of concentration versus time were approximately straight lines with slopes of -1.5 (Figure 10).

The tests conducted by *Garnier et al.* [1985] indicate that laboratory estimates of the matrix diffusion can be applied under field conditions. *Liu et al.* [2004] compiled the results

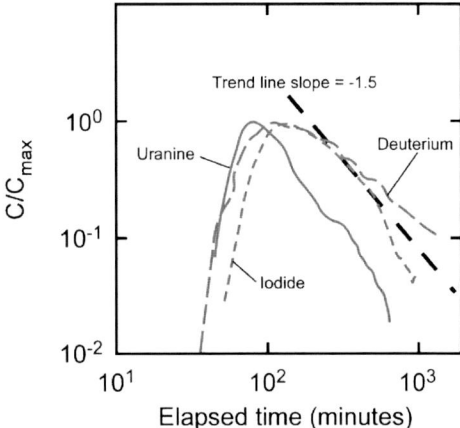

Figure 10. Breakthrough curves from the controlled tracer tests conducted by *Garnier et al.* [1985] (modified from *Moench*, 1995); the free-water diffusion coefficients for uranine, iodide, and deuterium are 1.4×10^{-2}, 5.0×10^{-2}, and 7.2×10^{-2} m^2/yr, respectively.

of tracer experiments conducted at different field sites over increasingly larger physical dimensions and showed a general increase in the effective diffusion coefficient with the physical scale of the experiment. In the following sections, the results from laboratory and field-scale tracer experiments at the Mirror Lake site are interpreted to estimate the effective diffusion coefficient, D_m, over increasingly larger physical dimensions.

5.2 Diffusion Experiments in Rock Cores

Wood et al. [1996] conducted diffusion experiments in samples of granite and schist collected from the Mirror Lake site. Cesium-137 (^{137}Cs) was used as the tracer in these experiments. ^{137}Cs is highly retarded by sorbing onto grain boundaries in the primary porosity of the rock. After accounting for the retardation of ^{137}Cs, the effective diffusion coefficient, D_m, was similar in magnitude to results of laboratory tracer experiments in similar types of rock reported in *Ohlsson and Neretnieks* [1995]. In general, D_m varies from approximately 10^{-4} to 10^{-3} square meters per year (m^2/yr).

5.3 In Situ Tracer Tests Over 10's of Meters

In situ tracer experiments were conducted in the fractured rock at the Mirror Lake site between boreholes in the FSE well field (Figure 1). In this section, the results of a series of tracer tests conducted between boreholes FSE9 and FSE6 are described. The tracer tests were conducted in one of the highly transmissive fracture intervals that intersect both boreholes, which are separated by approximately 35 m (Figure 7). The tests consisted of a converging tracer

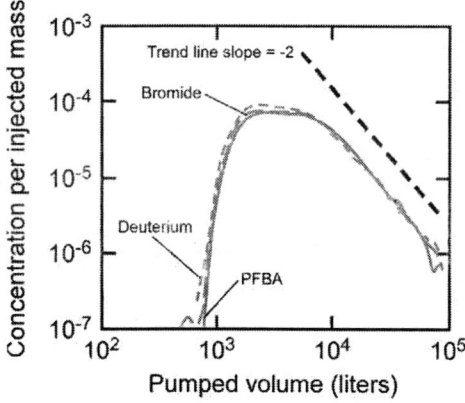

Figure 11. Breakthrough curves for multiple tracers at a pumped well (FSE6) from a pulse injection into FSE9 (modified from *Becker and Shapiro*, 2000); the free-water diffusion coefficients for pentafluorobenzoic acid (PFBA), bromide, and deuterium are 2.1 x 10^{-2}, 6.3 x 10^{-2}, and 7.2 x 10^{-2} m^2/yr, respectively.

experiment conducted by pumping continuously from a submersible pump placed between packers in FSE6 [*Becker and Shapiro*, 2000]. The tracer solutions were injected between packers in a hydraulically isolated interval in FSE9. The apparatus used to perform the injection of the tracer solution was designed to conduct a pulse injection of the tracer solution that strictly controlled the duration of the injection and the volume of the tracer solution injected into the fractures [*Shapiro and Hsieh*, 1996b]. Water samples were collected from the pumped water at FSE6 and analyzed for the concentration of the tracers. Pressure transducers were used to monitor the hydraulic responses in the wells of the FSE well field during the tracer test and the pumping rate in FSE6 was monitored using a flow meter.

Similar to the tests conducted by *Garnier et al.* [1985], the injection solution for the tests conducted in the FSE well field at the Mirror Lake site consisted of a suite of dissolved constituents with different free-water diffusion coefficients [*Becker and Shapiro*, 2000]. The results of one of the tracer experiments is shown in Figure 11, where the concentration of the tracers per mass of the tracer injected is plotted as a function of the pumped volume from FSE6. The pumped volume is a surrogate for time; dividing the pumped volume by the pumping rate is equivalent to time.

Unlike the experiments conducted by *Garnier et al.* [1985], the different tracers did not show a separation during the declining limb of their breakthrough curves. The three tracers had overlapping breakthrough curves when the concentration of each tracer was normalized with the mass of the tracer injected (Figure 11). In addition, the declining limbs of the breakthrough curves exhibited a straight-line behavior on log-log plots of concentration versus the pumped volume (Figure 11). The straight line on the declining limb of a breakthrough curve is usually indicative of chemical diffusion; however, the slope of the declining limb of the breakthrough curve on the log-log plot was –2, rather than the theoretical result of –1.5, as in the results of *Garnier et al.* [1985].

Additional tracer tests were also conducted between the same pair of wells in the FSE well field using a similar configuration to that used in the tracer test discussed above. These tracer tests were conducted using different pumping rates to alter the residence time of the tracer in the fractured rock [*Becker and Shapiro*, 2000]. If chemical diffusion into the primary porosity is responsible for the declining limbs of the breakthrough curves shown in Figure 11, then a longer residence time in the formation will result in an enhanced degree of diffusion. Figure 12 shows the results of 5 tracer tests conducted using bromide as the tracer. In this figure the concentration of bromide per mass of bromide injected is plotted as function of the pumped volume from FSE6; plot-

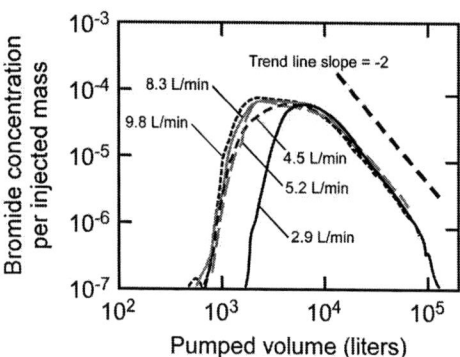

Figure 12. Breakthrough curves for bromide at a pumped well (FSE6) from the pulse injection into FSE9 conducted under different pumping rates (modified from *Becker and Shapiro*, 2000).

ting the breakthrough curves as a function of the pumped volume allows all of the tests to be superimposed. The pumping rates for these tests ranged from 2.9 to 9.8 L/min, which resulted in tests that lasted between 5 and 40 days for the tracer recovery.

The declining limb of the breakthrough curves from all of the tests shown in Figure 12 superimpose on one another, which indicates that diffusion into the rock matrix is not responsible for the character of the declining limbs of the breakthrough curves. In addition, the declining limbs of the breakthrough curves from these tests are again straight lines with a slope of –2, which is similar to the results shown in Figure 11 for the tracer test conducted with different tracers having different free-water diffusion coefficients. Furthermore, if the coefficients of advection, dispersion, and the effective matrix diffusion, D_m, are estimated from the breakthrough curves shown in Figures 11 and 12, the magnitude of D_m must be at least equal to or greater than the free-water diffusion coefficients associated with tracers used [*Becker and Shapiro*, 2000]. An effective diffusion coefficient that is equal to or greater than the free-water diffusion coefficient implies that processes other than chemical diffusion into the rock matrix must be responsible for the characteristics of the declining limbs of the breakthrough curves in these *in situ* tracer tests.

Becker and Shapiro [2000, 2003] and *Shapiro* [2001] hypothesized that diffusion into the rock matrix is ongoing in these tests; however, it is being dwarfed by the magnitude of a process that manifests itself similarly to diffusion over the dimensions of these tests. It is hypothesized that the declining limb of the breakthrough curves from these tests is an artifact of fluid advection ranging over several orders of magnitude. Usually, variability in the fluid velocity is attributed to mechanical dispersion; however, in fractured rock the hydraulic properties of fractures can vary over many orders of magnitude. The results from the single-hole hydraulic tests conducted on individual fractures show transmissivity to range over six orders of magnitude above the detection limit of the *in situ* testing equipment. With such a large range in the transmissivity of fractures, there is the potential for slow advection in the least transmissive fractures, resulting in the elongated tails of the breakthrough curves, rather than the symmetric breakthrough curve that would be associated with advection by a mean velocity and a Fickian model of mechanical dispersion.

Becker and Shapiro [2000, 2003] and *Shapiro* [2001] hypothesized that the tracer migrates rapidly through the most transmissive fractures and migrates much more slowly through the less transmissive fractures. For the tracer migration in the less transmissive fractures, eventually the tracer may enter a highly transmissive fracture and then be transported to the pumped well. While in the less transmissive fracture, the tracer has migrated only a short distance over an extended period of time before entering a more transmissive fracture. This phenomenon of migrating into a low permeability section of the formation and then back into a more transmissive section of the formation is analogous to the mass exchange between mobile and immobile fluids proposed by *Coats and Smith* [1964]. The result of this phenomenon is an apparent diffusion resulting in the elongated tails of breakthrough curves. The elongated tail is an artifact of the extreme variability in the fluid velocity. The extreme variability in the fluid velocity, however, cannot be incorporated into a Fickian interpretation of dispersion.

To further test the hypothesis that slow advection causes the elongated tails of the breakthrough curves exhibited in the tracers tests conducted in the FSE well field, *Becker and Shapiro* [2003] considered a simplified model of a fracture network shown in Figure 13. The model considered non-interconnecting channels of fluid migration, where within each channel a constituent migrates only as a result of advection and dispersion. The channels with the highest velocity will yield the most rapid breakthrough and the channels with the lowest velocity will yield a much more delayed breakthrough. The breakthrough curve associated with each channel is assumed to be governed by a transfer function, with the velocity in each channel being proportional to the square of the channel aperture [*Becker and Shapiro*, 2003]. The breakthrough curve associated with each channel yields a parabolic shape when plotted as the logarithm of concentration versus the logarithm of time. It is assumed that the tracer mass injected into each channel is proportional to the fluid flux through the channel, and the flux-averaged cumulative breakthrough curve at the mutual ending point of all channels is the superposition of the breakthrough curves from the individual channels. The results shown in Figure 13 were generated using a uniform distribution of channel apertures that yielded velocities ranging over 3 orders of magnitude. The superposition of the breakthrough curves results in a cumulative breakthrough curve with a declining limb having a slope equal to –2 on a log-log plot of concentration versus time [*Becker and Shapiro*, 2003]. The slope of the declining limb of the hypothetical breakthrough curve in Figure 13 is similar to the slopes of the declining limbs of the breakthrough curves from the tracer tests shown in Figures 11 and 12.

The tracer tests performed in the FSE well field were conducted within one of the highly transmissive fracture intervals that were identified from cross-borehole hydraulic tests conducted in the FSE well field. This interval is composed of multiple interconnected fractures, which most likely includes both steeply dipping and subhorizontal fractures. It is conceivable that the tracers introduced into the formation traversed fractures having a wide range of transmissivities. It is also possible that transport within individual fractures resulted in multiple flow paths (or channels) with a wide range of fluid velocities, which in combination with transport through multiple fractures resulted in the phenomenon of slow advection that gave rise to the elongated tails of the breakthrough curves.

The results of the *in situ* tracer tests conducted in the FSE well field are not necessarily contradictory to the results given by *Garnier et al.* [1985]. The results of *Garnier et al.* [1985] were conducted in a single, highly transmissive, subhorizontal fracture zone in a chalk aquifer. It is plausible that there was rather limited variability in the fluid velocity associated with this feature between the injection and recovery boreholes. In addition, the matrix porosity of chalk was estimated to be 0.36 [*Moench*, 1995], which is more than an order of magnitude greater than the matrix porosity of the granite and schist at the Mirror Lake site. The combination of a large matrix porosity and a fluid velocity that does not have significant variability gives rise to the dominance of the chemical diffusion, as opposed to the apparent diffusion that is an artifact of slow advection. In the interpretation of chemical migration and tracer tests conducted in fractured rock aquifers, we should not necessarily anticipate the results of *Garnier et al.* [1985], nor should we anticipate the results of the tracer tests conducted in the FSE well field. It is conceivable that there could be a continuum of responses, where the results of *Garnier et al.* [1985] represent one end member, where chemical diffusion dominates, and the results of the tests conducted in the FSE well field represent another extreme, where the elongated tails of breakthrough curves are an artifact of a wide range in the fluid advection.

5.4 Transport of Environmental Tracers over 100's of Meters

Over distances of 100's of meters to kilometers, controlled tracer tests cannot be conducted in fractured rock formations such as that at the Mirror Lake site. Over these distances in a complexly fractured formation, the ultimate location of a tracer that is introduced into the formation cannot necessarily be identified. In addition, with a sparse number of bedrock

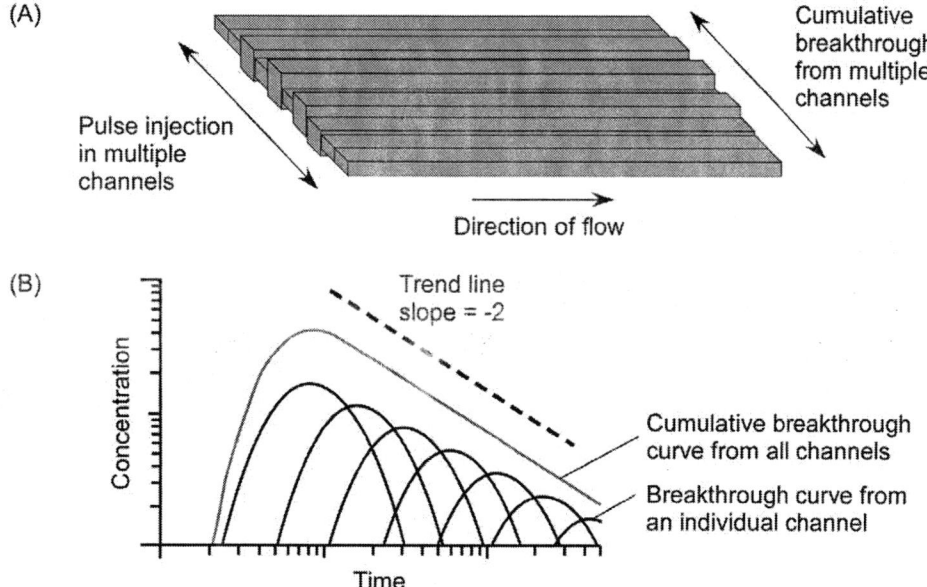

Figure 13. (A) Hypothetical model of unconnected, parallel channels, and (B) the breakthrough curves from each channel and the cumulative breakthrough curve from all channels (modified from *Becker and Shapiro*, 2003).

boreholes from which to sample, a quantitative interpretation of the tracer migration is unlikely to be successful because the majority of the tracer mass will not be recovered.

Over 100's of meters to kilometers, environmental tracers have been used to infer processes controlling chemical migration in many different geologic settings [*Busenberg and Plummer*, 1992; *Cook and Herczeg*, 2000]. Environmental tracers are either chemical species or dissolved gases in precipitation, which eventually recharge the ground water. With knowledge of the time-varying atmospheric concentrations, samples of the environmental tracers collected from wells can be interpreted to infer ground-water residence times. In relatively homogeneous porous media, fluid advection is assumed to control the spatial distribution of environmental tracers; the concentration of these tracers can then be directly translated to a ground-water age based on the assumed temporal variation of the tracer concentration in ground-water recharge. In fractured rock aquifers, however, diffusion into low-permeability environments, such as the matrix porosity, and dispersion, arising from the extreme variability in the fluid velocity, are likely to alter the concentration of the tracers as they migrate through the formation. Under such conditions, it is necessary to mathematically model the physical processes affecting chemical migration and not merely translate a tracer concentration into a ground-water age [*Shapiro*, 2002]

Shapiro [2001] used concentrations of dichlorodifluoromethane (CFC-12) and tritium (^{3}H) collected from discrete intervals of bedrock boreholes and piezometers in the unconsolidated glacial drift to infer the processes controlling chemical migration over the dimensions of the Mirror Lake watershed. The model of ground-water flow developed by *Tiedeman et al.* [1997] for the Mirror Lake watershed could not be used in the interpretation of CFC-12 and ^{3}H data. The bulk hydraulic properties of the fractured rock used in the ground-water flow model are sufficient in defining a water balance over the physical dimensions of the Mirror Lake watershed, but they are not capable of defining the intricacies of the flow regime that will affect the complex three-dimensional spatial distribution of CFC-12 and ^{3}H.

To interpret the CFC-12 and ^{3}H data from ground-water samples collected in the Mirror Lake watershed, *Shapiro* [2001] adopted a simplified interpretation of the flow regime and chemical transport. Ground-water flow lines were assumed to originate at the water table in the glacial drift and extend into the bedrock. Flow lines were assumed to move through a similar distance in the glacial deposits regardless of their starting location. In the bedrock, flow lines encounter fractures over a range of transmissivities, similar to the conceptual model of the rock underlying the FSE well field (Figure 7). Thus, over distances of more than 100 m in the bedrock, processes affecting chemical migration along one flow line were assumed to be similar to other flow lines. Therefore, sampling in the bedrock and glacial drift was regarded as sampling at locations along an ensemble of similar flow lines, which in turn, was equivalent to sampling one representative flow line at various distances from its origin [*Shapiro*, 2001]. The distance from the recharge location to the sampling point along the flow line, however, is unknown, and the relative distance between sampling locations also is unknown.

Because distance along a flow line to a sampling location is not known, the spatial distribution of CFC-12 and ^{3}H was removed from the procedure to estimate the parameters that control the chemical migration, in particular, the advection, dispersion, and matrix diffusion. The spatial distribution of the tracers is removed from the estimation procedure by taking advantage of multiple tracers moving simultaneously in the formation [*Shapiro*, 2001, 2002]. Concentrations from simulations of ^{3}H and CFC-12 transport along flow lines are plotted against each other, and model parameters are varied to reproduce the measured relation between ^{3}H and CFC-12 measured in drift piezometers and bedrock wells (Fig. 14).

Of interest in this investigation of chemical migration over dimensions of the Mirror Lake watershed was the magnitude of the effective diffusion coefficient, D_m, for comparison with the estimates of D_m from testing conducted over smaller volumes of rock. Other formation properties that control the chemical migration over dimensions of kilometers in the Mirror Lake watershed were also estimated from the ^{3}H and CFC-12 data. Details of the chemical-transport modeling, and the estimation of formation properties is given in *Shapiro* [2001].

Figure 14 shows the best fit curves for the transport modeling used to estimate the effective diffusion coefficient, D_m, from the concentrations of the environmental tracers measured at the Mirror Lake site. The value of D_m that best reproduces the measured relation between ^{3}H and CFC-12 is 1 m^2/yr; however, because of the variability in the data, values of D_m between 0.1 and 10 m^2/yr also qualitatively reproduce the data. The magnitude of these estimates exceeds the free-water diffusion coefficients for the environmental tracers by several orders of magnitude [*Shapiro*, 2001].

5.5 Diffusion in Fractured Rock From Cores to Kilometers

Figure 15 shows the effective diffusion coefficient, D_m, estimated over physical dimensions from cores to kilometers. The core experiments yield the theoretically expected result that D_m is less than the free-water diffusion coefficient, because of the tortuosity associated with the matrix porosity. From the interpretation of concentrations of tracers from *in*

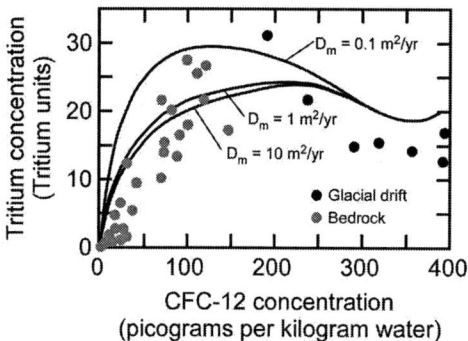

Figure 14. Measured concentrations of tritium and dichlorodifluoromethane (CFC-12) from piezometers in the glacial drift and isolated intervals from bedrock wells in the Mirror Lake watershed and its vicinity, along with best fit estimates for the effective diffusion coefficient, D_m (modified from *Shapiro*, 2001).

situ tests conducted over 10's of meters to kilometers, D_m increases with the physical dimension over which the data was interpreted.

Controlled tracer tests conducted over 10's of meters in the rock yield estimates of D_m that were greater than the free-water diffusion coefficient. From the interpretation of the environmental tracers over the dimensions of 100's of meters to kilometers, estimates of D_m were several orders of magnitude greater than the free-water diffusion coefficient.

Over dimensions of 10's of meters to kilometers, chemical diffusion into the rock matrix is ongoing. Laboratory experiments conducted by *Wood et al.* [1996] illustrated that ^{137}Cs diffused several millimeters into rock samples collected at the Mirror Lake site over a period of approximately 100 days. The magnitude of the chemical diffusion into the rock matrix, however, is being dwarfed over dimensions of 10's of meters to kilometers by a process that manifests itself similarly to diffusion.

The controlled tracer tests conducted over 10's of meters in the FSE well field showed the elongated declining limb of the breakthrough curves as a straight line when plotted on a log-log plot of concentration versus time. A straight line on a log-log plot of the declining limb of a breakthrough curve from a pulse injection of a tracer is usually attributed to diffusion. The slope of the log-log plot of the declining limb of the experimental breakthrough curves was -2, whereas, the theoretical result for diffusion from fractures to a porous matrix is -1.5.

It is hypothesized that the tracers introduced into rock migrate through fractures having a wide range in the fluid velocities, which results in the elongated tails observed in various tracer tests conducted in the FSE well field. The most transmissive fractures are responsible for the first arrival and peak concentration associated with the breakthrough curves. The least transmissive fractures are responsible for the tail of the breakthrough curve. If the least transmissive fractures connect with highly transmissive fractures before the tracer is recovered in the pumped well, then the least transmissive fractures cause the tracer to migrate a short distance in an extended period of time. This phenomenon is analogous to chemical mass diffusing into and out of a porous matrix. For diffusion into and out of a porous matrix, there is no travel distance associated with the time that the tracer is resident in the immobile fluid of the porous matrix before remerging into the mobile fluid of the fracture. Thus, there are differences between chemical diffusion and the slow advection arising from the tracer migration through the least transmissive fractures.

From the interpretation of the environmental tracers over dimensions of 100's of meters to kilometers in the Mirror Lake area, the effective diffusion coefficient, D_m, was several orders of magnitude greater than the free-water diffusion coefficient. Again this is attributed to the phenomenon of slow advection arising from the tracers migrating through fractures having a wide range of velocities. It is hypothesized that the magnitude of D_m from the interpretation of the environmental tracers is greater than the estimates of D_m from the tracer tests in the FSE well field, because the environmental tracers experience a greater range of the fluid velocities than *in situ* tracer tests conducted over 10's of meters. The tracer tests conducted in the FSE well field were conducted in one of the highly transmissive, subhorizontal fracture intervals identified from cross-borehole hydraulic testing. The highly transmissive intervals are likely composed of multiple interconnected fractures, and thus,

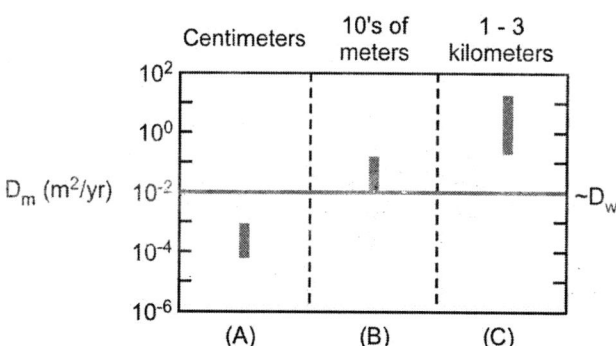

Figure 15. The effective diffusion coefficient, D_m, estimated over increasingly larger physical dimensions at the Mirror Lake site from (A) laboratory core experiments, (B) *in situ* tracer tests conducted over 10's of meters, and (C) the interpretation of the concentrations of environmental tracers over dimensions of 100's of meters to kilometers; D_w denotes the approximate magnitude of the free-water diffusion coefficient for most tracers.

the tracer most likely interrogated a number of fractures of different hydraulic properties, but not to the extent that the environmental tracers experienced in migrating through 100's of meters of rock.

From the estimates of D_m presented on Figure 15, one cannot tell whether the effective diffusion coefficient has achieved an asymptotic limit at the dimensions associated with the interpretation of the environmental tracers. If the environmental tracers encounter greater variability in the fluid velocity over dimensions larger than that considered in the interpretations presented by *Shapiro* [2001], then it is likely that estimates of the effective diffusion coefficient would increase further. If, however, the environmental tracers experienced the full range of the fluid velocity over the dimensions of the interpretation, then the results shown for interpretation of the environmental tracers in Figure 15 would be regarded as an asymptotic limit.

It is hypothesized that slow advection gives rise to the elongated tails of the tracer tests and the appearance of a chemical flux that resembles diffusion, even though it is an artifact of fluid advection. In fractured rock aquifers, slow advection should be anticipated when the fractures that are responsible for chemical migration have a wide range in their hydraulic conductivity. The hydraulic conductivity of fractures at the Mirror Lake site varies over six orders of magnitude. The hydraulic conductivity of the fracture data was censored by the detection limits of the *in situ* testing equipment. Therefore, it is likely that the hydraulic conductivity varied over a much greater range than that associated with the hydraulic conductivity measurements. In aquifers where fractures do not exhibit a wide range in fluid velocities, chemical diffusion is likely to dominate the declining limbs of breakthrough curves, and laboratory estimates of diffusion are likely to be sufficient in the characterization of field-scale chemical transport investigations.

6. FRACTURE CONTROLS ON GROUND-WATER FLOW AND CHEMICAL TRANSPORT AT THE MIRROR LAKE SITE

The hydraulic conductivity and the effective diffusion coefficient show different trends over increasingly larger physical dimensions [*Shapiro*, 2003]. With increasingly larger physical dimensions (up to kilometers), the hydraulic conductivity is bounded by the hydraulic conductivity associated with a background network of less transmissive fractures. The hydraulic conductivity associated with this background network of fractures at the Mirror Lake site is on the order of 10^{-7} m/s. Hydraulic tests conducted on individual fractures or closely spaced fractures, however, exhibits a range from 10^{-10} to 10^{-4} m/s, with the lower limit being associated with the detection limit of the *in situ* testing equipment. In contrast, the effective diffusion coefficient increases when interpreting chemical transport experiments from centimeters to kilometers. At the centimeter-scale of measurement, the effective diffusion coefficient is less than the free-water diffusion coefficient associated with the tracers used in the experiments. Over 10's of meters, the range of the effective diffusion coefficient is greater than the free-water diffusion coefficient, and over kilometers, the effective diffusion coefficient exceeds the free-water diffusion coefficient by several orders of magnitude.

The measurement and detailed mapping of fracture attributes on outcrops and road cuts, and in boreholes is integral in understanding the different trends that are observed in estimating the chemical transport and hydraulic properties over increasingly larger physical dimensions in the rock. Plots of fracture trends show a wide range of orientations from subhorizontal to moderate and steeply dipping (Figure 2). Fractures mapped on road cuts show poor connectivity with trace lengths that rarely exceed 10 m [*Barton*, 1996], which results in poor connectivity of the highly transmissive fractures. Results from cross-borehole hydraulic tests show that highly transmissive fractures are interconnected over lateral distances of approximately 20 m in the rock. The highly transmissive fractures are part of highly transmissive, subhorizontal zones composed of both moderate to steeply dipping fractures and subhorizontal sheeting fractures.

The poor connectivity of fractures results in the less transmissive fractures serving to connect the highly transmissive intervals of interconnected fractures. Consequently, the less transmissive fractures act as the "bottlenecks" that impede ground-water flow, and control the bulk hydraulic properties of the rock. The poor connectivity of the fractures is maintained over dimensions of several kilometers, as the bulk hydraulic conductivity of the rock as determined from ground-water flow modeling is again controlled by a background network of less transmissive fractures that impedes ground-water flow over these dimensions.

From investigations conducted in boreholes, road cuts, and outcrops, few faults were observed in the Mirror Lake area. The results from the ground-water flow modeling over the area of the Mirror Lake watershed implies that if other undetected faults are within the rock, they do not serve as highly transmissive features that affect the measured distribution of the hydraulic head or the ground-water discharge to streams. The results of the ground-water modeling conducted over dimensions of kilometers also implies that there is no interconnected "backbone" of highly transmissive features that results in an increase in the hydraulic conductivity with the scale of measurement. This is supported by the lack of a single dominant fracture trend in the orientation plots of Figure 2.

The poor connectivity of fractures had the opposite effect on the effective diffusion coefficient that was estimated over increasingly larger dimensions. The bulk hydraulic conductivity is controlled by the "bottlenecks" in the rock, and the effective diffusion coefficient is controlled by the range in the fluid velocity over dimensions from meters to kilometers. The poor connectivity and short areal extent of fractures give rise to a highly variable fluid velocity field. Consequently, as chemical constituents migrate in the subsurface they are forced to move into different fractures with a wide range of hydraulic properties.

Laboratory tracer tests conducted in cores are not affected by the fluid velocity in fractures. These tests yield estimates of the effective diffusion coefficient that are less than free-water diffusion coefficient associated with the tracers that were used. Over 10's of meters, *in situ* tracer tests in the FSE well field were conducted in one of the highly transmissive fracture intervals that was intersected in two adjacent boreholes. This highly transmissive interval is composed of multiple intersecting fractures. The range of the hydraulic conductivity from tests conducted in individual fractures varies over at least six orders of magnitude. Therefore, it is likely that the tracer interrogated multiple fractures with a wide range of hydraulic conductivities. The range in the hydraulic conductivity gives rise to the phenomena of "slow advection," which is attributed to the elongated tails in the breakthrough curves and the effective diffusion coefficient that is greater than the free-water diffusion coefficient associated with the tracers that were used in the *in situ* testing.

Over dimensions of kilometers, the interpretation of the environmental tracers yielded estimates of the effective diffusion coefficient that was orders of magnitude greater than the free-water diffusion coefficients. Over dimensions of kilometers, the tracers interrogated both highly transmissive fractures and the less transmissive background network of fractures. The range of the velocities encountered within the fractures over 100's of meters to kilometers was most likely greater than that encountered over 10's of meters between adjacent boreholes in the rock of the FSE well field.

7. SUMMARY

The characterization of fractured rock aquifers for issues of societal importance often requires the interpretation of formation properties that control ground-water flow and chemical migration over increasingly larger physical dimensions in these formations. Compilations of investigations from multiple field sites are often used to provide guidance on the effect that measurement scale has on the magnitude of chemical transport and hydraulic properties. While this information is informative about the general trends that could be anticipated in fractured rock aquifers, it may not be applicable to all fractured rock sites, because geologic conditions and fracture attributes that control ground-water flow and chemical migration vary from site to site.

A detailed investigation was undertaken to identify the effect of increasing physical dimensions on estimates of hydraulic conductivity and chemical diffusion in the granite and schist underlying the Mirror Lake watershed in central New Hampshire. Extensive geologic and fracture mapping was conducted over large rock exposures on road cuts, in boreholes, and at outcrops in the Mirror Lake area. The data on fracture orientations from road cuts, boreholes, and outcrops all showed moderate to steeply dipping and subhorizontal fractures. The similarity in fracture orientations between the detailed fracture mapping conducted on road cuts and boreholes and the fracture orientations observed from outcrops scattered over a 5 km radius in the Mirror Lake area suggests that fracture attributes identified from the detailed mapping over limited volumes of rock are maintained throughout the study area that covers more than 10 km^2.

The detailed fracture mapping on road cuts and in boreholes of the FSE and CO well fields illustrate that fractures in the Mirror Lake area are poorly connected, and the trace lengths of fractures rarely exceed 10 m. In general, the granitic intrusive rocks tend to be more fractured than the schist; however, the granite constitutes a smaller percentage of the rock volume than the schist and tends to be discontinuous. Therefore, the degree of fracturing in the granite does not translate into preferential flow in the formation.

From investigations conducted in boreholes, road cuts, and outcrops, only two fault features were observed over the Mirror Lake area. One fault was observed on the road cuts east of Mirror Lake and showed significant clay content. The hydraulic significance of the faults is unclear from its mapped features. The calibration of a ground-water model over dimensions of the Mirror Lake area was used to infer the hydraulic significance of large scale geologic features.

The hydraulic conductivity of the rock underlying the Mirror Lake watershed was estimated over physical dimensions ranging from meters to kilometers. In boreholes drilled into the bedrock, single-hole hydraulic tests were used to isolate individual fractures or closely space. fractures. Each single-hole test was conducted over 10's of minutes and was assumed to interrogate fractures within a few meters of the borehole. Over 200 single-hole hydraulic tests were conducted, which showed a range of hydraulic conductivity over approximately 6 orders of magnitude (10^{-10} to 10^{-4} m/s). The lower limit of that range was associated with the detection limit of the *in situ* testing equipment. Within individual boreholes, changes in the hydraulic conductivity of fractures (over several orders of magnitude) occurred over short distances, and the density of fracturing did not correlate with the magnitude of the hydraulic conductivity.

Cross-borehole hydraulic tests were conducted in clusters of wells with separation distances ranging from 10 to 100 m. These tests interrogated volumes of rock containing multiple fractures. The tests were designed to investigate the ambient conditions in the rock, which required packers to isolate various intervals in the bedrock boreholes. The results of these hydraulic tests showed the presence of highly transmissive subhorizontal fracture intervals of limited areal extent, with a hydraulic conductivity of approximately 10^{-4} m/s. The highly transmissive intervals were composed of multiple interconnected fractures, which was mostly likely a combination of both the steeply dipping and subhorizontal fractures that were mapped.

The results of the cross-borehole tests also showed that the highly transmissive intervals were connected only through a network of less transmissive fractures. These less transmissive fractures acted as "bottlenecks" that impeded ground-water flow. Consequently, over dimensions of 100 meters the bulk hydraulic conductivity of the fractured rock was controlled by the less conductive fractures, with a hydraulic conductivity of approximately 10^{-7} m/s.

Over dimensions of kilometers, ground-water flow modeling was used to estimate the bulk hydraulic conductivity of the rock. The measured hydraulic head and ground-water discharges to streams were used in the estimation procedure. Over dimensions of kilometers, the bulk hydraulic conductivity of the rock remained unchanged from the bulk hydraulic conductivity inferred from the cross-borehole hydraulic tests. The background network of less transmissive fractures appears to control the bulk hydraulic conductivity of the rock over an area greater than 10 km². Consequently, there is no interconnected "backbone" of highly transmissive features that would result in an increase in the hydraulic conductivity with increasingly larger physical dimensions. The poor connectivity of the rock and lengths of fractures that rarely exceed 10 m are the underlying cause for the poor connectivity of highly transmissive fractures and the stability in the hydraulic conductivity estimated over dimensions from 100 m to kilometers.

Estimates of chemical diffusion were made from the interpretation of tracer experiments conducted over increasingly larger physical dimensions. Estimates of diffusion coefficients in cores followed theoretical interpretations and were less than the free-water diffusion coefficient for the constituents under consideration. Controlled *in situ* tracer tests conducted over 10's of meters between boreholes in the rock yielded estimates of the diffusion coefficient that were greater than the free-water diffusion coefficient. The *in situ* tracer tests were conducted between boreholes that intersected one of the highly transmissive fracture intervals identified from the cross-borehole hydraulic tests. It is hypothesized that the tracer migrated through an interconnected network of fractures having a wide range of fluid velocities, which is attributed to the wide range in the hydraulic conductivities of fractures that were measured from single-hole hydraulic tests. The wide range in fluid velocities gives rise to elongated tails in the breakthrough curves of the tracer tests. The slow advection from the least transmissive fractures gives the appearance of a diffusive phenomenon over 10's of meters in the fractured rock that overwhelms the magnitude of the chemical diffusion that is ongoing between fractures and the rock matrix.

Over dimensions of kilometers, an interpretation of the concentrations of environmental tracers was used to estimate the magnitude of the effective diffusion coefficient in the fractured rock of the Mirror Lake site. The results over this scale showed an even larger estimate of the diffusion coefficient than that obtained from controlled *in situ* tracer tests conducted over 10's of meters. At the kilometer-scale, the large diffusion coefficient was again attributed to the wide range of fluid velocities and the slow advection, which manifests itself similarly to a diffusive phenomenon. The increase in the apparent diffusion coefficient with the physical dimension of the investigation is attributed to a larger variability in the fluid velocity encountered over larger dimensions, whereas tests conducted over 10's of meters did not encounter the full range in the fluid velocity encountered over kilometers.

The results of the effect of measurement scale on the magnitude of the chemical transport and hydraulic properties from the Mirror Lake site are not necessarily transferable to other geologic settings. The results of the investigations presented here point to the need to understand the underlying geologic structure and connectivity of fractures in formulating hypotheses regarding the magnitude of formation properties as a function of the measurement scale. For example, in bedded sedimentary sequences with highly transmissive, areally extensive, bedding plane partings, the bulk hydraulic conductivity over large volumes of rock may be controlled by the highly transmissive features, rather than a background network of less transmissive fractures.

Integral to the interpretations of chemical transport and hydraulic properties of the fractured rock over increasingly large physical dimensions at the Mirror Lake site was the integration of information from the geologic and fracture mapping. Attributes of fractures obtained from detailed mapping in boreholes and exposures of rocks on road cuts and outcrops provided a conceptual framework from which hypotheses were developed regarding the geologic controls on ground-water flow and chemical migration. Borehole geophysical logging tools that provided accurate imaging of fractures intersecting boreholes were important in verifying the similarity between fracture properties on surface exposures and at depth in the subsurface.

Acknowledgements. These investigations were conducted with the support of the Toxic Substances Hydrology and National Research Programs of the USGS. The authors gratefully acknowledge the logistical support of U.S. Forest Service personnel at the Hubbard Brook Experimental Forest, Grafton County, New Hampshire. The Hubbard Brook Experimental Forest is operated and maintained by the Northeastern Forest Experiment Station, U.S. Department of Agriculture Forest Service, Radnor, Pennsylvania. The authors also gratefully acknowledge the comments provided by Quanlin Zhou of Lawrence Berkeley Laboratory, the anonymous reviewer, and Frederick Day-Lewis of the USGS.

REFERENCES

Barton, C. C. (1996), Characterizing bedrock fractures in outcrop for studies of ground-water hydrology: An example from Mirror Lake, Grafton County, New Hampshire, *in* Morganwalp, D. W. and Aronson, D. A., eds., U.S. Geological Survey Toxic Substances Hydrology Program—Proceedings of the Technical Meeting, Colorado Springs, CO, September 20-24, 1993, U.S. Geological Survey Water-Resources Investigations Report 94-4015, vol. 1, p. 81–87.

Barton, C. C. (1997), Bedrock geologic map of Hubbard Brook Experimental Forest and maps of fractures and geology in road cuts along Interstate 93, Grafton County, New Hampshire, U.S. Geological Survey IMAP 2562, 1 sheet.

Bear, J. (1979), Hydraulics of Groundwater, McGraw-Hill, New York, 567 p.

Becker, M. W., and A. M. Shapiro (2000), Tracer transport in crystalline fractured rock: Evidence of non-diffusive breakthrough tailing, Water Resources Research, 36(7), 1677–1686.

Becker M. W., and A. M. Shapiro (2003), Interpreting tracer breakthrough tailing from different forced-gradient tracer experiment configurations in fractured bedrock, Water Resources Res., 39 (1), 1024, doi:10.1029/2001WR001190.

Birgersson, L., and I. Neretnieks (1990), Diffusion in the matrix of granitic rock: Field test in the Stripa Mine, Water Resources Research, 26(11), 2833–2842, 10.1029/90WR00822.

Burton, W. C., G. J. Walsh, and T. R. Armstrong (2000), Bedrock geologic map of the Hubbard Brook Experimental Forest, Grafton County, New Hampshire, U.S. Geological Survey Open-File Report 2000-45-A, CD.

Busenberg, E., and L. N. Plummer (1992), Use of chlorofluorocarbons (CCl_3F and CCl_2F_2) as hydrologic tracers and age-dating tools: The alluvium and terrace system of Central Oklahoma, Water Resources Research, 28(9), 2257–2283.

Clauser, C. (1992), Permeability of crystalline rocks, EOS, 73 (21), 233,237–238.

Coats, K. H., and B. D. Smith (1964), Dead-end pore volume and dispersion in porous media, Society of Petroleum Engineers Journal, 4, 73–84.

Cook, P., and A. L. Herczeg, eds. (2000), Environmental Tracers in Subsurface Hydrology, Kluwer Academic Publishers, Boston, 529 p.

Day-Lewis, F. D., P. A. Hsieh, and S. M. Gorelick (2000), Identifying fracture-zone geometry using simulated annealing and hydraulic-connection data, Water Resources Research, 36(7), 1707–1722, 10.1029/2000WR900073.

Garnier, J. M., N. Crampon, C. Préaux, G. Porel, and M. Vreulx (1985), Traçage par ^{13}C, ^2H, I- et uranine dans la nappe de la craie sénonienne en écoulement radial convergent (Béthune, France), Journal of Hydrology, 78, 379–392.

Gelhar, L. W., C. Welty, K. R. Rehfeldt (1992), A critical review of data on field-scale dispersion in aquifers, Water Resources Research, 28(7), 1955–1974.

Harte, P. T., and T. C. Winter (1996), Factors affecting recharge to crystalline rock in the Mirror Lake area, Grafton County, New Hampshire, *in* Morganwalp, D. W. and Aronson, D. A., eds., U.S. Geological Survey Toxic Substances Hydrology Program—Proceedings of the Technical Meeting, Colorado Springs, CO, September 20–24, 1993, U.S. Geological Survey Water-Resources Investigations Report 94-4015, vol. 1, p. 141–150.

Hsieh, P. A. (1998), Scale effects in fluid flow through fractured geologic media, *in* Sposito, G., ed., Scale Dependence and Scale Invariance in Hydrology, Cambridge University Press, p. 335–353.

Hsieh, P. A., R. L. Perkins, and D. O. Rosenberry (1996), Field instrumentation for multilevel monitoring of hydraulic head in fractured bedrock at the Mirror Lake site, Grafton County, New Hampshire, *in* Morganwalp, D. W. and Aronson, D. A., eds., U.S. Geological Survey Toxic Substances Hydrology Program—Proceedings of the Technical Meeting, Colorado Springs, CO, September 20–24, 1993, U.S. Geological Survey Water-Resources Investigations Report 94-4015, vol. 1, p. 137–140.

Hsieh, P. A., and A. M. Shapiro (1996), Hydraulic characteristics of fractured bedrock underlying the FSE well field at the Mirror Lake site, Grafton County, New Hampshire, *in* Morganwalp, D. W. and Aronson, D. A., eds., U.S. Geological Survey Toxic Substances Hydrology Program—Proceedings of the Technical Meeting, Colorado Springs, CO, September 20–24, 1993, U.S. Geological Survey Water-Resources Investigations Report 94-4015, vol. 1, p. 127–130.

Hsieh, P. A., A. M. Shapiro, C. C. Barton, F. P. Haeni, C. D. Johnson, C. W. Martin, F. L. Paillet, T. C. Winter and D. L. Wright (1993), Methods of characterizing fluid movement and chemical transport in fracture rock, *in* Cheney, J. T., and Hepburn, J. C., eds., Field Trip Guidebook for the Northeastern United States, Geological Society of America Annual Meeting, Boston, MA, University of Massachusetts, Department of Geology and Geography, Contribution no. 67, vol. 2, p. R1–R30.

Hsieh, P. A., A. M. Shapiro, and C. Tiedeman (1999), Computer simulation of fluid flow in fractured rocks at the Mirror Lake FSE well field, *in* Morganwalp, D. W. and Buxton, H. T., eds., U.S. Geological Survey Toxic Substances Hydrology Program—Proceedings of the Technical Meeting, Charleston, SC, March 8–12, 1999, U.S. Geological Survey Water-Resources Investigations Report 99-4018C, vol. 3, p. 777–781.

Jahns, R. H. (1943), Sheet structure in granites: Its origin and use as a measure of glacial erosion in New England, Journal of Geology, 51(2), 71–98.

Johnson, C. D. (1998), Subsurface lithology and fracture occurrence and their effects on hydraulic conductivity: Mirror Lake research site, Grafton County, New Hampshire, unpublished Master's Thesis, University of New Hampshire, Durham, NH, 159 p.

Johnson, C. D., and A. H. Dunstan (1998), Lithology and fracture characterization from drilling investigations in the Mirror Lake area, Grafton County, New Hampshire, U.S. Geological Survey Water-Resources Investigations Report 98-4183, 211 p.

Likens, G. E., ed. (1985), An Ecosystem Approach to Aquatic Ecology: Mirror Lake and Its Environment, Springer-Verlag, New York, 516 p.

Liu, H. H., G. S. Bodvarsson, and G. Zhang (2004), Scale dependency of the effective matrix diffusion coefficient, Vadose Zone Journal, 3, 312–315.

Mallard, G. E., and D. A. Aronson, eds. (1991), U.S. Geological Survey Toxic Substances Hydrology Program—Proceedings of the technical meeting, Monterey, California, March 11–15, 1991, U.S. Geological Survey Water-Resources Investigations Report 91-4034.

Maloszewski, P., and A. Zuber (1991), Influence of matrix diffusion and exchange reactions on radiocarbon ages in fissured carbonate aquifers, Water Resources Research, 27(8), 1937–1945.

Moench, A. F. (1995), Convergent radial dispersion in a double-porosity aquifer with fracture skin: Analytical solution and application to a field experiment in fractured chalk, Water Resources Research, 31(8), 1823–1835.

Morganwalp, D. W., and D. A. Aronson, eds. (1996), U.S. Geological Survey Toxic Substances Hydrology Program—Proceedings of the technical meeting, Colorado Springs, Colorado, September 20–24, 1993, U.S. Geological Survey Water-Resources Investigations Report 94-4015, vol. 1.

Morganwalp, D. W., and H. T. Buxton, eds. (1999), U.S. Geological Survey Toxic Substances Hydrology Program—Proceedings of the technical meeting, Charleston, South Carolina, March 8–12, 1999, U.S. Geological Survey Water-Resources Investigations Report 99-4018C, Volume 3.

Neretnieks, I., T. Eriksen, and P. Tähtinen (1982), Tracer movement in a single fissure in granitic rock: some experimental results and their interpretation, Water Resources Research, 18(4), 849–858.

Neuman, S. P. (1994), Generalized scaling of permeabilities: Validation and effect of support scale, Geophysical Research Letters, 21, 349–352.

Neuman S. P., and V. Di Federico (2003), Multifaceted nature of hydrogeologic scaling and its interpretation, Reviews in Geophysics, 41(3), 1014, doi:10.1029/2003RG000130.

Novakowski, K. S., and G. van der Kamp (1996), The radial diffusion method 2. A Semianalytical model for the determination of effective diffusion coefficients, porosity, and adsorption, Water Resources Research, 32(6), 1823–1830, 10.1029/95WR03720.

Ohlsson, Y., and I. Neretnieks (1995), Literature survey of matrix diffusion theory and of experiments and data including natural analogues, Swedish Nuclear Fuel and Waste Management company (SKB), Stockholm, Technical Report 95-12, 89 p.

Paillet, F. L. (1996), Use of well logs to prepare the way for packer strings and tracer tests: Lessons from the Mirror Lake study, in Morganwalp, D. W. and Aronson, D. A., eds., U.S. Geological Survey Toxic Substances Hydrology Program—Proceedings of the Technical Meeting, Colorado Springs, CO, September 20–24, 1993, U.S. Geological Survey Water-Resources Investigations Report 94-4015, vol. 1, p. 103–109.

Parker, B. L., R. W. Gillham, and J. A. Cherry (1994), Diffusive disappearance of immiscible-phase organic liquids in fractured geologic media, Ground Water, 32(5), 805–820.

Powers, C. J., K. Singha, and F. P. Haeni (1999), Integration of surface geophysical methods for fracture detection in bedrock at Mirror Lake, New Hampshire, in Morganwalp, D. W. and Buxton, H. T., eds., U.S. Geological Survey Toxic Substances Hydrology Program—Proceedings of the Technical Meeting, Charleston, SC, March 8–12, 1999, U.S. Geological Survey Water-Resources Investigations Report 99-4018C, vol. 3, p. 757–768.

Sánchez-Vila, X., J. Carrera, and J. P. Girardi (1996), Scale effects in transmissivity, Journal of Hydrology, 183, (1-2), 1–22.

Selvadurai, A. P. S., M. J. Boulon, and T. S. Nguyen (2005), The permeability of an intact granite, Pure and Applied Geophysics, 162, 373–407.

Shapiro, A. M. (2001), Effective matrix diffusion in kilometer-scale transport in fractured crystalline rock, Water Resources Research, 37(3), 507–522.

Shapiro, A. M. (2002), Modeling chemical transport over regional dimensions in bedrock aquifers: Hydrogeologic complexity doesn't necessarily warrant complex modeling, International Groundwater Modeling Center (IGWMC) Newsletter, XX(1), 5.

Shapiro, A. M. (2003), The effect of scale on the magnitude of formation properties governing fluid movement and chemical transport in fractured rock, in Krasny, J., Zbynek, H., and Bruthans, J., eds., Proceedings of the International Conference on Groundwater in Fractured Rocks, Prague, Czech Republic, September 15–19, 2003, International Association of Hydrogeologists, p. 13–14.

Shapiro, A. M. (2004), Borehole Testing System: U.S. Government Patent and Trademark Office, Washington, D.C., United States Patent Number 6,761,062, 12 p.

Shapiro, A. M., and P. A. Hsieh (1991), Research in fractured rock hydrogeology: Characterizing fluid movement and chemical transport in fracture rock at the Mirror Lake drainage basin, New Hampshire, in Mallard, G. E., and Aronson, D. A., eds., U.S. Geological Survey Toxic Substances Hydrology Program—Proceedings of the Technical Meeting, Monterey, CA, March 11–15, 1991, U.S. Geological Survey Water-Resources Investigations Report 91-4034, p. 155–161.

Shapiro, A. M., and P. A. Hsieh (1996a), Overview of research on use of hydrologic, geophysical and geochemical methods to characterize flow and chemical transport in fracture rock at the Mirror Lake site, New Hampshire, in Morganwalp, D. W. and Aronson, D. A., eds., U.S. Geological Survey Toxic Substances Hydrology Program—Proceedings of the Technical Meeting, Colorado Springs, CO, September 20–24, 1993, U.S. Geological Survey Water-Resources Investigations Report 94-4015, vol. 1, p. 71–80.

Shapiro, A. M., and P. A. Hsieh (1996b), A new method of performing controlled injection of traced fluid in fractured crystalline rock, in Morganwalp, D. W. and Aronson, D. A., eds., U.S. Geological Survey Toxic Substances Hydrology Program—Proceedings of the Technical Meeting, Colorado Springs, CO, September 20–24, 1993, U.S. Geological Survey Water-Resources Investigations Report 94-4015, vol. 1, p. 131–136.

Shapiro, A. M., and P. A. Hsieh (1998), How good are estimates of transmissivity from slug tests in fractured rock? Ground Water, 36(1), 37–48.

Shapiro, A. M., P. A. Hsieh, and F. P. Haeni (1999), Integrating multidisciplinary investigations in the characterization of fractured rock, in Morganwalp, D.W., and Buxton, H.T., eds., U.S. Geological Survey Toxic Substances Hydrology Program—Proceedings of the Technical Meeting, Charleston, South Carolina, March 8–12, 1999—Volume 3 of 3—Subsurface Contamination from Point Sources: U.S. Geological Survey Water-Resources Investigations Report 99-4018C, p. 669–680.

Shapiro, A. M., P. A. Hsieh, and T. C. Winter (1995), The Mirror Lake fractured rock research site—A multidisciplinary research effort in characterizing ground-water flow and chemical transport in fractured rock, U.S. Geological Survey Fact Sheet FS-138-95, 2 p.

Skoczylas, F., and J. P. Henry (1995), A study of the intrinsic permeability of granite to gas, Int. J. Rock Mech. Min. Sci. & Geomech. Abstr., 32(2), p. 171–179.

Tiedeman, C. R., D. J. Goode, and P. A. Hsieh (1997), Numerical simulation of ground-water flow through glacial deposits and crystalline bedrock in Mirror Lake area, Grafton County, New Hampshire, U.S. Geological Survey Professional Paper 1572, 50 p.

Tiedeman, C.R., D. J. Goode, and P. A. Hsieh (1998), Characterizing a ground water basin in a New England mountain and valley terrain, Ground Water, 36(4), 611–620.

Tiedeman, C. R., and P. A. Hsieh (2001), Assessing an open-well aquifer test in fractured crystalline rock, Ground Water, 39(1), 68–78.

Trimmer, D., B. Bonner, H. C. Heard, and A. Duba (1980), Effect of pressure and stress on water transport in intact and fractured gabbro and granite, Journal of Geophysical Research, 85, 7059–7071.

van der Kamp, G., D. R. Van Stempvoort, and L. I. Wassenaar (1996), The radial diffusion method 1. Using intact cores to determine isotopic composition, chemistry, and effective porosities for groundwater in aquitards, Water Resources Research, 32(6), 1815–1822, 10.1029/95WR03719.

Winter, T. C. (1984), Geohydrologic setting of Mirror Lake, West Thornton, New Hampshire, U.S. Geological Survey Water-Resources Investigations Report 84-4266, 61 p.

Wood, W. W., A. M. Shapiro, P. A. Hsieh, and T. B. Councell (1996), Observational, experimental and inferred evidence for solute diffusion in fractured granite aquifers: Examples from the Mirror Lake Watershed, Grafton County, New Hampshire, in Morganwalp, D. W. and Aronson, D. A., eds., U.S. Geological Survey Toxic Substances Hydrology Program—Proceedings of the Technical Meeting, Colorado Springs, CO, September 20–24, 1993, U.S. Geological Survey Water-Resources Investigations Report 94-4015, vol. 1, p. 167–170.

Allen M. Shapiro, U.S. Geological Survey, 12201 Sunrise Valley Drive, Mail Stop 431, Reston, VA 20192, USA, e-mail: ashapiro@usgs.gov

Accounting for Tomographic Resolution in Estimating Hydrologic Properties from Geophysical Data

Kamini Singha

Department of Geosciences, The Pennsylvania State University, University Park, Pennsylvania, USA

Frederick D. Day-Lewis

U.S. Geological Survey, Office of Ground Water, Branch of Geophysics, Storrs, Connecticut, USA

Stephen Moysey

School of the Environment, Clemson University, Clemson, South Carolina, USA

Geophysical measurements increasingly are being used in hydrologic field studies because of their ability to provide high-resolution images of the subsurface. In particular, tomographic imaging methods can produce maps of physical property distributions that have significant potential to improve subsurface characterization and enhance monitoring of hydrologic processes. In the tomographic imaging approach, geophysical images of the subsurface are converted to hydrologic property maps using petrophysical relations. In field studies, this transformation is complicated because measurement sensitivity and averaging during data inversion result in tomographic images that have spatially variable resolution (i.e., the estimated property values in the geophysical image represent averages of the true subsurface properties). Standard approaches to petrophysics do not account for variable geophysical resolution, and thus it is difficult to obtain quantitative estimates of hydrologic properties. We compare two new approaches that account for variable geophysical resolution: a Random Field Averaging (RFA) method and Full Inverse Statistical Calibration (FISt). The RFA approach uses a semi-analytical method whereas FISt calibration is based on a numerical solution to the problem.

1. INTRODUCTION

Data limitations represent the principal impediment to characterize and monitor subsurface hydrologic properties and processes at the field scale. Conventional hydrologic measurements (e.g., aquifer tests or fluid samples) commonly depend on direct access to the subsurface, making them expensive and sparse. Moreover, such measurements either sample conditions local to boreholes or integrate over large volumes of the subsurface. As a result, these measurements carry limited information about aquifer conditions between sampling locations or provide complex averages of properties, making it difficult to assess the distribution of heterogeneity throughout an aquifer. With recent advances in geophysical instrumentation and imaging algorithms, hydrologists increasingly are looking to geophysical imaging methods to help understand aquifer heterogeneity and monitor such processes as contaminant transport and seasonal dynamics in water content.

In recent years, the term "hydrogeophysics" has come to describe these kinds of interdisciplinary research efforts that bridge hydrology and geophysics. This new field is burgeoning, as evidenced by rapid growth of hydrogeophysics papers in the literature, as well as publication of recent texts on the subject [*Rubin and Hubbard*, 2005; *Vereecken et al.*, 2006]. In numerous studies, geophysical imaging has provided valuable data at spatial and temporal scales rarely attainable with standard hydrologic measurements. A conclusion common to much of this work is that the quantitative integration of geophysical and hydrologic data—through either coupled inversion or geostatistical techniques—results in characterization of the subsurface with higher resolution and greater reliability than is possible using conventional hydrologic measurements alone. Despite growing acceptance and clear evidence for the synergies that result from hydrogeophysical data integration, important research challenges remain. One of the foremost problems is determining how geophysical properties determined by a field survey (e.g., electrical conductivity, dielectric permittivity, or seismic velocity) are related to the properties that hydrologists are interested in, such as hydraulic conductivity, contaminant concentration, or water quantity. Petrophysical formulas that describe the relations between these properties are commonly used, and can be calibrated as site-specific conversions [e.g., *Alumbaugh et al.*, 2002] or based on theoretical or general empirical grounds [e.g., *Slater et al.*, 2002; *Singha and Gorelick*, 2005]. Petrophysical relations have been used extensively to convert geophysical images into two-dimensional (2D) maps or three-dimensional (3D) volumes of quantities such as saturation, concentration, porosity, or permeability [e.g., *Hubbard et al.*, 2001; *Slater et al.*, 2002; *Berthold and Masaki*, 2004].

One difficulty with using standard petrophysical relations to convert geophysical to hydrologic property values in tomographic studies is that the data or theory used to generate the relations may not fully capture conditions at the field scale. For example, a petrophysical relation may be based on data from a set of wells or cores. Because of subsurface heterogeneity, these data may be representative of only a small area near where they were collected; consequently, the calibrated petrophysical relation is most certain near the location where the data were collected, and reflects the particular support volume of the measurements at this location. Away from the sampling location, both the resolution of the geophysical survey and the type of material may change, causing the calibrated petrophysical relation to no longer apply. Additionally, the sensitivity of the geophysical methods and the effects of image reconstruction can contribute to the field-scale estimate of a geophysical property, thereby creating spatial dependence in the petrophysical relation. Consequently, relations between, for example, seismic velocity and hydraulic conductivity found at the laboratory scale may not be appropriate in the field. Direct estimation of hydrologic properties from geophysical tomograms at the field scale has been only moderately successful because 1) reconstructed tomograms are often highly uncertain and subject to inversion artifacts; 2) the range of subsurface conditions represented in calibration data sets is incomplete due to heterogeneity and the paucity of collocated well or core data; and 3) geophysical methods exhibit spatially variable sensitivity.

The uncertainty and non-uniqueness of petrophysical relations have led some to consider stochastic methods, such as co-simulation and co-kriging frameworks [e.g., *McKenna and Poeter*, 1995; *Cassiani et al.*, 1998; *Yeh et al.*, 2002; *Ramirez et al.*, 2005] or other geostatistical approaches for incorporating geophysical property estimates into hydrogeologic studies, such as the 1) correlation of site-specific soft geophysical data with collocated hard point data [*Doyen*, 1988; *McKenna and Poeter*, 1995; *Dietrich et al.*, 1998], 2) estimation of hydrologic properties from geophysical data based on probabilities of occurrence as mapped by geologists [*Carle and Ramirez*, 1999], and 3) estimation using geophysical methods for lithologic zonation [*Hyndman et al.*, 1994; *Hyndman and Gorelick*, 1996]. Some approaches to data integration include coupled inversion methods that consider hydrologic processes directly. For example, *Vanderborght et al.* [2005] used equivalent advection-dispersion equations and streamtube models to quantify breakthrough curves from synthetic 2D electrical resistivity tomography (ERT) inversions for estimating hydraulic conductivity and local-scale dispersivity values. *Kowalsky et al.* [2005] demonstrated coupled inversion of ground-penetrating radar (GPR) tomographic data and neutron probe data to monitor infiltration processes. Coupled inversion approaches may, in some cases, address the resolution issues related to geophysical imaging by directly integrating geophysical data in the hydrologic estimation problem to identify properties governing flow and transport. While the inherent limitations of geophysical method resolution cannot be fully circumvented, abandoning the idea of producing first a geophysical image may help in that the additional assumptions needed for this purpose (e.g., smoothness) are not necessary, and therefore do not "contaminate" the data inversion.

In this chapter, we work with inverted geophysical images, and highlight previous work regarding the estimation of petrophysical relations and the impacts of tomographic resolution for quantitatively integrating geophysical data into hydrogeologic estimation problems. By comparing emerging petrophysical methods that directly address geophysical resolution, we will offer suggestions for how to address these problems.

2. BACKGROUND

2.1. Traditional Petrophysical Approaches

Theoretical or empirical results are commonly used to develop petrophysical relations. For example, in theoretical studies, effective medium theory is typically used to predict the effective properties of a heterogeneous medium from the properties of its components. A well-known example of this approach to petrophysics was developed by *Hashin and Shtrikman* [1962; 1963] who estimated bounds on the effective magnetic and elastic properties of a composite medium based on the properties of the individual components. Other methods, such as differential effective medium approaches, have been developed when inclusions are sparse and do not form a connected network [e.g., *Berge et al.*, 1993].

At the field scale, interpretation of geophysical data in terms of hydrogeologic properties is often based on the linear regression of field data [e.g., *Kelly*, 1977; *Klimentos and McCann*, 1990; *Purvance and Andricevic*, 2000], or on theoretical relations [e.g., *Urish*, 1981; *Jorgensen*, 1988; *Blair and Berryman*, 1992; *Rubin et al.*, 1992; *Hubbard et al.*, 1997; *Gal et al.*, 1998; *Chan and Knight*, 1999; *Dunn et al.*, 1999; *Wang and Horne*, 2000]. Petrophysical relations can also be determined empirically at the laboratory scale. Two empirical petrophysical relations common in hydrogeophysics are 1) the Topp equation [*Topp et al.*, 1980], where laboratory measurements made on soils were used to fit a polynomial relation between the dielectric constant and water content of a soil, and 2) Archie's law [*Archie*, 1942], where well-log data were used to determine a relation between bulk electrical conductivity and porosity. Research has gone into applying theory to validate these relations, as can be seen in the work of *Hunt* [2004]. *Knight and Endres* [2005] give a comprehensive introduction to traditional petrophysical approaches relevant to hydrogeophysical studies.

Recent work has indicated that laboratory-scale petrophysical relations may not hold at the field scale. *Moysey and Knight* [2004] investigated the relation between dielectric constant and water content assuming that these properties could be represented by spatially correlated random fields. They found that when an electromagnetic wave produced by GPR averages over small-scale heterogeneities, the petrophysical relation at the measurement scale will be different from that defined at the scale of the property variations. Thus, they suggest that a petrophysical relation will be independent of measurement scale only when a medium is self-similar, given the same boundary conditions between the laboratory and field. Studies of this nature demonstrate that understanding how the subsurface is sampled is a critical aspect of developing appropriate petrophysical relations for field-scale problems.

2.2. Inversion and the "Geophysical Filter"

Each geophysical property value estimated in a field survey, such as a tomographic imaging experiment, represents an average of the true properties of the subsurface. Limitations in geophysical resolution mean that this average is rarely representative of only the region of the subsurface contained within the volume defined by a single voxel in the inverted model, leading to the smearing effect that is characteristic of geophysical images. Attempts at quantifying the resolution of the geophysical estimates date to *Backus and Gilbert* [1968], who consider each estimated parameter (or voxel) to depend upon some surrounding model space. Since this seminal work, quantifying tomographic resolution and geophysical measurement support has become an active area of geophysical research [e.g., *Menke*, 1984; *Ramirez et al.*, 1993; *Rector and Washbourne*, 1994; *Schuster*, 1996; *Oldenburg and Li*, 1999; *Alumbaugh and Newman*, 2000; *Friedel*, 2003; *Sheng and Schuster*, 2003; *Dahlen*, 2004]. Resolution has been found to be dependent on the measurement physics; survey design; measurement error; regularization criteria and inversion approach. In other words, the resolution of a target in the subsurface depends not only on the data collection, but on how the data are modeled and inverted. We conceptually refer to the cumulative effect of the factors that cause a loss in resolution between the true distribution of properties in the earth and that estimated by a geophysical survey as the *geophysical filter*.

We use an Occam's inversion, which is closely related to the Gauss-Newton approach (varying only in the explicit way in the which alpha is determined), for geophysical data inversion. We seek to identify the vector of model parameters that minimize an objective function, F, which consists of: 1) the least-squares, weighted misfit between observed and predicted measurements in the first term, and 2) a measure of solution complexity in the second term:

$$F = \left(\mathbf{d} - g(\hat{\mathbf{m}})\right)^T \mathbf{C}_D^{-1}\left(\mathbf{d} - g(\hat{\mathbf{m}})\right) + \alpha\left(\hat{\mathbf{m}} - \mathbf{m}_0\right)^T \mathbf{D}^T \mathbf{D}\left(\hat{\mathbf{m}} - \mathbf{m}_0\right) \quad (1)$$

where

- $\mathbf{d}$ is the vector of measurements;
- $g(\cdot)$ is the forward model;
- $\hat{\mathbf{m}}$ is the vector of parameter estimates, i.e., the calculated data;
- $\mathbf{m}_0$ is the model prior;
- $\mathbf{C}_D$ is the covariance matrix of measurement errors;
- α is a weight that determines the tradeoff between data misfit and regularization; and
- $\mathbf{D}$ is the model-weighting regularization matrix (e.g., a discretized second-derivative filter).

The model parameters are updated in an iterative fashion by repeated solution of a linear system of equations for $\Delta\hat{\mathbf{m}}$, the model update, at successive iterations such that

$$\left[\mathbf{J}^T\mathbf{C}_D^{-1}\mathbf{J} + \alpha\mathbf{D}^T\mathbf{D}\right]\Delta\hat{\mathbf{m}} = \mathbf{J}^T\mathbf{C}_D^{-1}\left(\mathbf{d} - g(\hat{\mathbf{m}}_{k-1})\right) - \alpha\mathbf{D}^T\mathbf{D}\left(\hat{\mathbf{m}}_{k-1} - \mathbf{m}_{prior}\right) \quad (2)$$

$$\hat{\mathbf{m}}_k = \hat{\mathbf{m}}_{k-1} + \Delta\hat{\mathbf{m}} \quad (3)$$

where

J is the Jacobian matrix, with elements $J_{ij} = \partial\hat{d}_i/\partial\hat{m}_j$;
$\hat{d}_i$ is the calculated value of measurement i;
$\hat{\mathbf{m}}_k$ is the vector of parameter estimates after updating in iteration k; and
$\Delta\hat{\mathbf{m}}$ is the vector of parameter updates for iteration k.

At each iteration of the inversion, a new Jacobian is calculated. A line search is performed to identify the α value such that the new model estimate from solution of equation (2) results in the expected root-mean squared (RMS) prediction error given the model of measurement errors. If such a value cannot be found, then the α that gives the lowest RMS error is taken, and the algorithm proceeds to the next iteration. The inversion continues until 1) the RMS error reaches the target RMS error, 2) the reduction in RMS error between successive iterations or the size of the objective function is less than a specified tolerance, or 3) a maximum number of iterations is reached. This approach is commonly referred to as an Occam inversion.

The model resolution matrix, which describes the degree to which model parameters can be determined independently from each other, can be calculated using the Jacobian matrix. The rows of the resolution matrix should sum to 1, and describe the smearing of the true model parameter; conceptually, the model resolution matrix is the lens or filter through which the inversion sees the study region:

$$\hat{\mathbf{m}} = \left[\mathbf{J}^T\mathbf{C}_D^{-1}\mathbf{J} + \alpha\mathbf{D}^T\mathbf{D}\right]^{-1}\mathbf{J}^T\mathbf{C}_D^{-1}\left(\mathbf{d} - \mathbf{J}\mathbf{m}_{prior}\right) + \mathbf{m}_{prior}$$
$$\approx \left[\mathbf{J}^T\mathbf{C}_D^{-1}\mathbf{J} + \alpha\mathbf{D}^T\mathbf{D}\right]^{-1}\mathbf{J}^T\mathbf{C}_D^{-1}\mathbf{J}\left(\mathbf{m}_{true} - \mathbf{m}_{prior}\right) + \mathbf{m}_{prior}, \quad (4)$$

where the model resolution matrix, **R**, is defined as

$$\mathbf{R} = \left[\mathbf{J}^T\mathbf{C}_D^{-1}\mathbf{J} + \alpha\mathbf{D}^T\mathbf{D}\right]^{-1}\mathbf{J}^T\mathbf{C}_D^{-1}\mathbf{J}. \quad (5)$$

Therefore,

$$\hat{\mathbf{m}} = \mathbf{R}\mathbf{m}_{true} + (\mathbf{I} - \mathbf{R})\mathbf{m}_{prior} \quad (6a)$$

where **I** is the identity matrix. Commonly, the prior model is spatially uniform. Under these conditions, equation 6a becomes

$$\hat{\mathbf{m}} = \mathbf{R}\mathbf{m}_{true}. \quad (6b)$$

For linear problems, where the elements of **J** are independent of the values of $\mathbf{m}_{true}$, **R** can be calculated prior to data collection. For non-linear problems, **R** can be calculated using the **J** and α from the last iteration of the inversion, and equation (6b) becomes approximate [e.g., *Alumbaugh and Newman*, 2000]. Since the model resolution matrix describes the loss of resolution incurred during a geophysical survey, it is a quantitative representation of the geophysical filter. According to equation (6), tomographic estimates can be interpreted as weighted averages of the true values of the imaged property, where the weights are described by the rows of **R**:

$$\hat{m}_i = \sum_{j=1}^{n} R_{ij} m_j^{true}. \quad (7)$$

In general, tomograms exhibit smaller variance and greater correlation lengths than the underlying property, and this distortion tends to be anisotropic and non-stationary (Plate 1). Because the correlation between point-scale measurements of hydrologic and geophysical properties is degraded by the inversion process (which may be quantifiable through **R** in certain circumstances), the use of standard petrophysical relations with field-scale tomograms may not be appropriate. *Cassiani et al.* [1998] noted correlation loss between tomographic estimates of seismic velocity and hydraulic conductivity in poorly resolved regions of tomograms. In an ERT study to monitor a fluid tracer in the unsaturated zone, *Binley et al.* [2002] applied locally derived petrophysical relations to convert resistivity tomograms to changes in moisture content; their analysis revealed a 50% mass balance error that was attributed to the poor sensitivity in the center of the image volume where the tracer was applied. In an effort to monitor tracer experiments with ERT, *Singha and Gorelick* [2005] noted the impact of regularization and inversion artifacts on the estimated tracer mass and spatial variance, and demonstrated that Archie's law failed to accurately reproduce solute concentrations without consideration of resolution.

2.3. New Approaches to Field-Scale Petrophysics

Because of issues associated with poor geophysical resolution and limited collocated data, numerous scientists have attempted to develop a correction that could be applied to their geophysical data. *McKenna and Poeter* [1995] noted weak correlation between tomographic estimates of seismic velocity and collocated measurements of hydraulic conductivity compared to the correlation seen for higher resolution sonic logs; they derived a correction based on regression

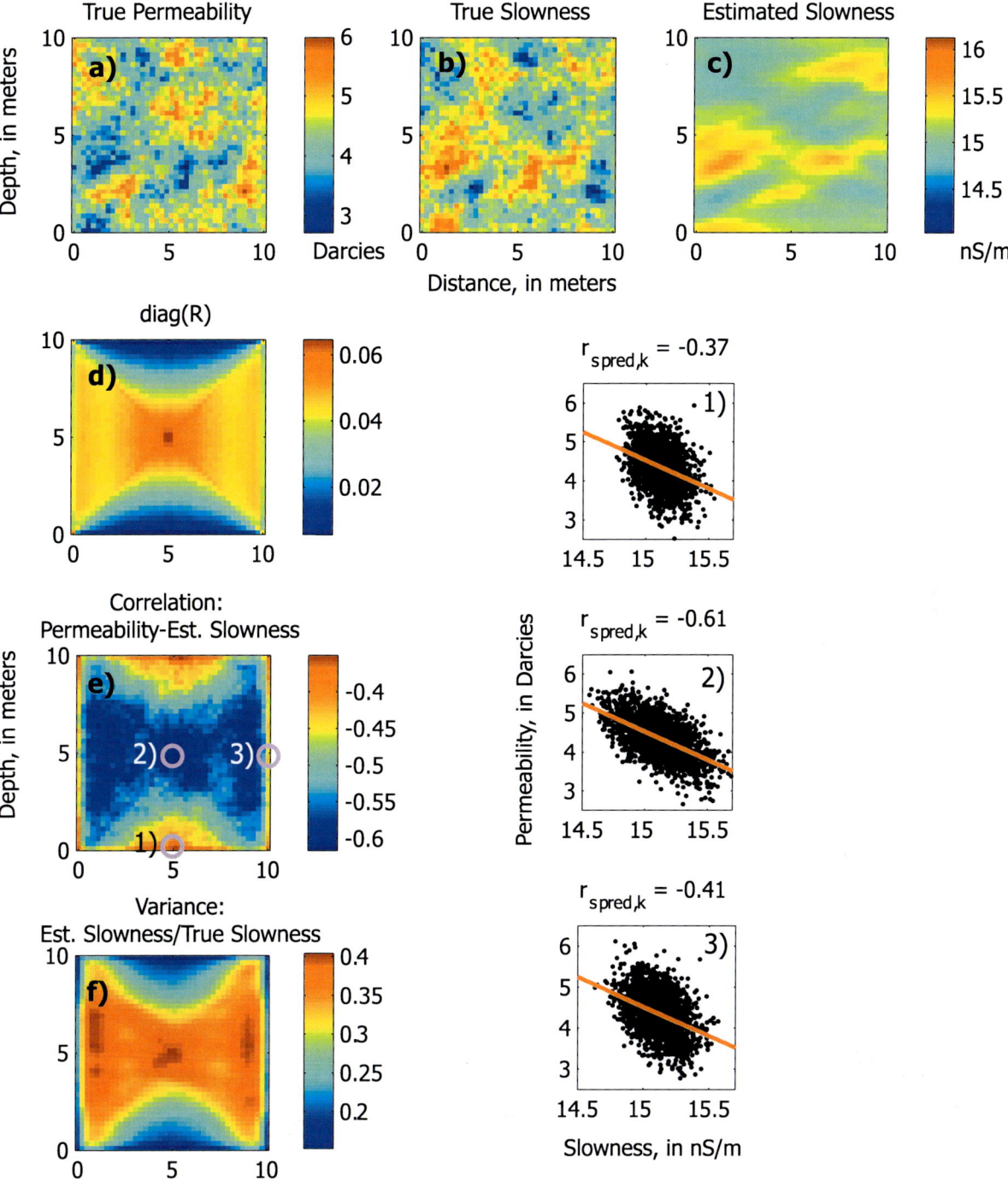

Plate 1. Radar travel time tomography for a field where a linear correlation with permeability is assumed. Cross sections of (a) true permeability; (b) true slowness; (c) inverted tomogram of slowness; (d) diagonal of the model resolution matrix; (e) the predicted correlation coefficient between true and estimated slowness; and (f) the predicted variance of the inverted tomogram normalized by the variance of the true slowness. Three pixels in the correlation coefficient matrix are highlighted [(1), (2), and (3)], showing that the estimated relation between permeability and slowness is spatially dependent.

and applied the correction uniformly over the tomogram to correct for the correlation loss between velocity and hydraulic conductivity. *Hyndman et al.* [2000] used an approach that combined geostatistical simulation, flow and transport simulation, and regression methods to calibrate a linear, field-scale relation between estimated seismic slowness and the logarithm of hydraulic conductivity. *Mukerji et al.* [2001] provided a framework for "statistical" petrophysics to account for conditions not explicitly represented in a data set with limited collocated measurements. In their approach, fluid saturations that were not directly observed at a well location but were likely to occur in the subsurface were included in the calibration data set. They did this by using Gassmann's relation [1951] to predict the change in seismic velocity for a given change in saturation. An assumption in these approaches is that the resulting petrophysical relations are not dependent on spatial location within the subsurface; recent theoretical work by *Day-Lewis and Lane* [2004], however, indicates that this assumption is often not appropriate.

2.3.1 Random field averaging approach. Day-Lewis and Lane [2004] developed an analytical method to determine the correlation loss between hydrological and geophysical measurements as a function of measurement physics, survey geometry, measurement error, spatial correlation structure of the subsurface, and regularization. This was accomplished by combining random field averaging (RFA) [*VanMarcke*, 1983], which allows calculation of the statistical properties of weighted averages of random functions, and the definition of the model resolution matrix (equation 6b).

As shown previously, tomographic estimates can be interpreted as weighted averages of point-scale properties. Estimating pixel values as a weighted average using equation (7), applying the random field average of *VanMarcke* [1983], making a Markov-type approximation [*Journel*, 1999], and calculating the cross-covariance between the geophysical parameter, m, and the hydrologic parameter of interest, p, we find:

$$\hat{\sigma}_{\hat{m}_i}^2 = \sum_{j=1}^{N}\sum_{k=1}^{N} R_{ij} R_{ik} \sigma_{m_j,m_k}, \qquad (8)$$

$$\hat{\sigma}_{\hat{m}_i,\hat{m}_k} = \sum_{j=1}^{N}\sum_{l=1}^{N} R_{ij} R_{kl} \sigma_{m_j,m_l}, \qquad (9)$$

$$\hat{r}_{\hat{m}_i,p_i} = \frac{\sigma_{\hat{m}_i,p_i}}{\sqrt{\sigma_{\hat{m}_i}^2 \sigma_{p_i}^2}} \approx r_{m,p}\hat{r}_{m_i,\hat{m}_i} = r_{m,p}\sum_{j=1}^{N} R_{ij}\sigma_{m_i,m_j} \Big/ \sqrt{\sigma_m^2 \hat{\sigma}_{\hat{m}_i}^2}. \qquad (10)$$

where $\sigma_{x,y}$ is the covariance between variables x and y where x or y may be m and $\hat{m}$ or p;

σ_x^2 is the variance of variable x; and
$r_{x,y}$ is the correlation coefficient between collocated values of x and y at the point scale.

These equations allow us to predict the variance reduction of the pixel-scale tomographic estimate ($\hat{m}_i$) compared to the point-scale property (m_i) (equation 8), the spatial covariance of the tomogram (equation 9), and the correlation loss between the tomographic estimate ($\hat{m}_i$) and collocated hydrologic property p_i (equation 10). Figure 1 shows a flowchart of this approach, which involves seven steps:

1. *Construction of "small-scale" hydrogeologic property realizations:* A realization of the geophysical parameter is generated, assuming a known covariance structure and second-order stationarity.
2. *Application of petrophysical relation:* Site-specific laboratory measurements and/or petrophysical theories are used to generate a field of the hydrologic property of interest, from the realization of the geophysical parameter.
3. *Geophysical forward modeling:* Synthetic geophysical data are calculated using an analytical or numerical model for the measurement physics and survey geometry. Given a model of expected measurement errors, random errors may be added to the data.
4. *Geophysical inverse modeling:* The synthetic measurements obtained via forward modeling in step 3 are inverted.
5. *Random Field Averaging (RFA)*: The RFA equations (Equations 8–10) are applied to predict the parameters describing the pixel-specific statistical distributions of the estimated geophysical parameter.
6. *Construct Bivariate Probability Distribution Functions (PDFs) between True and Estimated Geophysical Parameters:* Assuming Gaussian distributions, we construct bivariate probability distribution functions between the true and estimated geophysical parameters.
7. *Application of petrophysical relation:* Using the petrophysical relation from step 2, we convert the PDF from step 6 to a bivariate PDF between the estimated geophysical parameter and true hydrologic parameter.

For the simplified case of linear, straight-ray radar tomography and linear correlation between radar slowness (1/velocity) and the natural logarithm of permeability, *Day-Lewis and Lane* [2004] derived formulas to predict 1) how the inversion process degrades the correlation between imaged slowness and permeability, compared to point measurements; 2) how the variance of the estimated slowness compares to the variance of the true slowness; and 3) how the inversion alters the spatial covariance of the estimated slowness. This work was expanded in *Day-Lewis et al.* [2005] to consider non-linear tomographic inversion and non-linear petrophysi-

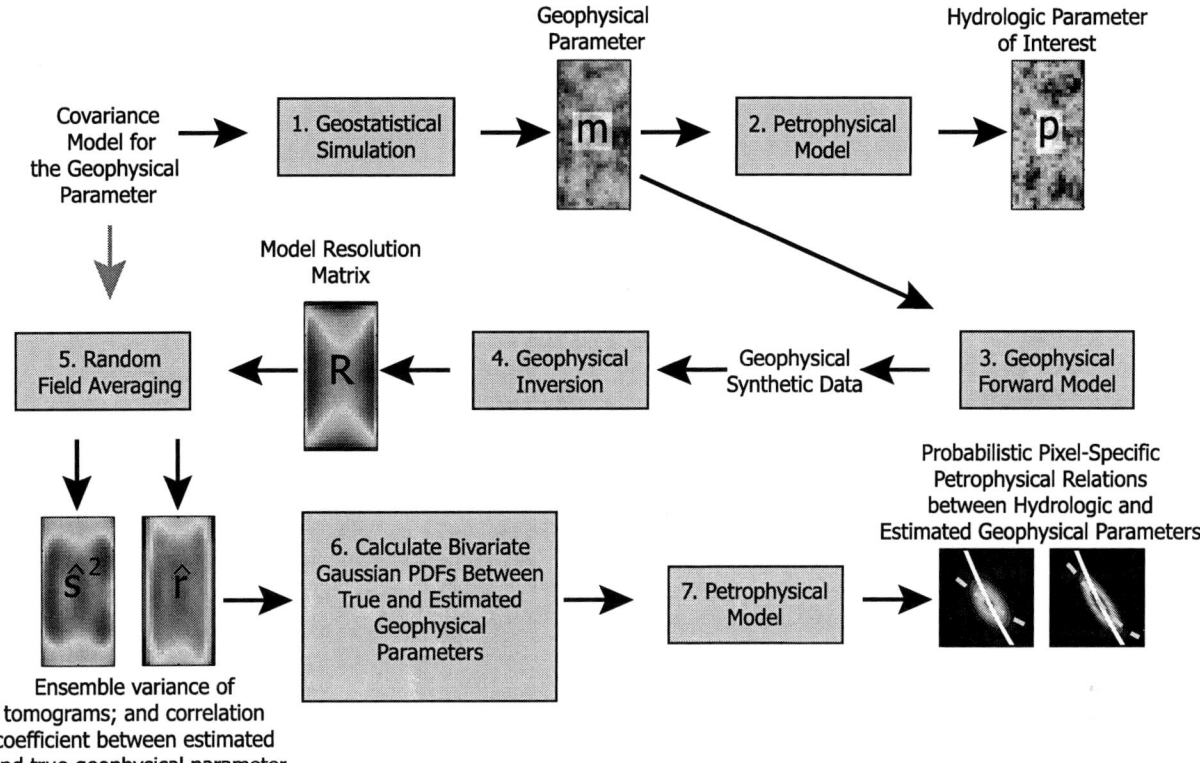

Figure 1. Flowchart for Random Field Averaging Analysis. Starting from an assumed covariance describing the spatial structure of the geophysical parameter, a realization is generated and converted using the petrophysical model to the hydrologic property of interest. Synthetic geophysical data are calculated in the next step. Then the data are inverted and the model resolution matrix calculated. Random field averaging is used to upscale the spatial covariance based on the model resolution matrix and to calculate (1) the ensemble variance of the estimated geophysical parameter and (2) the correlation coefficient between the estimated and true geophysical parameters. Based on these results, bivariate probability distribution functions between the true and estimated geophysical parameter are calculated. In the final step, the bivariate distributions are transformed using the petrophysical model to yield pixel-specific petrophysical relations. Adapted from *Day-Lewis et al.* [2005].

cal relations. In this later work, patterns of correlation loss and variance reduction for both ERT and fresnel-volume, or "fat ray", radar-traveltime tomography were investigated. In this case, because the petrophysical relations are non-linear, a more flexible approach was used, which considers the relation between the true and estimated geophysical parameter; the petrophysics was applied afterward to estimate hydrologic properties. The authors found that while ERT generally performs better near boreholes, where the electrodes were located, and radar-traveltime tomography performs better in the interwell region, the degradation in the relation between the geophysical and hydrologic property is a function of multiple factors: subsurface heterogeneity, the regularization used in the inverse problem, and the number and geometry of data collected. Consequently, imaging targets in the field is dependent on the distance of the targets from the electrodes, the number of data and the geometry with which they are collected, and the type of smoothing used to obtain convergence in the geophysical inverse problem.

The principal benefits of the RFA approach are that 1) it is semi-analytical and therefore provides clear insights into how choices of survey geometry, inversion parameters, or regularization criteria impact the use of tomograms for hydrologic estimation; and 2) it is no more CPU-intensive than the resolution modeling performed as part of a rigorous analysis of tomographic data.

Several key issues limit the applicability of the RFA approach. Whereas the RFA equations provide a semi-analytical way to estimate the degradation in the relation between hydrologic and geophysical parameters, the approach is based on a number of assumptions that may limit its utility to field applications, including: 1) the geophysical parameter is normally distributed, 2) both the geophysical and hydrologic properties share the same covariance structure, and 3) both properties are second-order stationary, i.e., the mean and variance are spatially uniform and the covariance between two points depends only on the separation between them.

2.3.2 Full inverse statistical calibration. An alternative approach to accounting for geophysical resolution in petrophysical relations is to generate a large number of subsurface property realizations, forward model and invert the tomographic experiment for each realization, and finally compare the resulting tomograms with the original realizations to assess the impact of the geophysical filter. *Moysey et al.* [2005] and *Singha and Moysey* [2006] give examples where this type of a numerical simulation approach is used to capture the spatial variability in resolution of geophysical surveys. The method is referred to as Full Inverse Statistical (FISt) calibration because a full forward and inverse simulation must be performed for each realization used in the calibration (or statistical inference) of the field-scale petrophysical relation. The approach used by these authors follows a six-step process, and is outlined schematically in a flowchart in Figure 2:

1. *Construction of "small-scale" hydrogeologic property realizations:* A set of realizations of the hydrogeologic property of interest is created using a technique that honors both the available data and conceptual model for the field site, e.g., geostatistical simulation or flow and transport modeling. These realizations should be simulated at an appropriate scale such that small-scale heterogeneities, should they be known, can be captured by effective parameters, but larger-scale heterogeneities affecting hydrologic (or geophysical) behavior are explicitly represented.
2. *Application of a traditional petrophysical relation:* Site-specific laboratory measurements and/or petrophysical theories are used to determine relations between the hydrogeologic and geophysical properties under investigation. Geophysical property realizations can then be obtained from the hydrogeologic property realizations generated in step 1 using this relation.
3. *Geophysical forward modeling:* A numerical analog to the experiment executed in the field is performed on each geophysical property realization from step 2. The numerical experiment should parallel as closely as possible the real field experiment in both experimental design (e.g., survey geometry, data acquisition parameters) and representation of the relevant physical processes.
4. *Geophysical inverse modeling:* The synthetic measurements obtained via forward modeling in step 3 are then inverted for each realization. The forward and inverse geophysical model need not use the same grid; in fact, doing so assumes that there are no subgrid-scale heterogeneities impacting the data. The inversion of the measurements into tomograms mimics the inversion of the field data, including the parameterization (i.e., model grid) and selection of regularization criteria. The goal is to reproduce the processing and inversion steps that have been applied to the field measurements.
5. *Generation of field-scale hydrogeologic property realizations:* Each hydrogeologic property realization is upscaled to the model grid selected in step 4 using an appropriate spatial weighting function. For example, if hydrogeologic properties of interest are volumetric properties, e.g., water content, this step can be carried out using volumetric averaging.
6. *Development of "apparent" or field-scale petrophysical relations:* The sets of field-scale hydrogeologic analogs from step 5 and geophysical analogs from step 4 are used to calculate the apparent petrophysical relation at every location in space (as defined by a pixel or voxel). The petrophysical relations should also be updated for each observation time during a monitoring experiment. The resulting relations can then be used to post-process the real-world geophysical properties to obtain an estimate of the hydrogeologic properties for the actual field site.

One final practical consideration in implementing FISt calibration is the decision of how to find the best-fit line between the geophysical and hydrologic property for each pixel. There are many different means of obtaining a best-fit line. The most obvious choice, least-squares regression, produces a result that depends on which variable is considered independent—the best-fit linear relation of permeability versus velocity may be quite different from that of velocity vs. permeability. Ideally with FISt, we want to find the relation that allows the estimated and "true" property, permeability in this case, to fall on a 1:1 line. One way to do this is to regress both the x and y axes and minimize the sum of squares of the perpendicular distances from the line, i.e., major-axis regression. Another possibility is to consider a distribution transformation:

$$Y = \sigma_Y \left(\frac{X - \mu_X}{\sigma_X} \right) + \mu_Y \qquad (11)$$

In theory, the field-scale petrophysical relation determined at each spatial location could be a non-parametric estimate of the joint probability density function (PDF) between the geophysical and hydrologic parameters of interest. In practice, such inference would require a large number of simulations to be performed. When the full joint PDF is not needed, e.g., if a multi-modal PDF is not of concern, it is more practical to fit a simple model (e.g., linear relation) at each spatial location using a limited number of realizations.

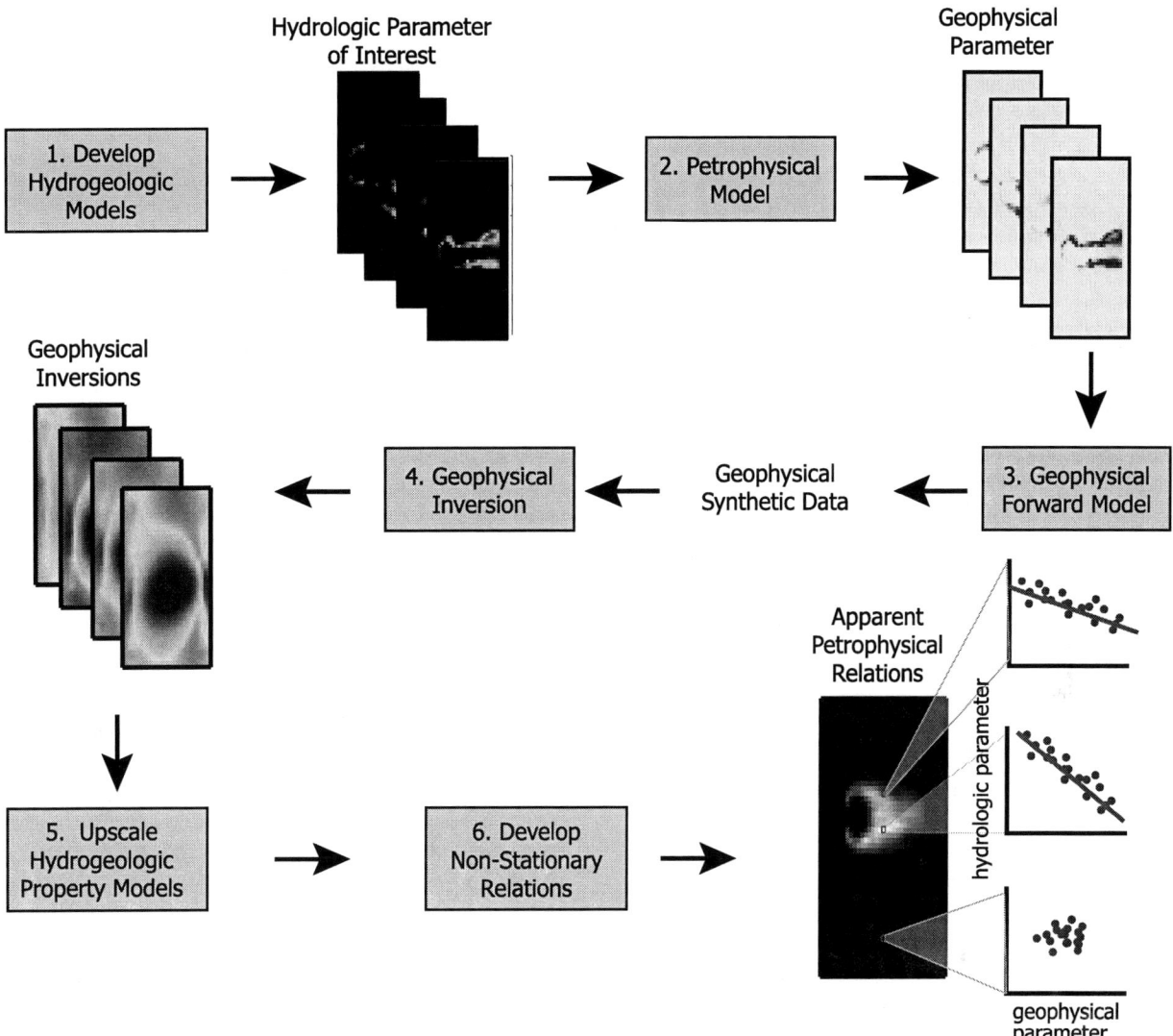

Figure 2. Flowchart for FISt calibration. Realizations of hydrologic properties can be generated by geostatistical calibration, flow and transport models, or from conceptual understanding of field sites. These realizations are converted to geophysical properties through a petrophysical relation. Following this step, forward and inverse geophysical simulations are conducted. By considering multiple realizations, relations are then built between the inverted geophysical parameter and the hydrologic parameter at every pixel to account for spatially variable resolution. These relations are applied to an inversion of field geophysical data for a better estimation of hydrologic properties than otherwise attainable. Adapted from *Moysey et al.* [2005] and *Singha and Gorelick* [2006].

In 2D synthetic examples, *Moysey et al.* [2005] and *Singha and Moysey* [2006] used FISt calibration to improve estimates of water content and solute concentration for radar traveltime and ERT experiments, respectively. Extending this work to a 3D transient system in the field, *Singha and Gorelick* [2006] applied a similar approach to ERT monitoring of a tracer test performed at the Massachusetts Military Reservation, Cape Cod, Massachusetts. The tracer concentrations and total solute mass estimated from the ERT survey were in better agreement with multi-level sampler results when the authors used field-scale petrophysical relations rather than Archie's law to convert resistivity to concentration. *Singha and Gorelick* [2006] also demonstrated that rather than develop "apparent" petrophysical relations between a hydrologic and geophysical property, FISt can also be used to "correct" tomograms by building relations between the true and estimated geophysical property. The benefit of comparing the true and estimated geophysical parameter, rather than converting to a hydrologic parameter, is that the relation between them is likely to be well-described by a bivariate normal distribution that can be captured in a linear relation.

One benefit of FISt calibration is that it is conceptually straightforward to implement. Another strength of this method is that it is easy to account for different kinds of uncertainty within the Monte Carlo-type framework of the method. For example, if the hydrologic conceptual model at a site is considered uncertain, it is straightforward to implement FISt calibration using realizations based on different conceptual models or generated using different simulation techniques. This degree of flexibility makes the method potentially quite powerful. A third advantage of the FISt approach is the capability to condition to secondary information, e.g., direct point measurements of the geophysical or hydrologic properties under investigation.

There are at least three key issues that should be considered when using FISt calibration: 1) a meaningful relation between the geophysical and hydrologic parameters, 2) the appropriateness of the realizations used in the simulations with respect to the field-site hydrogeology, and 3) the ability of the numerical models used in the forward simulations to adequately capture 'real-world' processes. The lack of geophysical data sensitivity to the hydrologic properties of interest, i.e., non-informative data, is a general problem that will defeat any approach to data integration. The second and third issues, however, are important considerations that can potentially lead to biased estimates of the resulting field-scale petrophysical relation, and therefore, inaccurate estimates of hydrologic properties. An additional consideration in using FISt calibration is that it is computationally expensive, especially for 3D transient problems.

Both FISt calibration and the RFA-based approach are similar to traditional petrophysics in that they use a mathematical (or numerical) model to describe how geophysical measurements sample the subsurface. In contrast to traditional approaches, however, these methods also account for the impacts of inversion on geophysical resolution rather than focusing on how a single measurement averages the subsurface. Both the RFA method and FISt calibration can determine the petrophysical relation between a geophysical and hydrologic variable as a statistical association, captured by a PDF that is explicitly dependent on spatial location, therefore inherently accounting for the spatially varying resolution of geophysical surveys. The overall conceptual similarity between the two approaches is made apparent by comparing the flowcharts in Figures 1 and 2. The main difference between the two approaches is that RFA is a semi-analytical approach that relies on an assumption of second-order stationary distributions, whereas FISt calibration is a non-parametric, numerical approach that allows for any model of spatial variability. In summary, the RFA method will typically be more computationally efficient, but the flexibility of FISt calibration makes it more generally applicable.

3. EXAMPLE

We demonstrate the utility of FISt and RFA for a synthetic example where radar slowness is considered to be linearly related to permeability. Although FISt has been used in the past to convert tomograms to hydrologic estimates, RFA has not; RFA has been used only to predict the loss of information arising from limited survey geometry, regularization criteria, measurement errors, and other factors that affect tomographic resolution. Here, we demonstrate how the field-scale petrophysical relations generated with RFA can be used to produce more reliable hydrologic estimates from tomograms.

Our example assumes a linear relation with a correlation coefficient of -0.9 between radar slowness and permeability. There have been examples where similar relations were assumed to be valid [e.g., *Hubbard et al.*, 1997; *Linde et al.*, 2006]; we note, however, that such a linear model may only be hypothetical, with limited realism in many or most field scenarios, and should be applied in the field only with great care. Correlated permeability and slowness fields (Figure 3a, b, respectively) are generated using sequential Gaussian simulation with an exponential covariance model, assuming an isotropic correlation length of 2.5 m, mean $\ln k$ [darcies] of 4.35, variance of $\ln k$ of 0.25, mean radar slowness of 15.12 ns/m, and variance of radar slowness of 0.0973 ns^2/m^2. Radar traveltimes are calculated assuming straight raypaths, for 1600 measurements with antenna spacings of 0.25 m along each 10-m deep borehole. The grids for forward simulation and tomographic inversion are identical, with a discretization of 0.25-m square pixels. For simplicity, we assume straight rays for forward modeling traveltimes. More sophisticated eikonal-solver forward models have been considered in other applications [*Day-Lewis et al.*, 2005; *Moysey et al.*, 2005]; however, for our present purpose of comparing FISt and RFA for linear problems, straight rays are appropriate and sufficient. The measurement errors are assumed to be normally distributed with zero mean and standard deviation of 2.0 ns (1.5%); this standard error is large relative to common sampling periods, but is intended to represent the combined effects of errors in traveltime picking, inaccurate borehole deviation and antenna positions, and modeling errors arising from the straight-ray approximation. It should be noted that errors are considered independent in this synthetic example, but are likely correlated in real field data. The inverse model

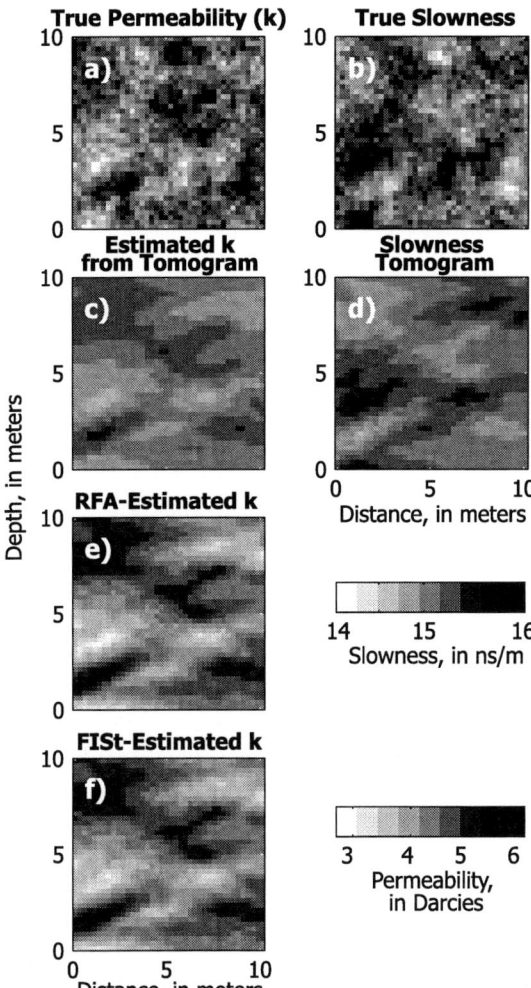

Figure 3. Application of FISt and RFA to a synthetic example of estimating permeability from crosswell radar data, where (a) is the true permeability and (b) is the true slowness. Shown are the (c) estimated permeability and (d) slowness distributions obtained using inversion and the application of our assumed relation; (e) the estimation of permeability using RFA; and (f) the estimation of permeability using FISt.

is linear and amounts to a single iteration of equations (2–3), with J_{ij} equal to the length of raypath i in pixel j.

Compared to the true slowness field (Figure 3b), the tomogram (Figure 3d) is smoother and shows less variation in slowness. The tomography resolves only large-scale structures and even these are smeared or blunted. The differences between true and estimated slowness arise because of measurement errors, the limited data collection geometry, and regularization; therefore, after tomographic reconstruction, the relation between true permeability and estimated slowness varies within the tomogram as a function of the spatial variability in resolution. Consequently, any estimation of permeability from radar data is biased, even in this simple synthetic example where a strong correlation between permeability and slowness exists. To demonstrate the degree of bias, we use the linear relation between true slowness and permeability to convert the tomogram to a cross-section of permeability estimates (Figure 3c), and plot the permeability estimates against true, synthetic permeability (Figure 4a). High values of permeability are underestimated and low values are overestimated; the loss of extreme values is clearly evident in quantile-quantile comparison between estimated and true permeability (Figure 4d).

Correction of the tomogram with FISt and RFA shows that accounting for spatial variability can improve the estimation of permeability values without greatly impacting the fit to the data (Figure 3e, f; Table 1). To create the slope and intercept maps for both FISt and RFA, we transformed the distribution of slowness to the natural logarithm of permeability assuming both to be normal—while this assumption may not always be appropriate in the field, it is applicable for this example. The updated cross sections of permeability show increased variance; areas of high and low slowness (and permeability) are better captured in the RFA and FISt estimates (Figure 3e, f) than in the original tomogram. The RMS error and overall correlation between true and estimated slowness are slightly poorer than in the original

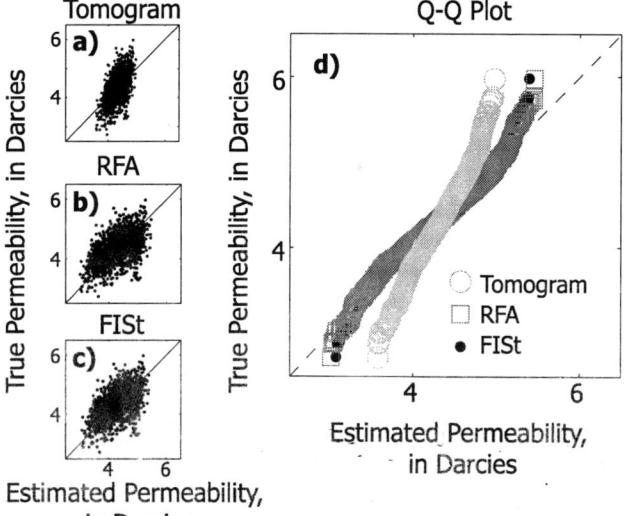

Figure 4. Plots of true versus estimated permeability illustrate the accuracy of the estimates obtained using: (a) tomographic inversion and application of "true" relation, (b) RFA, and (c) FISt; accurate estimates fall along the 1:1 line. The (d) Q-Q plot, which compares the quantiles of the distributions of the true versus estimated permeabilities, shows that the estimate from tomographic inversion significantly underestimates the range of true permeability. Both RFA and FISt match the permeabilities in the central quartiles, but still underestimate the extremes.

Table 1. Estimates of correlation between true and estimated permeability over the entire tomogram, the RMS error associated with the slowness tomograms versus the original field, the forward-modeled data misfit associated with the slowness tomogram, and the fraction of the variance in the slowness tomogram compared to the original slowness field.

	k-k̂ Correlation	RMS Error	Data Misfit	Fraction of True Variance
Original Tomogram	0.577	0.408	0.996	0.299
FISt-corrected	0.562	0.460	1.03	0.94
RFA-corrected	0.560	0.476	1.05	1.07

tomogram; however, these metrics may be less important to a hydrogeologist than capturing the tails of the permeability distribution. While the correlation between the true and estimated permeability does not markedly increase (Figure 4b, c), the scatter better fits the 1:1 line than the original tomogram, and the variance in the permeability is greatly improved, as illustrated in quantile-quantile plots (Figure 4d). Improved mapping of the high and low permeabilities in the field, should a relation between permeability and slowness exist, would be critical to understanding flow and transport processes in a given setting.

4. DISCUSSION

FISt calibration and RFA provide improved methods for estimating hydrologic properties (in this case, permeability) from geophysical data, should the relation between the geophysics and hydrogeology exist. Both perform equally well for the example presented; however, both have important limitations when being applied to field data. First and foremost, both require that a relation between the geophysical and hydrologic property exists. While this relation is relatively clear in some circumstances (changes in water content or total dissolved solids and electrical resistivity, for instance), estimation of static properties such as permeability or the concentration of a dissolved non-aqueous phase liquid at some time may be impossible for geophysical methods to image directly. An assumption of correlation between hydrologic and geophysical parameters that does not exist may lead one to believe that geophysical methods can be used to estimate hydrologic processes when the geophysical data are actually uninformative. Another major issue is the sensitivity of the tomogram appearance to the final RMS error. Both methods may overpredict the hydrologic property if the data are overfitted, which could happen with the translation of geophysical tomograms to hydrologic data given standard petrophysical relations. It is similarly important that all the inversions for the numerical analogs used in FISt and the field data converge to similar error levels. If the data are over- or underfitted, the final tomograms exhibit different degrees of spatial variation; consequently the apparent petrophysical relations may not be meaningful.

The RFA approach depends only on the resolution matrix and covariance models to estimate correlation loss. Numerous studies, however, have indicated that two-point statistics, or covariance models, are inappropriate for generating realistic connectivity in highly heterogeneous media [e.g., *Western et al.*, 1998; *Caers et al.*, 2003; *Zinn and Harvey*, 2003; *Knudby and Carrera*, 2005]. While RFA is more mathematically rigorous than FISt, it is unlikely to be used to correct tomograms in the field, as the assumption of second-order stationarity is unlikely to be valid. FISt therefore has a wider applicability in real-world scenarios, where objects such as plumes, which are not easily described by two-point statistics, may be the target of interest. Additionally, RFA may prove to be limited in cases of highly nonlinear physics. The calculation of **R** in nonlinear methods is local to the final solution; for mild non-linearity, the resolution will be relatively stable, but this may not be the case in the presence of high velocity contrasts. This limitation has the potential to impact the applicability of RFA to some scenarios.

While FISt has the advantage of not being limited by a covariance model, the method is computationally expensive, and like other Monte Carlo methods may require many hundreds or thousands of inversions to obtain good results [*Peck et al.*, 1988]. FISt is also model-dependent, and requires an estimate of the subsurface properties to build the calibration relations. While we often have more data about our field sites than are used in inversion (a geologist's rendition of the field area, for instance), we must attempt to minimize the introduction of features that are unlikely to exist—a poor choice for the underlying hydrogeology in construction of the realizations can cause spurious results [*Singha and Gorelick*, 2006]. Another issue with FISt is that the pixels or voxels are also assumed to be independent, which is not true; they are dependent upon the resolution of the inverse procedure. Each estimated parameter is dependent upon the surrounding model space as dictated by the Backus and Gilbert averaging kernel [*Backus and Gilbert*, 1968]. This is not a concern for RFA, which uses the entire resolution matrix when calculating relations. A strength of FISt, however, is that it remains applicable with nonlinear inversion [e.g., *Singha and Gorelick*, 2006; *Singha and Moysey*, 2006].

While these methods can be used in the "design" phase of a survey to assess a priori the nature of the resulting, space-dependent, constitutive models, they can also be used to estimate hydrologic properties a posteriori from field data by developing petrophysical models specific to given acquisition geometries and local geology. Despite the complications above, both of these methods provide improved estimates of hydrologic properties and processes when applied with care, and also allow for quantification of the correlation loss between geophysical properties measured in the field, and hydrologic properties of interest, either before entering the field, or once data have been collected.

5. CONCLUSIONS

Geophysical tomograms are plagued by spatially variable resolution, making the estimation of hydrologic properties from them a difficult task. We have presented two methods currently available for mitigating these problems: FISt and RFA. Both have distinct advantages and disadvantages. While RFA provides a semi-analytical way to quantify spatially variable correlation loss, it is limited by the requirement of a known covariance model. FISt, on the other hand, is applicable in situations where two-point statistics may not be valid (the movement of contaminant plumes or infiltration), but is computationally expensive because of the number of realizations that must be considered; FISt (as described here) assumes a linear model between parameters and pixel independence. Nevertheless, both of these methods provide a manner for estimating spatially variable petrophysical relations applicable in field settings, to improve quantification of hydrologic properties and processes from geophysical data.

Acknowledgments. Aspects of this work were funded by the National Science Foundation (awards EAR-0229896 and EAR-012462) and the U.S. Geological Survey Toxics Substances Hydrology Program. The authors are grateful for useful comments from John W. Lane, Jr., James Irving, and Adam Pidlisecky. Any opinions, findings, and conclusions or recommendations expressed in this material are those of the authors and do not necessarily reflect the views of the National Science Foundation.

REFERENCES

Alumbaugh, D. L., P. Y. Chang, L. Paprocki, J. R. Brainard, R. J. Glass and C. A. Rautman, Estimating moisture contents in the vadose zone using cross-borehole ground penetrating radar: a study of accuracy and repeatability, *Water Resources Research,* 38(12): 1309, doi: 10.1029/2001/WR000754, 2002.

Alumbaugh, D. L. and G. A. Newman, Image appraisal for 2-D and 3-D electromagnetic inversion, *Geophysics,* 65(5): 1455–1467, 2000.

Archie, G. E., The electrical resistivity log as an aid in determining some reservoir characteristics, *Transactions of the American Institute of Mining, Metallurgical and Petroleum Engineers,* 146: 54–62, 1942.

Backus, G. E. and J. F. Gilbert, The resolving power of gross earth data, *Geophysical Journal of the Royal Astronomical Society,* 16: 169–205, 1968.

Berge, P. A., J. G. Berryman and B. P. Bonner, Influence of microstructure on rock elastic properties, *Geophysical Research Letters,* 20: 2619–2622, 1993.

Berthold, S. and L. R. H. Masaki, Integrated hydrogeological and geophysical study of depression-focused groundwater recharge in the Canadian prairies, *Water Resources Research,* 40(6): doi:10.1029/2003WR002982, 14p., 2004.

Binley, A., G. Cassiani, R. Middleton and P. Winship, Vadose zone flow model parameterisation using cross-borehole radar and resistivity imaging, *Journal of Hydrology,* 267: 147–159, 2002.

Blair, S. C. and J. G. Berryman, Permeability and relative permeability in rocks, *Fault Mechanics and Transport Properties of Rocks*, Academic Press Ltd.: . 169–186, 1992.

Caers, J., S. Strebelle and K. Payrazyan, Stochastic integration of seismic data and geologic scenarios, *The Leading Edge,* 22(3): 192–196, 2003.

Carle, S. and A. Ramirez, *Integrated Subsurface Characterization Using Facies Models, Geostatistics, and Electrical Resistance Tomography* Lawrence Livermore National Laboratory, UCRL-JC-136739, Livermore CA, 1999.

Cassiani, G., G. Böhm, A. Vesnaver and R. Nicolich, A Geostatistical Framework for Incorporating Seismic Tomography Auxiliary Data into Hydraulic Conductivity Estimation, *Journal of Hydrology,* 206(1-2): 58–74, 1998.

Chan, C. Y. and R. J. Knight, Determining water content and saturation from dielectric measurements in layered materials, *Water Resources Research,* 35(1): 85–93, 1999.

Dahlen, F. A., Resolution limit of traveltime tomography, *Geophysical Journal International,* 157: 315–331, doi:10.1111/j.1365-246X.2004.02214.x, 2004.

Day-Lewis, F. D. and J. W. Lane, Jr., Assessing the resolution-dependent utility of tomograms for geostatistics, *Geophysical Research Letters,* 31: L07503, doi:10.1029/2004GL019617, 2004.

Day-Lewis, F. D., K. Singha and A. M. Binley, Applying petrophysical models to radar traveltime and electrical-resistivity tomograms: resolution-dependent limitations, *Journal of Geophysical Research,* 110: B08206, doi:10.1029/2004JB003569, 2005.

Dietrich, P., T. Fechner, J. Whittaker and G. Teutsch, *An integrated hydrogeophysical approach to subsurface characterization*, GQ 98 conference, Tubingen, Federal Republic of Germany, [Louvain] : International Association of Hydrological Sciences, 513–519, 1998.

Doyen, P. M., Porosity from seismic data; a geostatistical approach, *Geophysics,* 53(10): 1263–1276, 1988.

Dunn, K.-J., G. A. LaTorraca and D. J. Bergman, Permeability relation with other petrophysical parameters for periodic porous media, *Geophysics,* 64(2): 470–478, 1999.

Friedel, S., Resolution, stability, and efficiency of resistivity tomography estimated from a generalized inverse approach, *Geophysical Journal International,* 153: 305–316, 2003.

Gal, D., J. Dvorkin and A. Nur, A physical model for porosity reduction in sandstones, *Geophysics,* 63(2): 454–459, 1998.

Gassmann, F., Elastic waves through a packing of spheres, *Geophysics,* 16(673–685), 1951.

Hashin, Z. and S. Shtrikman, A variational approach to the theory of effective magnetic permeability of multiphase materials, *Journal of Applied Physics,* 33: 3125–3131, 1962.

Hashin, Z. and S. Shtrikman, A variational approach to the elastic behavior of multiphase materials, *J. Mech. Phys. Solids,* 11: 127–140, 1963.

Hubbard, S. S., J. Chen, J. Peterson, E. L. Majer, K. H. Williams, D. J. Swift, B. Mailloux and Y. Rubin, Hydrogeological characterization of the South Oyster Bacterial Transport Site using geophysical data, *Water Resources Research,* 37(10): 2431–2456, 2001.

Hubbard, S. S., J. E. Peterson, Jr., E. L. Majer, P. T. Zawislanski, K. H. Williams, J. Roberts and F. Wobber, Estimation of permeable pathways and water content using tomographic radar data, *The Leading Edge,* 16(11): 1623–1628, 1997.

Hunt, A. G., Continuum percolation theory and Archie's Law, *Geophysical Research Letters,* 31(19): L19503, 10.1029/2004GL020817, 2004.

Hyndman, D. W. and S. M. Gorelick, Estimating lithologic and transport properties in three dimensions using seismic and tracer data; the Kesterson Aquifer, *Water Resources Research,* 32(9): 2659–2670, 1996.

Hyndman, D. W., J. M. Harris and S. M. Gorelick, Coupled seismic and tracer test inversion for aquifer property characterization, *Water Resources Research,* 30(7): 1965–1977, 1994.

Hyndman, D. W., J. M. Harris and S. M. Gorelick, Inferring the relation between seismic slowness and hydraulic conductivity in heterogeneous aquifers, *Water Resources Research,* 36(8): 2121–2132, 2000.

Jorgensen, D. G., Estimating permeability in water-saturated formations, *Log Analyst,* 29(6): 401–409, 1988.

Journel, A. G., Markov models for cross-covariances, *Mathematical Geology,* 31: 955–964, 1999.

Kelly, W. E., Geoelectric sounding for estimating aquifer hydraulic conductivity, *Ground Water,* 15(6): 420–425, 1977.

Klimentos, T. and C. McCann, Relationships among compressional wave attenuation, porosity, clay content, and permeability in sandstones, *Geophysics,* 55(8): 998–1014, 1990.

Knight, R. J. and A. L. Endres, Physical properties of near-surface materials and approaches to geophysical determination of properties, *Near-Surface Geophysics, Volume 1*, editors. Tulsa, OK, Society of Exploration Geophysicists: 723 pp., 2005.

Knudby, C. and J. Carrera, On the relationship between indicators of geostatistical, flow, and transport connectivity, *Advances in Water Resources,* 28: 405–421, 2005.

Kowalsky, M. B., S. Finsterle, J. Peterson, S. Hubbard, Y. Rubin, E. Majer, A. Ward and G. Gee, Estimation of field-scale soil hydraulic and dielectric parameters through joint inversion of GPR and hydrological data, *Water Resources Research,* 41(W11425, doi:10.1029/2005WR004237), 2005.

Linde, N., S. Finsterle and S. Hubbard, Inversion of tracer test data using tomographic constraints, *Water Resources Research,* 42(4): W04410, 10.1029/2004WR003806, 2006.

McKenna, S. A. and E. P. Poeter, Field example of data fusion in site characterization, *Water Resources Research,* 31(12): 3229–3240, 1995.

Menke, W., *Geophysical data analysis; discrete inverse theory,* New York-London-Toronto, Academic Press, 289, 1984.

Moysey, S. and R. J. Knight, Modeling the field-scale relationship between dielectric constant and water content in heterogeneous systems, *Water Resources Research,* 40: W03510, doi:10.1029/2003WR002589, 2004.

Moysey, S., K. Singha and R. Knight, A framework for inferring field-scale rock physics relationships through numerical simulation, *Geophysical Research Letters,* 32: L08304, doi:10.1029/2004GL022152, 2005.

Mukerji, T., I. Takahashi and E. F. Gonzalez, Statistical rock physics: combining rock physics, information theory, and geostatistics to reduce uncertainty in seismic reservoir characterization, *The Leading Edge,* 20(3): 313–319, 2001.

Oldenburg, D. W. and Y. Li, Estimating the depth of investigation in dc resistivity and IP surveys, *Geophysics,* 64: 403–416, 1999.

Peck, A., S. Gorelick, G. d. Marsily, S. Foster and V. Kovalevsky, *Consequences of spatial variability in aquifer properties and data limitations for groundwater modelling practice,* Oxfordshire UK, IAHS Press, 272 p., 1988.

Purvance, D. T. and R. Andricevic, On the electrical-hydraulic conductivity correlation in aquifers, *Water Resources Research,* 36(10): 2905–2913, 2000.

Ramirez, A., W. Daily, D. LaBrecque, E. Owen and D. Chesnut, Monitoring an underground steam injection process using electrical resistance tomography, *Water Resources Research,* 29(1): 73–87, 1993.

Ramirez, A. L., J. J. Nitao, W. G. Hanley, R. Aines, R. E. Glaser, S. K. Sengupta, K. M. Dyer, T. L. Hickling and W. D. Daily, Stochastic inversion of electrical resistivity changes using a Markov Chain Monte Carlo approach, *Journal of Geophysical Research,* 110: B02101, doi:10.1029/2004JB003449, 2005.

Rector, J. W. and J. K. Washbourne, Characterization of resolution and uniqueness in crosswell direct-arrival traveltime tomography using the Fourier projection slice theorem, *Geophysics,* 59(11): 1642–1649, 1994.

Rubin, Y. and S. S. Hubbard, Eds., *Hydrogeophysics,* Springer, 2005.

Rubin, Y., G. Mavko and J. Harris, Mapping permeability in heterogeneous aquifers using hydrologic and seismic data, *Water Resources Research,* 28(7): 1809–1816, 1992.

Schuster, G. T., Resolution limits for crosswell migration and traveltime tomography, *Geophysical Journal International,* 127: 427–440, 1996.

Sheng, J. and G. T. Schuster, Finite-frequency resolution limits of wave path traveltime tomography for smoothly varying velocity models, *Geophysical Journal International,* 152(3): 669–676, 2003.

Singha, K. and S. M. Gorelick, Saline tracer visualized with electrical resistivity tomography: field scale moment analysis, *Water Resources Research,* 41: W05023, doi:10.1029/2004WR003460, 2005.

Singha, K. and S. M. Gorelick, Hydrogeophysical tracking of 3D tracer migration: the concept and application of apparent petrophysical relations, *Water Resources Research,* 42 W06422, doi:10.1029/2005WR004568, 2006.

Singha, K. and S. Moysey, Accounting for spatially variable resolution in electrical resistivity tomography through field-scale rock physics relations, *Geophysics,* 71(4): A25–A28, 2006.

Slater, L., A. Binley, R. Versteeg, G. Cassiani, R. Birken and S. Sandberg, A 3D ERT study of solute transport in a large experimental tank, *Journal of Applied Geophysics,* 49: 211–229, 2002.

Topp, G. C., J. L. Davis and A. P. Annan, Electromagnetic determination of soil water content: measurements in coaxial transmission lines, *Water Resources Research,* 16(3): 574–582, 1980.

Urish, D. W., Electrical resistivity-hydraulic conductivity relationships in glacial outwash aquifers, *Water Resources Research,* 17(5): 1401–1407, 1981.

Vanderborght, J., A. Kemna, H. Hardelauf and H. Vereecken, Potential of electrical resistivity tomography to infer aquifer transport characteristics from tracer studies: A synthetic case study, *Water Resources Research,* 41: W06013, doi:10.1029/2004WR003774, 23 pp., 2005.

VanMarcke, R., *Random Fields: Analysis and Synthesis,* Cambridge, MA, MIT Press, 1983.

Vereecken, H., A. Binley, G. Cassiani, A. Revil and K. Titov, Eds., *Applied Hydrogeophysics,* NATO Science Series: IV: Earth and Environmental Sciences, Springer-Verlag, 2006.

Wang, P. and R. Horne, N., *Integrating resistivity data with production data for improved reservoir modelling,* SPE Asia Pacific Conference of Integrated Modelling for Asset Management, Yokohama, Japan, Society of Petroleum Engineers, 2000.

Western, A. W., G. Bloschl and R. B. Grayson, How well do indicator variograms capture connectivity of soil moisture?, *Hydrological Processes,* 12: 1851–1868, 1998.

Yeh, T. C. J., S. Liu, R. J. Glass, K. Baker, J. R. Brainard, D. L. Alumbaugh and D. LaBrecque, A geostatistically based inverse model for electrical resistivity surveys and its applications to vadose zone hydrology, *Water Resources Research,* 38(12): 14-1:14-13, 2002.

Zinn, B. and C. F. Harvey, When good statistical models of aquifer heterogeneity go bad: a comparison of flow, dispersion, and mass transfer in connected and multivariate Gaussian hydraulic conductivity fields, *Water Resources Research,* 39(3): doi: 10.1029/2001WR001146, 2003.

F. D. Day-Lewis, U.S. Geological Survey, Office of Ground Water, Branch of Geophysics, 11 Sherman Place, Unit 5015, Storrs, CT 06269, USA. (daylewis@usgs.gov)

S. Moysey, School of the Environment, Clemson University, P.O. Box 340919, Clemson, SC 29634-0919, USA. (smoysey@clemson.edu)

K. Singha, Department of Geosciences, Pennsylvania State University, 311 Deike Building, University Park, PA 16802, USA. (ksingha@psu.edu)

A Probabilistic Perspective on Nonlinear Model Inversion and Data Assimilation

Dennis McLaughlin

Department of Civil and Environmental Engineering, Massachusetts Institute of Technology, Cambridge, Massachusetts, USA

Environmental data assimilation and related applications of inverse theory seek to characterize uncertain variables by combining measurements and model predictions. The need for efficient and reliable environmental characterization methods is increasing as the quantity of in situ and remotely sensed measurements increases and as numerical models become more complex. Bayesian probability theory provides a convenient framework for analyzing the properties of available alternatives. In practical applications this theory typically focuses on the estimation of particular properties (e.g. the mean, mode, covariance, etc.) of the conditional distributions of uncertain model input and output variables, given available measurements. Difficult-to-justify assumptions about normality and linearity are needed to derive the most popular environmental estimation methods (3DVAR and 4DVAR variational algorithms and ensemble Kalman estimators) from Bayesian theory. When the required assumptions do not hold, the point estimates provided by these methods are not necessarily equal, or even close, to any of the true conditional distributional properties of interest in Bayesian theory. This result raises questions about the probabilistic significance of estimates produced by commonly-used nonlinear estimation methods. It also suggests that the emphasis in large nonlinear environmental estimation problems should be on achieving robust rather than optimal solutions.

1. INTRODUCTION

The closely-related related fields of nonlinear inverse theory and environmental data assimilation seem at first glance to have generated a diverse collection of methods, assumptions, and numerical algorithms that have little common rationale or justification (see *Banks and Kunisch* [1989], *Daley* [1991], *Bennett* [1992], *Sun* [1995], *Wunsch* [1996], *Kalnay* [2002], and *Tarantola* [2005]). The exception is the very well-developed theory for linear systems. Unfortunately, many important physical phenomena observed in the earth sciences are direct manifestations of nonlinearity and cannot be reproduced, at least over extensive time periods and spatial regions, with linear models. We cannot generally assume that linear estimation methods will perform well for nonlinear problems, even though they may be successful in certain situations.

Considering the importance of nonlinearity and the possible limitations of a linear approach to environmental estimation it is useful to examine the assumptions and capabilities of the most popular approaches to inverse estimation and data assimilation. In order to provide a common framework, we adopt a Bayesian probabilistic perspective, which concedes that models, measurements, and related estimates of environmental variables are fundamentally imperfect and uncertain. This perspective can be used to consider the properties of a variety of different estimation methods, including ordinary nonlinear least-squares, regularized least-squares, variational data assimilation, and ensemble estimation. For context, we begin with a deterministic formulation of the inverse problem and indicate why a probabilistic reformulation is useful. We then present the basic concepts of Bayesian estimation theory. This leads us to consider the assumptions required to derive the most popular nonlinear data assimilation methods from a general Bayesian solution. We conclude by examining the implications of our brief survey of environmental estimation.

2. DETERMINISTIC INVERSION

Environmental model inversion and data assimilation problems can be formulated at various levels of generality. The essential elements are: 1) a set of measurements taken at various times and locations and characterized by different scales, errors, and coverage; 2) a model that describes the physical system of interest; and 3) another model that describes the measurement process. The objective is to characterize uncertain system variables by combining measurements and model predictions. Depending on one's perspective, this process can be viewed as using measurements to constrain the models or as using the models to interpret and enhance observations. In either case the hope is that measurements and models will together provide better descriptions of environmental conditions than either taken alone.

In environmental applications the system model is often most naturally expressed in terms of a set of coupled partial differential equations, or state equations. We suppose that these equations have been discretized in space and time in order to obtain an approximate numerical solution for a vector y of output variables defined at all discretized times and locations, given a vector u of discretized input variables. The numerical solution for y is then written as:

$$y = f(u) \qquad (1)$$

Known inputs are not explicitly included as arguments.

The output variables (y) of interest in environmental applications typically include system states (the dependent variables in the differential equations) as well as other variables derived from the states. In subsurface hydrology, a classic nonlinear example is unsaturated flow. Here the system state, which is included in y, is the pressure of the system, expressed through Richard's equation as a function of time and location. Other outputs of interest (such as saturation, infiltration, or runoff) can be derived from the pressure through constitutive and mass balance relationships and also included in y. The inputs (u) in this example include boundary and initial values for the states (or their gradients) and various coefficients that describe the dependence of saturation and hydraulic conductivity on pressure. In this case, the discretized solutions to the system equations (Richards equation and the associated constitutive and mass balance relationships) implicitly define the function f in the input-output relationship $y = f(u)$.

It is easiest to appreciate the issues involved in nonlinear estimation problems if we adopt a batch approach and assemble all measurements used for estimation in a single large vector z. These measurements may only provide indirect information about the system variables. In the unsaturated flow example relevant measurements include remotely sensed passive microwave brightness temperature, *in situ* measurements of soil saturation, and remote or *in situ* precipitation measurements. All of these can be incorporated into the measurement vector z. Postulated relationships between such measurements and the system inputs and outputs constitute a measurement model, which can be written compactly as:

$$z = g(y, u) = g[f(u), u] = h(u) \qquad (2)$$

The relationship $z = h(u)$ defined in (2) is frequently called a "forward model." In our unsaturated flow example, the function g might be expressed as a product $g(y, u) = g(y) = G y$ of the output vector y and a selection matrix G that identifies which outputs are measured (e.g. a subset of the discretized soil saturation values). In this special case the forward model would be $z = h(u) = Gf(u)$. Note that the forward model depends on both the system and measurement models.

The batch description provided above is convenient for our present purposes but does not exploit the temporal structure found in most dynamic problems. It is frequently helpful in applications to distinguish two types of time-dependent estimation: 1) smoothing problems, which seek estimates of output variables throughout a specified measurement time interval, and 2) filtering problems, which seek estimates of these variables only at the end of the measurement interval [*Gelb*, 1974]. Smoothing is typically used for retrospective analysis of historical data while filtering is used in real-time control and to initialize forecasts. Both types of problems are frequently solved recursively, with new measurements processed as they become available. Recursive algorithms are usually more computationally efficient than batch estimation algorithms. But a batch problem formulation provides the most general way to compare different estimation approaches.

Inverse methods derive estimates of the uncertain input u from (2), with z given. Formally, this can be viewed as an inversion of the forward model $z = h(u)$. Once an estimate of u is obtained it is typically substituted into (1) to give a corresponding estimate for y. Direct algebraic inversion of (2) is usually possible only in special situations. If there are more inputs than measurements the problem is under determined and will generally not have a unique solution. In this case it may be possible to reproduce observations exactly with an unrealistic model that has no predictive value. If there are more measurements than inputs the problem is over determined and direct inversion of (2) is feasible only if the model and measurements are perfect. In practice, it is necessary to generalize the concept of inversion to deal with data limitations, imperfect models, and measurement errors. The most common generalization is the least-squares approach.

2.1. Classical Least-Squares

Measurement and model errors can be accounted for in an aggregate way by modifying (2) to include a vector v of unknown measurement-model deviations:

$$z = g(y,u) + v = g[f(u),u] + v = h(u) + v \quad (3)$$

In this case we seek the value of u that yields a "best fit" rather than a "perfect fit" to the measurements. We can measure goodness-of-fit in terms of the mean-squared error between z and $h(u)$:

$$J_{ls}(u) = [z - h(u)]^T W_v [z - h(u)] \quad (4)$$

where W_v is a positive definite matrix that can be used to give different weights to different measurement–model deviations. Note that this need not be an inverse covariance matrix (although that is a possibility discussed below). In a deterministic context the weighting matrix is chosen for convenience. The resulting "best fit" input u_{ls} is the solution to the following minimization problem:

$$u_{ls} = \underset{u}{\mathrm{argmin}}\, J_{ls}(u) = \underset{u}{\mathrm{argmin}}\, [z - h(u)]^T W_v [z - h(u)] \quad (5)$$

There can be a unique solution only if the problem is not underdetermined (i.e. the number of distinct measurements n_z must be at least as large as the number n_u of unknown inputs). When $h(u)$ is linear in u this is equivalent to requiring that the rank of the constant matrix $\partial h/\partial u$ is at least n_u.

There is a rich and extensive literature on the analysis of least-squares estimates derived from (5) [*Bard*, 1974]. In order to provide an assessment of estimation accuracy least-squares theory treats the actual measurement z and the measurement-model deviation v as samples of random variables which are, with some abuse of notation, also written as z and v. The true input u is assumed to be unknown but not random. The probability density $p_v(v)$ of the error v is usually assumed to be normal, perhaps with a mean and covariance that are inferred from the actual observations. Given the additive form of (3) it follows that the measurement density $p_z(z)$ is also normal. Most classical least-squares analyses of model significance, parameter confidence intervals, and estimation accuracy are based on these normality assumptions and the additional assumption that $h(u)$ is linear. Analyses of nonlinear non-normal problems generally rely on Monte Carlo simulations [*Tarantola*, 2005].

Least-squares analysis raises some important conceptual issues. One is the ambiguity associated with the definition of the random measurement-model deviation v, which accounts for the aggregate effects of both model and measurement error. Since these two sources of error have different origins and enter the model and measurement equations in different ways it is difficult to objectively specify a $p_v(v)$ that properly describes their aggregate effect. In practice, this probability density is typically assumed to be normal with a mean and covariance that do not depend on the model state. In many applications, such as situations where the output variables are confined to limited ranges or measurement errors depend on the magnitude of the signal, this assumption is difficult to justify.

Another issue that arises, particularly in nonlinear least-squares, is the difficulty of finding a unique set of least-squares estimates. Even when the problem is not under determined many combinations of different parameters may give nearly the same mean-squared error. Unfortunately, these more or less indistinguishable inverse solutions may yield very different predictions for times/locations that extend beyond the observation period/region.

The limitations of the classical least-squares approach have prompted the development of modified versions that distinguish model and measurement errors and constrain the set of permissible solutions to deal with non-uniqueness.

2.2. Regularization and Prior Information

The least-squares approach can be extended if the objective function in (4) is augmented with terms that bias the solution towards input values with certain desirable properties [*Tikhonov and Arsenin*, 1977; *Banks and Kunisch*, 1989]. One common option is to add a "regularization term" that penalizes deviations from a specified "first guess" or "prior" value u_f as follows

$$J_{reg} = [z - h(u)]^T W_v [z - h(u)] + [u - u_f]^T W_u [u - u_f] \quad (6)$$

The first term in this regularized objective can be viewed as a penalty on measurement error (weighted by W_v) while the second can be viewed as a penalty on input error (weighted by W_u). When measurements are more uncertain W_v is reduced and larger measurement deviations are tolerated. Similarly, when the first guess input is more uncertain W_u is reduced and larger input deviations are tolerated (if W_u is zero the classical least-squares objective is recovered). This tradeoff between measurement and model error is fundamental in data assimilation theory.

Regularization can give more physically plausible estimates by effectively constraining the set of possible solutions and forcing underdetermined problems to have unique solutions [*Tikhonov and Arsenin*, 1977]. However, regularization raises its own conceptual issues. First, it sacrifices the objective concept of "best fit" for a more subjective interpretation of plausibility. The final input estimates will not generally give as good a fit to the data as a classical least-squares procedure. Second, regularization theory does not provide, in itself, any objective way to select the weights and first guess values that control the final solution. Nearly any desired solution can be obtained by sufficient manipulation of the regularization term. These difficulties have prompted efforts to give regularized least-squares, particularly the regularizing weights and first guess values, a more fundamental foundation. For the most part, these efforts draw on Bayesian estimation theory. Further details are discussed in the following section.

3. BAYESIAN ESTIMATION

Bayesian estimation treats unknown input and output variables and measurements as random variables [*Jazwinski*, 1970; *Miller et al.*, 1999]. Most Bayesian theory assumes that the structure of the function $f(.)$ defined in (1) is known perfectly so that all system model uncertainties enter through the random inputs assembled in the vector u, which has a specified prior probability density $p_u(u)$. In addition, it is common to assume that (3) describes the relationship between the measurements and system variables, that the function $g(.)$ is known perfectly, and that v is an additive random variable independent of u with a specified probability density $p_v(v)$.

If f and $p_u(u)$ are known the prior (or unconditional) output density $p_y(y)$ can be derived from (1). For general nonlinear problems this derivation must usually be carried out numerically, with a Monte Carlo simulation. Measurement information is incorporated through the conditional densities $p_{u/z}(u \mid z)$ and $p_{y/z}(y \mid z)$, which may be derived from the unconditional densities and the measurement model [*Miller et al.*, 1999]. These conditional probability densities convey everything we know about the uncertain inputs and outputs, given a particular set of measurements. In environmental applications the densities of interest are usually too large to be derived or visualized directly, except for certain special cases (e.g. when all the relevant variables are jointly normal). Consequently, practical Bayesian estimation methods typically focus on particular properties of $p_{u/z}(u \mid z)$ and $p_{y/z}(y \mid z)$, such as their means, modes, covariances, quantiles, and univariate marginal densities (e.g. the density of a particular scalar component of y).

It is useful to distinguish two approaches for deriving the statistical properties of the conditional probability densities of u and y. These can be broadly identified as parameter and state estimation and are illustrated in Figure 1. Parameter estimation derives $p_{u/z}(u \mid z)$ from the following version of Bayes theorem:

$$p_{u \mid z}(u \mid z) = c p_{z \mid u}(z \mid u) p_u(u) = c p_v[z - h(u)] p_u(u) \qquad (7)$$

where $p_{z/u}(z \mid u)$ is the input likelihood function (treated as a function of the variable u for a given z) and c is a normalization constant. Subscripts are added to the probability densities to clarify the difference between the random variable described by the density (in the subscript) and the argument of the density function (in parentheses). The second equality in (7) follows because the additive measurement error assumption of (3) allows the likelihood to be derived directly from p_v, which is specified *a priori*. The conditional output density $p_{y/z}(y \mid z)$ or its properties can, in principle, be obtained from $p_{u/z}(u \mid z)$ using derived distribution tech-

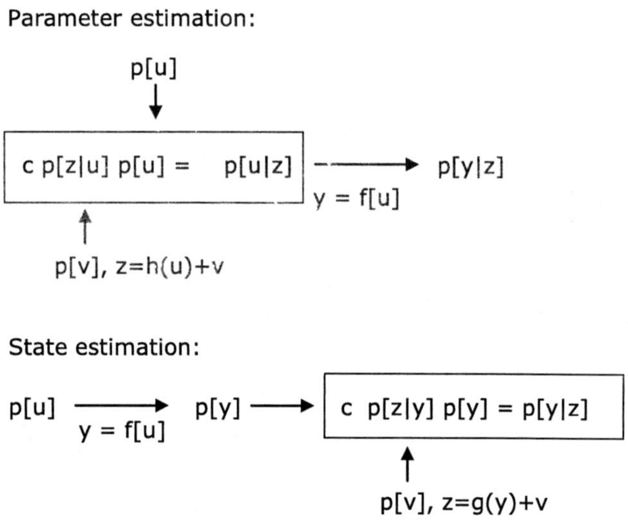

Parameter estimation:

p[u]
↓
c p[z|u] p[u] = p[u|z] ——→ p[y|z]
↑ y = f[u]
p[v], z=h(u)+v

State estimation:

p[u] ——→ p[y] ——→ c p[z|y] p[y] = p[y|z]
y = f[u] ↑
p[v], z=g(y)+v

Figure 1. Comparison of Bayesian parameter and state estimation approaches to data assimilation problems.

niques (e.g. Monte Carlo simulation) but this is rarely practical for large nonlinear problems.

The alternative state estimation approach focuses directly on $p_{y/z}(y \mid z)$, which can be obtained from the following version of Bayes theorem:

$$p_{y \mid z}(y \mid z) = c p_{z \mid y}(z \mid y) p_y(y) \qquad (8)$$

where $p_{z/y}(z \mid y)$ is the output likelihood function (treated as a function of the variable y, for a given z). Derivation of this likelihood is simplified if the measurement function $g(y, u) = g(y)$ depends only on output variables. This requirement can be met if the output vector is redefined to include all measured inputs, generally at the cost of introducing additional sources of nonlinearity. The result is:

$$p_{y \mid z}(y \mid z) = c p_v[z - g(y)] p_y(y) \qquad (9)$$

Note that the state estimation approach does not provide information about $p_{u/z}(u|z)$. The parameter and state estimation versions of Bayes theorem ultimately yield the same $p_{y/z}(y/z)$ when their assumptions are compatible but they require different information and may have different computational requirements.

In order to apply Bayesian estimation we must specify the appropriate prior density $p_v(v)$ as well as either $p_u(u)$ or $p_y(y)$. For parameter estimation the most common approach is to assume that u and v have normal prior densities. Then the natural log of the conditional density for u may be written as:

$$\ln p_{u \mid z}(u \mid z) = \ln c -$$
$$\frac{1}{2}[z - h(u)]^T C_v^{-1}[z - h(u)] + \qquad (10)$$
$$\frac{1}{2}[u - \bar{u}]^T C_u^{-1}[u - \bar{u}]$$

where C_v and C_u are the covariances of v and u, $\bar{u}$ is the mean of u, and v is assumed, without loss of generality, to have a zero mean. Note that $p_{u/z}(u|z)$ is generally not normal if $h(u)$ is nonlinear, even if the u and v priors are normal. In this case the mean, covariance, and most other statistics of $p_{u/z}(u \mid z)$ are generally difficult to derive. However, the mode can be found by minimizing $-\ln p_{u/z}(u|z) - 2\ln c$, which can be written as a least-squares objective function:

$$J_{pe}(u) =$$
$$[z - h(u)]^T C_v^{-1}[z - h(u)] \qquad (11)$$
$$+ [u - \bar{u}]^T C_u^{-1}[u - \bar{u}]$$

This mode is equal to u_{reg}, the regularized estimate that minimizes (6), if the least-squares first guess is taken to be $u_f = \bar{u}$ and the weighting matrices W_v and W_u are taken to be the corresponding inverse covariances. In this respect, Bayesian estimation theory appears to provide the desired fundamental justification for regularization. The regularized estimate now represents a balance between best fit and first guess that is based on our relative confidence in measured data and prior information.

If we focus on minimization of (11) and only find the mode of $p_{u/z}(u \mid z)$ it is not generally possible to derive any of the conditional statistics of y, since we do not know how u is distributed around the mode. Although it is possible to evaluate the point estimate $y = f(u_{mode})$ this estimate is not generally equal to the mode or any other common statistical property of $p_{y/z}(y \mid z)$ in the nonlinear case (see the example presented below).

As an alternative, we can pursue the state estimation approach and assume that y and v (rather than u and v) are normal. In this case, the natural log of the conditional density for y can be written:

$$\ln p_{y \mid z}(y \mid z) = \ln c -$$
$$\frac{1}{2}[z - g(y)]^T C_v^{-1}[z - g(y)] + \qquad (12)$$
$$\frac{1}{2}[y - \bar{y}]^T C_y^{-1}[y - \bar{y}]$$

where C_y is the covariance of y, and $\bar{y}$ is the mean of y. Note that $p_{y/z}(y \mid z)$ is generally not normal if $g(u)$ is nonlinear, even if the y and v priors are normal. Also, the assumption that u is normal (used in (11)) and the assumption that y is normal (used in (12)) are generally not compatible if $f(u)$ is nonlinear. So the normal (quadratic) versions of the parameter and state estimation approaches actually solve different problems and will generally give different results in nonlinear applications.

By analogy with the parameter estimation formulation the mode of $p_{y/z}(y|z)$ can be found by minimizing $-\ln p_{y|z}(y|z) - 2\ln c$, which can be written as a least-squares objective function that depends on y rather than u:

$$J_{se}(y) =$$
$$[z - g(y)]^T C_v^{-1}[z - g(y)] + \qquad (13)$$
$$[y - \bar{y}]^T C_y^{-1}[y - \bar{y}]$$

Although the state estimation objective of (13) looks similar to the parameter estimation objective of (11) the normality assumptions required are different.

The Bayesian estimation concepts outlined are illustrated in Figure 2 for a simple nonlinear estimation example. In this case, the state y is derived from a lognormally distributed input η as follows:

$$y = f(\eta) = \frac{\eta^2}{a^2 + \eta^2} \qquad (14)$$

This is a saturation function similar to those frequently used to describe sorption and other limiting processes. Figure 2a shows the saturation function for $a = 2.0$, together with the lognormal density for η, which is generated from a normal random variable $u = \ln(\eta)$ with mean -1.0 and standard deviation 2.0. For these parameters the most probable values of η and $y = f(\eta)$ are near zero. In order to maintain consistency with the normality assumptions of (11) it is convenient to formulate the estimation problem in terms of the normally distributed input u rather than η. Figure 2b shows the unconditional probability distributions of u and y., derived from a Monte Carlo simulation with 10^6 replicates. The $p(y)$ plot confirms that small values of y are most probable.

Now suppose that we have a measurement $z = y + v$, where v is normally distributed with mean 0.0 and standard deviation 0.1. We seek the conditional probability densities $p(u|z)$ and $p(y|z)$, conditioned on z. This estimation problem may be solved by using (7) to obtain $p(u|z)$ and (9) to obtain $p(y|z)$. The conditional densities for this example with $z = 0.7$ are shown in Figure 2c. These were derived with by using the closed form normal likelihood function obtained from $p(v)$ to weight the Monte Carlo replicates for $p(y)$, as specified in (9). The unconditional mean and mode of y and the conditional means and modes of u and y are noted on the plots.

The conditional mode of u, $mode[u|z] = 0.93$ is the point estimate of u that is obtained by minimizing (11). The point estimate $f[exp(mode(u|z))] = 0.62$ obtained when the conditional mode of u is substituted into the saturation function is much larger than the actual conditional mode, which is $mode(y|z) = 0.01$. Both $f[exp(mode(u|z))]$ and the actual conditional expectation $E[y|z]$ are in the midst of an extended region of relatively low probability corresponding to the transition portion of the saturation function. Conditioning on the noisy measurement $z = 0.7$ somewhat increases the probability of values in this region but the actual mode of the y conditional density remains near zero.

The results shown in this particular example are dramatic but illustrative of what may happen when relatively simple nonlinearities are combined with non-Gaussian densities. It should be noted that the values of $f[exp(mode(u|z))]$ and $E[y|z]$ gradually move toward the actual conditional mode as the measurement value in this example moves toward zero. Also, as the measurement error standard deviation decreases the actual conditional mode moves closer to $f[exp(mode(u|z))]$ and $E[y|z]$, which become clustered near the measured value. In both cases, the point estimate that minimizes (11) gives a better estimate of the mode of y when substituted into the saturation function. On the other hand, it is not difficult to construct other examples that give an even larger spread between point estimates and actual values of the mean and mode. The differences can be especially large if the function $f(\eta)$ is non-monotonic. The real lesson to take from this example is that conventional point estimates can be relatively uninformative in nonlinear problems since the most likely values of uncertain variables may differ significantly from these estimates. We return to this topic at the end of the paper.

4. VARIATIONAL METHODS

Bayesian estimation techniques are widely used in fields as diverse as hydrology, economics, petroleum engineering, and meteorology. In meteorology and oceanography it has become common to call these methods variational algorithms, in recognition of the fact that the solutions are often derived from variational calculus (see *Daley* [1991], *Bennett* [1992], *Wunsch* [1996], and *Kalnay* [2002]). Meteorologists typically distinguish three-dimensional variational (3DVAR) and four-dimensional variational (4DVAR) algorithms, depending on how time is handled [*Courtier*, 1997]. In most cases variational algorithms are based on (11) and (13) and rely on the normality assumptions that are implicit in these equations if a probabilistic perspective is adopted.

Using the terminology introduced above, 3DVAR is a static state estimation procedure. Suppose that a state estimate y_t at time t is to be derived solely from the current measurement z_t, a corresponding measurement error density $p(v_t)$, and a specified prior state probability density $p(y_t)$. 3DVAR methods nearly always adopt normality assumptions and use the mode of $p(y_t | z_t)$ as a point estimate of y_t. This mode minimizes an instantaneous version of (13):

$$J_{se,t}(y_t) = [z_t - g_t(y_t)]^T C_{vt}^{-1} [z_t - g_t(y_t)] + \qquad (15)$$
$$[y_t - \bar{y}_t]^T C_{yt}^{-1} [y_t - \bar{y}_t]$$

The necessary condition for a local minimum is that the gradient $\partial J_{se,t}(y_t) / \partial y_t$ of this objective must equal zero. When $g_t(y_t)$ is linear this condition yields a convenient closed form solution that is also the global minimum (since $J_{se,t}(y_t)$ is then convex). When $g_t(y_t)$ is not linear it is generally necessary to solve the minimization problem with a gradient-based numerical search algorithm [*Tarantola*, 2005].

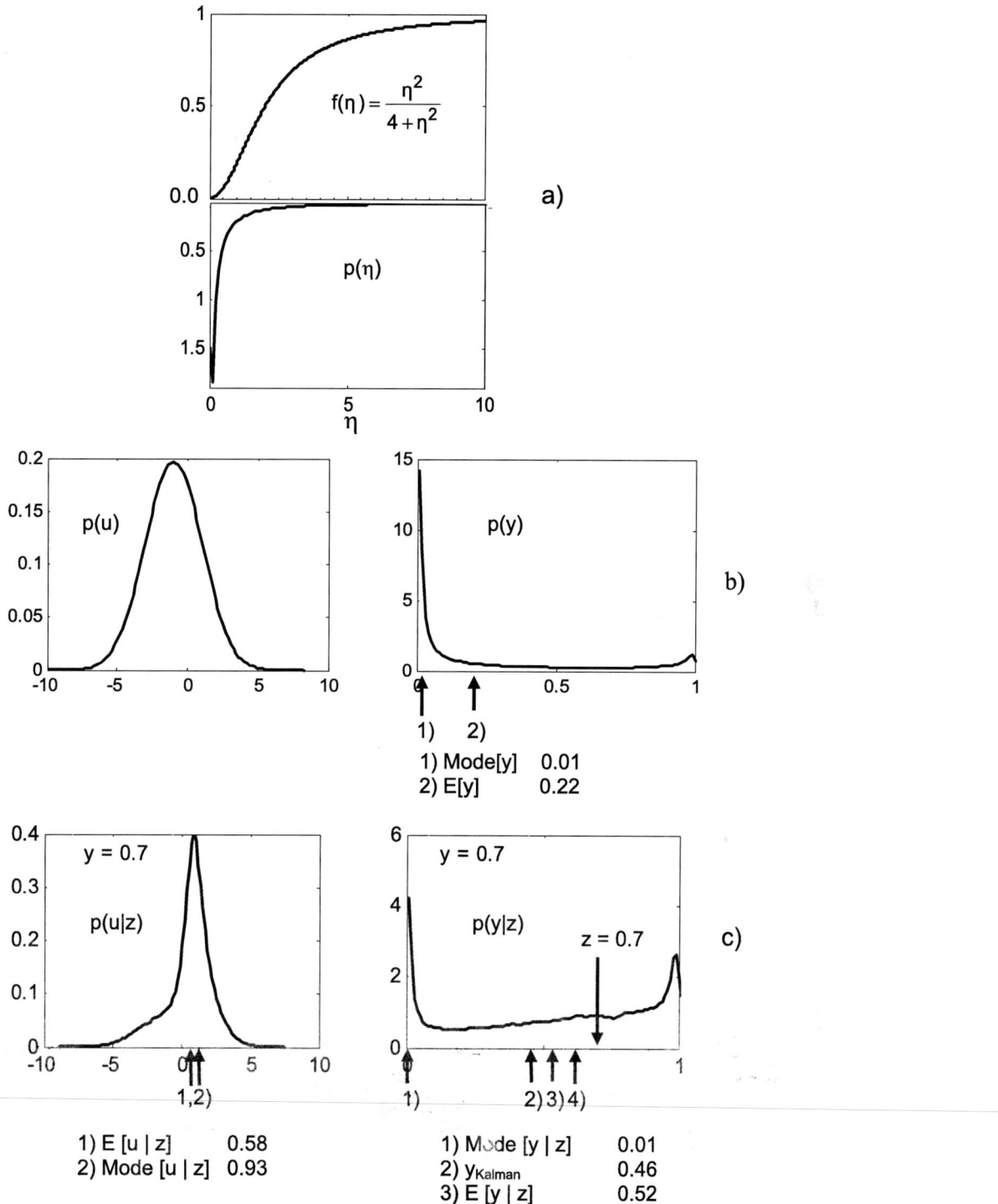

Figure 2. Specified and derived probability densities for the example: a) Forward model $f(\eta)$ and unconditional $p(\eta)$, b) unconditional $p(u)$ and $p(y)$, c) conditional $p(u|z)$ and $p(y|z)$. Measurement value $z = 0.7$.

In practice, 3DVAR typically updates state estimates recursively, using new prior statistics and measurements at each update time. The updated state estimate is often used to initialize a real-time forecast derived from the system model. In our unsaturated flow example a 3DVAR approach could be used at a single time t to estimate pressures (y_t) throughout the soil column from radiobrightness measurements (z_t). The resulting estimates could then serve as initial conditions for a Richards equation solution that predicts the state at times greater than t.

The 4DVAR approach is an extension of 3DVAR that is, from a Bayesian perspective, fundamentally a parameter estimation technique. In a 4DVAR algorithm measurements span an extended time period and inputs and states are typically estimated throughout this period. In meteorology the uncertain input of most interest is often the state at the beginning of the measurement period, which serves as an initial condition for the model solution over this period. In hydrogeology and petroleum engineering the most important uncertain input is often permeability. 4DVAR methods nearly always adopt normality assumptions and use the mode of $p(u|z)$ as a point estimate of u. This mode is obtained by minimizing (11). As in 3DVAR, gradient-based search algorithms are typically used to find the minimum. Adjoint methods for computing the objective function gradient often provide a significant computational advantage over finite difference methods and are generally used in operational 4DVAR applications [*Courtier et al.*, 1993].

Most 4DVAR algorithms derive a point estimate $\hat{y}$ of the system state by substituting *mode(u)* directly into (1) to obtain $\hat{y} = fmode(u)$. In any applications the state estimate at the end of the measurement period is used to initialize a forecast, as in 3DVAR. As mentioned earlier, the state estimate *fmode(u)* has no particular probabilistic significance for nonlinear problems. However, it satisfies the model equations for a set of inputs which are deemed the most probable, subject to all the assumptions made in the analysis. The fact that 4DVAR state estimates always satisfy the model equations has been one of the attractions of the approach. "Weak constraint" variational formulations challenge this viewpoint by including a random additive model error term in (1) [*Zupanski*, 1997]. In this case, the system model equations do not need to be satisfied exactly since they are no longer believed to be perfect. From a mathematical viewpoint the weak constraint variational formulation is equivalent to extending the uncertain input vector to include error variables that are not part of the original system model. When this is done the problem can be written in the form given in (11).

In order for the variational approach to properly balance measurement and prior information the prior means and covariances must be physically reasonable. This is particularly important in 3DVAR, where the prior mean and background covariance are the only sources of information about the evolution of the state from the previous update time. Both of these statistics should be dynamically consistent (i.e. they should be consistent with conservation laws that relate the various state variables). This requirement is sometimes difficult to enforce in complex problems. In 4DVAR the input covariance should properly describe spatial and temporal relationships among the uncertain inputs. For example, in groundwater hydrology or reservoir engineering this covariance should reflect the effects of anisotropy and connectedness in the permeability field. The assumption that the prior input density is normal limits the flexibility available for such descriptions since variability must be characterized entirely by the first two moments of u.

The variational approach to model inversion and data assimilation is currently the method of choice for large operational applications. This is reflected, for example, in the activities of the major weather forecasting centers, which use variational algorithms to issue routine operational products. The greatest conceptual limitations of the variational approach are 1) its dependence on normality assumptions and covariance-based descriptions of variability, 2) its reliance on point state estimates that generally do not have any particular statistical significance, and 3) its inability to provide distributional information about uncertain variables. These limitations have prompted interest in ensemble methods that are more flexible and are may be able to provide more complete statistical information about the system states.

5. ENSEMBLE STATE ESTIMATION

Ensemble state estimation methods seek approximate solutions for the conditional state density $p_{y|z}(y|z)$ or its statistical properties [*Arulampalam*, 2002; *Evensen*, 2003, 2004]. From a probabilistic perspective these methods are ultimately based on (9), although they are often derived without specific reference to their Bayesian origins. In order to understand the approximations typically used in large problems it is helpful to consider the widely-studied special case of normally distributed y, v and z and linear $g(y) = Gy$. In this case, $p_{y|z}(y|z)$ is multivariate normal and completely determined by the conditional mean and covariance, which may be written in closed form as [*Tarantola*, 2005]:

$$E[y \mid z] = \bar{y} + C_{yz} C_{zz}^{-1} [z - G\bar{y}] \quad (16)$$

$$\text{Cov}[y \mid z] = C_{yy} - C_{yz} C_{zz}^{-1} C_{yz}^T \quad (17)$$

where the cross-covariance $C_{yz} = C_{yy}G^T$ and the measurement covariance $C_{zz} = GC_{yy}G^T + C_{vv}$. Note that the conditional mean is a linear function of the measurements. Also, the conditional covariance depends on G, and the prior covariances of y and v but not on the measurements themselves. When u is also normal and $f(u) = Fu$ is linear $C_{yy} = FC_{uu}F^T$ and the Bayesian state and parameter estimation problems are equivalent. These batch estimation equations have the same form as a cokriging algorithm with a known mean [*Marsily*, 1986]. Recursive versions of (16) and (17) yield the Kalman smoother or, when estimates are desired only at the end of the measurement interval, the Kalman filter [*Gelb*, 1974]. Also, the conditional mean estimate of (16) is the same as the conditional mode estimate obtained from 3DVAR for the linear normal case. The linear normal problem has such a straightforward solution because it is able to use covariances to describe physical relationships between inputs, states and measurements.

The convenient closed form expressions for $p_{y/z}(y/z)$ and its moments obtained in the linear normal case do not apply when y is non-normal and/or $g(y)$ is nonlinear. Large non-normal nonlinear problems must generally be solved numerically with a Monte Carlo procedure. In this case the desired probability densities are approximated with ensembles of randomly generated replicates [*Arulampalam*, 2002; *Evensen*, 2003, 2004]. Each replicate is assigned a value and a discrete probability. The prior input replicates u^i ($i = 1, ..., N$) are obtained by selecting N equally likely samples from the specified prior input density. These may be substituted into (1) and (2) to give a corresponding set of equally likely prior state replicates y^i and measurement prediction replicates $z^i = g(y^i)$. The conditional state density is approximated by updated replicates $y^{i/z}$ derived from the prior replicates and the measurements. There are a number of different state updating approaches characterized by different assumptions and computational requirements.

Particle-based updating methods generally adjust only replicate probabilities and leave replicate values unchanged. The simplest of these uses (9) to derive the updated discrete probability for replicate i [*Arulampalam*, 2002]:

$$p_y(y^i \mid z) = cp_v(z - z^i)p_y(y^i) = cp_v(z - z^i)/N \quad (18)$$

The complete ensemble of state replicates and their updated probabilities may be used to estimate distributional properties such as the conditional mean, mode, covariance, etc. The particle approach has the virtue of providing a truly probabilistic description of the state without making restrictive assumptions about the forms of the prior densities of u and v. Its primary disadvantage is computational. As the number of measurements and unknown inputs increases the number of replicates required to prevent the collapse of the ensemble to a single replicate (degeneracy) grows rapidly, making the approach infeasible for large problems [*Arumpalam*, 2002].

A practical alternative is to focus on updates of the replicate values rather than their probabilities. This is the approach taken in ensemble methods that rely on linear normal theory. An example is the ensemble Kalman smoother and its filter counterpart, the ensemble Kalman filter [*Evensen*, 2003, 2004]. The ensemble Kalman smoother assumes that the conditional density $p(y|z)$ has the same mean and variance as the classical linear normal Kalman smoother. Consequently, the updated replicates $y^{i/z}$ are selected so that their sample mean $E[y|z]$ and covariance $Cov[y|z]$ correspond to (16) and (17), with the prior covariances C_{yy}, C_{yz}, and C_{zz} replaced by sample estimates computed from the y^i and z^i replicates. There are many ways to generate updated replicates that converge to the desired $E[y|z]$ and $Cov[y|z]$. These alternatives all lead to somewhat different versions of the ensemble Kalman smoother. In the version proposed by *Evensen* [2004] the mean-removed updated replicates are linear combinations of the mean-removed prior replicates, with weights that depend on the sample estimates of C_{yy}, C_{yz}, and C_{zz}.

The updated ensemble produced by the ensemble Kalman smoother can be proven to converge to the exact conditional density $p(y|z)$ as the ensemble grows, provided that the linear normal assumptions are valid. It is reasonable to ask why one might want to use an ensemble Kalman updating procedure in the more general case when linear normal assumptions do not hold. The honest answer is that the Kalman update is a very convenient approximation that has been observed to work well, even in situations where the prior and conditional densities of interest are highly non-normal (e.g. skewed or multi-modal) [*Zhou et al.*, 2006]. Despite this, there is no reason to expect that the distributional properties of the ensemble obtained from a Kalman update will be correct in any given nonlinear problem. In the example presented in Figure 2 the Kalman conditional mean estimate is reasonably close to the true conditional mean, primarily because the measured value is also reasonably close. This may not always be the case. Care should be used in applying the Kalman update in nonlinear problems since it is possible that the updated ensemble values can actually move away from the true value, especially if the measurement error is large.

Ensemble estimation is becoming popular in a variety of data assimilation applications, largely because it is so easy to use. It has the advantage of providing information about the distribution of likely states, rather than just a single point estimate. Like the variational approach, ensemble estimation is general in concept but limited in practice. At present, the computational demands of particle methods have prevented their application to large environmental data assimilation problems. Ensemble Kalman filtering and smoothing are more efficient but rely on normality assumptions that are even stronger than those required in variational methods. That is, the ensemble filter tries to duplicate (17) and (18), which assumes that all conditional and unconditional densities are normal, while the variational approach only assumes that the unconditional densities $p(v)$ and either $p(y)$ or $p(u)$ are normal. The implicit normality assumptions of the ensemble Kalman filter may be acceptable in some situations but can be problematic in others where variables are highly skewed or even multi-modal.

Even when the linear normal assumptions are satisfactory ensemble methods may be compromised by sampling error issues. This is especially true when the number of replicates is small compared to the number of estimated unknowns (as is usually the case in real-world applications). Sample covariances derived from a small ensemble can lead to nonphysical artifacts that can be difficult to detect and remove without creating other problems. Many of these difficulties may be resolved in the future, as better techniques are developed for generating and updating replicates.

6. CONCLUSIONS

A probabilistic perspective on nonlinear model inversion and data assimilation problems prompts us to rethink existing approaches. It is striking that the most popular state-of-the-art data assimilation methods cannot be proven to yield accurate estimates of any of the distributional properties of interest in Bayesian estimation theory. In particular, we have no reason to believe that either variational methods or ensemble Kalman estimators will give reliable estimates of the conditional mean, mode, covariance, or quantiles of the system state, except in the special linear normal case discussed above.

Perhaps the point state estimates produced by variational methods and the sample statistics produced by an ensemble Kalman estimator are useful approximations of reality in certain practical nonlinear applications. But it is difficult to assign any probabilistic significance to these state estimates and statistics or to compute credible nonprobabilistic indicators of the "usefulness" or "accuracy" of these approximations. Similarly, it is difficult to compare estimation alternatives without an objective performance measure. *Post-hoc* or off-line performance measures, such as the "forecast skill" used to assess the match between meteorological forecasts and observations, can be argued to be useful alternatives to more classical distributional measures. But there is no reason to believe that a variational method designed to minimize (11) or (13) or an ensemble Kalman estimator designed to estimate conditional moments subject to normality assumptions will necessarily maximize forecast skill. There is also no reason to believe that a skill-maximizing estimator would be better in other respects than existing alternatives.

Although variational and ensemble methods will undoubtedly both improve over time it seems appropriate to ask at this point whether the classical preoccupation with "optimal" point estimates has any place in nonlinear data assimilation. What we really seek in most practical applications is a robust estimation strategy that nearly always improves on our prior knowledge, even when our assumptions and inputs are imperfect. It would be helpful to have a theory of robust estimation that properly acknowledges the possible deficiencies of the models and measurement sources used in nonlinear data assimilation problems. In this respect it is worth noting that operational data assimilation in meteorology and oceanography is largely concerned with creating robust estimators, using information gained when so-called optimal estimators fail. In operational applications the biases, artifacts, and various other errors frequently observed in practice provide valuable clues about possible deficiencies in current estimation methods. These clues may point the way to new developments in robust environmental estimation theory.

It is likely that the need to describe and monitor environmental change will become more pressing, that new instruments and data sources will continue to become available, and that the capabilities and resolution of numerical earth system models will improve. All of these developments point to the need for a realistic theory of environmental data assimilation that can adequately deal with more complex problems. This theory needs to take nonlinear behavior as a given and to focus more on robustness than optimality.

Acknowledgments. The author is grateful to Sara Friedman for her help with the example and for a number of useful discussions. Partial support for this work was provided by US National Science Foundation grants 0003361, 0121182, 0530851, and 0540259.

REFERENCES

Arulampalam, M. S., A tutorial on particle filters for online nonlinear/nongaussian Bayesian tracking, *IEEE Trans. Signal Proc.*, 50(2), 174–188, 2002.

Banks, H. and K. Kunisch, *Estimation Techniques for Distributed Parameter Systems*, Birkhauser, Boston, 315 pp., 1989.

Bard, Y., *Nonlinear Parameter Estimation,* Academic Press, New York, NY, 1974.

Bennett, A., *Inverse Methods in Physical Oceanography*, Cambridge Univ. Press, Cambridge, UK, 1992.

Courtier, P., A dual formulation of four-dimensional variational assimilation, *Quart.. J. Royal Meteor. Soc.*, 120(519), 1367–1387, 1997.

Courtier, P., J. Derber, R. M. Errico, J.-F. Louis, and T. Vukićević, Review of the use of adjoint, variational methods, and Kalman filters in meteorology. *Tellus*, 45A, 343–357, 1993.

Daley, R., *Atmospheric Data Analysis*, Cambridge Univ. Press, Cambridge, UK, 457pp, 1991.

Evensen, G., The Ensemble Kalman Filter: theoretical formulation and practical implementation, *Ocean Dynamics*, 53(4), 343–367, 2003.

Evensen, G., Sampling strategies and square root analysis schemes for the ensemble Kalman filter, *Ocean Dynamics*, 54(6), 539–560, 2004.

Gelb, A. (Ed.), *Applied Optimal Estimation*, MIT Press, Cambridge, MA, 374 pp., 1974.

Jazwinski, A. H., *Stochastic Processes and Filtering Theory*, Academic, San Diego, CA, 376 pp., 1970.

Kalnay, E., *Atmospheric Modeling, Data Assimilation, and Predictability*, 341pp., Cambridge Univ. Press, Cambridge, UK, 2002.

Marsily, G. de, *Quantitative Hydrogeology: Groundwater Hydrology for Engineers*, Academic Press, 1986.

Miller, R. N., E. F. Carter, and S. T. Blue, Data assimilation into nonlinear stochastic models, *Tellus*, 51A, 167–194, 1999.

Tarantola, A., *Inverse Problem Theory and Model Parameter Estimation*. Albert Tarantola, SIAM, Philadelphia, 2005.

Tikhonov, A. N. and V. I. F. A. F. Arsenin, *Solutions of Ill-posed Problems*, Scripta series in mathematics, Winston, Washington, DC, 258 pp., 1977.

Wunsch, C., *The Ocean Inverse Problem*, Cambridge Univ. Press, Cambridge, UK, 1996.

Zhou, Y., D. McLaughlin, and D. Entekhabi, Assessing the performance of the ensemble Kalman filter for land surface data assimilation, *Mon. Weather Rev.*, 134 (8), 2128–2142, 2006.

Zupanski, D., A General Weak Constraint Applicable to Operational 4DVAR Data Assimilation Systems, *Mon. Weather Rev.*, 125 (9), 2274–2292, 1997.